# Lecture Notes in Computer Science 6492

*Commenced Publication in 1973*
Founding and Former Series Editors:
Gerhard Goos, Juris Hartmanis, and Jan van Leeuwen

Ron Kimmel Reinhard Klette
Akihiro Sugimoto (Eds.)

# Computer Vision – ACCV 2010

10th Asian Conference on Computer Vision
Queenstown, New Zealand, November 8-12, 2010
Revised Selected Papers, Part I

Springer

Volume Editors

Ron Kimmel
Department of Computer Science
Technion – Israel Institute of Technology
Haifa 32000, Israel
E-mail: ron@cs.technion.ac.il

Reinhard Klette
The University of Auckland
Private Bag 92019, Auckland 1142, New Zealand
E-mail: r.klette@auckland.ac.nz

Akihiro Sugimoto
National Institute of Informatics
Chiyoda, Tokyo 1018430, Japan
E-mail: sugimoto@nii.ac.jp

ISSN 0302-9743
ISBN 978-3-642-19314-9
DOI 10.1007/978-3-642-19315-6
e-ISSN 1611-3349
e-ISBN 978-3-642-19315-6

Springer Heidelberg Dordrecht London New York

Library of Congress Control Number: 2011921594

CR Subject Classification (1998): I.4, I.5, I.2.10, I.2.6, I.3.5, F.2.2

LNCS Sublibrary: SL 6 – Image Processing, Computer Vision, Pattern Recognition, and Graphics



*Typesetting:* Camera-ready by author, data conversion by Scientific Publishing Services, Chennai, India
*Coverpicture:* Lake Wakatipu and the The Remarkables, from 'Skyline Queenstown' where the conference dinner took place.

Printed on acid-free paper

Springer is part of Springer Science+Business Media (www.springer.com)

# Preface

The 2010 Asian Conference on Computer Vision took place in the southern hemisphere, in "The Land of the Long White Cloud" in Maori language, also known as New Zealand, in the beautiful town of Queenstown. If we try to segment the world we realize that New Zealand does not belong officially to any continent. Similarly, in computer vision we often try to define outliers while attempting to segment images, separate them to well-defined "continents" we refer to as objects. Thus, the ACCV Steering Committee consciously chose this remote and pretty island as a perfect location for ACCV2010, to host the computer vision conference of the most populated and largest continent, Asia. Here, on South Island we studied and exchanged ideas about the most recent advances in image understanding and processing sciences.

Scientists from all well-defined continents (as well as ill-defined ones) submitted high-quality papers on subjects ranging from algorithms that attempt to automatically understand the content of images, optical methods coupled with computational techniques that enhance and improve images, and capturing and analyzing the world's geometry while preparing for higher-level image and shape understanding. Novel geometry techniques, statistical-learning methods, and modern algebraic procedures rapidly propagate their way into this fascinating field as we witness in many of the papers one can find in this collection.

For this 2010 issue of ACCV, we had to select a relatively small part of all the submissions and did our best to solve the impossible ranking problem in the process. We had three keynote speakers (Sing Bing Kang lecturing on modeling of plants and trees, Sebastian Sylwan talking about computer vision in production of visual effects, and Tim Cootes lecturing about modelling deformable object), eight workshops (Computational Photography and Esthetics, Computer Vision in Vehicle Technology, e-Heritage, Gaze Sensing and Interactions, Subspace, Video Event Categorization, Tagging and Retrieval, Visual Surveillance, and Application of Computer Vision for Mixed and Augmented Reality), and four tutorials. Three Program Chairs and 38 Area Chairs finalized the decision about the selection of 35 oral presentations and 171 posters that were voted for out of 739, so far the highest number of ACCV, submissions. During the reviewing process we made sure that each paper was reviewed by at least three reviewers, we added a rebuttal phase for the first time in ACCV, and held a three-day AC meeting in Tokyo to finalize the non-trivial acceptance decision-making process.

Our sponsors were the Asian Federation of Computer Vision Societies (AFCV), NextWindow–Touch-Screen Technology, NICTA–Australia's Information and Communications Technology (ICT), Microsoft Research Asia, Areograph–Interactive Computer Graphics, Adept Electronic Solutions, and 4D View Solutions.

Finally, the *International Journal of Computer Vision* (IJCV) sponsored the Best Student Paper Award.

We wish to acknowledge a number of people for their invaluable help in putting this conference together. Many thanks to the Organizing Committee for their excellent logistical management, the Area Chairs for their rigorous evaluation of papers, the Program Committee members as well as external reviewers for their considerable time and effort, and the authors for their outstanding contributions.

We also wish to acknowledge the following individuals for their tremendous service: Yoshihiko Mochizuki for support in Tokyo (especially also for the Area Chair meeting), Gisela Klette, Konstantin Schauwecker, and Simon Hermann for processing the 200+ Latex submissions for these proceedings, Kaye Saunders for running the conference office at Otago University, and the volunteer students during the conference from Otago University and the *.enpeda..* group at The University of Auckland. We also thank all the colleagues listed on the following pages who contributed to this conference in their specified roles, led by Brendan McCane who took the main responsibilities.

ACCV2010 was a very enjoyable conference. We hope that the next ACCV meetings will attract even more high-quality submissions.

November 2010

Ron Kimmel
Reinhard Klette
Akihiro Sugimoto

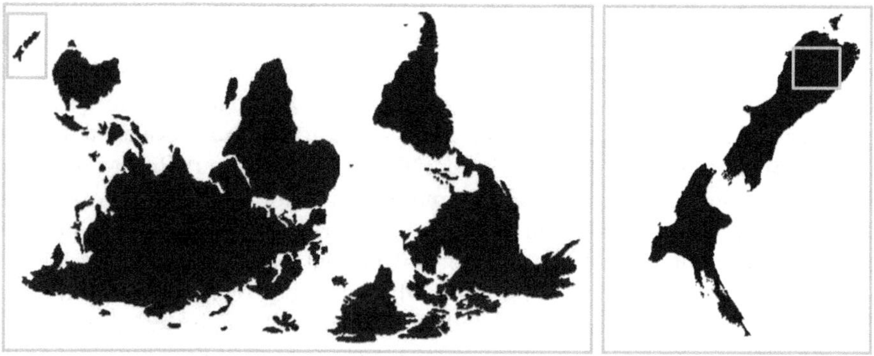

# Organization

## Steering Committee

| | |
|---|---|
| Katsushi Ikeuchi | University of Tokyo, Japan |
| Tieniu Tan | Institute of Automation, Chinese Academy of Science, China |
| Chil-Woo Lee | Chonnam National University, Korea |
| Yasushi Yagi | Osaka University, Japan |

## Honorary Chairs

| | |
|---|---|
| P. Anandan | Microsoft Research India |
| Richard Hartley | Australian National University, NICTA |

## General Chairs

| | |
|---|---|
| Brendan McCane | University of Otago, New Zealand |
| Hongbin Zha | Peking University, China |

## Program Chairs

| | |
|---|---|
| Ron Kimmel | Israel Institute of Technology |
| Reinhard Klette | University of Auckland, New Zealand |
| Akihiro Sugimoto | National Institute of Informatics, Japan |

## Local Organization Chairs

| | |
|---|---|
| Brendan McCane | University of Otago, New Zealand |
| John Morris | University of Auckland, New Zealand |

## Workshop Chairs

| | |
|---|---|
| Fay Huang | Ilan University, Yi-Lan, Taiwan |
| Reinhard Koch | University of Kiel, Germany |

## Tutorial Chair

| | |
|---|---|
| Terrence Sim | National University of Singapore |

## Demo Chairs

| | |
|---|---|
| Kenji Irie | Lincoln Ventures, New Zealand |
| Alan McKinnon | Lincoln University, New Zealand |

## Publication Chairs

| | |
|---|---|
| Michael Cree | University of Waikato, New Zealand |
| Keith Unsworth | Lincoln University, New Zealand |

## Publicity Chairs

| | |
|---|---|
| John Barron | University of Western Ontario, Canada |
| Domingo Mery | Pontificia Universidad Católica de Chile |
| Ioannis Pitas | Aristotle University of Thessaloniki, Greece |

## Area Chairs

| | |
|---|---|
| Donald G. Bailey | Massey University, Palmerston North, New Zealand |
| Horst Bischof | TU Graz, Austria |
| Alex Bronstein | Technion, Haifa, Israel |
| Michael S. Brown | National University of Singapore |
| Chu-Song Chen | Academia Sinica, Taipei, Taiwan |
| Hui Chen | Shandong University, Jinan, China |
| Laurent Cohen | University Paris Dauphine, France |
| Daniel Cremers | Bonn University, Germany |
| Eduardo Destefanis | Technical University Cordoba, Argentina |
| Hamid Krim | North Carolina State University, Raleigh, USA |
| Chil-Woo Lee | Chonnam National University, Gwangju, Korea |
| Facundo Memoli | Stanford University, USA |
| Kyoung Mu Lee | Seoul National University, Korea |
| Stephen Lin | Microsoft Research Asia, Beijing, China |
| Kai-Kuang Ma | Nanyang Technological University, Singapore |
| Niloy J. Mitra | Indian Institute of Technology, New Delhi, India |
| P.J. Narayanan | International Institute of Information Technology, Hyderabad, India |
| Nassir Navab | TU Munich, Germany |
| Takayuki Okatani | Tohoku University, Sendai City, Japan |
| Tomas Pajdla | Czech Technical University, Prague, Czech Republic |
| Nikos Paragios | Ecole Centrale de Paris, France |
| Robert Pless | Washington University, St. Louis, USA |
| Marc Pollefeys | ETH Zürich, Switzerland |
| Mariano Rivera | CIMAT Guanajuato, Mexico |
| Antonio Robles-Kelly | National ICT, Canberra, Australia |
| Hideo Saito | Keio University, Yokohama, Japan |

| | |
|---|---|
| Yoichi Sato | The University of Tokyo, Japan |
| Nicu Sebe | University of Trento, Italy |
| Stefano Soatto | University of California, Los Angeles, USA |
| Nir Sochen | Tel Aviv University, Israel |
| Peter Sturm | INRIA Grenoble, France |
| David Suter | University of Adelaide, Australia |
| Robby T. Tan | University of Utrecht, The Netherlands |
| Toshikazu Wada | Wakayama University, Japan |
| Yaser Yacoob | University of Maryland, College Park, USA |
| Ming-Hsuan Yang | University of California, Merced, USA |
| Hong Zhang | University of Alberta, Edmonton, Canada |
| Mengjie Zhang | Victoria University of Wellington, New Zealand |

## Program Committee Members

Abdenour, Hadid
Achard, Catherine
Ai, Haizhou
Aiger, Dror
Alahari, Karteek
Araguas, Gaston
Arica, Nafiz
Ariki, Yasuo
Arslan, Abdullah
Astroem, Kalle
August, Jonas
Aura Vese, Luminita
Azevedo-Marques, Paulo
Bagdanov, Andy
Bagon, Shai
Bai, Xiang
Baloch, Sajjad
Baltes, Jacky
Bao, Yufang
Bar, Leah
Barbu, Adrian
Barnes, Nick
Barron, John
Bartoli, Adrien
Baust, Maximilian
Ben Hamza, Abdessamad
BenAbdelkader, Chiraz
Ben-ari, Rami
Beng-Jin, AndrewTeoh
Benosman, Ryad
Berkels, Benjamin
Berthier, Michel
Bhattacharya, Bhargab
Biswas, Prabir
Bo, Liefeng
Boerdgen, Markus
Bors, Adrian
Boshra, Michael
Bouguila, Nizar
Boyer, Edmond
Bronstein, Michael
Bruhn, Andres
Buckley, Michael
Cai, Jinhai
Cai, Zhenjiang
Calderón, Jesús
Camastra, Francesco
Canavesio, Luisa
Cao, Xun
Carlo, Colombo
Carlsson, Stefan
Caspi, Yaron
Castellani, Umberto
Celik, Turgay
Cham, Tat-Jen
Chan, Antoni
Chandran, Sharat
Charvillat, Vincent

Chellappa, Rama
Chen, Bing-Yu
Chen, Chia-Yen
Chen, Chi-Fa
Chen, Haifeng
Chen, Hwann-Tzong
Chen, Jie
Chen, Jiun-Hung
Chen, Ling
Chen, Xiaowu
Chen, Xilin
Chen, Yong-Sheng
Cheng, Shyi-Chyi
Chia, Liang-Tien
Chien, Shao-Yi
Chin, Tat-Jun
Chuang, Yung-Yu
Chung, Albert
Chunhong, Pan
Civera, Javier
Coleman, Sonya
Cootes, Tim
Costeira, JoaoPaulo
Cristani, Marco
Csaba, Beleznai
Cui, Jinshi
Daniilidis, Kostas
Daras, Petros
Davis, Larry
De Campos, Teofilo
Demirci, Fatih
Deng, D. Jeremiah
Deng, Hongli
Denzler, Joachim
Derrode, Stephane
Diana, Mateus
Didas, Stephan
Dong, Qiulei
Donoser, Michael
Doretto, Gianfranco
Dorst, Leo
Duan, Fuqing
Dueck, Delbert
Duric, Zoran
Dutta Roy, Sumantra
Ebner, Marc
Einhauser, Wolfgang
Engels, Christopher
Eroglu-Erdem, Cigdem
Escolano, Francisco
Esteves, Claudia
Evans, Adrian
Fang, Wen-Pinn
Feigin, Micha
Feng, Jianjiang
Ferri, Francesc
Fite Georgel, Pierre
Flitti, Farid
Frahm, Jan-Michael
Francisco Giro Martín, Juan
Fraundorfer, Friedrich
Frosini, Patrizio
Fu, Chi-Wing
Fuh, Chiou-Shann
Fujiyoshi, Hironobu
Fukui, Kazuhiro
Fumera, Giorgio
Furst, Jacob
Fusiello, Andrea
Gall, Juergen
Gallup, David
Gang, Li
Gasparini, Simone
Geiger, Andreas
Gertych, Arkadiusz
Gevers, Theo
Glocker, Ben
Godin, Guy
Goecke, Roland
Goldluecke, Bastian
Goras, Bogdan
Gross, Ralph
Gu, I
Guerrero, Josechu
Guest, Richard
Guo, Guodong
Gupta, Abhinav
Gur, Yaniv
Hajebi, Kiana
Hall, Peter

Hamsici, Onur
Han, Bohyung
Hanbury, Allan
Harit, Gaurav
Hartley, Richard
HassabElgawi, Osman
Havlena, Michal
Hayes, Michael
Hayet, Jean-Bernard
He, Junfeng
Hee Han, Joon
Hiura, Shinsaku
Ho, Jeffrey
Ho, Yo-Sung
Ho Seo, Yung
Hollitt, Christopher
Hong, Hyunki
Hotta, Kazuhiro
Hotta, Seiji
Hou, Zujun
Hsu, Pai-Hui
Hua, Gang
Hua, Xian-Sheng
Huang, Chun-Rong
Huang, Fay
Huang, Kaiqi
Huang, Peter
Huang, Xiangsheng
Huang, Xiaolei
Hudelot, Celine
Hugo Sauchelli, Víctor
Hung, Yi-Ping
Hussein, Mohamed
Huynh, Cong Phuoc
Hyung Kim, Soo
Ichimura, Naoyuki
Ik Cho, Nam
Ikizler-Cinbis, Nazli
Il Park, Jong
Ilic, Slobodan
Imiya, Atsushi
Ishikawa, Hiroshi
Ishiyama, Rui
Iwai, Yoshio
Iwashita, Yumi
Jacobs, Nathan
Jafari-Khouzani, Kourosh
Jain, Arpit
Jannin, Pierre
Jawahar, C.V.
Jenkin, Michael
Jia, Jiaya
Jia, JinYuan
Jia, Yunde
Jiang, Shuqiang
Jiang, Xiaoyi
Jin Chung, Myung
Jo, Kang-Hyun
Johnson, Taylor
Joshi, Manjunath
Jurie, Frederic
Kagami, Shingo
Kakadiaris, Ioannis
Kale, Amit
Kamberov, George
Kanatani, Kenichi
Kankanhalli, Mohan
Kato, Zoltan
Katti, Harish
Kawakami, Rei
Kawasaki, Hiroshi
Keun Lee, Sang
Khan, Saad-Masood
Kim, Hansung
Kim, Kyungnam
Kim, Seon Joo
Kim, TaeHoon
Kita, Yasuyo
Kitahara, Itaru
Koepfler, Georges
Koeppen, Mario
Koeser, Kevin
Kokiopoulou, Effrosyni
Kokkinos, Iasonas
Kolesnikov, Alexander
Koschan, Andreas
Kotsiantis, Sotiris
Kown, Junghyun
Kruger, Norbert
Kuijper, Arjan

Kukenys, Ignas
Kuno, Yoshinori
Kuthirummal, Sujit
Kwolek, Bogdan
Kwon, Junseok
Kybic, Jan
Kyu Park, In
Ladikos, Alexander
Lai, Po-Hsiang
Lai, Shang-Hong
Lane, Richard
Langs, Georg
Lao, Shihong
Lao, Zhiqiang
Lauze, Francois
Le, Duy-Dinh
Le, Triet
Lee, Jae-Ho
Lee, Soochahn
Leistner, Christian
Leonardo, Bocchi
Leow, Wee-Kheng
Lepri, Bruno
Lerasle, Frederic
Li, Chunming
Li, Hao
Li, Hongdong
Li, Stan
Li, Yongmin
Liao, T.Warren
Lie, Wen-Nung
Lien, Jenn-Jier
Lim, Jongwoo
Lim, Joo-Hwee
Lin, Huei-Yung
Lin, Weisi
Lin, Wen-Chieh(Steve)
Ling, Haibin
Lipman, Yaron
Liu, Cheng-Lin
Liu, Jingen
Liu, Ligang
Liu, Qingshan
Liu, Qingzhong
Liu, Tianming
Liu, Tyng-Luh
Liu, Xiaoming
Liu, Yuncai
Loog, Marco
Lu, Huchuan
Lu, Juwei
Lu, Le
Lucey, Simon
Luo, Jiebo
Macaire, Ludovic
Maccormick, John
Madabhushi, Anant
Makris, Dimitrios
Manabe, Yoshitsugu
Marsland, Stephen
Martinec, Daniel
Martinet, Jean
Martinez, Aleix
Masuda, Takeshi
Matsushita, Yasuyuki
Mauthner, Thomas
Maybank, Stephen
McHenry, Kenton
McNeill, Stephen
Medioni, Gerard
Mery, Domingo
Mio, Washington
Mittal, Anurag
Miyazaki, Daisuke
Mobahi, Hossein
Moeslund, Thomas
Mordohai, Philippos
Moreno, Francesc
Mori, Greg
Mori, Kensaku
Morris, John
Mueller, Henning
Mukaigawa, Yasuhiro
Mukhopadhyay, Jayanta
Muse, Pablo
Nagahara, Hajime
Nakajima, Shin-ichi
Nanni, Loris
Neshatian, Kourosh
Newsam, Shawn

Niethammer, Marc
Nieuwenhuis, Claudia
Nikos, Komodakis
Nobuhara, Shohei
Norimichi, Ukita
Nozick, Vincent
Ofek, Eyal
Ohnishi, Naoya
Oishi, Takeshi
Okabe, Takahiro
Okuma, Kenji
Olague, Gustavo
Omachi, Shinichiro
Ovsjanikov, Maks
Pankanti, Sharath
Paquet, Thierry
Paternak, Ofer
Patras, Ioannis
Pauly, Olivier
Pavlovic, Vladimir
Peers, Pieter
Peng, Yigang
Penman, David
Pernici, Federico
Petrou, Maria
Ping, Wong Ya
Prasad Mukherjee, Dipti
Prati, Andrea
Qian, Zhen
Qin, Xueyin
Raducanu, Bogdan
Rafael Canali, Luis
Rajashekar, Umesh
Ramalingam, Srikumar
Ray, Nilanjan
Real, Pedro
Remondino, Fabio
Reulke, Ralf
Reyes, EdelGarcia
Ribeiro, Eraldo
Riklin Raviv, Tammy
Roberto, Tron
Rosenhahn, Bodo
Rosman, Guy
Roth, Peter
Roy Chowdhury, Amit
Rugis, John
Ruiz Shulcloper, Jose
Ruiz-Correa, Salvador
Rusinkiewicz, Szymon
Rustamov, Raif
Sadri, Javad
Saffari, Amir
Saga, Satoshi
Sagawa, Ryusuke
Salzmann, Mathieu
Sanchez, Jorge
Sang, Nong
Sang Hong, Ki
Sang Lee, Guee
Sappa, Angel
Sarkis, Michel
Sato, Imari
Sato, Jun
Sato, Tomokazu
Schiele, Bernt
Schikora, Marek
Schoenemann, Thomas
Scotney, Bryan
Shan, Shiguang
Sheikh, Yaser
Shen, Chunhua
Shi, Qinfeng
Shih, Sheng-Wen
Shimizu, Ikuko
Shimshoni, Ilan
Shin Park, You
Sigal, Leonid
Sinha, Sudipta
So Kweon, In
Sommerlade, Eric
Song, Andy
Souvenir, Richard
Srivastava, Anuj
Staiano, Jacopo
Stein, Gideon
Stottinge, Julian
Strecha, Christoph
Strekalovskiy, Evgeny
Subramanian, Ramanathan

Sugaya, Noriyuki
Sumi, Yasushi
Sun, Weidong
Swaminathan, Rahul
Tai, Yu-Wing
Takamatsu, Jun
Talbot, Hugues
Tamaki, Toru
Tan, Ping
Tanaka, Masayuki
Tang, Chi-Keung
Tang, Jinshan
Tang, Ming
Taniguchi, Rinichiro
Tao, Dacheng
Tavares, João Manuel R.S.
Teboul, Olivier
Terauchi, Mutsuhiro
Tian, Jing
Tian, Taipeng
Tobias, Reichl
Toews, Matt
Tominaga, Shoji
Torii, Akihiko
Tsin, Yanghai
Turaga, Pavan
Uchida, Seiichi
Ueshiba, Toshio
Unger, Markus
Urtasun, Raquel
van de Weijer, Joost
Van Horebeek, Johan
Vassallo, Raquel
Vasseur, Pascal
Vaswani, Namrata
Wachinger, Christian
Wang, Chen
Wang, Cheng
Wang, Hongcheng
Wang, Jue
Wang, Yu-Chiang
Wang, Yunhong
Wang, Zhi-Heng
Wang, Zhijie
Wolf, Christian
Wolf, Lior
Wong, Kwan-Yee
Woo, Young
Wook Lee, Byung
Wu, Jianxin
Xue, Jianru
Yagi, Yasushi
Yan, Pingkun
Yan, Shuicheng
Yanai, Keiji
Yang, Herbert
Yang, Jie
Yang, Yongliang
Yi, June-Ho
Yilmaz, Alper
You, Suya
Yu, Jin
Yu, Tianli
Yuan, Junsong
Yun, Il Dong
Zach, Christopher
Zelek, John
Zha, Zheng-Jun
Zhang, Cha
Zhang, Changshui
Zhang, Guofeng
Zhang, Hongbin
Zhang, Li
Zhang, Liqing
Zhang, Xiaoqin
Zheng, Lu
Zheng, Wenming
Zhong, Baojiang
Zhou, Cathy
Zhou, Changyin
Zhou, Feng
Zhou, Jun
Zhou, S.
Zhu, Feng
Zou, Danping
Zucker, Steve

## Additional Reviewers

Bai, Xiang
Collins, Toby
Compte, Benot
Cong, Yang
Das, Samarjit
Duan, Lixing
Fihl, Preben
Garro, Valeria
Geng, Bo
Gherardi, Riccardo
Giusti, Alessandro
Guo, Jing-Ming
Gupta, Vipin
Han, Long
Korchev, Dmitriy
Kulkarni, Kaustubh
Lewandowski, Michal
Li, Xin
Li, Zhu
Lin, Guo-Shiang
Lin, Wei-Yang
Liu, Damon Shing-Min
Liu, Dong
Luo, Ye
Magerand, Ludovic
Molineros, Jose
Rao, Shankar
Samir, Chafik
Sanchez-Riera, Jordy
Suryanarayana, Venkata
Tang, Sheng
Thota, Rahul
Toldo, Roberto
Tran, Du
Wang, Jingdong
Wu, Jun
Yang, Jianchao
Yang, Linjun
Yang, Kuiyuan
Yuan, Fei
Zhang, Guofeng
Zhuang, Jinfeng

## ACCV2010 Best Paper Award Committee

| | |
|---|---|
| Alfred M. Bruckstein | Technion, Israel Institute of Techonlogy, Israel |
| Larry S. Davis | University of Maryland, USA |
| Richard Hartley | Australian National University, Australia |
| Long Quan | The Hong Kong University of Science and Technology, Hong Kong |

## Sponsors of ACCV2010

| | |
|---|---|
| Main Sponsor | The Asian Federation of Computer Vision Societies (AFCV) |
| Gold Sponsor | NextWindow – Touch-Screen Technology |
| Silver Sponsors | Areograph – Interactive Computer Graphics<br>Microsoft Research Asia<br>Australia's Information and Communications Technology (NICTA)<br>Adept Electronic Solutions |
| Bronze Sponsor | 4D View Solutions |
| Best Student Paper Sponsor | *The International Journal of Computer Vision* (IJCV) |

## Best Paper Prize ACCV 2010

Context-Based Support Vector Machines for Interconnected Image Annotation
Hichem Sahbi, Xi Li.

## Best Student Paper ACCV 2010

Fast Spectral Reflectance Recovery Using DLP Projector
Shuai Han, Imari Sato, Takahiro Okabe, Yoichi Sato

## Best Application Paper ACCV 2010

Network Connectivity via Inference Over Curvature-Regularizing Line Graphs
Maxwell Collins, Vikas Singh, Andrew Alexander

## Honorable Mention ACCV 2010

Image-Based 3D Modeling via Cheeger Sets
Eno Toeppe, Martin Oswald, Daniel Cremers, Carsten Rother

## Outstanding Reviewers ACCV 2010

Philippos Mordohai
Peter Roth
Matt Toews
Andres Bruhn
Sudipta Sinha
Benjamin Berkels
Mathieu Salzmann

# Table of Contents – Part I

## Keynote



## Geometry and Correspondence



## Computational Photography and Low Level Vision

## Detection and Recognition I



## Keynote



## Applied Computer Vision



## Tracking and Categorization

## Image Sensing



## Keynote



## Segmentation and Texture



## Detection and Recognition II

## Matching and Similarity

# Table of Contents – Part II

## Posters on Day 1 of ACCV 2010

# Table of Contents – Part III

## Posters on Day 2 of ACCV 2010

# Table of Contents – Part IV

## Posters on Day 3 of ACCV 2010

# Deformable Object Modelling and Matching

Tim F. Cootes

The University of Manchester, UK

**Abstract.** Statistical models of the shape and appearance of deformable objects have become widely used in Computer Vision and Medical Image Analysis. Here we give an overview of such models and of two efficient algorithms for matching such models to new images (Active Shape Models and Active Appearance Models). We also describe recent work on automatically constructing such models from minimally labelled training images.

## 1 Statistical Shape Models

Many objects of interest in computer vision can be considered to be some deformed version of an "average" shape. For instance, most human faces have two eyes, a nose and a mouth in similar relative positions, and good approximations to each face can be generated by modest distortions of a standard template. Similarly many anatomical structures (such as human bones, or the heart) have broadly similar shapes across a population.

Statistical shape models seek to represent such objects. Since their introduction Point Distribution Models [1], which represent shapes as a linear combination of modes of variation about the mean, have found wide application.

These represent a shape using a set of points $\{x_i, y_i\}$, $(i = 1..n)$, which define particular positions on the object of interest. They are placed at consistent positions on every example in a training set (see Figure 1).

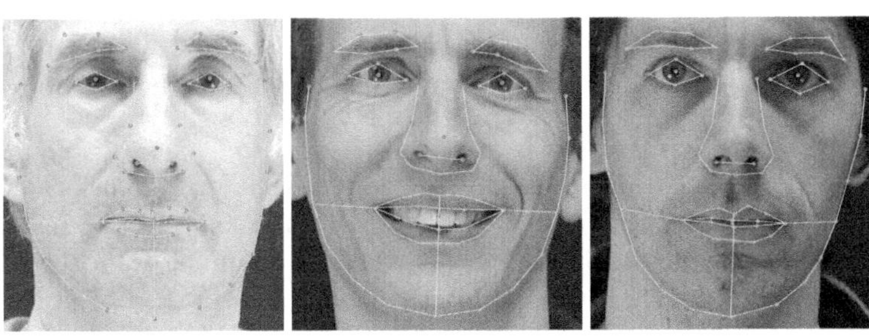

**Fig. 1.** Examples of faces with 68 points annotated on each, defining correspondences across the set

By applying Procrustes Analysis [2] the examples can be aligned into a common co-ordinate frame. Principal Component Analysis is then used to build a linear model of the variation over the set as

R. Kimmel, R. Klette, and A. Sugimoto (Eds.): ACCV 2010, Part I, LNCS 6492, pp. 1–10, 2011.

$$\boldsymbol{x} = \hat{\boldsymbol{x}} + \boldsymbol{P}\boldsymbol{b} \tag{1}$$

where $\boldsymbol{x} = (x_1, y_1, ..., x_n, y_n)^T$, $\hat{\boldsymbol{x}}$ is the mean shape, $\boldsymbol{P}$ is a set of eigenvectors of the covariance matrix describing the modes of variation, and $\boldsymbol{b}$ is a vector of shape parameters (see [1] for details). For instance, Figure 2 shows the effect of varying the first three parameters of a face shape model[1].

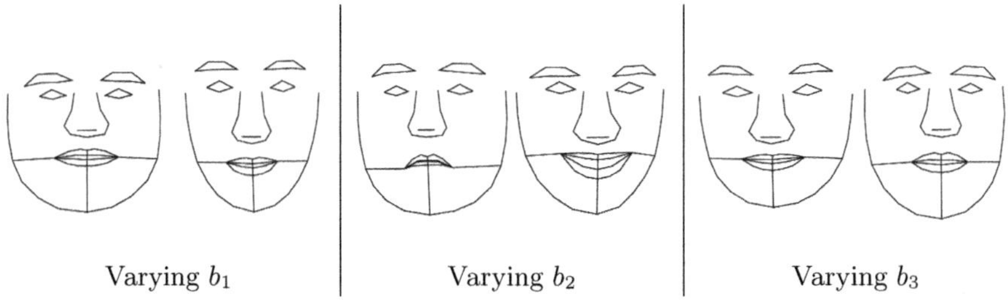


**Fig. 2.** First three modes of shape variation of a face model

Such models typically have far fewer parameters (elements of $\boldsymbol{b}$) than points, as they take advantage of the correlation between points on the shape. For instance, nearby points on a boundary are usually correlated, and symmetries of the shape reduce the degrees of freedom even further.

Statistical shape models can be used to analyse the differences in shapes between populations, or can be used to help locate structures in new images.

## 2 Active Shape Models

An Active Shape Model (ASM) is a method of matching a statistical shape model of the form described above to a new image. The positions of the points in the image are given by the equation

$$\boldsymbol{X} = T_\theta(\hat{\boldsymbol{x}} + \boldsymbol{P}\boldsymbol{b}) \tag{2}$$

where $T_\theta(\boldsymbol{x})$ applies a transformation (typically a similarity transformation) to the set of points encoded in the vector $\boldsymbol{x}$, with parameters $\theta$ (for instance rotation, scaling and translation). $T_\theta$ thus defines the mapping from the reference frame to the target image frame, giving the global pose of the object.

For an ASM we require a method of locating a good candidate position for each model point in a region. Typically this involves building a statistical model of the image patch about each point from the training set, then searching the region for the best match using this model [1,3].

The Active Shape Model matches using a simple alternating algorithm:

[1] C++ source code for building shape models will be made available in VXL (vxl.sourceforge.net) in the `contrib/mul/msm` library.

1. Search around each current point position $\boldsymbol{X}_i$ with the associated local model to find a better position, $\boldsymbol{Z}_i$
2. Find the shape and pose parameters $\{\boldsymbol{b}, \theta\}$ to best fit the model to the found points $\{\boldsymbol{Z}_i\}$

Though in early work the local search was along profiles normal to the model boundaries [1], this is naturally generalised to searching regions around each point [3]. The local models can either be simple Gaussian models of the image patch [1,4], or more sophisticated classifiers [5,6]. Alternatively regression techniques can be used to directly predict the movement of each point [7].

## 3 Active Appearance Models

Rather than search for each point independently, as the ASM does, the Active Appearance Model (AAM) approach is to predict an update to the model parameters directly from samples of the image. This takes into account correlations between the image patterns across the shape.

The original formulation [8,9] was designed to efficiently fit a statistical appearance model to a new image. Such a model combines a statistical shape model with a model of image texture in a normalised reference frame. It is a generative model, in that it can create synthetic images of the objects on which it has been trained [10,11]. A texture model is used to generate the intensity pattern in the reference frame, which is then warped to the target image frame using a deformation defined by the shape model parameters. Figure 3 shows the first three appearance modes of a face model, demonstrating how shape and texture vary together.

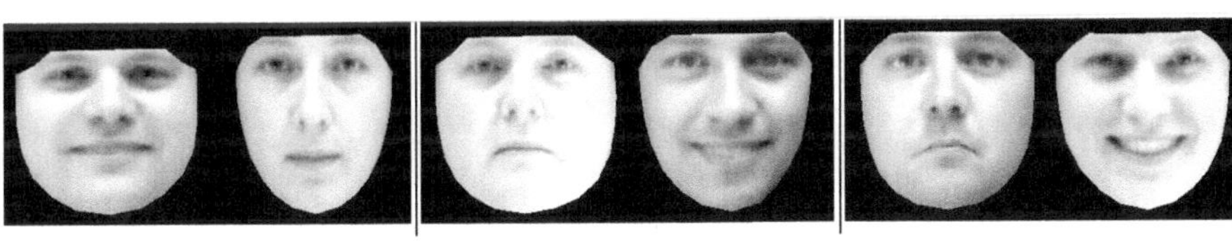

**Fig. 3.** First three modes of appearance variation of a face model

Matching such a model to a new image is a difficult optimisation problem. If $\boldsymbol{p}$ represents all the model parameters, $\boldsymbol{t}(\boldsymbol{p})$ the texture generated by the current model, and $\boldsymbol{s}(\boldsymbol{p})$ a vector of image samples taken at the current position, then a simple approach is to seek the parameters which minimise

$$E(\boldsymbol{p}) = |\boldsymbol{s}(\boldsymbol{p}) - \boldsymbol{t}(\boldsymbol{p})|^2 \tag{3}$$

the sum of squared errors between model and image.

By differentiating this equation and making certain approximations, it is possible to show [12] that a good estimate of the optimal update to the parameters is given by a simple linear equation,

$$\delta \boldsymbol{p} = \boldsymbol{R}(\boldsymbol{s}(\boldsymbol{p}) - \boldsymbol{t}(\boldsymbol{p})) \tag{4}$$

where the update matrix $\boldsymbol{R}$ is the pseudo-inverse of a Jacobian which can be estimated from training data.

The AAM matching algorithm then simply repeats this process, updating the model parameters based on the difference between the model and the image samples. Usually a coarse-to-fine approach is used, in which low resolution models are used during the early parts of the search, and later refined with more detailed models.

Although the derivation of the update matrix $\boldsymbol{R}$ as the pseudo-inverse of the Jacobian is elegant, it turns out that often it is better to treat the task as a regression problem, in which we seek the matrix which gives the best parameter updates. If we generate many random parameter displacements $\delta\boldsymbol{p}$ on a training set, and for each compute the residual error $\boldsymbol{r} = \boldsymbol{s}(\boldsymbol{p} + \delta\boldsymbol{p}) - \boldsymbol{t}(\boldsymbol{p} + \delta\boldsymbol{p})$, we can then use regression methods to estimate the matrix $\boldsymbol{R}$ [9,13]. This has been shown to lead to more accurate results than using the inverse of the Jacobian [13,14].

This general approach has proved very effective, and has been developed in many directions including

- a more elegant compositional update scheme [15]
- the inclusion of constraints [16]
- the use of other image features for improved robustness [17,18]
- methods of dealing with occlusion [19]
- combining 2D and 3D models for face tracking [20]
- more sophisticated update schemes [21,22,14]

amongst many others.

Though often used for face model matching, the AAM has been widely applied in medical image interpretation. For instance, modelling the heart [23], hand [24], brain [25] or knee [26].

## 4 Automatic Model Construction

The shape and appearance models used in the above methods rely on training sets containing points annotated on each of a representative set of images. Manually annotating such data is time consuming, and is particularly difficult for three dimensional volume images, widely used in medical imaging.

Thus there has been a long history of work attempting to automate the model building process. The points on each image define the correspondences between the objects viewed. If such correspondences can be estimated automatically, a model can be constructed with minimal human intervention.

If we have the boundary of each 2D object (or surface of each 3D object), effective correspondence can be found using techniques which optimise the ability of the resulting model to encode the shapes - Minimum Description Length methods [27,28].

Where we only have images, with no annotation, the problem is more challenging. A common approach is to use non-rigid registration or optical flow methods to find the correspondences between each image and a reference image. Some of the earliest work in this vein was by Vetter, Jones and Poggio [29,30] who used a combination of model fitting and optical flow to estimate dense vector fields across sets of objects, to build 'morphable models' - statistical models of shape and appearance.

Registering images to build an average 'atlas' is a widely used technique in medical image analysis. For instance, Guimond *et al.* [31] used Thirion's 'Demons' algorithm [32] to register sets of images, describing how to iteratively update the group mean. Frangi *et al.* used non-rigid B-spine registration [33] to correspond 3D images and build statistical shape models [34]. Joshi *et al.* [35,36] demonstrate how to simultaneously estimate the reference shape and image in the case of large scale diffeomorphisms, where linear approximations to averages break down.

Our early work in this field focused on estimating diffeomorphisms (smooth invertable mappings) from a reference frame to each target image in a set. We assume that each image in a set should contain the same structures, and hence there should be a unique and invertible one-to-one correspondence between all points on each pair of images - a diffeomorphism (see [37]).

For any two diffeomorphisms $f(\boldsymbol{x})$, $g(\boldsymbol{x})$, their composition $(f \circ g)(\boldsymbol{x}) \equiv f(g(\boldsymbol{x}))$ is also a diffeomorphism. We can thus construct a wide class of diffeomorphic functions by repeated compositions of a basis set of simple diffeomorphisms. In [38] we describe a coarse-to-fine algorithm for estimating the correspondences across a group using such compositions.

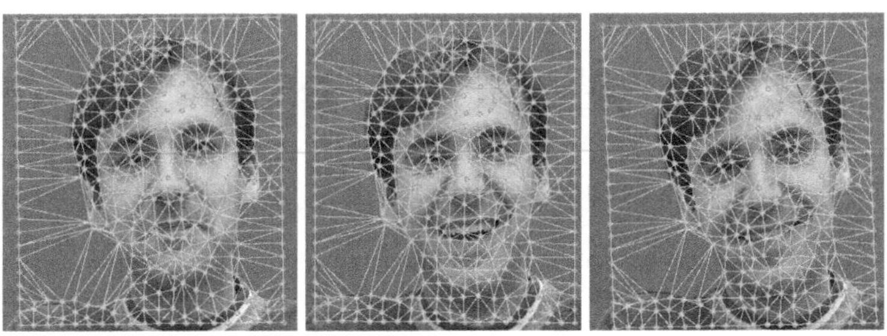

**Fig. 4.** Examples from a training set with resulting control points [39]

However, although the compositional approach can generate deformation fields which are guaranteed to be invertable, actually computing the inverse can be difficult, as it usually involves a non-linear optimisation. A more pragmatic alternative is to represent deformation fields using a triangulated mesh (in 2D) or a tetrahedral mesh (in 3D) guided by the control points at the nodes. Affine interpolation can be used to compute the transformation inside the mesh elements, leading to a piece-wise affine representation of the warp. As long as the triangles/tetrahedra do not 'flip', the transformation is invertable - one simply swaps the source and destination control points [39]. Although not smooth, as there are discontinuities in the derivative at the element boundaries, such representations are accurate enough for many applications (Figure 4).

In [12,39] we describe a groupwise image registration algorithm in which such meshes are used in a Minimum Description Length optimisation framework. A related method is described by Baker *et al.* [40].

Figure 5 shows the first three appearance modes of a model constructed from 300 face images of different people, with the correspondences computed automatically using the groupwise algorithm from [41].

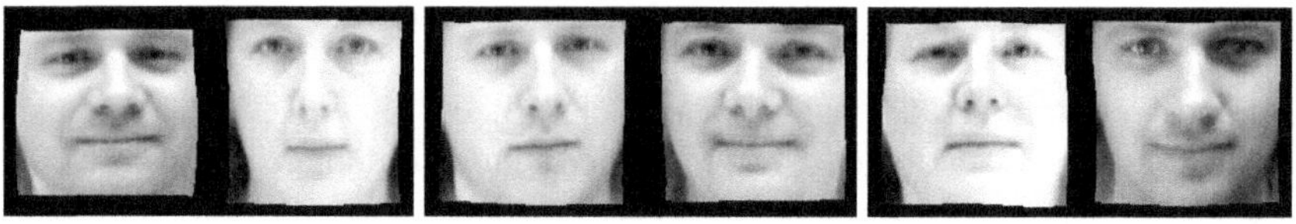

Mode 1 Mode 2 Mode 3

**Fig. 5.** First three modes of appearance variation of a face model constructed from automatically computed correspondences

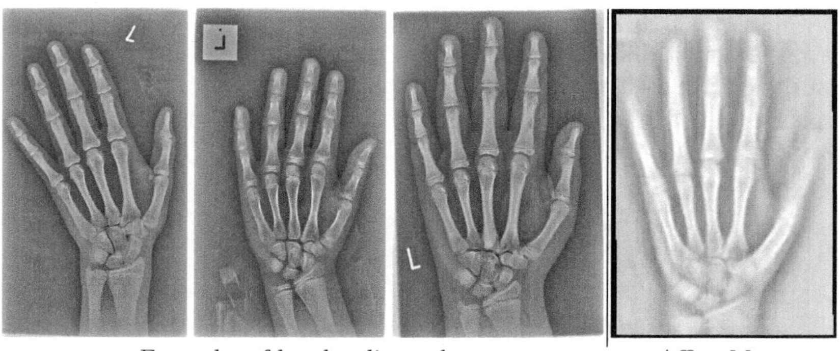

Examples of hand radiographs Affine Mean

**Fig. 6.** Examples of hand radiographs, which display considerable shape variation, and the resulting affine mean (from [42])

### 4.1 Initialising Groupwise Registration

The methods described above work well, given a good enough initialisation. As they involve local optimisation, if poorly initialised, they fall into local minima. Typically an affine transformation is estimated as the initialisation. However, for objects with significant shape variation this may not be sufficient. For example when registering hand radigraphs (Figure 6), the resulting affine mean is poor, and further registration does not significantly improve the result as it has fallen into a local minima.

To overcome this, more accurate initialisation is required. In [43] we show that by manually annotating a small number of key points on a single image we can construct a parts+geometry model from the whole training set, which can accurately locate those points on all the images. Such points give a sparse corresondence, which can be used to initialise a denser groupwise registration.

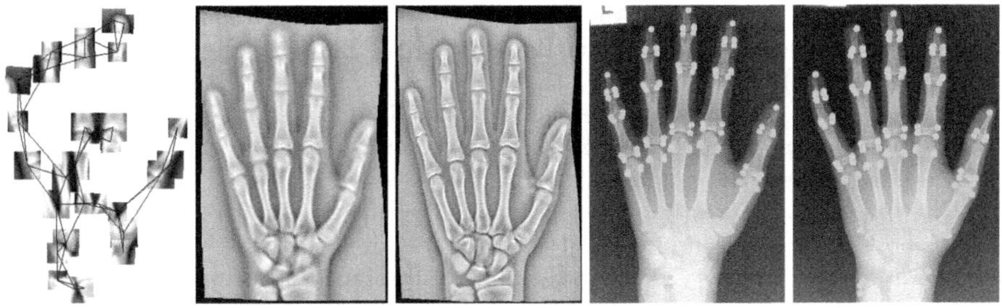


**Fig. 7.** Automatically generated parts+geometry models and results of dense groupwise registration [42]

In [42] this approach is extended to automatically select a sparse set of points which can be accurately located across the whole set. We use a variant of the Genetic Algorithm to select good subsets from a large pool of candidate part models. For instance, Figure 7 shows the application of this approach to hand radiographs.

## 5 Conclusions

Statistical models of shape and appearance are powerful tools for interpretting images. They can be matched to new images using the Active Shape Model or Active Appearance Model algorithms or their variants. Such models are built from annotated training images. Though significant progress has been made in developing algorithms to automatically estimate suitable correspondences, more work is required to make such algorithms sufficiently robust on challenging data.

**Acknowledgements.** Thanks to all to collaborators within ISBE and beyond who have contributed to this work.

## References

1. Cootes, T.F., Taylor, C.J., Cooper, D., Graham, J.: Active shape models - their training and application. Computer Vision and Image Understanding 61, 38–59 (1995)
2. Goodall, C.: Procrustes methods in the statistical analysis of shape. Journal of the Royal Statistical Society B 53, 285–339 (1991)
3. Milborrow, S., Nicolls, F.: Locating facial features with an extended active shape model. In: Forsyth, D., Torr, P., Zisserman, A. (eds.) ECCV 2008, Part IV. LNCS, vol. 5305, pp. 504–513. Springer, Heidelberg (2008), http://www.milbo.users.sonic.net/stasm
4. Cootes, T.F., Page, G., Jackson, C., Taylor, C.J.: Statistical grey-level models for object location and identification. Image and Vision Computing 14, 533–540 (1996)

5. van Ginneken, B., Frangi, A.F., Stall, J.J., ter Haar Romeny, B.: Active shape model segmentation with optimal features. IEEE Trans. Medical Imaging 21, 924–933 (2002)
6. Wimmer, M., Stulp, F., Tschechne, S.J., Radig, B.: Learning robust objective functions for model fitting in image understanding applications. In: Proc. British Machine Vision Conference, vol. 3, pp. 1159–1168 (2006)
7. Cristinacce, D., Cootes, T.: Boosted active shape models. In: Proc. British Machine Vision Conference, vol. 2, pp. 880–889 (2007)
8. Edwards, G., Taylor, C.J., Cootes, T.F.: Interpreting face images using active appearance models. In: 3rd International Conference on Automatic Face and Gesture Recognition 1998, Japan, pp. 300–305 (1998)
9. Cootes, T.F., Edwards, G.J., Taylor, C.J.: Active appearance models. In: Burkhardt, H., Neumann, B. (eds.) ECCV 1998. LNCS, vol. 1407, pp. 484–498. Springer, Heidelberg (1998)
10. Lanitis, A., Taylor, C.J., Cootes, T.F.: Automatic interpretation and coding of face images using flexible models. IEEE Trans. on Pattern Analysis and Machine Intelligence 19, 743–756 (1997)
11. Edwards, G.J., Taylor, C.J., Cootes, T.F.: Learning to identify and track faces in image sequences. In: 8th British Machine Vision Conference, Colchester, UK, pp. 130–139 (1997)
12. Cootes, T.F., Edwards, G.J., Taylor, C.J.: Active appearance models. IEEE Trans. Pattern Analysis and Machine Intelligence 23, 681–685 (2001)
13. Donner, R., Reiter, M., Langs, G., Peloschek, P., Bischof, H.: Fast active appearance model search using canonical correlation analysis. IEEE Trans. Pattern Analysis and Machine Intelligence 28, 1690–1694 (2006)
14. Tresadern, P., Sauer, P., Cootes, T.: Additive update predictors in active appearance models. In: British Machine Vision Conference. BMVA Press (2010)
15. Matthews, I., Baker, S.: Active appearance models revisited. International Journal of Computer Vision 60, 135–164 (2004)
16. Cootes, T.F., Taylor, C.J.: Constrained active appearance models. In: 8th International Conference on Computer Vision, vol. 1, pp. 748–754. IEEE Computer Society Press, Los Alamitos (2001)
17. Cootes, T.F., Taylor, C.J.: On representing edge structure for model matching. Computer Vision and Pattern Recognition 1, 1114–1119 (2001)
18. Scott, I.M., Cootes, T.F., Taylor, C.J.: Improving appearance model matching using local image structure. In: Taylor, C.J., Noble, J.A. (eds.) IPMI 2003. LNCS, vol. 2732, pp. 258–269. Springer, Heidelberg (2003)
19. Gross, R., Matthews, I., Baker, S.: Constructing and fitting active appearance models with occlusion. In: Proceedings of the IEEE Workshop on Face Processing in Video (2004)
20. Xiao, J., Baker, S., Matthews, I., Kanade, T.: Real-time combined 2D+3D active appearance models. In: Computer Vision and Pattern Recognition, vol. 2, pp. 535–542. IEEE, Los Alamitos (2004)
21. Saragih, J., Goecke, R.: Iterative error bound minimisation for AAM alignment. In: Proc. ICPR, vol. 2, pp. 1192–1195 (2006)
22. Saragih, J., Goecke, R.: A non-linear discriminative approach to AAM fitting. In: Proc. ICCV 2007 (2007)
23. Mitchell, S., Lelieveldt, B., van der Geest, R., Schaap, J., Reiber, J., Sonka, M.: Segmentation of cardiac MR images: An active appearance model approach. In: SPIE Medical Imaging (2000)

24. Thodberg, H.H.: Hands-on experience with active appearance models. In: SPIE Medical Imaging (2002)
25. Babalola, K., Cootes, T., Twining, C., Petrovic, V., Taylor, C.: 3D brain segmentation using active appearance models and local regressors. In: Metaxas, D., Axel, L., Fichtinger, G., Székely, G. (eds.) MICCAI 2008, Part I. LNCS, vol. 5241, pp. 401–408. Springer, Heidelberg (2008)
26. Vincent, G., Wolstenholme, C., Scott, I., Bowes, M.: Fully automatic segmentation of the knee joint using active appearance models. In: Medical Image Analysis for the Clinic: A Grand Challenge (2010)
27. Davies, R., Twining, C., Cootes, T., Taylor, C.: A minimum description length approach to statistical shape modelling. IEEE Trans. on Medical Imaging 21, 525–537 (2002)
28. Davies, R., Twining, C., Cootes, T., Waterton, J., Taylor, C.: 3D Statistical Shape Models Using Direct Optimisation of Description Length. In: Heyden, A., Sparr, G., Nielsen, M., Johansen, P. (eds.) ECCV 2002. LNCS, vol. 2352, pp. 3–20. Springer, Heidelberg (2002)
29. Vetter, T., Jones, M., Poggio, T.: A bootstrapping algorithm for learning linear models of object classes. In: Computer Vision and Pattern Recognition Conference 1997, pp. 40–46 (1997)
30. Jones, M.J., Poggio, T.: Multidimensional morphable models. In: 6th International Conference on Computer Vision, pp. 683–688 (1998)
31. Guimond, A., Meunier, J., Thirion, J.P.: Automatic computation of average brain models. In: Wells, W.M., Colchester, A.C.F., Delp, S.L. (eds.) MICCAI 1998. LNCS, vol. 1496, pp. 631–640. Springer, Heidelberg (1998)
32. Thirion, J.P.: Image matching as a diffusion process: an analogy with Maxwell's demons. Medical Image Analysis 2, 243–260 (1998)
33. Rueckert, D., Sonoda, L.I., Hayes, C., Hill, D.L.G., Leach, M.O., Hawkes, D.J.: Non-rigid registration using free-form deformations: Application to breast MR images. IEEE Trans. Medical Imaging 18, 712–721 (1999)
34. Frangi, A., Rueckert, D., Schnabel, J., Niessen, W.: Automatic 3D ASM construction via atlas-based landmarking and volumetric elastic registration. In: Insana, M.F., Leahy, R.M. (eds.) IPMI 2001. LNCS, vol. 2082, pp. 78–91. Springer, Heidelberg (2001)
35. Joshi, S., Davis, B., Jomier, M., Gerig, G.: Unbiased diffeomorphic atlas construction for computational anatomy. NeuroImage 23, S151–S160 (2004)
36. Lorenzen, P., Davis, B., Joshi, S.: Unbiased atlas formation via large deformations metric mapping. In: Duncan, J.S., Gerig, G. (eds.) MICCAI 2005. LNCS, vol. 3750, pp. 411–418. Springer, Heidelberg (2005)
37. Twining, C., Marsland, S., Taylor, C.: Measuring geodesic distances on the space of bounded diffeomorphisms. In: Rosin, P.L., Marshall, D. (eds.) 13th British Machine Vision Conference, vol. 2, pp. 847–856. BMVA Press (2002)
38. Cootes, T., Marsland, S., Twining, C., Smith, K., Taylor, C.: Groupwise diffeomorphic non-rigid registration for automatic model building. In: Pajdla, T., Matas, J(G.) (eds.) ECCV 2004. LNCS, vol. 3024, pp. 316–327. Springer, Heidelberg (2004)
39. Cootes, T., Twining, C., Petrović, V., Schestowitz, R., Taylor, C.: Groupwise construction of appearance models using piece-wise affine deformations. In: 16th British Machine Vision Conference, vol. 2, pp. 879–888 (2005)
40. Baker, S., Matthews, I., Schneider, J.: Automatic construction of active appearance models as an image coding problem. IEEE Trans. on Pattern Analysis and Machine Intelligence 26, 1380–1384 (2004)

41. Cootes, T.F., Twining, C.J., Petrović, V.S., Babalola, K.O., Taylor, C.J.: Computing accurate correspondences across groups of images. IEEE Trans. Pattern Analysis and Machine Intelligence 32 (to appear, 2010)
42. Zhang, P., Cootes, T.F.: Learning sparse correspondences for initialising groupwise registration. In: Jiang, T., Navab, N., Pluim, J.P.W., Viergever, M.A. (eds.) MICCAI 2010. LNCS, vol. 6362, pp. 635–642. Springer, Heidelberg (2010)
43. Adeshina, S., Cootes, T.F.: Constructing part-based models for groupwise registration. In: Proc. ISBI (2010)

# New Efficient Solution to the Absolute Pose Problem for Camera with Unknown Focal Length and Radial Distortion

Martin Bujnak[1], Zuzana Kukelova[2], and Tomas Pajdla[2]

[1] Bzovicka 24, 85107, Bratislava, Slovakia
[2] Center for Machine Perception, Czech Technical University in Prague

**Abstract.** In this paper we present a new efficient solution to the absolute pose problem for a camera with unknown focal length and radial distortion from four 2D-to-3D point correspondences. We propose to solve the problem separately for non-planar and for planar scenes. By decomposing the problem into these two situations we obtain simpler and more efficient solver than the previously known general solver. We demonstrate in synthetic and real experiments significant speedup as our new solvers are about 40× (non-planar) and 160× (planar) faster than the general solver. Moreover, we show that our two solvers can be joined into a new general solver, which gives comparable or better results than the existing general solver for of most planar as well as non-planar scenes.

## 1 Introduction

The Perspective-n-Point (PnP) problem, i.e. the problem of determining the absolute position and orientation of a camera given its intrinsic parameters and a set of $n$ 2D-to-3D point correspondences, is one of the most important problems in computer vision with a broad range of applications in structure from motion [1, 21] or recognition [16, 17]. One of the oldest papers considering this problem dates back to 1841 [11]. Recently a huge number of solutions to the calibrated PnP problems for three and more than three points have been published [4, 9, 13, 18, 19, 24, 25].

The minimal number of points needed to estimate the camera position and orientation is three, resp. six, for a fully calibrated, resp. a fully uncalibrated, camera. The linear solution to the problem of estimating absolute position and orientation together with five inner calibration parameters of a fully uncalibrated camera from six 2D-3D point correspondences is known as Direct Linear Transform (DLT) [2, 20].

Modern digital cameras have square pixels and the principal point close to the center of the image [12]. Therefore, for most of the applications this prior knowledge can be used and four out of the five internal calibration parameters can be safely set to some prior value (the skew to 0, the pixel aspect ratio to 1 and the principal point to the center of the image).

Adopting these calibration constraints has several advantages. First, the minimal number of points needed to solve the absolute pose of a camera is reduced. Secondly, since fewer parameters are estimated, the results are more stable.

R. Kimmel, R. Klette, and A. Sugimoto (Eds.): ACCV 2010, Part I, LNCS 6492, pp. 11–24, 2011.

In this paper we use this prior calibration knowledge and provide efficient solution to the minimal problem of estimating the absolute pose of a camera with unknown focal length and radial distortion from images of four known 3D points.

This solution is non-iterative and based on Gröbner basis methods [8] for solving systems of polynomial equations.

The problem of estimating absolute pose of a camera together with its focal length for image points without radial distortion was firstly solved by Abidi and Chandra [3]. In this paper authors formulated the problem using areas of triangular subdivisions of a planar quadrangle and arrived to a closed form solution which works only for planar scenes.

The first solution to this focal length problem, which works for non-planar scenes, was presented by Triggs in [23]. This solution uses calibration constraints arising from using dual image of the absolute quadric and solves resulting polynomial equations using multivariate resultants [8]. The solution works for non-planar scenes but fails for the planar ones. In this paper authors also proposed a solution which handles both planar and non-planar points and is based on eigendecomposition of multiplication matrices, however this solution is numerically unstable and not practical. The paper [23] also provided a solution to the problem of estimating absolute pose of a camera with unknown focal length and unknown principal point from five 2D-to-3D correspondences.

A solution working for both planar and non-planar scenes was proposed only recently in [5]. This solution is based on Euclidean rigidity constraint and results in a system of four polynomial equations in four unknowns which are solved using the Gröbner basis method [8].

The problem of estimating absolute pose with unknown focal length from four 2D-to-3D correspondences is not minimal and one additional calibration parameter can be handled in this problem. In [14] authors included the radial distortion to the problem and proposed a method for solving absolute pose problem for a camera with radial distortion and unknown focal length from four point correspondences based on Gröbner bases. In this paper authors show that in many real applications the consideration of radial distortion brings a significant improvement. The presented solution uses quaternions to parametrize rotations and one parameter division model for the radial distortion [10] and results in five equations in five unknowns. These equations are quite complex and therefore the Gröbner basis method results in relatively large solver (1134 x 720 matrix) which runs about $70ms$. Therefore, the proposed solver is not really practical in real-time applications.

In this paper we propose two new solutions to this minimal problem of determining absolute pose of a camera with unknown focal length and radial distortion. One solver works for non-planar scenes and one for the planar ones. By decomposing the general problem to the non-planar and the planar case we obtain much simpler systems of polynomial equations and therefore also much simpler and more practical solutions.

The most significant improvement is in speedup, since our new solvers are about 40× (non-planar) and 160× (planar) faster than the general solver presented in [14].

Both our solutions are based on the Gröbner basis method for solving systems of polynomial equations [8]. Our new solution to the non-planar case requires to perform G-J elimination of significantly smaller matrix of size $136 \times 152$ than [14] and eigenvalue computation of a $16 \times 16$ matrix. The planar solver is even simpler and requires G-J elimination of only $12 \times 18$ matrix. Moreover, the proposed solvers return less solutions, 16 and 6 compared to 24 in [14], and their run-times are about $1ms$ which is important for real-time applications and RANSAC.

We show in experiments that our two new specialized solvers can be joined to a one general solver, which gives comparable or better results than the existing general solver [14] for most scenes, including the near-planar ones.

Next we provide our formulation of the presented problem and its solutions for both non-planar and planar scenes. We compare our new solutions with the only existing general solution [14]. By evaluating our solutions on synthetic and real data we show that our solutions are stable and efficient and that the joined solver works well in real situations.

## 2 Problem Formulation

Let us assume the standard pinhole camera model [12]. In this model the image projection $\mathbf{u}_i$ of a 3D reference point $\mathbf{X}_i$ can be written as

$$\lambda_i \, \mathbf{u}_i = \mathtt{P} \, \mathbf{X}_i, \tag{1}$$

where $\mathtt{P}$ is a $3 \times 4$ projection matrix, $\lambda_i$ is an unknown scalar value and points $\mathbf{u}_i = [u_i, v_i, 1]^\top$ and $\mathbf{X}_i \, [x_i, y_i, z_i, 1]^\top$ are represented by their homogeneous coordinates.

The projection matrix $\mathtt{P}$ can be written as

$$\mathtt{P} = \mathtt{K} \, [\mathtt{R} \,|\, \mathbf{t}], \tag{2}$$

where

$\mathtt{R} = [r_{ij}]_{i,j=1}^{3}$ is a $3 \times 3$ rotation matrix, $\mathbf{t} = [t_x, t_y, t_z]^\top$ contains the information about camera position and $\mathtt{K}$ is the calibration matrix of the camera.

As described in Introduction we assume that the only unknown parameter from the calibration matrix $\mathtt{K}$ is the focal length. Therefore, the calibration matrix $\mathtt{K}$ has the form $diag\,[f, f, 1]$. Since the projection matrix is given only up to scale we can equivalently write $\mathtt{K} = diag\,[1, 1, w]$

for $w = 1/f$.

Using these assumptions the projection equation (1) can be written as

$$\lambda_i \, \mathbf{u}_i = \begin{bmatrix} r_{11} & r_{12} & r_{13} & t_x \\ r_{21} & r_{22} & r_{23} & t_y \\ wr_{31} & wr_{32} & wr_{33} & wt_z \end{bmatrix} \mathbf{X}_i. \tag{3}$$

In our problem we assume that the image points are affected by some amount of radial distortion. Here we model the radial distortion by the one-parameter division model proposed by Fitzgibbon [10]. This model is given by formula

$$\mathbf{p}_u \sim \mathbf{p}_d / \left(1 + k r_d^2\right), \tag{4}$$

where $k$ is the distortion parameter, $\mathbf{p}_u = [u_u, v_u, 1]^\top$, resp. $\mathbf{p}_d = [u_d, v_d, 1]^\top$, are the corresponding undistorted, resp. distorted, image points, and $r_d$ is the radius of $\mathbf{p}_d$ w.r.t. the distortion center. We assume that the distortion center is in the center of the image. Therefore $r_d^2 = u_d^2 + v_d^2$ and we have

$$\mathbf{u}_i = \left[u_i, v_i, 1 + k\left(u_i^2 + v_i^2\right)\right]^\top. \tag{5}$$

We can eliminate the scalar values $\lambda_i$ from the projection equation (3) by multiplying it with the skew symmetric matrix $[\mathbf{u}_i]_\times$.

Since $[\mathbf{u_i}]_\times \mathbf{u_i} = 0$ we obtain the matrix equation

$$\begin{bmatrix} 0 & -1 - k\,r_i^2 & v_i \\ 1 + k\,r_i^2 & 0 & -u_i \\ -v_i & u_i & 0 \end{bmatrix} \begin{bmatrix} r_{11} & r_{12} & r_{13} & t_x \\ r_{21} & r_{22} & r_{23} & t_y \\ w r_{31} & w r_{32} & w r_{33} & w t_z \end{bmatrix} \begin{bmatrix} x_i \\ y_i \\ z_i \\ 1 \end{bmatrix} = 0 \tag{6}$$

for $\mathbf{X}_i = [x_i, y_i, z_i, 1]^\top$.

This matrix equation results in three polynomial equations from which only two are linearly independent. This is caused by the fact that the skew symmetric matrix $[\mathbf{u}_i]_\times$ has rank two.

In the case of the image points not affected by the radial distortion, i.e. when $k = 0$, the projection equation (6) gives us for each point correspondence two linear homogeneous equations in 12 elements of the projection matrix $\mathtt{P}$. For $N$ 2D-to-3D point correspondences these equations can be written as $\mathtt{Mp} = 0$, where $\mathtt{M}$ is a $2N \times 12$ coefficient matrix and $\mathtt{p}$ is the vector consisting of 12 elements of the projection matrix $\mathtt{P}$. Therefore, the projection matrix can be written as a linear combination of the $12 - 2N$ null space basis vectors $\mathtt{P}_i$ of the matrix $\mathtt{M}$

$$\mathtt{P} = \sum_{i=1}^{12-2N} \alpha_i \mathtt{P}_i, \tag{7}$$

where $\alpha_i$ are unknown parameters from which one can be set to 1. In this way the projection matrix $\mathtt{P}$ can be parameterized using $11 - 2N$ unknowns. This parameterization was for example used in [23] for solving absolute pose problem for camera with unknown focal length and works only for non-planar scenes.

Unfortunately this parameterization cannot be used in the case of image points affected by the radial distortion (5). Therefore, we will next provide two different parameterizations of the projection matrix $\mathtt{P}$ which are applicable also to image points affected by the radial distortion (5). Both parametrization are very similar, the first one works for non-planar scenes and the second for the planar ones.

### 2.1 Absolute Pose for a Camera with Unknown Focal Length and Radial Distortion for Non-planar Scene

Let us denote the elements of the projection matrix $\mathtt{P}$ as $p_{ij}$, where $p_{ij}$ is the element from the $i^{th}$ row and $j^{th}$ column of the matrix $\mathtt{P}$. The equation corresponding to the third row of the matrix equation (6) can be then written as

$$-v_i\,(p_{11}\,x_i + p_{12}\,y_i + p_{13}\,z_i + p_{14}) + u_i\,(p_{21}\,x_i + p_{22}\,y_i + p_{23}\,z_i + p_{24}) = 0. \quad (8)$$

This is a homogeneous linear equation in eight unknowns $p_{11}, p_{12}, p_{13}, p_{14}, p_{21}, p_{22}, p_{23}$ and $p_{24}$. Since we have four 2D-to-3D point correspondences we have four such equations. These four equations can be rewritten in the matrix form

$$\mathtt{M}\,\mathbf{v} = 0, \quad (9)$$

where $\mathtt{M}$ is a $4 \times 8$ coefficient matrix and $\mathbf{v} = [p_{11}, p_{12}, p_{13}, p_{14}, p_{21}, p_{22}, p_{23}, p_{24}]^\top$ is a $8 \times 1$ vector of unknowns. Therefore we can write our eight unknowns in $\mathbf{v}$ as a linear combination of the four null space basis vectors $\mathbf{n}_i$ of the matrix M

$$\mathbf{v} = \sum_{i=1}^{4} \alpha_i\,\mathbf{n}_i, \quad (10)$$

where $\alpha_i$ are new unknowns from which one can be set to one, e.g. $\alpha_4 = 1$.

In this way we obtain parametrization of the first two rows of the projection matrix $\mathtt{P}$ with three unknowns $\alpha_1, \alpha_2$ and $\alpha_3$.

To parametrize the third row of the projection matrix $\mathtt{P}$ we use one from the remaining two equations from the projection equation (6). When $u_i = 0$ we use the equation corresponding to the first row of (6) and when $v_i = 0$ the equation corresponding to the second row. In all remaining situations, which are most common, we can select arbitrarily from these two equations, e.q. the equation corresponding to the second row. This equation has the form

$$\left(1 + k\,r_i^2\right)(p_{11}\,x_i + p_{12}\,y_i + p_{13}\,z_i + p_{14}) - u_i\,(p_{31}\,x_i + p_{32}\,y_i + p_{33}\,z_i + p_{34}) = 0. \quad (11)$$

Equation (11) contains elements $p_{31}, p_{32}, p_{33}$ and $p_{34}$ from the third row of the projection matrix and elements $p_{11}, p_{12}, p_{13}$ and $p_{14}$ from the first row of $\mathtt{P}$ which are already parametrized with $\alpha_1, \alpha_2$ and $\alpha_3$. We again have four equations of the form (11). Using (10) we can rewrite these equations as

$$\mathtt{A}\,[p_{31}, p_{32}, p_{33}, p_{34}]^\top = \mathtt{B}\,[\alpha_1, \alpha_2, \alpha_3, k\,\alpha_1, k\,\alpha_2, k\,\alpha_3, k, 1]^\top, \quad (12)$$

where $\mathtt{A}$ and $\mathtt{B}$ are coefficient matrices, $\mathtt{A}$ of size $4 \times 4$ and $\mathtt{B}$ of size $4 \times 8$.

If the matrix $\mathtt{A}$ has full rank, i.e. points $X_1, X_2, X_3$ and $X_4$ are not coplanar, we can write

$$[p_{31}, p_{32}, p_{33}, p_{34}]^\top = \mathtt{A}^{-1}\,\mathtt{B}\,[\alpha_1, \alpha_2, \alpha_3, k\,\alpha_1, k\,\alpha_2, k\,\alpha_3, k, 1]^\top. \quad (13)$$

This gives us a parametrization of the third row of the projection matrix $\mathtt{P}$ with four unknowns, $\alpha_1, \alpha_2, \alpha_3$ and $k$. Together with the parametrization of the first two rows (10) we obtain a parametrization of the whole projection matrix $\mathtt{P}$ with these four unknowns $\alpha_1, \alpha_2, \alpha_3$ and $k$.

With this parameterization of the projection matrix $\mathtt{P}$ in hand we can now solve the absolute pose problem for the camera with unknown focal length and radial distortion.

To solve this problem we use constraints that the three rows of the $3 \times 3$ submatrix of the projection matrix $\mathtt{P}$ are perpendicular and that the first two rows of this submatrix have the same norm. These constraints results from the fact that the $3 \times 3$ submatrix of the projection matrix $\mathtt{P}$ has the form $\mathtt{K}\,\mathtt{R}$, where $\mathtt{R}$ is a rotation matrix. In this way we obtain four equations in four unknowns $\alpha_1, \alpha_2, \alpha_3, k$ (two from them quadratic and two cubic)

$$p_{11}p_{21} + p_{12}p_{22} + p_{13}p_{23} = 0, \tag{14}$$

$$p_{31}p_{11} + p_{32}p_{12} + p_{33}p_{13} = 0, \tag{15}$$

$$p_{31}p_{21} + p_{32}p_{22} + p_{33}p_{23} = 0, \tag{16}$$

$$p_{11}^2 + p_{12}^2 + p_{13}^2 - p_{21}^2 - p_{22}^2 - p_{23}^2 = 0. \tag{17}$$

To solve these four polynomial equations in four unknowns we use the Gröbner basis method [8]. This method was recently used to solve several minimal computer vision problems [5, 14, 22] and the automatic generator of the Gröbner basis solvers is available online [15]. For more details about this Gröbner basis method for solving systems of polynomial equations see for example [6, 8, 15].

Using this automatic generator we have obtained solver for our equations consisting of one G-J elimination of a $136 \times 152$ matrix and the eigenvalue computation of a $16 \times 16$ matrix. This solver gives us 16 solutions for $\alpha_1, \alpha_2, \alpha_3$ and $k$ from which we can create the projection matrix $\mathtt{P}$ using (10) and (13).

Finally we can use the constraint that the squared norm of the first row of the $3 \times 3$ submatrix of the projection matrix $\mathtt{P}$ multiplied by $w^2$ is equal to the squared norm of the third row of this submatrix

$$w^2\,p_{11}^2 + w^2\,p_{12}^2 + w^2\,p_{13}^2 - p_{31}^2 - p_{32}^2 - p_{33}^2 = 0, \tag{18}$$

This is a quadratic equation in $w = 1/f$ from which the positive root give us a solution for the focal length $f$.

### 2.2 Absolute Pose for a Camera with Unknown Focal Length and Radial Distortion for Planar Scene

In the planar case, i.e. when all four 3D points are on the plane, we can not directly use the parametrization presented in the Section 2.1. However, we can use a similar parametrization.

Without loss of generality let us assume that all four 3D points $X_i$ have the fourth coordinate $z_i = 0$. In this case the equation (8) corresponding to the third row of the matrix equation (6) can be written as

$$-\,v_i\,(p_{11}\,x_i + p_{12}\,y_i + p_{14}) + u_i\,(p_{21}\,x_i + p_{22}\,y_i + p_{24}) = 0. \tag{19}$$

This is a homogeneous linear equation in only six unknowns $p_{11}, p_{12}, p_{14}, p_{21}, p_{22}$ and $p_{24}$. Since we have four 2D-to-3D point correspondences we have four such equations which can be again rewritten in the matrix form $\mathtt{M}\,\mathbf{v} = 0$, where $\mathtt{M}$ is a $4 \times 6$ coefficient matrix and $\mathbf{v} = [p_{11}, p_{12}, p_{14}, p_{21}, p_{22}, p_{24}]^\top$ is a $6 \times 1$ vector of unknowns. Therefore, in this case we can write our unknowns in $\mathbf{v}$ as a linear combination of the two null space basis vectors $\mathbf{n}_1$ and $\mathbf{n}_2$ of the matrix M

$$\mathbf{v} = \beta_1\,\mathbf{n}_1 + \mathbf{n}_2, \tag{20}$$

where $\beta_1$ is a new unknown. Using (20) we obtain a parametrization of the first two rows of the matrix $\mathtt{P}$ (without the third column) with one unknown $\beta_1$.

To parametrize the third row we again use one from the remaining two equations from the projection equation (6). Let's again consider the equation corresponding to the second row of the projection equation (6). In this planar case has this equation the form

$$\left(1 + k\,r_i^2\right)(p_{11}\,x_i + p_{12}\,y_i + p_{14}) - u_i\,(p_{31}\,x_i + p_{32}\,y_i + p_{34}) = 0. \tag{21}$$

This equation contains elements from the first row of $\mathtt{P}$ which are already parametrized with $\beta_1$ and three elements $p_{31}, p_{32}$ and $p_{33}$ from the third row which we want to parametrize. We again have four equations of the form (11). However, we will now use only three of them, e.q. equations corresponding to the first three 2D-to-3D point correspondences. Using (20) we can rewrite these three equations as

$$\mathtt{C}\,[p_{31}, p_{32}, p_{34}]^\top = \mathtt{D}\,[\beta_1, k\,\beta_1, k, 1]^\top, \tag{22}$$

where $\mathtt{C}$ and $\mathtt{D}$ are coefficient matrices, $\mathtt{C}$ of size $3 \times 3$ and $\mathtt{B}$ of size $3 \times 4$.

If the matrix $\mathtt{C}$ has full rank, i.e. points $X_1, X_2$ and $X_3$ are not collinear, we can write

$$[p_{31}, p_{32}, p_{34}]^\top = \mathtt{C}^{-1}\,\mathtt{D}\,[\beta_1, k\,\beta_1, k, 1]^\top. \tag{23}$$

In this way we obtain parametrization of the third row (without the third column) of the projection matrix $\mathtt{P}$ with two unknowns, $\beta_1$ and $k$. Together with (20) we have parametrized the first, second and fourth column of the projection matrix $\mathtt{P}$ with $\beta_1$ and $k$.

In this case we can not use constraints (14)-(17) on the rows of the projection matrix. It is because we do not have information about the third column of $\mathtt{P}$.

However, we can use constraints that the columns of the rotation matrix are perpendicular and of the same norm. These constraints in this case gives us two equations of degree four in three unknowns $\beta_1, k$ and $w = 1/f$

$$w\,p_{11}\,w\,p_{12} + w\,p_{21}\,w\,p_{22} + p_{31}\,w\,p_{32} = 0, \tag{24}$$

$$w^2\,p_{11}^2 + w^2\,p_{21}^2 + p_{31}^2 - w^2\,p_{12}^2 - w^2\,p_{22}^2 - p_{32}^2 = 0. \tag{25}$$

Moreover, we have one more equation of the form (21), for the fourth 2D-3D point correspondence, which was not used in (22). This equation has the form

$$\left(1 + k\,r_4^2\right)(p_{11}\,x_4 + p_{12}\,y_4 + p_{14}) - u_4\,(p_{31}\,x_4 + p_{32}\,y_4 + p_{34}) = 0 \tag{26}$$

and after using parametrization (20) and (23) of the unknowns $p_{11}, p_{12}, p_{14}, p_{31}$, $p_{32}$ and $p_{34}$ it results in one quadratic equation in two unknowns $\beta_1$ and $k$.

Equation (26) together with equations (24) and (25) give us three equations in three unknowns $\beta_1, k$ and $w = 1/f$ which can be again solved using Gröbner basis method [8] and automatic generator [15]. In this case the resulting solver results in one G-J elimination of relatively small $12 \times 18$ matrix and gives up to 6 real solutions to $\beta_1, k$ and $f$.

The third column of the projection matrix $\mathtt{P}$ can be finally easily obtained from its structure and the properties that the columns of the rotation matrix are perpendicular and of the same norm.

## 3 Experiments

In this section we compare our two new solutions (non-planar and planar) to the absolute pose problem of a camera with unknown focal length and radial distortion presented in Sections 2.1 and 2.2 with the general solution to this problem proposed in [14]. We compare these solutions on synthetically generated scenes and show that all solvers return comparable results.

Then we study the performance of our two specialized solvers on near-planar scenes and show that these solvers can be joined into a new general solver, which gives comparable or better results then the existing general solver [14] for most scenes, including these near-planar ones.

Finally we show the performance of this new joined general solver on real datasets and compare it with the general solver from [14].

### 3.1 Synthetic Datasets

In the following synthetic experiments we use synthetically generated ground-truth 3D scenes. These scenes were generated using 3D points randomly distributed on a plane or in a 3D cube depending on the testing configuration. Each 3D point was projected by a camera with random feasible orientation and position and random or fixed focal length. Then the radial distortion using the division model [10] was added to all image points to generate noiseless distorted points. Finally, Gaussian noise with standard deviation $\sigma$ was added to the distorted image points assuming a $1000 \times 1000$ pixel image.

**Numerical stability.** In the first experiment we have studied the behavior of both presented solvers on noise free data to check their numerical stability.

In this experiment 1500 random scenes and feasible camera poses were generated. The radial distortion parameter was randomly drawn from the interval $k \in [-0.45, 0]$ and the focal length from the interval $f \in [0.5, 2.5]$.

Figure 1 shows results of our new non-planar solver on non-planar scenes (Top) and of the planar solver on planar ones (Bottom). In both cases we compare our solvers (Red) with the general solver from [14] (Blue). The $log_{10}$ relative error of the focal length $f$ obtained by selecting the real root closest to the ground truth value is on the left and the $log_{10}$ absolute error of the radial distortion parameter on the right.

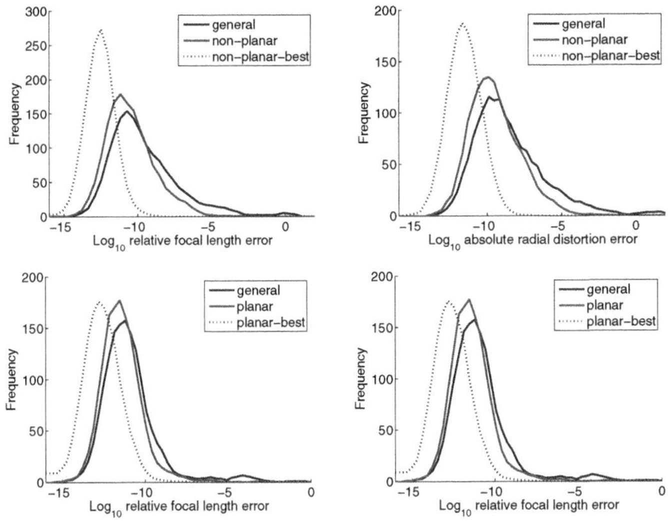


**Fig. 1.** $Log_{10}$ relative error of the focal length $f$ (Left) and $Log_{10}$ absolute error of the radial distortion parameter $k$ obtained by selecting the real root closest to the ground truth value for the non-planar solver (Top) and the planar solver (Bottom)

As it can be seen both our new algorithms give similar results to the general algorithm from [14]. The small difference is in the number of results with error greater than $10^{-5}$. In our new solutions such results occur in about 1% of cases while in the general solver from [14] in about 4.5%. This "failure" will be also visible in the near-planar and real experiments.

Note that the general solution from [14] uses techniques for improving numerical stability of Gröbner basis solvers based on changing basis and QR decomposition [6] while our solutions use standard Gröbner basis method [8] without these improvements. We therefore believe that such techniques can further improve numerical stability of our solvers. It is partially visible also from Figure 1 where the results denoted as non-planar-best and planar-best (Dashed black) correspond to the most precise results obtained by our solvers by permuting the input points. This permutation of input data is in some sense similar to the permutation of columns of a coefficient matrix in Gröbner basis solver and therefore it is also similar to the changing of basis used in [14]. However, these techniques for improving numerical stability [6] are little bit expensive and as it will be shown in real experiments in the case of our solvers also unnecessary.

**Noise test.** In the next experiment we have tested behavior of our non-planar and planar solvers in the presence of noise added to image points. We again compare both presented specialized solvers with the general solver from [14].

Since both our specialized solvers have similar numerical stability as the general solver from [14], these solvers should behave similarly also in the presence of noise. This is visible also from Figure 2 which shows results for our non-planar solver (Red) and the general solver from [14] (Blue).

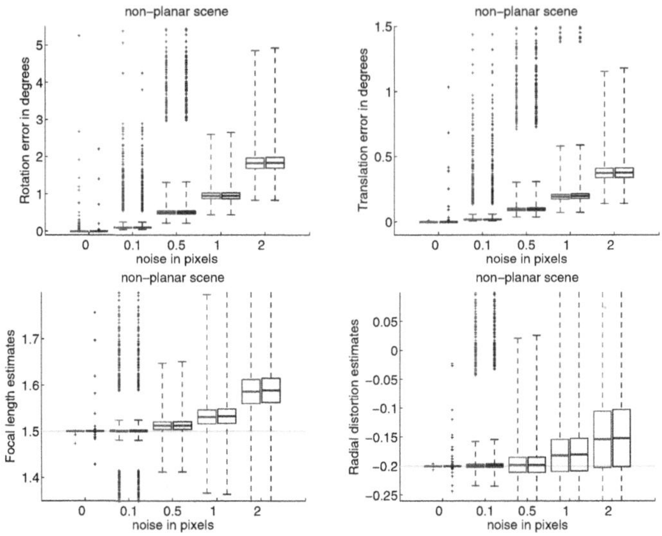


**Fig. 2.** Error of rotation (Top left), translation (Top right), focal length estimates (Bottom left) and radial distortion estimates (Bottom right) in the presence of noise for our non-planar solver (Red) and the general solver from [14] (Blue)

In this experiment for each noise level, from 0.0 to 2 pixels, 2000 estimates for random scenes and camera positions, focal length $f_{gt} = 1.5$ and radial distortion $k_{gt} = -0.2$, were made.

Results in Figure 2 are represented by the Matlab boxplot function which shows values 25% to 75% quantile as a box with horizontal line at median. The crosses show data beyond 1.5 times the interquartile range. In this case the rotation error (Top left) was measured as the rotation angle in the angle-axis representation of the relative rotation $RR_{gt}^{-1}$ and the translation error (Top right) as the angle between ground-truth and estimated translation vector.

It can be seen that our new non-planar solver provides quite precise results even for larger noise levels. Similar results were obtained also for our planar solver and therefore we are not showing them here.

**Computational complexity.** The most significant improvement of our new specialized solvers over the general solver from [14] is in speedup, since our solvers are about 40× (non-planar) and 160× (planar) faster than the general solver [14].

This is caused by the fact that our new solvers results in much simpler systems of polynomial equations and therefore also in much simpler and practical solvers. While the general solver [14] requires to perform LU decomposition of a $1134 \times 720$ matrix, QR decomposition of a $56 \times 56$ matrix and eigenvalue computations of a $24 \times 24$ matrix, our non-planar solver requires only one G-J elimination of a $136 \times 152$ matrix and eigenvalue computations of a $16 \times 16$ matrix. The planar solver is even simpler and requires one G-J elimination of only $12 \times 18$ matrix. Moreover, our two solvers return less solutions, 16 and 6 compared to 24 in [14].

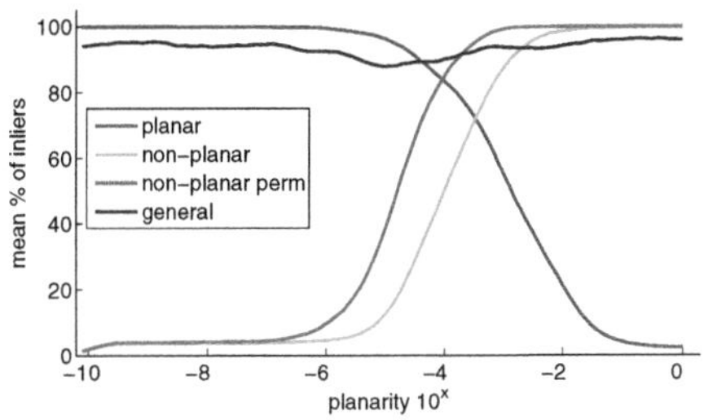


**Fig. 3.** Results of experiment on the near-planar scene

All these facts are important in RANSAC and real applications in which the general solver from [14] was due to its speed impractical.

**Near planar test.** In this experiment we have studied the behavior of our non-planar and planar solvers and the general solver from [14] on planar, general and non-planar scenes. We have focused primarily to near-planar scenes in order to show how to build fast joined general solver by composing our two new specialized solvers. For this purpose we created a synthetic scene where we could control scene planarity by a scalar value $a$. Given planarity $a$ we have constructed the scene as follows: Assume that we have a synthetic scene generated as described above in Subsection 3.1. Now let's denote by $\rho$ a plane created from the first three non-collinear 3D points and by $s$ a normalizing scale. We calculate the scale $s$ as the distance of the furthest point from these three points from their center of gravity $CG$. Then, we randomly generate the fourth point at the distance $s\,a$ from the plane $\rho$ and such that it is not further than $s$ from the center of gravity $CG$. Note that for planarity $a = 0$ we get four points on the plane and for $a = 1$ we obtain a well defined non-planar four-tuple of 3D points.

In this experiment we did not contaminate the image points corresponding to these four 3D points by a noise. We added noise with deviation of 0.5 pixels only to the remaining image points. In this way we created a scene with one uncontaminated four-tuple of 2D-to-3D point correspondences for which we can control planarity by a scalar value $a$.

Next, for each given planarity value $a$ we created a scene and calculated camera pose from the four-tuple of correspondences using the planar, the non-planar and the general solver [14]. Note that this four-tuple is not affected by a noise and hence the only deviation from the ground truth solutions comes from the numerical instability of the solvers itself. To evaluate the impact of the instability on the solution we used the estimated camera, focal length and radial distortion to project all 3D points to the image plane. Then we measured how many points were projected closer than one pixel to its corresponding 2D image - we call them inliers.

Figure 3 shows results for all three examined solvers i.e. the planar (Red), the non-planar (Cyan) and the general (Blue) together with the results obtained from the non-planar solver by permuting the input points (Magenta). Here, for each given planarity value $a$ we created 100 random scenes and evaluated algorithms.

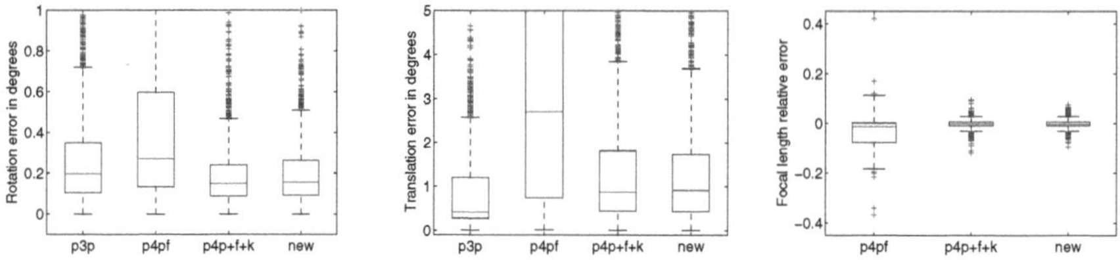


**Fig. 4.** Real experiment: Comparison of estimated results. The new joined planar-non-planar solver (new) behaves similar to the general solver from [14] (p4pf+k). Radial distortion boxplots are almost equal for both of these solvers and therefore are omitted.

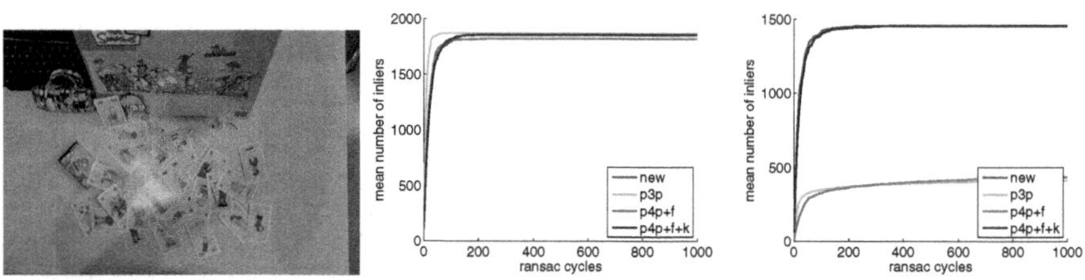


**Fig. 5.** Example of an input image (left), RANSAC sampling history for image without (middle) and with (right) strong radial distortion

Interesting points in this Figure 3 are the intersections of the planar and the general solvers and also the general and the non-planar solvers. Ideally one would combine planar, general and non-planar solvers to gain maximal precision at maximal speed. Reasonable thresholds for our planar solver is planarity less than $10^{-4.2}$ and for the non-planar solver greater than $10^{-2.8}$. General solver from [14] should be used in-between. However, our further experiments show that using only our new planar and non-planar solvers with splitting threshold $a = 10^{-3.2}$ is sufficient in practice.

### 3.2 Real Data

For the real data experiment we created a simple scene with two dominant planar structures (Figure 5 left). Our intention was to show behavior of our new joined planar-non-planar solver in a real scene with sampling on the plane, near the plane and off the plane points. We used new joined planar and non-planar solver with spitting planarity threshold $a = 10^{-3.2}$.

First, we captured around 20 photos with the cell phone and the digital camera to get images with different distortions. Then we used Phototurism-like [21] pipeline to create the 3D reconstruction, 2D-3D correspondences and to get the ground truth reference for the camera poses. Note that 2D correspondences are positions of detected feature points and not ideal projections of 3D points and hence there is a natural noise. Since all our 2D-3D correspondences coming from the reconstruction pipeline are inliers we randomly modify 50% of 2D measurements to get 50% of outliers.

We plugged all examined solvers, the calibrated P3P solver [9] (Cyan), the P4P+f solver for camera with unknown focal length [5] (Magenta), the general P4P+f+k solver from [14] (Blue) and our new joined general solver (Red) to locally optimized RANSAC estimator [7]. Then, we calculated the camera pose of each camera using given 2D-3D correspondences.

Figure 4 shows boxplots obtained by collecting results from 1000 executions of RANSAC for each camera. Boxplot shows that joined planar-non-planar solver returns very competitive results comparing to the general solver from [14].

Note, that we did not calibrate radial distortion before calling P3P and P4P+f. Since many of images have strong radial distortion one cannot expect good results without correcting the distortion. On the other hand these results show that radial distortion solvers are useful in practice. Figure 5 (Right) shows difference in RANSAC convergence when using solvers with and without radial distortion estimation and Figure 5 (Middle) results for not so distorted image.

## 4 Conclusion

In this paper we have proposed a new efficient solver to the absolute pose problem for camera with unknown focal length and radial distortion from four 2D-to-3D point correspondences. The presented solver is obtained by joining two specialized solvers, one for non-planar scenes and one for planar ones. By decomposing the problem into these two cases we obtain a simpler and more efficient solver than previously known general solver [14].

We have demonstrated in synthetic and real experiments significant speedup of our solvers over the general solver from [14]. Moreover, we have shown that our new joined general solver gives comparable or better results than the existing general solver for of most planar as well as non-planar scenes. – Matlab source codes of the presented solvers are available online at `http://cmp.felk.cvut.cz/minimal/`.

**Acknowledgment.** This work has been supported by EC project FP7-SPACE-218814 PRoVisG, by Czech Government research program MSM6840770038 and by Grant Agency of the CTU Prague project SGS10/072/OHK4/1T/13.

## References

1. 2D3. Boujou, `http://www.2d3.com`
2. Abdel-Aziz, Y., Karara, H.: Direct linear transformation from comparator to object space coordinates in close-range photogrammetry. In: ASP Symp. Close-Range Photogrammetry, pp. 1–18 (1971)
3. Abidi, M.A., Chandra, T.: A new efficient and direct solution for pose estimation using quadrangular targets: Algorithm and evaluation. IEEE PAMI 17, 534–538 (1985)
4. Ameller, M.A., Quan, M., Triggs, L.: Camera pose revisited – new linear algorithms. In: ECCV (2000)

5. Bujnak, M., Kukelova, Z., Pajdla, T.: A general solution to the P4P problem for camera with unknown focal length. In: CVPR (2008)
6. Byröd, M., Josephson, K., Åström, K.: Fast and Stable Polynomial Equation Solving and Its Application to Computer Vision. Int. J. Computer Vision 84, 237–255 (2009)
7. Chum, O., Matas, J., Kittler, J.: Locally optimized RANSAC. In: DAGM, pp. 236–243 (2003)
8. Cox, D., Little, J., O'Shea, D.: Using Algebraic Geometry, 2nd edn., vol. 185. Springer, Berlin (2005)
9. Fischler, M.A., Bolles, R.C.: Random sample consensus: a paradigm for model fitting with applications to image analysis and automated cartography. Comm. ACM 24, 381–395 (1981)
10. Fitzgibbon, A.: Simultaneous linear estimation of multiple view geometry and lens distortion. In: CVPR, pp. 125–132 (2001)
11. Grunert, J.A.: Das pothenot'sche problem, in erweiterter gestalt; nebst bemerkungen über seine anwendung in der geodäsie. Archiv der Mathematik und Physik 1, 238–248 (1841)
12. Hartley, R., Zisserman, A.: Multiple View Geometry in Computer Vision. Cambridge University Press, Cambridge (2003)
13. Moreno-Noguer, F., Lepetit, V., Fua, P.: Accurate non-iterative o(n) solution to the pnp problems. In: ICCV (2007)
14. Josephson, K., Byröd, M., Aström, K.: Pose Estimation with Radial Distortion and Unknown Focal Length. In: CVPR (2009)
15. Kukelova, Z., Bujnak, M., Pajdla, T.: Automatic Generator of Minimal Problem Solvers. In: Forsyth, D., Torr, P., Zisserman, A. (eds.) ECCV 2008, Part III. LNCS, vol. 5304, pp. 302–315. Springer, Heidelberg (2008)
16. Leibe, B., Cornelis, N., Cornelis, K., Gool, L.J.V.: Dynamic 3d scene analysis from a moving vehicle. In: CVPR (2007)
17. Leibe, B., Schindler, K., Gool, L.J.V.: Coupled detection and trajectory estimation for multi-object. In: ICCV (2007)
18. Quan, L., Lan, Z.-D.: Linear n-point camera pose determination. IEEE PAMI 21, 774–780 (1999)
19. Reid, G., Tang, J., Zhi, L.: A complete symbolic-numeric linear method for camera pose determination. In: ISSAC, pp. 215–223 (2003)
20. Slama, C.C. (ed.): Manual of Photogrammetry. American Society of Photogrammetry and Remote Sensing, Falls Church, Virginia (1980)
21. Snavely, N., Seitz, S.M., Szeliski, R.: Photo tourism: Exploring photo collections in 3D. ACM Trans. Graphics Proc. 25 (2006)
22. Stewénius, H., Engels, C., Nistér, D.: Recent developments on direct relative orientation. ISPRS J. Photogrammetry Remote Sensing 60, 284–294 (2006)
23. Triggs, B.: Camera pose and calibration from 4 or 5 known 3d points. In: ICCV, pp. 278–284 (1999)
24. Wu, Y., Hu, Z.: PNP problem revisited. J. Math. Imaging Vision 24, 131–141 (2006)
25. Zhi, L., Tang, J.: A complete linear 4-point algorithm for camera pose determination. In: AMSS, Academia Sinica (2002)

# Efficient Large-Scale Stereo Matching

Andreas Geiger[1], Martin Roser[1], and Raquel Urtasun[2]

[1] Dep. of Measurement and Control, Karlsruhe Institute of Technology
geiger@kit.edu, martin.roser@kit.edu
[2] Toyota Technological Institute at Chicago
rurtasun@ttic.edu

**Abstract.** In this paper we propose a novel approach to binocular stereo for fast matching of high-resolution images. Our approach builds a prior on the disparities by forming a triangulation on a set of support points which can be robustly matched, reducing the matching ambiguities of the remaining points. This allows for efficient exploitation of the disparity search space, yielding accurate dense reconstruction without the need for global optimization. Moreover, our method automatically determines the disparity range and can be easily parallelized. We demonstrate the effectiveness of our approach on the large-scale Middlebury benchmark, and show that state-of-the-art performance can be achieved with significant speedups. Computing the left and right disparity maps for a one Megapixel image pair takes about one second on a single CPU core.

## 1 Introduction

Estimating depth from binocular imagery is a core subject in low-level vision as it is an important building block in many domains such as multi-view reconstruction. In order to be of practical use for applications such as autonomous driving, disparity estimation methods should run at speeds similar to other low-level visual processing techniques, e.g. edge extraction or interest point detection. Since depth errors increase quadratically with the distance [1], high-resolution images are needed to obtain accurate 3D representations. While the benefits of high resolution imagery are already exploited exhaustively in structure-from-motion, object recognition and scene classification, only few binocular stereo methods deal efficiently with large images.

Stereo algorithms based on local correspondences [2,3] are typically fast, but require an adequate choice of window size. As illustrated in Fig. 1 this leads to a trade-off between low matching ratios for small window sizes and border bleeding artifacts for larger ones. As a consequence, poorly-textured and ambiguous surfaces cannot be matched consistently.

Algorithms based on global correspondences [4,5,6,7,8,9] overcome some of the aforementioned problems by imposing smoothness constraints on the disparities in the form of regularized energy functions. Since optimizing such MRF-based energy functions is in general NP-hard, a variety of approximation algorithms have been proposed, e.g., graph cuts [4,5] or belief propagation [6]. However, even

R. Kimmel, R. Klette, and A. Sugimoto (Eds.): ACCV 2010, Part I, LNCS 6492, pp. 25–38, 2011.

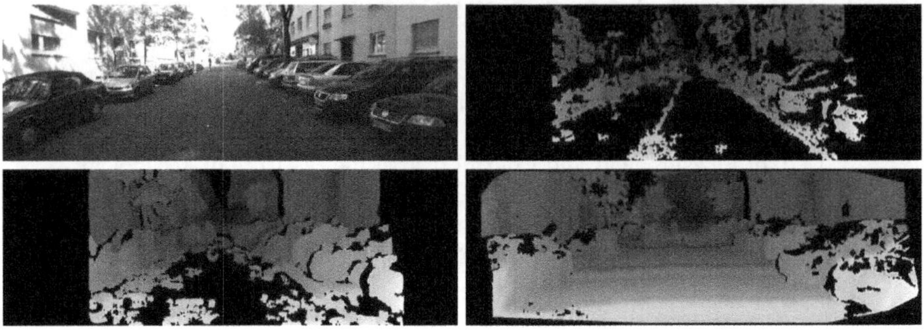

**Fig. 1.** Low-textured areas often pose problems to stereo algorithms. Using local methods one faces the trade-off between low matching ratios (top-right, window size $5 \times 5$) and border bleeding effects (bottom-left, window size $25 \times 25$). Our method is able to combine small window sizes with high matching ratios (bottom-right).

on low-resolution imagery, they generally require large computational efforts and high memory capacities. For example, storing all messages of a one Megapixel image pair requires more than 3 GB of RAM [10]. In these approaches, the disparity range usually has to be known in advance, and a good choice of the regularization parameters is crucial. Furthermore, when increasing image resolution, the widely used priors based on binary potentials fail to reconstruct poorly-textured and slanted surfaces, as they favor fronto-parallel planes. Recently developed methods based on higher-order cliques [7] overcome these problems, but are even more computationally demanding.

In this paper we propose a generative probabilistic model for stereo matching, called *ELAS (Efficient LArge-scale Stereo)*[1], which allows for dense matching with small aggregation windows by reducing ambiguities on the correspondences. Our approach builds a prior over the disparity space by forming a triangulation on a set of robustly matched correspondences, named 'support points'. Since our prior is piecewise linear, we do not suffer in the presence of poorly-textured and slanted surfaces. This results in an efficient algorithm that reduces the search space and can be easily parallelized. As demonstrated in our experiments, our method is able to achieve state-of-the-art performance with significant speedups of up to three orders of magnitude when compared to prevalent approaches; we obtain 300 MDE/s (million disparity evaluations per second) on a single CPU core.

## 2 Related Work

In the past few years much progress has been made towards solving the stereo problem, as evidenced by the excellent overview of Scharstein et al. [2]. Local methods typically aggregate image statistics in a small window, thus imposing smoothness implicitly. Optimization is usually performed using a winner-takes-all strategy, which selects for each pixel the disparity with the smallest value

[1] C++ source code, Matlab wrappers and videos online at **http://www.cvlibs.net**

under some distance metric [2]. Weber et al. [3] achieved real-time performance using the Census transform and a GPU implementation. However, as illustrated by Fig. 1, traditional local methods [11] often suffer from border bleeding effects or struggle with correspondence ambiguities. Approaches based on adaptive support windows [12,13] adjust the window size or adapt the pixel weighting within a fixed-size window to improve performance, especially close to border discontinuities. Unfortunately, since for each pixel many weight factors have to be computed, these methods are much slower than fixed-window ones [13].

Dense and accurate matching can be obtained by global methods, which enforce smoothness explicitly by minimizing an MRF-based energy function which can be decomposed as the sum of a data fitting term and a regularization term. Since for most energies of practical use such an optimization is NP-hard, approximate algorithms have been proposed, e.g. graph-cuts [4,5], belief propagation [6]. Klaus et al. [14] extend global methods to use mean-shift color segmentation, followed by belief propagation on super-pixels. In [15], a parallel VLSI hardware design for belief propagation that achieves real time performance on VGA imagery was proposed . The application of global methods to high-resolution images is, however, limited by their high computational and memory requirements, especially in the presence of large disparity ranges. Furthermore, models based on binary potentials between pixels favor fronto-parallel surfaces which leads to errors in low-textured slanted surfaces. Higher order cliques can overcome these problems [7], but they are even more computationally demanding.

Hirschmüller proposed semi-global matching [16], an approach which extends polynomial time 1D scan-line methods to propagate information along 16 orientations. While reducing streaking artifacts and improving accuracy compared to traditional methods based on dynamic programming, computational complexity increases with the number of computed paths. 'ground control points' are used in [17] to improve the occlusion cost sensitivity of dynamic programming algorithms. In [18,19] disparities are 'grown' from a small set of initial correspondence seeds. Though these methods produce accurate results and can be faster than global approaches, they do not provide dense matching and struggle with textureless and distorted image areas. Approaches to reduce the search space have been investigated for global stereo methods [10,20]. However, they mainly focus on memory requirements and start with a full search using local methods first. Furthermore, the use of graph-cuts imposes high computational costs particularly for large-scale imagery.

In contrast, in this paper we propose a Bayesian approach to stereo matching that is able to compute accurate disparity maps of high resolution images at frame rates close to real time without the need for global optimization. The remainder of this paper is structured as follows: In Section 3 we describe our approach to efficient large-scale stereo matching. Experimental results on real-world datasets and comparisons to a variety of other methods on large-scale versions of the Middlebury benchmark images are reported in Section 4. Finally, Section 5 gives our conclusions and future work.

## 3 Efficient Large-Scale Stereo Matching

In this section we describe our approach to efficient stereo matching of high-resolution images. Our method is inspired from the observation that despite the fact that many stereo correspondences are highly ambiguous, some of them can be robustly matched. Assuming piecewise smooth disparities, such reliable 'support points' contain valuable prior information for the estimation of the remaining ambiguous disparities. Our approach proceeds as follows: First, the disparities of a sparse set of support points are computed using a full disparity range. The image coordinates of the support points are then used to create a 2D mesh via Delaunay triangulation. A prior is computed to disambiguate the matching problem, making the process efficient by restricting the search to plausible regions. In particular, this prior is formed by computing a piecewise linear function induced by the support point disparities and the triangulated mesh. For simplicity of the presentation, we will assume rectified input images, such that correspondences are restricted to the same line in both images.

### 3.1 Support Points

As support points, we denote pixels which can be robustly matched due to their texture and uniqueness. While a variety of methods for obtaining stable correspondences are available [17,21,22], we find that matching support points on a regular grid using the $\ell_1$ distance between vectors formed by concatenating the horizontal and vertical Sobel filter responses of $9 \times 9$ pixel windows to be both efficient and effective. In all of our experiments we used Sobel masks of size $3 \times 3$ and a grid with fixed step-size of 5 pixels. A large disparity search range of half the input image width was employed to impose no restrictions on the disparities. We also experimented with sparse interest point descriptors such as SURF [23], but found that they did not improve matching accuracy while being slower to compute.

For robustness we impose consistency, i.e., correspondences are retained only if they can be matched from left-to-right and right-to-left. To get rid of ambiguous matches, we eliminate all points whose ratio between the best and the second best match exceeds a fixed threshold, $\tau = 0.9$ . Spurious mismatches are removed by deleting all points which exhibit disparity values dissimilar from all surrounding support points. To cover the full image, we add additional support points at the image corners whose disparities are taken to be the ones of their nearest neighbors.

### 3.2 Generative Model for Stereo Matching

We now describe our probabilistic generative model which, given a reference image and the support points, can be used to draw samples from the other image. More formally, let $\mathbf{S} = \{\mathbf{s}_1, ..., \mathbf{s}_M\}$ be a set of robustly matched support points. Each support point, $\mathbf{s}_m = (u_m, v_m, d_m)^T$, is defined as the concatenation of its image coordinates, $(u_m, v_m) \in \mathbb{N}^2$, and its disparity, $d_m \in \mathbb{N}$. Let $\mathbf{O} = \{\mathbf{o}_1, ..., \mathbf{o}_N\}$ be a set of image observations, with each observation $\mathbf{o}_n = (u_n, v_n, \mathbf{f}_n)^T$ formed as the concatenation of its image coordinates,

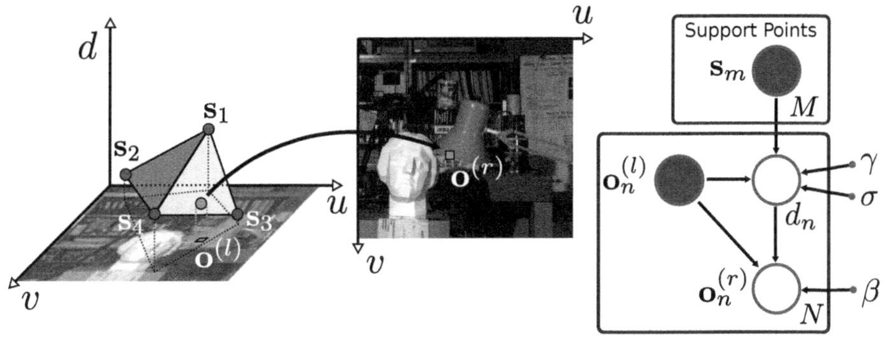


(a) Sampling process and graphical model

(b) Left image (c) Sample mean (d) Right image

**Fig. 2. Illustration of the sampling process.** (a) Graphical model and sampling process: Given support points $\{\mathbf{s}_1, ..., \mathbf{s}_M\}$ and an observation in the left image $\mathbf{o}_n^{(l)}$, a disparity $d$ is drawn. Given the observation on the left image and the disparity, we can draw an observation in the right image $\mathbf{o}_n^{(r)}$. (c) Repeating this process 100 times for each pixel and (d) computing the mean results in a blurred version of the right image.

$(u_n, v_n) \in \mathbb{N}^2$, and a feature vector, $\mathbf{f}_n \in \Re^Q$, e.g., the pixel's intensity or a low-dimensional descriptor computed from a small neighborhood. We denote $\mathbf{o}_n^{(l)}$ and $\mathbf{o}_n^{(r)}$ as the observations in the left and right image respectively. Without loss of generality, in the following we consider the left image as the reference image.

Assuming that the observations $\{\mathbf{o}_n^{(l)}, \mathbf{o}_n^{(r)}\}$ and support points $\mathbf{S}$ are conditionally independent given their disparities $d_n$, the joint distribution factorizes

$$p(d_n, \mathbf{o}_n^{(l)}, \mathbf{o}_n^{(r)}, \mathbf{S}) \propto p(d_n|\mathbf{S}, \mathbf{o}_n^{(l)}) p(\mathbf{o}_n^{(r)}|\mathbf{o}_n^{(l)}, d_n) \tag{1}$$

with $p(d_n|\mathbf{S}, \mathbf{o}_n^{(l)})$ the prior and $p(\mathbf{o}_n^{(r)}|\mathbf{o}_n^{(l)}, d_n)$ the image likelihood. The graphical model of our approach is depicted in Fig. 2(a). In particular, we take the prior to be proportional to a combination of a uniform distribution and a sampled Gaussian

$$p(d_n|\mathbf{S}, \mathbf{o}_n^{(l)}) \propto \begin{cases} \gamma + \exp\left(-\frac{(d_n - \mu(\mathbf{S}, \mathbf{o}_n^{(l)}))^2}{2\sigma^2}\right) & \text{if } |d_n - \mu| < 3\sigma \ \vee \ d_n \in N_{\mathbf{S}} \\ 0 & \text{otherwise} \end{cases} \tag{2}$$

with $\mu(\mathbf{S}, \mathbf{o}_n^{(l)})$ a mean function linking the support points and the observations, and $N_\mathbf{S}$ the set of all support point disparities in a small $20 \times 20$ pixel neighborhood around $(u_n^{(l)}, v_n^{(l)})$. We gain efficiency by excluding all disparities farther than $3\sigma$ from the mean. The condition $d_n \in N_\mathbf{S}$ enables the prior to locally extend its range to better handle disparity discontinuities in places where the linearity assumption might be violated.

We express $\mu(\mathbf{S}, \mathbf{o}_n^{(l)})$ as a piecewise linear function, which interpolates the disparities using the Delaunay triangulation computed on the support points. For each triangle, we thus obtain a plane defined by

$$\mu_i(\mathbf{o_n}^{(l)}) = a_i u_n + b_i v_n + c_i \tag{3}$$

where $i$ is the index of the triangle the pixel $(u_n, v_n)$ belongs to, and $\mathbf{o}_n = (u_n, v_n, \mathbf{f}_n)^T$ is an observation. For each triangle, the plane parameters $(a_i, b_i, c_i)$ are easily obtained by solving a linear system. Hence, the mode of the proposed prior, $\mu$, is a linear interpolation between support point disparities, serving as a coarse representation.

We express the image likelihood as a constrained Laplace distribution

$$p(\mathbf{o}_n^{(r)} | \mathbf{o}_n^{(l)}, d_n) \propto \begin{cases} \exp(-\beta \, \|\mathbf{f}_n^{(l)} - \mathbf{f}_n^{(r)}\|_1) & \text{if} \begin{pmatrix} u_n^{(l)} \\ v_n^{(l)} \end{pmatrix} = \begin{pmatrix} u_n^{(r)} + d_n \\ v_n^{(r)} \end{pmatrix} \\ 0 & \text{otherwise} \end{cases} \tag{4}$$

where $\mathbf{f}_n^{(l)}$,$\mathbf{f}_n^{(r)}$ are feature vectors in the left and right image respectively, and $\beta$ is a constant. The if-condition ensures that correspondences are located on the same epipolar line and matched via the disparity $d_n$. In our experiments, the features $\mathbf{f}_n$ are taken as the concatenation of image derivatives in a $5 \times 5$ pixel neighborhood around $(u_n, v_n)$, computed from Sobel filter responses, leading to $2 \times 5 \times 5 = 50$-dimensional feature vectors. Note that in [14] similar features were shown to be robust to illumination changes (i.e., additive bias). Empirically, we found features based on Sobel responses to work noticeably better than features based on Laplacian-of-Gaussian (LoG) filters. We do not make use of any color information, although this could be incorporated with little additional effort. We refer the reader to [2,24] for a comprehensive study of dissimilarity metrics for stereo matching.

An advantage of having a generative model is that we can use it to draw samples, see Fig. 2(a) for an illustration. Given the support points and an observation in the left image, samples from the corresponding observation in the right image can be obtained as follows:

1. Given $\mathbf{S}$ and $\mathbf{o}_n^{(l)}$ draw a disparity $d_n$ from $p(d_n | \mathbf{S}, \mathbf{o}_n^{(l)})$
2. Given $\mathbf{o}_n^{(l)}$ and $d_n$ draw an observation $\mathbf{o}_n^{(r)}$ from $p(\mathbf{o}_n^{(r)} | \mathbf{o}_n^{(l)}, d_n)$

Fig. 2(b-d) depicts the left input image, as well as the mean of the samples drawn from the right image given the left image and the support points. In order to obtain a comprehensive visualization in Fig. 2, here we use pixel intensities as features and draw 100 samples for each pixel. As expected, the sample mean corresponds to a blurred version of the right image.

### 3.3 Disparity Estimation

In the previous section we have proposed a prior and an image likelihood for stereo matching. We have also shown how to draw samples of the right image given support points and observations in the left image. At inference, however, we are interested in estimating the disparity map given the left and right images. We rely on maximum a-posteriori (MAP) estimation to compute the disparities

$$d_n^* = \operatorname{argmax}\ p(d_n | \mathbf{o}_n^{(l)}, \mathbf{o}_1^{(r)}, ..., \mathbf{o}_N^{(r)}, \mathbf{S}) \ , \tag{5}$$

where $\mathbf{o}_1^{(r)}, ..., \mathbf{o}_N^{(r)}$ denotes all observations in the right image which are located on the epipolar line of $\mathbf{o}_n^{(l)}$. The posterior can be factorized as

$$p(d_n | \mathbf{o}_n^{(l)}, \mathbf{o}_1^{(r)}, ..., \mathbf{o}_N^{(r)}, \mathbf{S}) \propto p(d_n | \mathbf{S}, \mathbf{o}_n^{(l)}) p(\mathbf{o}_1^{(r)}, ..., \mathbf{o}_N^{(r)} | \mathbf{o}_n^{(l)}, d_n) \ . \tag{6}$$

The observations along the epipolar line on the right image are structured, i.e., given a disparity associated with $\mathbf{o}_n^{(l)}$, there is a deterministic mapping to which observations have non-zero probability on the line. We capture this property by modeling the distribution over all the observations along the epipolar line as

$$p(\mathbf{o}_1^{(r)}, ..., \mathbf{o}_N^{(r)} | \mathbf{o}_n^{(l)}, d_n) \propto \sum_{i=1}^{N} p(\mathbf{o}_i^{(r)} | \mathbf{o}_n^{(l)}, d_n). \tag{7}$$

Note that from Eq. (4), there is only one observation with non-zero probability for each $d_n$. Plugging Eq. (2) and (4) into Eq. (6) and taking the negative logarithm yields an energy function that can be easily minimized

$$E(d) = \beta \ \|\mathbf{f}^{(l)} - \mathbf{f}^{(r)}(d)\|_1 - \log \left[\gamma + \exp\left(-\frac{[d - \mu(\mathbf{S}, \mathbf{o}^{(l)})]^2}{2\sigma^2}\right)\right] \tag{8}$$

with $\mathbf{f}^{(r)}(d)$ the feature vector located at pixel $(u^{(l)} - d, v^{(l)})$. Note that from the definition of the image likelihood, the energy $E(d)$ is required to be evaluated only if $|d - \mu| < 3\sigma$, or $d$ is an element of the neighboring support point disparities. A dense disparity map can be obtained by minimizing Eq. (8). Importantly, this can be done in parallel for each pixel as the support points decouple the different observations.

Although in this section we have focused on obtaining the disparity map of the right image, similarly one can obtain the left disparity map. In practice, we apply our approach to both images, and perform a left/right consistency check to eliminate spurious mismatches and disparities in occluded regions. Following [16] we also remove small segments with an area smaller than 50 pixels.

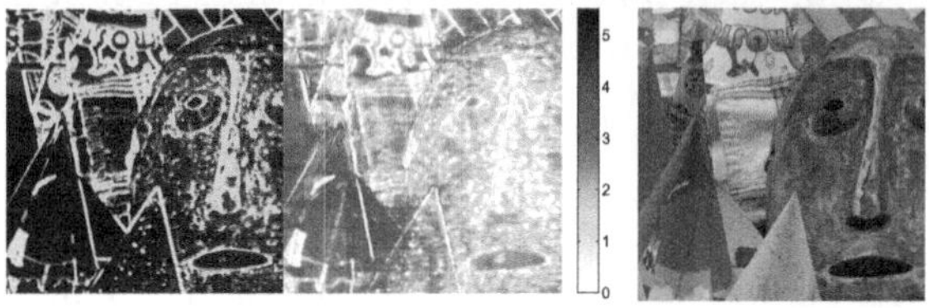

(a) Entropy (uniform vs. our prior) (b) Missed disp.

| Cones (900x750) | Time |
|---|---|
| Filtering | 27 ms |
| Support points | 118 ms |
| Triangulation | 7 ms |
| Matching | 359 ms |
| L/R consistency | 33 ms |
| Postprocessing | 125 ms |
| Total time | **669 ms** |

(c) Running times

**Fig. 3.** (a) Entropy of the posterior in Eq. (5) in nats using a uniform prior and our method for a subset of the *Cones* image pair. (b) Pixels for which our proposal distribution does not contain the ground truth disparity are shown in red. (c) Running time of the individual parts of our algorithm for an image of size $900 \times 750$ pixels on a single CPU core.

# 4 Experimental Evaluation

In this section we compare our approach to state-of-the-art methods in terms of accuracy and running time. Throughout all experiments we set $\beta = 0.03$, $\sigma = 3$, $\gamma = 15$ and $\tau = 0.9$ which were found to empirically perform well. We employed the library 'Triangle' [25] to compute the triangulations. All experiments were conducted on a single i7 CPU core running at 2.66 GHz. Since our goal is to achieve near real time rates on high resolution imagery, we compare all approaches on large images in the Middlebury dataset. This is in contrast to the $450 \times 375$ resolution typically used in the literature.

## 4.1 Entropy Reduction

We first evaluate the quality of our prior by evaluating the entropy of the posterior in Eq. (6). We expect a good prior to reduce the matching entropy since it gets rid of ambiguous matches. Fig. 3a shows the posterior entropy in nats on the *Cones* image set from the Middlebury dataset. Our approach disambiguates the problem as the matching entropy is significantly lower compared to using a uniform prior. As shown in Fig. 3b, the pixels for which our prior is erroneous mainly occur in occluded regions. Notably, our algorithm is able to compute the disparity maps for the left and right image in 0.6 s. An overview of the running times of the different parts of our algorithm is shown in Fig. 3c. Post-processing refers to the aforementioned removal of small segments and constant interpolation of missing disparities in order to obtain dense disparity maps.

## 4.2 Accuracy and Running Time for the Middlebury Dataset

We compare our approach to a wide range of baselines on medium to high resolution images from the Middlebury benchmark. In particular, we compare against two global methods [5,6] and two seed-and-grow algorithms [18,19], using their publicly available implementations[2]. We also compare our prior to a *uniform*

[2] http://people.cs.uchicago.edu/~pff/, http://www.cs.ucl.ac.uk/staff/V.Kolmogorov/, http://cmp.felk.cvut.cz/~stereo/, http://cmp.felk.cvut.cz/cechj/GCS/

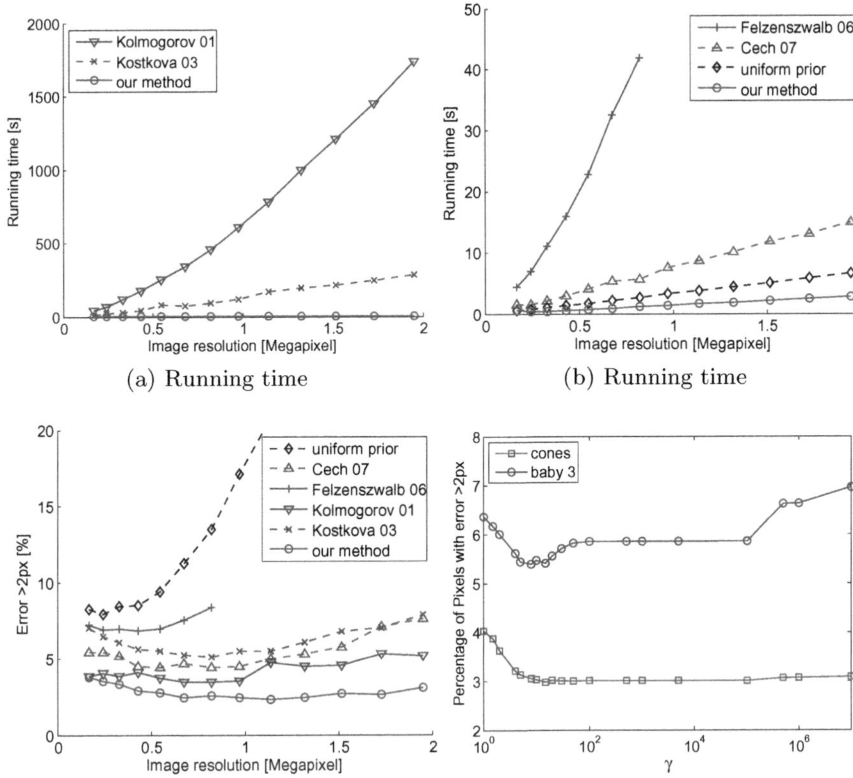


(c) Accuracy (d) Sensitivity with respect to $\gamma$

**Fig. 4. Comparison to state-of-the-art and sensitivity** on the *Cones* image pair. (a,b) Running time and (c) accuracy as a function of the image resolution. (d) Sensitivity of our algorithm wrt. $\gamma$.

*prior* that uses the same image likelihood as our method, and selects disparities using a winner-takes-all strategy over the whole disparity range.

For graph-cuts with occlusion handling [5] we use $K = 60$, $\lambda_1 = 40$, $\lambda_2 = 20$, the $l_2$ norm with an intensity threshold of 8 in combination with the Birchfield-Tomasi dissimilarity measure. We iterate until convergence to obtain best results. These parameters are set based on suggestions in [5] and adaptations to the large-scale imagery we use. We run efficient hierarchical belief propagation [6] using 4 scales, 10 iterations per scale, $\lambda = 0.1$ and $\sigma = 1$. For the seed-and-grow methods [18,19] we set $\alpha = \beta = 0$, which empirically gave best performance. For all baselines which are not able to estimate the disparity search range automatically, we set it manually to the largest disparity in the ground truth. We use the aforementioned large search space for our method and [18,19]. To obtain dense results for all methods, missing disparities are interpolated using a piecewise constant function on the smallest valid neighbor in the same image line.

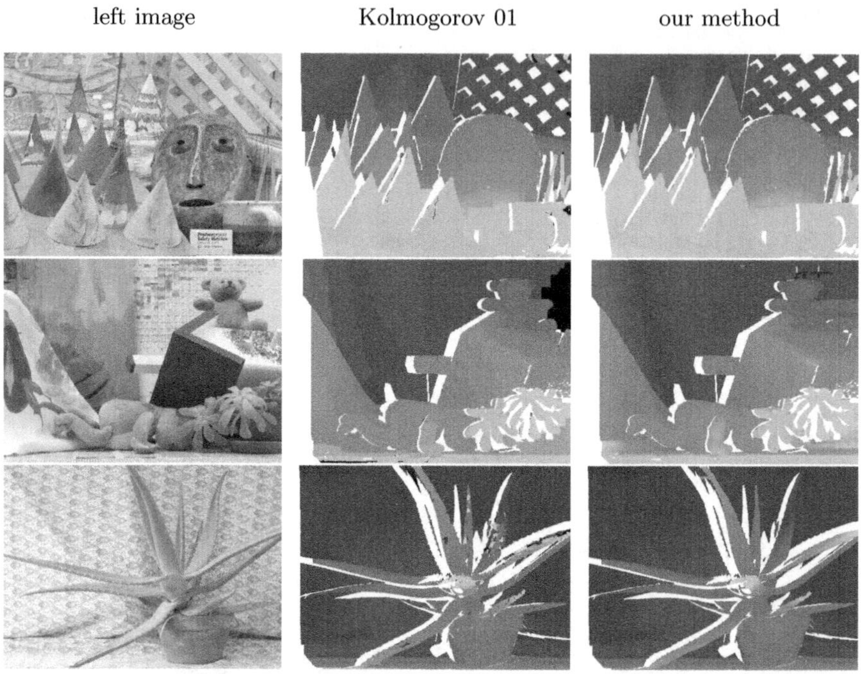


| | Cones | Teddy | Art | Aloe | Dolls | Baby3 | Cloth3 | Lamp2 | Rock2 |
|---|---|---|---|---|---|---|---|---|---|
| Image width | 900 | 900 | 1390 | 1282 | 1390 | 1312 | 1252 | 1300 | 1276 |
| Image height | 750 | 750 | 1110 | 1110 | 1110 | 1110 | 1110 | 1110 | 1110 |
| Support points | 3236 | 3095 | 6164 | 6268 | 8241 | 5901 | 6805 | 4424 | 6670 |
| Correct points | 99.2% | 99.1% | 99.1% | 99.8% | 99.4% | 98.7% | 100.0% | 98.9% | 100.0% |
| Triangles | 6376 | 6128 | 12237 | 12417 | 16353 | 11689 | 13473 | 8769 | 13215 |
| Missed pixels | 0.7% | 4.0% | 5.3% | 1.6% | 1.1% | 1.0% | 0.4% | 9.7% | 0.6% |
| Non-occluded pixels: Error > 1 | | | | | | | | | |
| uniform prior | 18.0% | 37.5% | 43.0% | 12.8% | 33.1% | 49.4% | 7.6% | 74.9% | 7.8% |
| Felzenszwalb 06 | 15.2% | 18.7% | 23.3% | 12.8% | 20.9% | 13.0% | 6.1% | 32.0% | 7.6% |
| Kolmogorov 01 | 8.2% | 16.5% | 30.3% | 13.5% | 28.7% | 26.2% | 4.3% | 65.7% | 10.4% |
| Cech 07 | 7.2% | 15.8% | 18.8% | 9.2% | 19.8% | 17.4% | 2.8% | 36.7% | 3.6% |
| Kostkova 03 | 7.2% | 13.5% | 17.9% | 7.2% | 14.4% | 14.2% | 2.7% | 31.5% | 3.0% |
| **our method** | **5.0%** | **11.5%** | **13.3%** | **5.0%** | **11.0%** | **10.8%** | **1.4%** | **17.5%** | **1.9%** |
| Non-occluded pixels: Error > 2 | | | | | | | | | |
| uniform prior | 16.4% | 35.0% | 41.1% | 11.3% | 29.6% | 46.9% | 7.3% | 74.2% | 7.3% |
| Felzenszwalb 06 | 7.8% | 11.4% | 16.5% | 7.8% | 10.5% | 7.0% | 3.5% | 26.0% | 3.1% |
| Kolmogorov 01 | 4.1% | 8.1% | 21.0% | 8.1% | 17.0% | 19.0% | 1.8% | 60.7% | 6.0% |
| Cech 07 | 4.4% | 10.2% | 11.2% | 4.8% | 10.6% | 9.7% | 1.8% | 27.1% | 2.1% |
| Kostkova 03 | 5.3% | 10.1% | 13.0% | 4.8% | 8.2% | 8.2% | 2.2% | 26.7% | 2.2% |
| **our method** | **2.7%** | **7.3%** | **8.7%** | **3.0%** | **5.3%** | **4.5%** | **0.9%** | **10.4%** | **1.0%** |

**Fig. 5. Results on Middlebury data set.** White regions highlight occluded areas.

We first evaluate running time and accuracy for varying resolutions ranging from 0.16 Megapixel (Middlebury benchmark size) to 2 Megapixels on the *Cones* image pair. For best performance, we adjust the smoothing parameter of the global methods to adapt linearly with the image scale. Since hallucinating occluded areas is not the focus of this paper, we evaluate the error in all non-occluded regions, i.e. all non-white pixels in Fig. 5. As shown in Fig. 4(a,b), our method, which takes about three seconds to process a 2 Megapixel image, runs up to three orders of magnitude faster than global methods. Note that while when using a uniform prior the error rate increases quickly with image size due to an increase in the number of ambiguities, as depicted by Fig. 4c, our method's performance remains constant. Also note that due to memory limitations, we were only able to process images up to 0.8 Megapixels for [6]. Fig. 4d shows the sensitivity of our method with respect to the choice of $\gamma$ on the *Cones* and *Baby3* image pairs. While our approach is insensitive to the precise value of $\gamma$, we observed best performance for $\gamma = 15$, especially for poorly-textured images.

We also compare the accuracy of our approach to the baselines in a variety of stereo images from the Middlebury data set [2], i.e., *Cones*, *Teddy*, *Art*, *Aloe*, *Dolls*, *Baby3*, *Cloth3*, *Lampshade2* and *Rock2*. As before, we use the non-occluded pixels to compute error rates. The top row of Fig. 5 depicts the left camera image, the disparity maps created using Kolmogorov's graph-cuts [5] and our method. The upper rows of the table depict statistics of the input images as well as of our prior. We refer as 'correct points' the ratio of correctly matched support points and 'missed pixels' as the amount of ground truth disparities not contained in the prior; this is a lower bound on the error of our method. Note that, for most of the images, more than 98% of the correct disparities are included in our prior. A comparison to the baselines is depicted in the bottom two rows of the table. Due to memory limitations, we downscaled the images bi-cubically by $\frac{2}{3}$ for [6]. The final disparity maps were up-sampled again, using the nearest neighbor disparities. Note that even though our method mainly aims for efficient matching, for all images we perform competitively to global methods. As expected, smallest errors are achieved for highly textured objects: *Cloth3* and *Rock2*. Worst performance is obtained for the *Lamp2* image pair, as it is poorly textured. The bad performance of the uniform prior is mostly due to its inability to match textureless regions. In contrast, our approach handles them via the proposed prior. This is particularly evidenced in the *Dolls*, *Baby3* and *Lamp2* image pairs. Note that even though the graph-cuts baseline is able to capture disparity discontinuities more accurately than our method, it is not able to perform very well overall. This is mainly due to the inability of the Potts model to truthfully recover poorly-textured slanted surfaces, as shown in Fig. 5.

### 4.3 Urban Sequences and Face Image Set

We now demonstrate the effectiveness of our approach on challenging real-world matching problems using high resolution imagery (1382 × 512 pixels) recorded from a moving vehicle. Typical challenges are poorly-textured regions and sensor saturation. While some regions are unmatched due to half occlusions

**Fig. 6. Challenging urban scene and face dataset.** We evaluate our algorithm on an urban video sequence at $1382 \times 512$ pixels resolution. By computing disparity maps at 2 fps and using visual odometry with sparse features, we were also able to obtain a 3D scene reconstruction in real time (see **http://www.cvlibs.net**). The bottom row shows a one Megapixel face image from the dataset of [26] and the corresponding disparity map obtained using our algorithm. Best viewed in color.

(i.e., disparities depicted in black), most of the scene is accurately estimated by our approach. For this dataset, we obtain frame rates of $\geq 2$ fps on a single core. Real time 3D scene reconstruction is made possible by computing one disparity map every 0.5 seconds, or by exploiting the parallel nature of our algorithm.

The bottom row of Fig. 6 shows the disparity map of a 1 Megapixel face image [26], generated by our algorithm. Importantly, note that all computations took only 1 second. Fine facial structures are clearly visible, while only few ambiguous regions could not be matched.

## 5 Conclusion and Future Work

In this paper we have proposed a Bayesian approach to stereo matching that is able to compute accurate disparity maps of high resolution images at frame rates close to real time. We have shown that a prior distribution estimated from robust support points can decrease stereo matching ambiguities. Our experiments on the Middlebury benchmark and real-world imagery show that our approach performs comparably to state-of-the-art approaches, while being orders of magnitude faster. Importantly, exploiting the parallel nature of our algorithm e.g., in a GPU implementation, will enable real time stereo matching at resolutions above 1 Megapixel. To increase the robustness of our method and account for small surfaces, we intend to study the use of adaptive support windows to estimate support points. We further plan to incorporate scene segmentation to better model disparity discontinuities.

**Acknowledgement.** We thank the reviewers for their feedback, Thabo Beeler for providing the face test image and the *Karlsruhe School of Optics and Photonics* and the *Deutsche Forschungsgemeinschaft* for supporting this work.

## References

1. Gallup, D., Frahm, J.M., Mordohai, P., Pollefeys, M.: Variable baseline/resolution stereo. In: CVPR, pp. 1–8 (2008)
2. Scharstein, D., Szeliski, R.: A taxonomy and evaluation of dense two-frame stereo correspondence algorithms. Int. Journal of Computer Vision 47, 7–42 (2002)
3. Weber, M., Humenberger, M., Kubinger, W.: A very fast census-based stereo matching implementation on a graphics processing unit. In: IEEE Workshop on Embedded Computer Vision (2009)
4. Boykov, Y., Veksler, O., Zabih, R.: Markov random fields with efficient approximations. In: CVPR, pp. 648–655 (1998)
5. Kolmogorov, V., Zabih, R.: Computing visual correspondence with occlusions using graph cuts. In: International Conference on Computer Vision, pp. 508–515 (2001)
6. Felzenszwalb, P., Huttenlocher, D.: Efficient belief propagation for early vision. International Journal of Computer Vision 70, 41–54 (2006)
7. Woodford, O., Torr, P., Reid, I., Fitzgibbon, A.: Global stereo reconstruction under second-order smoothness priors. PAMI 31, 2115–2128 (2009)
8. Cheng, L., Caelli, T.: Bayesian stereo matching. Computer Vision and Image Understanding 106, 85–96 (2007)

9. Kong, D., Tao, H.: Stereo matching via learning multiple experts behaviors. In: BMVC, pp. 97–106 (2006)
10. Wang, L., Jin, H., Yang, R.: Search space reduction for MRF stereo. In: Forsyth, D., Torr, P., Zisserman, A. (eds.) ECCV 2008, Part I. LNCS, vol. 5302, pp. 576–588. Springer, Heidelberg (2008)
11. Konolige, K.: Small vision system. hardware and implementation. In: International Symposium on Robotics Research, pp. 111–116 (1997)
12. Kanade, T., Okutomi, M.: A stereo matching algorithm with an adaptive window: Theory and experiment. In: ICRA (1994)
13. Yoon, K.j., Member, S., Kweon, I.S.: Adaptive support-weight approach for correspondence search. PAMI 28, 650–656 (2006)
14. Klaus, A., Sormann, M., Karner, K.: Segment-based stereo matching using belief propagation and a self-adapting dissimilarity measure. In: ICPR (2006)
15. Liang, C.K., Cheng, C.C., Lai, Y.C., Chen, L.G., Chen, H.H.: Hardware-efficient belief propagation. In: Computer Vision and Pattern Recognition (2009)
16. Hirschmueller, H.: Stereo processing by semiglobal matching and mutual information. PAMI 30, 328–341 (2008)
17. Bobick, A.F., Intille, S.S.: Large occlusion stereo. International Journal of Computer Vision 33, 181–200 (1999)
18. Cech, J., Sára, R.: Efficient sampling of disparity space for fast and accurate matching. In: Computer Vision and Pattern Recognition (2007)
19. Kostkova, J., Sara, R.: Stratified dense matching for stereopsis in complex scenes. In: BMVC (2003)
20. Veksler, O.: Reducing search space for stereo correspondence with graph cuts. In: British Machine Vision Conference (2006)
21. Xiaoyan Hu, P.M.: Evaluation of stereo confidence indoors and outdoors. In: CVPR (2010)
22. Sára, R.: Finding the largest unambiguous component of stereo matching. In: Heyden, A., Sparr, G., Nielsen, M., Johansen, P. (eds.) ECCV 2002. LNCS, vol. 2352, pp. 900–914. Springer, Heidelberg (2002)
23. Bay, H., Tuytelaars, T., Gool, L.V.: Surf: Speeded up robust features. In: Leonardis, A., Bischof, H., Pinz, A. (eds.) ECCV 2006. LNCS, vol. 3951, pp. 404–417. Springer, Heidelberg (2006)
24. Hirschmueller, H., Scharstein, D.: Evaluation of cost functions for stereo matching. In: CVPR, pp. 1–8 (2007)
25. Shewchuk, J.R.: In: Lin, M.C., Manocha, D. (eds.) FCRC-WS 1996 and WACG 1996. LNCS, vol. 1148, pp. 203–222. Springer, Heidelberg (1996)
26. Beeler, T., Bickel, B., Beardsley, P., Sumner, B., Gross, M.: High-quality single-shot capture of facial geometry. In: SIGGRAPH, vol. 29 (2010)

# Towards Full 3D Helmholtz Stereovision Algorithms

Amaël Delaunoy[1], Emmanuel Prados[1], and Peter N. Belhumeur[2]

[1] INRIA Grenoble / LJK
[2] Columbia University

**Abstract.** Helmholtz stereovision methods are limited to binocular stereovision or depth maps reconstruction. In this paper, we extend these methods to recover the full 3D shape of the objects of a scene from multi-view Helmholtz stereopsis. Thus, we are able to reconstruct the complete three-dimensional shape of objects made of any arbitrary and unknown bidirectional reflectance distribution function. Unlike previous methods, this can be achieved using a full surface representation model. In particular occlusions (self occlusions as well as cast shadows) are easier to handle in the surface optimization process. More precisely, we use a triangular mesh representation which allows to naturally specify relationships between the geometry of a point of the scene and its surface normal. We show how to implement the presented approach using a coherent gradient descent flow. Results and benefits are illustrated on various examples.

## 1 Introduction

Reconstructing shape and appearance of objects from images is still one of the major problems in computer vision and graphics. In this work, we are interested in recovering a full and dense representation of the object's three-dimensional shape. Multi-view reconstruction systems are commonly used to estimate such a model, as they provide information from many viewpoints around the object of interest. Among these approaches, variational methods have been popular because they can be used to solve a wide variety of vision problems. The idea is to minimize an energy functional that depends on the considered object surface and on the input images, whose minima is reached at the object of interest. Many methods have been proposed in order to solve this problem, but these approaches are often limited in the kind of appearance they can handle. To overcome these limitations, we exploit Helmholtz reciprocity and propose a single framework for normal estimation and normal integration using a triangular mesh-based deformable model.

**A Reconstruction Approach for Real World Objects.** Most of the multiview reconstruction algorithms rely on image correspondences (as done for instance in multiview stereo [1]) or shading (using the normal information in multiview shape from shading [2,3] or multiview photometric stereo [4]). When texture information (stereo case) is good enough or Lambertian assumption is

R. Kimmel, R. Klette, and A. Sugimoto (Eds.): ACCV 2010, Part I, LNCS 6492, pp. 39–52, 2011.

sufficiently verified, those methods have been proved to give good results with surfaces that are nearly. They then obtain either accurate correspondences or accurate normal estimates. But when the scene is not Lambertian, which is the case for most (if not all) real world scenes, such cues are not valid and algorithms fail to reconstruct accurately the surface. In order to solve this problem, many alternatives have been proposed. Some authors consider specular highlights as outliers [4], and consider a large number of images in order to compensate. Some others modify the input images to have specular free images, and be photometric invariant [5,6]. Another approach is to have a robust similarity measurement [7,8]. All these methods try to compensate the non-Lambertian components in order to run reconstruction algorithms designed for the Lambertian case.

Some authors consider another strategy: they propose to use more general reflectance / radiance parametric models; see (e.g.) [9,10,11]. In practice, these approaches suffer from several limitations. First of all the reflectance model has to be known in advance and this constrains scene to be composed by materials consistent with the chosen reflectance model. Such algorithms tend to solve nonlinear systems of thousands of variables (one reflectance / radiance model per surface point), or need additional assumptions (single or fixed number of materials, single specular component, etc.). Those models are difficult to be optimized and generally require to alternatively estimate the reflectance and the shape. They are numerically unstable and easily tend to get stuck in local minima [3]. Moreover, the algorithms are generally ill-posed, and then require strong regularization which over-smooth the obtained results. Finally, the reflectance and illumination models need also to be approximated. Although such algorithms show reasonable results for perfect synthesized scenes, their application to real-world scenes is complex and requires accurate camera and light calibration.

A different approach introduced by [12] uses radiance samples of reflectance exemplars to match it with the observed images. A direct matching allows to estimate the surface normal in order to reconstruct the shape by normal integration. Although it can deal with many materials and anisotropic BRDF, material samples are needed. This can be a restrictive assumption in applications.

During the last decade, some authors have proposed to use Helmholtz reciprocity in order to perform 3D reconstruction [13,14,15,16,17]. In practice, Helmholtz reciprocity is exploited by taking a pair of images under a single light source, where camera centers and light positions are exchanged at each shot. It uses the fact that in this particular setup, for a single reciprocal pair, the relationship between two radiances of a single surface point is independent of the reflectance. Contrary to works described in previous paragraphs, methods based on Helmholtz reciprocity [13,14,15,16,17] allow to accurately estimate the normals at one point, independently of the reflectance model. This can be used to obtain a 3D surface. In this context, modeling the reflectance, having material samples or being photometric invariant is not required. Nevertheless, contrary to most of multiview stereovision algorithms, the state of the art in Helmholtz reconstruction is limited to depth map reconstructions. In this paper, we push the envelope by proposing a Full 3D multiview Helmholtz stereovision method.

**A Surface-Based Approach.** Until now, all the previous Helmholtz reconstruction methods were camera-view centered. On the contrary, we propose to change the surface representation and to adopt an object-centered strategy. Thus, instead of using a 2.5D surface as it used to be done previously, we represent the object's surface by a closed and dense 2D manifold embedded in the euclidean 3D space. The interest of this choice is two-fold. First this allows to naturally recover a full 3D surface when the other approaches only recover a depth map (or a needle map). Secondly, this allow to easily and properly handle visibility (shadows as well as self occlusions). Figure 1 sheds light on these advantages: the picture on the left illustrates the case of conventional Helmholtz Stereovision methods, that recover a surface based on a virtual view (here in red). The drawing on the right illustrates the proposed approach which is surface-based instead of view-based reconstruction. The red contours show the optimal surface each method can recover. Clearly, the conventional approaches are confronted with difficulties with visibility since it generates discontinuities in the depth maps and the needle maps. Also, this difficulty spreads at the integration steps which behave badly in presence of discontinuities.

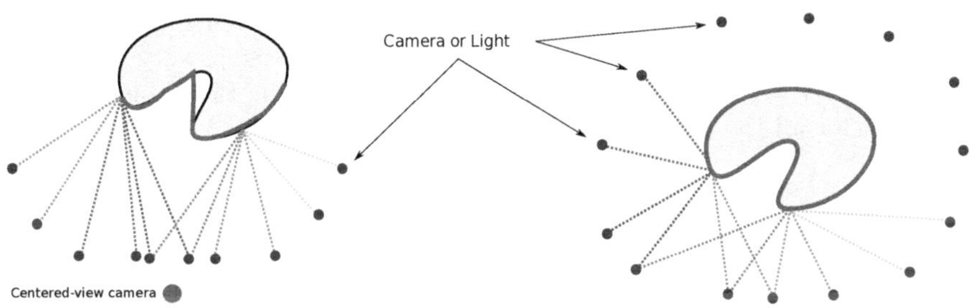


**Fig. 1.** Recovered surfaces by conventional Helmholtz Stereopsis versus proposed approach (surface in red). Left: case of conventional Helmholtz Stereo methods which recovers the shape based on a virtual view (in red). Right: the proposed approach.

In this work, as we will discuss in section 3, we represent the 2D surface using a triangle mesh. In our context, where the surface normals play a key role, this representation offers significant advantages. In particular, it intrinsically links a depth to its surface normal which is, in the case of piecewise planar surfaces like triangular meshes, well defined on each facet. Also, this allows us to naturally combine both the normal integration and normal estimation when these were two separate stages in most of previous approaches [13,14,15].

**Contributions.** This article presents several improvements to state-of-the-art 3D reconstruction techniques exploiting Helmholtz reciprocity. First, we model the problem as an energy minimization problem which is completely surface-based (when all previous methods were camera based). Secondly, we present a method for solving this problem and show how to implement it on triangular surface meshes by using a coherent discrete gradient flow. Finally, since we optimize a full surface, the method is able handle visibility. This also allows to fully reconstruct dense 3D surfaces of complex objects in a single framework.

## 2 Helmholtz Stereopsis - Variational Formulation

As described previously, Helmholtz reciprocity exploits the fact that BRDFs are generally symmetric and therefore, for any incoming angle $\hat{\imath}$ and outgoing direction $\hat{o}$, we have $\beta(\hat{\imath}, \hat{o}) = \beta(\hat{o}, \hat{\imath})$ where $\beta$ is the BRDF function. By interchanging light and camera positions, one can exploit the constraint on $\beta$ in radiance equations in order to solve 3D reconstruction problems.

Given a camera – light pair, one can write the radiance equation $I_c$ of a scene seen from a camera $c$ as:

$$I_c = \alpha\beta(\mathbf{v}_c, \mathbf{v}_l)\frac{\mathbf{v}_l \cdot \mathbf{n}}{\|\mathbf{v}_l\|^3} , \tag{1}$$

where $\mathbf{v}_c$ is the vector from the camera center to the point, $\mathbf{v}_l$ is the vector from the light center to the view-point and $\alpha$ is a constant. $\mathbf{n}$ is the surface normal of the considered point and $\beta(\mathbf{v}_c, \mathbf{v}_l)$ is the BRDF at the surface point (See Figure 2). The same radiance equation can be written for modeling the radiance $I_l$ with the BRDF $\beta(\mathbf{v}_l, \mathbf{v}_c)$. Using Helmholtz reciprocity allows us to write $\beta(\mathbf{v}_c, \mathbf{v}_l) = \beta(\mathbf{v}_l, \mathbf{v}_c)$. This equality then defines the Helmholtz stereopsis constraint for all the point of the surface $\mathcal{S}$:

$$\left[ I_c \, \frac{\mathbf{v}_c}{\|\mathbf{v}_c\|^3} - I_l \, \frac{\mathbf{v}_l}{\|\mathbf{v}_l\|^3} \right] \cdot \mathbf{n} = 0 . \tag{2}$$

Now, we are going to formulate this constraint in the variational framework via a weighted area functional defined over the surface of the object.

We denote by $\pi_c(\mathbf{x})$ (resp. $\pi_l(\mathbf{x})$) the projection of a point in space $\mathbf{x}$ in the camera $c$ (light $l$ respectively), and $I_c$ (or $I_l$) its corresponding intensity value in the image. For more clarity, we also denote

$$\mathbf{h}(\mathbf{x}) = \left[ I_c(\pi(\mathbf{x}))\frac{\mathbf{v}_c}{\|\mathbf{v}_c\|^3} - I_l(\pi(\mathbf{x}))\frac{\mathbf{v}_l}{\|\mathbf{v}_l\|^3} \right] . \tag{3}$$

In this case, the surface that "best" verifies Equation (2) can be obtained by minimizing the following energy functional defined over the surface, with respect to the surface itself:

$$E_{HS}(\mathcal{S}) = \int_{\mathcal{S}} \left(\mathbf{h}(\mathbf{x}) \cdot \mathbf{n}(\mathbf{x})\right)^2 \, \nu_{\mathcal{S},c,l}(\mathbf{x}) \mathrm{d}s , \tag{4}$$

where $\nu_{\mathcal{S},c,l}$ is the characteristic function such that $\nu_{\mathcal{S},c,l}(\mathbf{x}) = 1$ if $\mathbf{x}$ is visible from both images, or 0 otherwise. $\mathrm{d}s$ is the element of area of the surface. This problem formulation allows thus to naturally integrate both multiview geometry and normal constraint. The functional (4) constrains the surface normals to be on the orthogonal plane of $\mathbf{h}(\mathbf{x})$. This is an ill-posed problem since there is an unlimited choice for the normal. In this context several reciprocal pairs are needed in order to better pose the problem and Energy (4) has to be adapted to multiview settings. At the end, we then consider the energy to minimize as the sum of all energies for all Helmholtz pairs $i$:

$$E_{HS}(\mathcal{S}) = \sum_i \int_{\mathcal{S}} (\mathbf{h}_i(\mathbf{x}) \cdot \mathbf{n}(\mathbf{x}))^2 \ \nu_{\mathcal{S},c_i,l_i}(\mathbf{x}) \mathrm{d}s \ , \tag{5}$$

where $i$ is the $i^{\text{th}}$ camera / light pair.

In the next section, we show how to minimize this energy via gradient descent when the surface is represented by a triangle mesh. To simplify the notations, we will just consider and compute the gradient of the functional (4). The gradient of (5) is then obtained by summing the gradients of all the camera / light pairs.

## 3 Optimization for Triangle Mesh Representation

**Choice of Representation.** In section 1 we show the interest of using an intrinsic and full surface representation. In practice, there exists several possibilities for such representation. We propose to minimize energy (5) using a gradient descent algorithm. The choice of the representation must then be consistent with the surface evolution technique which is the base of gradient descent methods. Here we chose to use the Lagrangian framework and to represent the surface by a triangle mesh. Several reasons motivate this choice. In recent years, Lagrangian methods have taken advantage of significant advances in mesh processing allowing these methods to enjoy practical properties such as topological changes [18,19]. In Lagrangian methods, the gradient is computed directly from the discrete representation, whereas in the Eulerian framework the continuous gradient is computed and then discretized. Performing gradient descent in the context of discrete representations allows to make the minimization coherent with the handled numerical object. In other respects, the visibility of a point from a vantage point is well defined and easy to compute with a mesh representation (by using graphic hardware). In practice, it is easier to check the visibility with such a representation than with a level-set representation. All these reasons make nowadays Lagrangian methods more and more popular. Also these methods have recently proved their strong potential for 3D applications [8,20,21].

**Shape Gradient and Evolution Algorithm.** From now, we consider that the surface is a piecewise planar triangular mesh. Let $\mathbf{X} = \{\mathbf{x}_1 \dots \mathbf{x}_n\}$ be a discrete mesh, $\mathbf{x}_k$ being the $k^{th}$ vertex of $\mathbf{X}$, and let $\mathcal{S}_j$ be the $j^{\text{th}}$ triangle of $\mathbf{X}$. With such a representation, functional (4) can be rewritten as:

$$E_{HS}(\mathcal{S}) = \sum_j \int_{\mathcal{S}_j} (\mathbf{h}(\mathbf{x}) \cdot \mathbf{n}_j)^2 \ \nu_{\mathcal{S},c,l}(\mathbf{x}) \mathrm{d}s_j \ , \tag{6}$$

where $\mathbf{n}_j$ is the normal to $\mathcal{S}_j$ and where the sum is over all the triangles of the mesh $\mathbf{X}$. Figure 2 illustrates these notations.

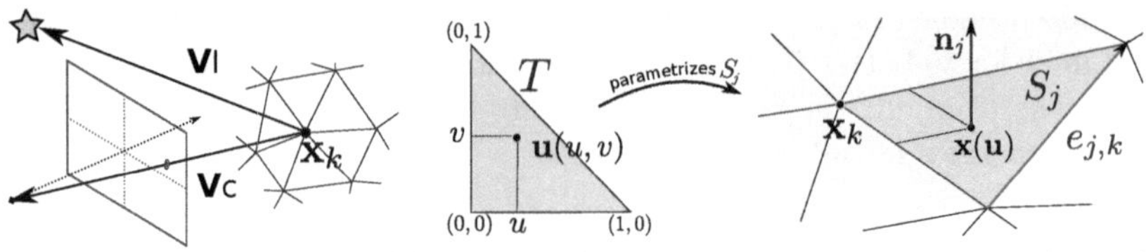


**Fig. 2.** Parametrization of discrete surface representation into a triangle mesh

We propose to optimize $E_{HS}$ with respect to $\mathcal{S}$ using the shape gradient [22]. Let $\mathbf{V}$ be a vector field defined on all the vertices $\mathbf{x}$ of the mesh $\mathbf{X}$ representing the surface deformation. Let us consider the evolution of $E_{HS}$ according to the deformation $\mathbf{V}$. In other words, we assume that the vertices $\mathbf{x}_k[t]$ of $\mathbf{X}[t]$ are moving according to $\mathbf{x}_k[t] = \mathbf{x}_k^0 + t\mathbf{V}_k$. The method for computing the gradient of $E_{HS}$ with respect to $\mathcal{S}$ consists in computing the directional derivative of $E(\mathcal{S})$ for this deformation $\mathbf{V}$, i.e., $\frac{d}{dt}E(\mathcal{S}[t])\Big|_{t=0}$, and then in rewriting it as a scalar product of $\mathbf{V}$, i.e. as $\langle \mathbf{V}, \mathbf{G} \rangle = \sum_k \mathbf{V}_k \cdot \mathbf{G}_k$. The obtained vector $\mathbf{G}$ is called the *gradient* and the energy necessarily decreases when deforming the surface according to its opposite direction $-G$. Indeed, for $\mathbf{x}_k[t] = \mathbf{x}_k^0 - t\mathbf{G}$, we have $\frac{d}{dt}E(\mathcal{S}[t])\Big|_{t=0} = -\langle \mathbf{G}, \mathbf{G} \rangle \leq 0$, see [20].

To obtain the gradient of our energy $E_{HS}$, we have then to calculate the expression of $\frac{d}{dt}E_{HS}(\mathcal{S}[t])\Big|_{t=0}$ and express it as a scalar product of $\mathbf{V}$. In appendix 5, we have detailed this calculus in the general case where a functional $\int_{\mathcal{S}} g(\mathbf{x}, \mathbf{n}(\mathbf{x}))\mathrm{d}s$ is minimized. By replacing $g(\mathbf{x}, \mathbf{n}(\mathbf{x}))$ by $(\mathbf{h}_i(\mathbf{x}) \cdot \mathbf{n}(\mathbf{x}))^2$, we get:

$$\frac{d}{dt}E(\mathcal{S}[t])\Big|_{t=0} = \sum_j \sum_{k \in K_j} \mathbf{V}_k \cdot \left\{ \frac{\mathbf{e}_{j,k}}{A_j} \wedge \int_{\mathcal{S}_j} \Big( (\mathbf{h} \cdot \mathbf{n}_j)^2 \, \mathbf{n}_j - 2\,(\mathbf{h} \cdot \mathbf{n}_j)\, \mathbf{h} \Big)\, \mathrm{d}s_j - \int_{\mathcal{S}_j} 2(\mathbf{h} \cdot \mathbf{n}_j) \nabla_{\mathbf{x}}(\mathbf{h} \cdot \mathbf{n}_j)\; \phi_k(\mathbf{x}) \mathrm{d}s_j \right\}, \quad (7)$$

where $A_j$ is the area of $\mathcal{S}_j$, $\mathbf{e}_{j,k}$ is the edge of $\mathcal{S}_j$ that is at the opposite of vertex $k$; $K_j$ is the set of the indexes of the three vertices of the triangle $\mathcal{S}_j$ and $\phi_k : \mathcal{S} \to \mathbf{R}$ is the piecewise linear interpolating basis function such that $\phi_k(\mathbf{x}_k) = 1$ and $\phi_k(\mathbf{x}_i) = 0$ if $i \neq k$.

Then the $L^2$ gradient descent flow using the triangular mesh uses the gradient $\nabla E(\mathbf{X}) = M^{-1}\frac{\partial E}{\partial \mathbf{X}}(\mathbf{X})$, where $M$ is the *mass* matrix and $\frac{\partial E}{\partial \mathbf{X}}(\mathbf{X})$ is directly given by the part in braces of Equation (7). One classically approximates $M$ by the diagonal mass lumping $\tilde{M}$, where $\tilde{M}_{ii}$ is the area of the Voronoi dual cell of $\mathbf{x}_i$ times the identity matrix $Id_3$; for more details, see e.g. [20,23]. For some initial mesh $\mathbf{X}^0$, the evolution algorithm used here is as follows:

$$\begin{cases} \mathbf{X}[0] = \mathbf{X}^0 \,, \\ \mathbf{X}[t+1] = \mathbf{X}[t] - dt\; \tilde{M}^{-1} \dfrac{\partial E}{\partial \mathbf{X}}(\mathbf{X}[t]) \,. \end{cases} \quad (8)$$

## 4 Experimental Results

In this section we present results using the gradient developed in section 3 (Equation (7)) that directly corresponds to the minimization of our original Energy (Equation (6)).

Let us emphasize here that the gradient descent flow used to obtain the result is exactly the one described previously. In particular, it does not require additional terms or parameters such as surface smoothing present in most variational formulations. Of course, adding such a term would help in being more robust to noise and calibration errors. The only parameters used by us are for numerical computations, like for instance the numerical integration over the triangles. All those can be easily estimated automatically. Since this is a gradient descent approach, one needs some reasonable initialization of the surface such as the visual hull to avoid local minima.

The experiments were implemented in C++ and OpenGL using the CGAL library for mesh computation running on a standard 2.4GHz linux machine – and the topology adaptive meshes of [18]. The optimization starts from an initial condition which is the visual hull in our case. A coarse-to-fine approach is applied to help prevent from local minima. The rendered results use only one constant normal per facet (flat shading).

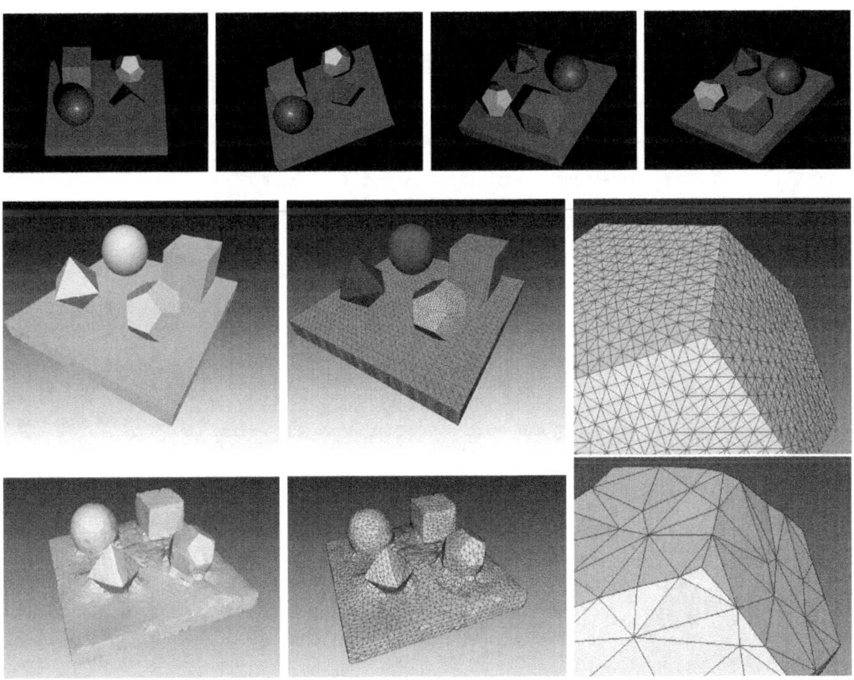

**Fig. 3.** A simple synthetic example. Top row: 2 camera / light pairs out of the 32. Middle row: ground truth surface; ground truth mesh representation; Bottom row: final result; mesh representation of the result; details on the mesh.

We first apply our method to synthesized data. Figure 3 show an example of simple objects disposed on a plane, where images were generated using non-Lambertian reflectance. This dataset is composed of 32 reciprocal pairs with images of resolution $800 \times 600$, placed all around the object of interest. This example shows that our method is able to recover the surface whereas previous Helmholtz stereo methods, where visibility is not accounted for, would fail. In order to solve this problem, they would need to cluster the camera positions to find several view-centered cameras, integrate multiple normal maps and then merge the final reconstructions into a single surface. Our method is simpler in the sense that it works without additional steps.

Even though the number of vertices is low, the gradient flows tends to place them in their correct location. In particular triangle edges perfectly match the one in the images. Again, this is made possible because the discrete gradient is computed with respect to the discrete representation. Also the approach is suitable for reconstructing objects with sharp edges, having depth discontinuities or self-occlusions.The example in figure 4 shows that our method can be applied to reconstruct full and high quality object shapes. Even though the input resolution of the images is low ($1024 \times 768$), the recovered surface nicely matches the ground truth model. The images were generated using a mixture of different

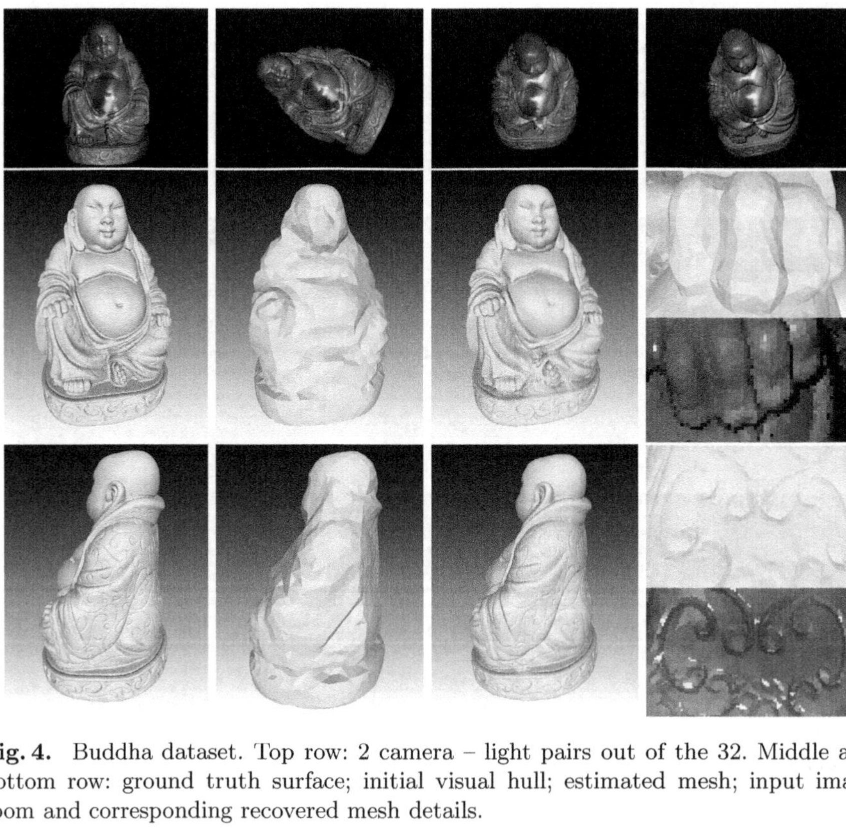

**Fig. 4.** Buddha dataset. Top row: 2 camera – light pairs out of the 32. Middle and bottom row: ground truth surface; initial visual hull; estimated mesh; input image zoom and corresponding recovered mesh details.

specular models so that it looks realistic and non Lambertian. Details are well recovered, and the quality of the mesh is good enough for further object relighting.

Following the evaluation presented in [24], we perform a quantitative evaluation of examples from Figures 3 and 4 where we consider the object in a 2m diameter bounding box. For Figure 3, the completeness at 10mm is 87.51%, and the accuracy at 95% is 9.301mm. For the Buddha, the completeness at 10mm is 95.766%, and the accuracy at 95% is 5.29mm.

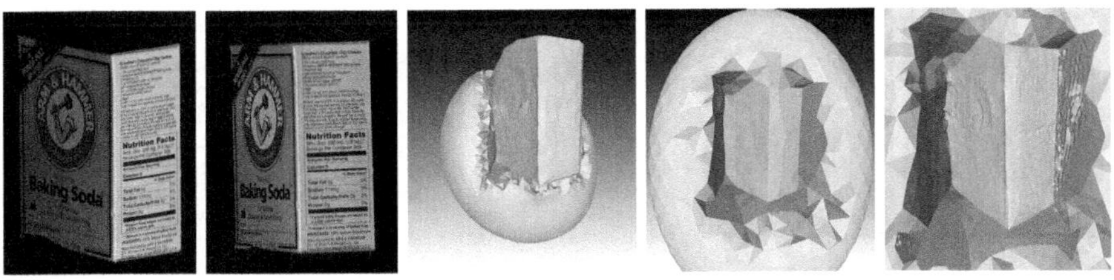

**Fig. 5.** Small box data: results of the proposed method with different initial conditions and mesh resolution. From left to right: 1 camera – light pair out of the 8 pairs; result from a small non-encompassing initial surface; result from an encompassing initial surface; result from an encompassing initial surface with a more dense resolution.

The next two figures 5 and 6 show real data used in previous work [13,14], that have been capture only from one side. We then cannot reconstruct the full surface of the object since images behind the object are missing. Figure 5 shows the 3D reconstruction of a textured box. Since we can choose the mesh resolution, having large triangles compared to the image resolution allows to integrate the gradient over the triangle and have a correct gradient flow for the vertices of the triangle. Then we can reconstruct objects with textured or rough surfaces using Helmholtz reciprocity similarly as in [15]. This figure also show different optimization taken from different initial conditions that finally give similar results.

Figure 6 show results for two real datasets, the mask containing 18 reciprocal pairs, and the mannequin containing 8 reciprocal pairs.

Finally, the approach was tested on real dataset on full 3D objects (Figure 7). It consists of 18 reciprocal pairs (using $1104 \times 828$ resolution images) taken on a ring around the object slightly on top of it. The two data show a "Fish" reconstruction highly specular with fine changing surface structure, whereas the second one "Dragon" has strong self-occlusions and complex shape. Starting from the visual hull, we are able to recover details on the surface even though the input image resolution is not too high. Results are illustrated in Figure 7 and present the recovered full 3D surface. Some parts of the surface are not visible from the images and thus cannot be recovered. Camera calibration was performed using a checker board without distortion corrections and light positions were empirically positioned and calibrated. These datasets will be made available for comparisons purposes. Images are taken around the object so some parts are occluded from images and objects also contain self-occlusions. For those reasons, previous

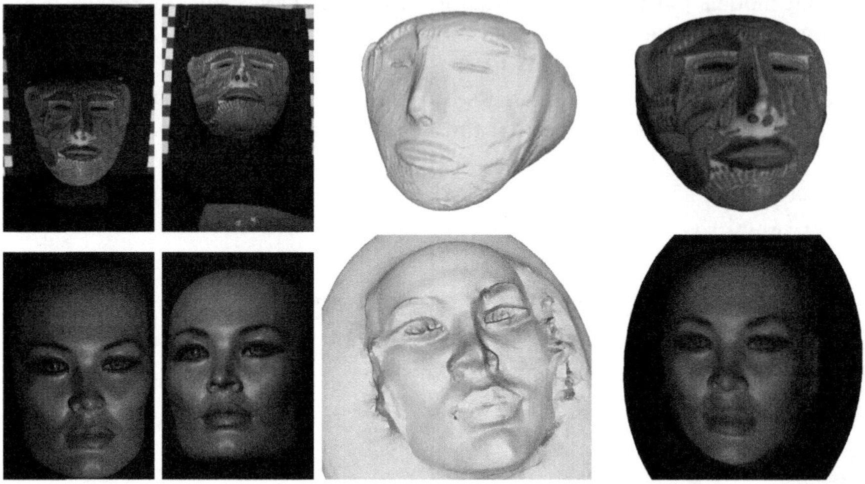

**Fig. 6.** Mannequin (8 reciprocal pairs) and Mask (18 reciprocal pairs) data: results of the proposed method for two real dataset containing varying complex appearance. From left to right: 1 camera / light pair out of the 18 (respectively 8); final result; final mesh textured with the mean of the reprojected colors from the cameras.

approaches using a camera-center view do not apply – since it requires depth continuity – whereas our surface-based approach can recover the full 3D shape.

**Discussion.** All these examples illustrated the advantage of having a mesh representation. It allows to preserve edges, and we show that having a coherent discrete gradient flow allows vertices to be placed at their correct locations.

Second order minimization is known to recover well the higher frequency of the surface, but poorly recover the lower ones. Also, due to the integration process, the optimization might be slow if the surface is too far from the solution. A good initial condition and a coarse to fine approach significantly help to prevent these problems, as illustrated in the experiments. In particular, the gradient flow tends to shrink the surface and introduce a minimal surface bias.

However, since the method minimizes a weighted area functional defined over the surface, one of its global minima (in addition to the real surface) is the empty set. To prevent from this choice, one can add an additional term, start closer to the solution, or fix boundary conditions. A more elegant way is to see the problem as a reprojection error. Instead of minimizing a weighted area functional, one can reformulate the problem (2) by minimizing the following energy functional:

$$E(\mathcal{S}) = \int_{\mathcal{I}_c} \left( I_c - \left( I_l \circ \pi_{\mathcal{S},I_l} \circ \pi^{-1}_{\mathcal{S},I_c} \right) \frac{\hat{\mathbf{v}}_l \cdot \left( \mathbf{n} \circ \pi^{-1}_{\mathcal{S},I_c} \right)}{\hat{\mathbf{v}}_c \cdot \left( \mathbf{n} \circ \pi^{-1}_{\mathcal{S},I_c} \right)} \right)^2 \mathrm{d}\mathbf{u} \,, \qquad (9)$$

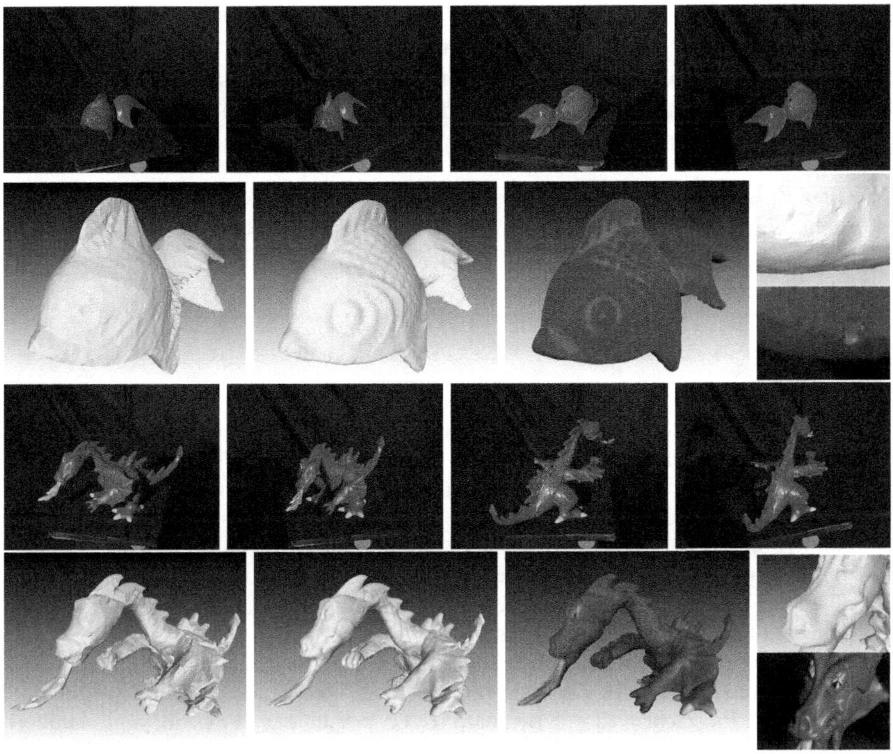

**Fig. 7.** Fish and Dragon data (18 reciprocal pairs of 1104 × 828): Results of the proposed method for two full 3D real world datasets. Top and third row: 2 camera / light pairs out of the 36. Second and last row: initial visual hull; recovered 3D mesh; final mesh textured with the mean of the reprojected colors from the cameras; input image zoom and corresponding recovered mesh details.

where $\pi_{\mathcal{S},I_c}^{-1}(\mathbf{u})$ is the reprojection of an image point $\mathbf{u}$ of the image $I_c$ on the surface $\mathcal{S}$. This formulation as been presented by [17], but has not been minimized as a reprojection error, like for instance in the case of [25] for continuous surfaces, and of [21] using deformable meshes. In fact one may rewrite Equation (9) as an energy over the visible surface [25] instead of the image in order to optimize the surface. Such formulation lets appear an additional term, that turns out to behave like a visual hull constrain on the silhouette occluding contours, or as a contour matching term for self occlusions. Such a term will give boundary conditions to prevent from shrinkage, and will help the method to be more robust to initial conditions.

# 5 Conclusion

In this paper we have presented a surface-based method to estimate the 3D shape of objects from multiview Helmholtz stereo pairs. As far as we know, this is the first time Helmholtz stereopsis can be used to recover dense and full 3D models into a single framework. This is made possible thanks to the compact surface representation that allow to easily compute surface point visibility. Moreover, the mesh based representation allows to naturally exploit geometric relationships between a point of the scene and its surface normal. Tests on synthetic and real datasets demonstrate the benefit of our approach.

**Acknowledgement.** Research was supported by the Agence Nationale pour la Recherche within the Flamenco project (Grant ANR-06-MDCA-007). The authors would like to thank Todd Zickler for providing Helmholtz stereo data used in Figures 5 and 6.

## References

1. Goesele, M., Curless, B., Seitz, S.M.: Multi-view stereo revisited. In: IEEE Conference on Computer Vision and Pattern Recognition, vol. 02, pp. 2402–2409 (2006)
2. Jin, H., Cremers, D., Yezzi, A.J., Soatto, S.: Shedding light on stereoscopic segmentation. In: IEEE Conference on Computer Vision and Pattern Recognition, vol. 01, pp. 36–42 (2004)
3. Jin, H., Cremers, D., Wang, D., Prados, E., Yezzi, A., Soatto, S.: 3-D reconstruction of shaded objects from multiple images under unknown illumination. International Journal of Computer Vision 76 (2008)
4. Hernandez, C., Vogiatzis, G., Cipolla, R.: Multiview photometric stereo. IEEE Transactions on Pattern Analysis and Machine Intelligence 30, 548–554 (2008)
5. Mallick, S.P., Zickler, T., Kriegman, D.J., Belhumeur, P.N.: Beyond lambert: Reconstructing specular surfaces using color. In: IEEE Conference on Computer Vision and Pattern Recognition, vol. 2, pp. 619–626 (2005)
6. Zickler, T., Mallick, S.P., Kriegman, D.J., Belhumeur, P.: Color subspaces as photometric invariants. International Journal of Computer Vision (2008)
7. Pons, J.P., Keriven, R., Faugeras, O.: Multi-view stereo reconstruction and scene flow estimation with a global image-based matching score. The International Journal of Computer Vision 72, 179–193 (2007)
8. Vu, H., Keriven, R., Labatut, P., Pons, J.P.: Towards high-resolution large-scale multi-view stereo. In: IEEE Conference on Computer Vision and Pattern Recognition (2009)
9. Jin, H., Soatto, S., Yezzi, A.J.: Multi-view stereo beyond lambert. In: IEEE Computer Society Conference on Computer Vision and Pattern Recognition, vol. 1, p. 171 (2003)
10. Yu, T., Xu, N., Ahuja, N.: Recovering shape and reflectance model of non-lambertian objects from multiple views. In: IEEE Conference on Computer Vision and Pattern Recognition, pp. 226–233 (2004)
11. Yoon, K.J., Prados, E., Sturm, P.: Joint estimation of shape and reflectance using multiple images with known illumination conditions. International Journal of Computer Vision (2009) (to appear)
12. Hertzmann, A., Seitz, S.: Example-based photometric stereo: Shape reconstruction with general, varying brdfs. IEEE Trans. Pattern Anal. Mach. Intell. 27, 1254–1264 (2005)
13. Zickler, T., Belhumeur, P.N., Kriegman, D.J.: Helmholtz stereopsis: Exploiting reciprocity for surface reconstruction. International Journal of Computer Vision 49, 215–227 (2002)

14. Zickler, T., Ho, J., Kriegman, D., Ponce, J., Belhumeur, P.: Binocular helmholtz stereopsis. In: Proceedings Ninth IEEE International Conference on Computer Vision, vol. 2, pp. 1411–1417 (2003)
15. Guillemaut, J.Y., Drbohlav, O., Sara, R., Illingworth, J.: Helmholtz stereopsis on rough and strongly textured surfaces. In: 3DPVT 2004: Proceedings of the 3D Data Processing, Visualization, and Transmission, 2nd International Symposium, Washington, DC, USA, pp. 10–17. IEEE Computer Society, Los Alamitos (2004)
16. Zickler, T.: Reciprocal image features for uncalibrated helmholtz stereopsis. In: IEEE Computer Society Conference on Computer Vision and Pattern Recognition, vol. 2, pp. 1801–1808 (2006)
17. Tu, P., Mendon, P.R.S.: Surface reconstruction via helmholtz reciprocity with a single image pair. In: IEEE Computer Society Conference on Computer Vision and Pattern Recognition, vol. 1, p. 541 (2003)
18. Pons, J.P., Boissonnat, J.D.: Delaunay deformable models: Topology-adaptive meshes based on the restricted delaunay triangulation. In: IEEE Conference on Computer Vision and Pattern Recognition, Minneapolis, USA (2007)
19. Zaharescu, A., Boyer, E., Horaud, R.: Transformesh: a topology-adaptive mesh-based approach to surface evolution. In: Proceedings Asian Conference on Computer Vision, Tokyo, Japan (2007)
20. Eckstein, I., Pons, J.P., Tong, Y., Kuo, C.C.J., Desbrun, M.: Generalized surface flows for mesh processing. In: Eurographics Symposium on Geometry Processing (2007)
21. Delaunoy, A., Prados, E., Gargallo, P., Pons, J.P., Sturm, P.: Minimizing the multi-view stereo reprojection error for triangular surface meshes. In: British Machine and Vision Conference, Leeds, UK (2008)
22. Debreuve, É., Gastaud, M., Barlaud, M., Aubert, G.: Using the shape gradient for active contour segmentation: from the continuous to the discrete formulation. Journal of Mathematical Imaging and Vision (2007)
23. Meyer, M., Desbrun, M., Schröder, P., Barr, A.H.: Discrete differential-geometry operators for triangulated 2-manifolds (2002)
24. Seitz, S.M., Curless, B., Diebel, J., Scharstein, D., Szeliski, R.: A comparison and evaluation of multi-view stereo reconstruction algorithms. In: IEEE Conference on Computer Vision and Pattern Recognition, pp. 519–528 (2006)
25. Gargallo, P.: Contributions to the Bayesian approach to Multi-view Stereo. PhD thesis, Institut National Polytechique de Grenoble, France (2008)

## Appendix: Gradient Flows of Weighted Area Functionals Using Triangular Meshes

Let $\mathcal{S}_j$ be the $j^{th}$ triangle of the mesh and let us consider a parametrization of the triangle $\mathcal{S}_j$ such that $\mathbf{x}(\mathbf{u}) = x_k^j + u\ \overrightarrow{x_k^j x_1^j} + v\ \overrightarrow{x_k^j x_2^j}$ where $x_k^j$ and $x_1^j$ and $x_2^j$ are the three vertices associated to the triangle $\mathcal{S}_j$ and where $\mathbf{u}(u,v) \in T = \{(u,v) | u \in [0,1] \text{ and } v \in [0,u]\}$. (See Figure 2)

In this appendix, we detail the calculus of $\frac{d}{dt}E(\mathcal{S}_j[t])\Big|_{t=0}$ for energy functionals of the form:

$$E(\mathcal{S}) = \sum_i \int_{\mathcal{S}_j} g(\mathbf{x}, \mathbf{n}(\mathbf{x}))\, d\sigma = \sum_j 2\, A_j \int_T g(\mathbf{x}(\mathbf{u}), \mathbf{n}_j)\, du\ , \tag{10}$$

where $A_j$ is then the area of the triangle $\mathcal{S}_j$ and $\mathbf{n}_j$ the outward surface normal of triangle $\mathcal{S}_j$.
Let us focus on the evolution of $E(S)$ under the induced velocity $\mathbf{V}$ on triangle $\mathcal{S}_j$ only (In other words when a vertex $\mathbf{x}_k$ is moved according to $\mathbf{x}_k[t] = \mathbf{x}_k^0 + t\mathbf{V}_k$). We have:

$$E(\mathcal{S}_j[t]) = 2\,A_j[t] \int_T g(\mathbf{x}(\mathbf{u}) + t\mathbf{V}(\mathbf{x}(\mathbf{u})), \mathbf{n}_j[t])\,du \;, \tag{11}$$

where $A_j[t]$ is the area of the triangle $\mathcal{S}_j[t]$, $\mathbf{n}_j[t]$ is its normal and $\mathbf{V}(\mathbf{x})$ is the piecewise linear extension of the $\mathbf{V}$ vector field on the whole surface $\mathcal{S}$. By simple derivation we get:

$$\begin{aligned} \frac{d}{dt}E(\mathcal{S}_j[t])\Big|_{t=0} = \; & \frac{A'_j[0]}{A_j} \int_{\mathcal{S}_j} g(\mathbf{x}, \mathbf{n}_j)\,\mathrm{d}s_j \\ & + \int_{\mathcal{S}_j} \nabla_{\mathbf{x}} g(\mathbf{x}, \mathbf{n}_j) \cdot \mathbf{V}(\mathbf{x})\,\mathrm{d}s_j + \int_{\mathcal{S}_j} \nabla_{\mathbf{n}} g(\mathbf{x}, \mathbf{n}_j) \cdot \mathbf{n}'_j[0]\,\mathrm{d}s_j \;, \end{aligned} \tag{12}$$

where $A'_j[0] = \frac{d}{dt}A_j[t]\Big|_{t=0}$ and $\mathbf{n}'_j[0] = \frac{d}{dt}\mathbf{n}_j[t]\Big|_{t=0}$ are detailed below. In order to rewrite $\frac{d}{dt}E(\mathcal{S}_j[t])\Big|_{t=0}$ as a scalar production of $\mathbf{V}$, we then need to detail $A'_j[0]$ and $\mathbf{n}'_j[0]$. Since for all $\mathbf{x} \in \mathcal{S}_j$, $\mathbf{V}(\mathbf{x}) = \sum_{k \in K_j} \mathbf{V}_k \phi_k(\mathbf{x})$ ($K_j$, $\phi_k$ and $\mathbf{e}_{j,k}$ being defined in section 3), one can show that: $A'_j[0] = \sum_{k \in K_j} \frac{1}{2}\,(\mathbf{n}_j \wedge \mathbf{e}_{j,k}) \cdot \mathbf{V}_k$ , and

$$\mathbf{n}'_j[0] = \frac{1}{2\,A_j} \left( \left( \sum_{k \in K_j} \mathbf{e}_{j,k} \wedge \mathbf{V}_k \right) - \left( (\sum_{k \in K_j} \mathbf{e}_{j,k} \wedge \mathbf{V}_k) \cdot \mathbf{n}_j \right) \mathbf{n}_j \right) . \tag{13}$$

It follows that

$$\begin{aligned} \frac{d}{dt}E(\mathcal{S}_j[t])\Big|_{t=0} = \sum_{k \in K_j} \mathbf{V}_k \cdot \Bigg\{ & \frac{\mathbf{n}_j \wedge \mathbf{e}_{j,k}}{2A_j} \int_{\mathcal{S}_j} g(\mathbf{x}, \mathbf{n}_j)\,\mathrm{d}s_j \\ & + \int_{\mathcal{S}_j} \nabla_{\mathbf{x}} g(\mathbf{x}, \mathbf{n}_j) \phi_k(\mathbf{x})\,\mathrm{d}s_j - \frac{\mathbf{e}_{j,k}}{2\,A_j} \wedge \int_{\mathcal{S}_j} g_{\mathbf{n}}(\mathbf{x}, \mathbf{n}_j)\,\mathrm{d}s_j \Bigg\} . \end{aligned} \tag{14}$$

Above, we have denoted $g_{\mathbf{n}} = \nabla_{\mathbf{n}} g(\mathbf{x}, \mathbf{n}_j) - \langle \nabla_{\mathbf{n}} g(\mathbf{x}, \mathbf{n}_j), \mathbf{n}_j \rangle\, \mathbf{n}_j$, where $\nabla_{\mathbf{n}} g(\mathbf{x}, \mathbf{n}_j)$ is the gradient of $g$ with respect to the second variable (i.e. $\mathbf{n} \in \mathbf{Re}^3$).
It then immediately follows that:

$$\begin{aligned} \frac{d}{dt}E(\mathcal{S}[t])\Big|_{t=0} = \sum_k \mathbf{V}_k \cdot \sum_{j \in J_k} & \Bigg\{ \int_{\mathcal{S}_j} \nabla_{\mathbf{x}} g(\mathbf{x}, \mathbf{n}_j) \phi_k(\mathbf{x})\,\mathrm{d}s_j \\ & - \frac{\mathbf{e}_{j,k}}{2A_j} \wedge \int_{\mathcal{S}_j} g(\mathbf{x}, \mathbf{n}_j)\mathbf{n}_j + g_{\mathbf{n}}(\mathbf{x}, \mathbf{n}_j)\,\mathrm{d}s_j \Bigg\} , \end{aligned} \tag{15}$$

where $J_k$ is the set of triangles containing vertex $\mathbf{x}_k$. By replacing $g(\mathbf{x}, \mathbf{n}(\mathbf{x}))$ by $(\mathbf{h}_i(\mathbf{x}) \cdot \mathbf{n}(\mathbf{x}))^2$ in the equations below, where $\mathbf{h}_i(\mathbf{x})$ is the vector defined in the paper, we have:

$$\begin{aligned} \nabla_{\mathbf{x}} g(\mathbf{x}, \mathbf{n}_j) &= 2(\mathbf{h}(\mathbf{x}) \cdot \mathbf{n}_j) \nabla_{\mathbf{x}} (\mathbf{h}(\mathbf{x}) \cdot \mathbf{n}_j) \;, \\ g(\mathbf{x}, \mathbf{n}_j)\mathbf{n}_j &= (\mathbf{h}(\mathbf{x}) \cdot \mathbf{n}_j)^2\, \mathbf{n}_j \;, \\ \text{and} \;\; g_{\mathbf{n}}(\mathbf{x}, \mathbf{n}_j) &= 2\,(\mathbf{h}(\mathbf{x}) \cdot \mathbf{n}_j)\, \mathbf{h}(\mathbf{x}) - 2\,(\mathbf{h}(\mathbf{x}) \cdot \mathbf{n}_j)^2\, \mathbf{n}_j \;. \end{aligned} \tag{16}$$

# Image-Based 3D Modeling via Cheeger Sets

Eno Töppe[1,2], Martin R. Oswald[1], Daniel Cremers[1], and Carsten Rother[2]

[1] Technische Universität München, Germany
[2] Mircosoft Research, Cambridge, UK

**Abstract.** We propose a novel variational formulation for generating $3D$ models of objects from a single view. Based on a few user scribbles in an image, the algorithm automatically extracts the object silhouette and subsequently determines a $3D$ volume by minimizing the weighted surface area for a fixed user-specified volume. The respective energy can be efficiently minimized by means of convex relaxation techniques, leading to visually pleasing smooth surfaces within a matter of seconds. In contrast to existing techniques for single-view reconstruction, the proposed method is based on an implicit surface representation and a transparent optimality criterion, assuring high-quality $3D$ models of arbitrary topology with a minimum of user input.

## 1 Introduction

**Single-View Reconstruction.** Generating models of the three-dimensional world from sets of images is at the heart of Computer Vision. An interesting limiting case is the problem of single view reconstruction – a highly ill-posed problem where stereo and multiview concepts like point correspondence and photo-consistency cannot be applied. Nevertheless, it is an important problem: In many applications we may only have a single image of the scene, and yet we may want to interactively extract solid $3D$ models of respective objects for virtual and augmented reality applications, or we may want to simply render the same scene from a novel vantage point or with different illumination based on estimates of the geometric structure.

Human observers have an excellent ability to generate plausible $3D$ models of objects around them – even from a single image. To this end, they partially rely on prior knowledge about the geometric structures and primitives in their world. Yet, they also generate plausible models of objects they have never seen before. It is beyond the scope of this work to contemplate on the multitude of criteria the human visual system may be employing for solving the single view reconstruction problem. Instead, we will demonstrate that for a large variety of real-world images very simple extremality assumptions give rise to convincing $3D$ models. The key idea is to compute a silhouette-consistent weighted minimal surface for a user-specified volume. In this sense, the proposed formulation is closely related to the concept of *Cheeger sets* – sets which minimize the ratio of area over volume [1].

R. Kimmel, R. Klette, and A. Sugimoto (Eds.): ACCV 2010, Part I, LNCS 6492, pp. 53–64, 2011.

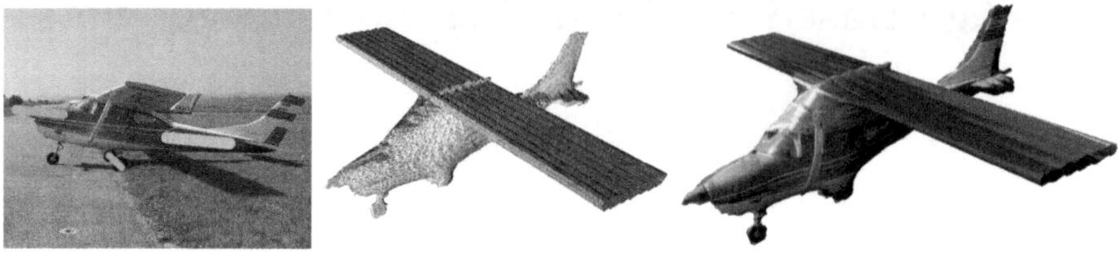


**Fig. 1.** The proposed method generates convincing *3D* models from a single image computed by fixed volume weighted minimal surfaces. Colored lines in the input image mark user input, which locally alters the surface smoothness. Red marks low, yellow marks high smoothness (see section 4.4 for details).

**Related Work.** Existing work on single view reconstruction and on interactive *3D* modeling can be grouped into two classes based on the choice of mathematical surface representation, namely explicit surface representation and implicit surface representations. Some of the pioneering works on single view reconstruction are those of Criminisi and coworkers [2,3] on generating three-dimensional models of architecture from single images by exploiting the perspective structure of parallel lines and other aspects of man-made environments. Different aspects of the reconstruction have been emphasized among related works. Horry et al. [4] aim for pleasant 3D visual effects that do not result in high quality meshes. Hoiem et al. [5] are similar in this respect but they try to fully automate the reconstruction process. Also related to the field are easy-to-use tools like *Teddy* [6] and *FiberMesh* [7] that have pioneered sketch based modeling but are not image-based. Note that there are also approaches that seek to reconstruct height fields [8] and are therefore not suited for getting closed *3D* surfaces.

All of the above works are using explicit surface representation – while surface manipulation is often straight forward and a variety of cues are easily integrated leading to respective forces or constraints on the surface, there are two major limitations: Firstly numerical solutions are generally not independent of the choice of parameterization. And secondly, parametric representations are not easily extended to objects of varying topology. While Prasad et al. [9] were able to extend their approach to surfaces with one or two holes, the generalization to objects of arbitrary topology is by no means straight forward. Similarly, topology-changing interaction in the FiberMesh system requires a complex remeshing of the modeled object leading to computationally challenging numerical optimization schemes.

A first effort in single view reconstruction using an implicit representation was recently proposed by Oswald et al. [10]. There the authors combined a minimal surface constraint with a data term that favored the object thickness to be proportional to the distance to the boundary.

Despite a number of convincing results, the latter work suffers from several drawbacks: Firstly, imposing a thickness proportional to the distance from the curve is very strong and not always a correct assumption. Secondly, the modeling required a large number of not necessarily intuitive parameters controlling the data term. Our work is different from [10] in that we impose exact volume consistency and do not require additional tuning parameters.

All cited works on single view reconstruction have in common that they revert to inflation heuristics in order to avoid surface collapsing. These techniques boil down to fixing absolute depth values, which undesirably restrict the solution space. A precursor to volume constraints are the volume inflation terms pioneered for deformable models by Cohen and Cohen [11]. However, no constant volume constraints were considered and no implicit representations were used.

**Contribution.** In this paper, we revisit the problem of single view reconstruction. We will show that one can compute silhouette-consistent weighted minimal surfaces for a user-prescribed volume using convex relaxation techniques. To this end, we revert to an implicit representation of the surface given by the indicator function of its interior (sometimes referred to as voxel-occupancy). In this representation, the weighted minimal surface problem is a convex functional and relaxation of the binary function leads to an overall convex problem. In addition, we will show that the volume constraint amounts to a convex constraint which is easily integrated in the reconstruction process. We show that the relaxed indicator function can be binarized so that we obtain a surface which firstly has exactly the user-specified volume and secondly is within a computable energetic bound of the optimal combinatorial solution.

The convex optimization is solved by a recently proposed provably convergent primal-dual algorithm enabling *interactive* reconstruction within seconds. We show on a variety of real-world images that the simple extremality condition of a *fixed-volume minimal surface* gives rise to convincing $3D$ models for a large variety of real-world images, comparing favorably to alternative approaches.

To the best of our knowledge this is the first work on convex shape optimization with guaranteed volume preservation.

## 2 Variational Formulation

Assume we are given the silhouette of an object in an image as returned by an interactive segmentation tool[1]. The goal is then to obtain a smooth $3D$ model of the object which is consistent with the silhouette. How should we select the correct $3D$ model among the infinitely many that match the silhouette? Clearly, we need to impose additional information, at the same time we want to keep this information at a minimum since user interaction is always tedious and slow. In the following, we will show that merely specifying the object's volume and computing a minimal surface of given volume is sufficient to give rise to a family of plausible $3D$ models.

We propose a solution that comes in two flavors: one is formulated with a soft the other with a hard volume constraint. We then go into detail on the fast

[1] For brevity and since it is not part of our contribution we will not detail the graph cut based interactive segmentation algorithm we use. Instead we refer to representative work in the field [12].

optimization of the resulting energy which finally leads to an interactive user interface for single view reconstruction.

### 2.1 Implicit Weighted Variational Surfaces

We are given an image plane $\Omega$ which contains the input image and lies in $\mathbb{R}^3$. As part of the image we also have an object silhouette $\Sigma \subset \Omega$. Now, we are seeking to compute reconstructions as minimal weighted surfaces $S \subset \mathbb{R}^3$ that have a certain target volume $V_t$ and are compliant with the object silhouette $\Sigma$:

$$min \int_S g(s)ds \tag{1}$$

$$\text{subject to} \qquad \pi(S) = \Sigma \tag{2}$$

$$Vol(S) = V_t \tag{3}$$

where $\pi : \mathbb{R}^3 \rightarrow \Omega$ is the orthographic projection onto the image plane $\Omega$, $g : \mathbb{R}^3 \rightarrow \mathbb{R}^+$ is a smoothness weighting function, $Vol(S)$ denotes the volume enclosed by the surface $S$ and $s \in S$ is a surface element. In the following we will gradually derive an implicit representation for the above problem.

We begin by replacing the surface $S$ with its implicit binary indicator function $u \in BV(\mathbb{R}^3; \{0,1\})$, where $BV$ denotes the functions of bounded variation [13]. The desired minimal weighted surface area is then given by minimizing the total variation over a suitable set $U$ of feasible functions $u$:

$$\min_{u \in U} \int g(x)|\nabla u(x)| d^3x \tag{4}$$

where $\nabla u$ denotes the derivative in the distributional sense. Eq. (4) favors smooth solutions. However, smoothness is locally affected by the function $g(x)$ : $\mathbb{R}^3 \rightarrow \mathbb{R}^+$ which will be used later for modeling.

How does the set $U$ of feasible functions look like? For simplicity, we assume the silhouette to be enclosed by the surface. Then all surface functions that are consistent with the silhouette $\Sigma$ must be in the set

$$U_\Sigma = \left\{ u \in BV(\mathbb{R}^3; \{0,1\}) \;\middle|\; u(x) = \begin{cases} 0, & \pi(x) \notin \Sigma \\ 1, & x \in \Sigma \end{cases} \right\} \tag{5}$$

Still, solving (4) with respect to the set $U_\Sigma$ of silhouette consistent functions will result in the silhouette itself. In the following section we will show a way to avoid this trivial solution.

### 2.2 Volume Constraint

In order to inflate the solution of (4) we propose to use a constraint on the size of the volume enclosed by the minimal surface. We formulate this both as a soft- and as a hard constraint and discuss the two approaches in the following.

**Hard Constraint.** By further constraining the feasible set $U_\Sigma$ one can force the reconstructed surface to have a specific target volume $V_t$. We regard the problem

$$\min_{u \in U_\Sigma \cap U_V} E(u) \qquad \text{where} \quad E(u) = \int g(x)|\nabla u(x)| d^3x \tag{6}$$

$$\text{and} \quad U_V = \left\{ u \in BV(\mathbb{R}^3; \{0,1\}) \;\middle|\; \int u(x) d^3x = V_t \right\} \tag{7}$$

where $U_V$ denominates all reconstructions with bounded variation that have the specific volume $V_t$.

**Soft Constraint.** For the sake of completeness we also consider the soft formulation of the volume constraint. One can add a ballooning term to (4):

$$E_V(u) = \lambda \left( \int u(x) d^3x - V_t \right)^2 \tag{8}$$

The integral quadratically punishes the deviation of the surface volume from a certain target volume $V_t$. In contrast to the constant volume constraint above, this formulation comes with an extra parameter $\lambda$ which is why in the following we will focus on (6) instead.

Different approaches to finding $V_t$ can be considered. In the implementation the optimization domain is naturally bounded. We choose $V_t$ to be a fraction of the volume of this domain. In a fast interactive framework the user can then adapt the target volume with the help of instant visual feedback. Most importantly, as opposed to a data term driven model volume constraints do not dictate where inflation takes place.

### 2.3 Fast Minimization

In order to convexify the problem in (6) we make use of a relaxation technique [14]. To this end we relax the binary range of functions $u$ in (5) and (7) to the interval $[0, 1]$. In other words we replace $U_V$ and $U_\Sigma$ with their respective convex hulls $U_V^r$ and $U_\Sigma^r$. The corresponding optimization problem is then convex:

**Proposition 1.** *The relaxed set* $U^r := U_\Sigma^r \cap U_V^r$ *is convex.*

*Proof.* The constraint in the definition of $U_V$ is clearly linear in $u$ and therefore $U_V^r$ is convex. The same argument holds for $U_\Sigma$. Being an intersection of two convex sets $U^r$ is convex as well.

One standard way of finding the globally optimal solution to this problem is gradient descent, which is known to converge very slowly. Since optimization speed is an integral part of an interactive reconstruction framework, we employ a recently proposed significantly faster and provably convergent primal-dual algorithm published in [15]. The scheme is based on the weak formulation of the total variation:

$$\min_{u \in U^r} \int g(x)|\nabla u| d^3x = \min_{u \in U^r} \sup_{|\xi(x)|_2 \le g(x)} \left\{ \int -u \operatorname{div}\xi \; d^3x \right\} \tag{9}$$

Optimization is done by alternating a gradient descent with respect to the function $u$ and a gradient ascent for the dual variable $\xi \in C_c^1(\mathbb{R}^3; \mathbb{R}^3)$ interlaced with an over-relaxation step on the primal variable:

$$\begin{cases} \xi^{k+1} = \Pi_{|\xi(x)|_2 \leq g(x)}(\xi^k + \tau \cdot \nabla \bar{u}^k) \\ u^{k+1} = \Pi_{U^r}(u^k + \sigma \cdot \text{div}\xi^{k+1}) \\ \bar{u}^{k+1} = 2u^{k+1} - u^k \end{cases} \tag{10}$$

where $\Pi_A$ denotes the projection onto the set $A$. Projection of $\xi$ is done by simple clipping while that of the primal variable $u$ will be detailed in the next paragraph. The scheme (10) is numerically attractive since it avoids division by the potentially zero-valued gradient-norm which appears in the Euler-Lagrange equation of the TV-norm. Moreover, it is parallelizable and we therefore implemented it on the GPU. On a volume of 63x47x60 voxels the computation takes only 0.47 seconds.

**Projection Scheme.** The projection $\Pi_{U^r}$ in (10) needs to ensure three constraints on $u$: Silhouette consistency, constant volume and $u \in [0,1]$. In order to maintain silhouette consistency (5) of the solution we restrict updates to those voxels which project onto the silhouette interior excluding the silhouette itself.

Still we need to enforce the other two constraints. An iterative algorithm which computes the Euclidean projection of a point onto the intersection of arbitrary convex sets is the one of Boyle and Dykstra [16]. It is fast for a low number of convex constraints and converges provably to the projection point. In our case step $i$ of this algorithm reduces to two seperate projections for volume and range

$$\begin{cases} u_V^i = u_R^{i-1} - v_V^{i-1} + \frac{V_d}{N} \\ v_V^i = u_V^i - (u_R^{i-1} - v_V^{i-1}) \end{cases} \tag{11}$$

$$\begin{cases} u_R^i = \Pi_{[0,1]}(u_V^i - v_R^{i-1}) \\ v_R^i = u_R^i - (u_V^i - v_R^{i-1}) \end{cases} \tag{12}$$

where we initialize $u_R$ with the current $u^k$ in (10) and $v_R, v_V$ with zero. $\Pi_{[0,1]}(u)$ simply clips the value of $u$ to the unit interval and $V_d$ is the difference between the target volume $V_t$ and the current volume of the values $u_R^{i-1} - v_V^{i-1}$. $N$ is the number of voxels in the discrete implementation.

### 2.4 Optimality Bounds

Having computed a global optimal solution $u_{opt}$ of (9), the question remains how we obtain a binary solution and how the two solutions relate to one another energetically. Unfortunately no thresholding theorem holds, which would imply energetic equivalence of the relaxed optimum and its thresholded version for arbitrary thresholds.

Nevertheless we can construct a binary solution $u_{bin}$ as follows:

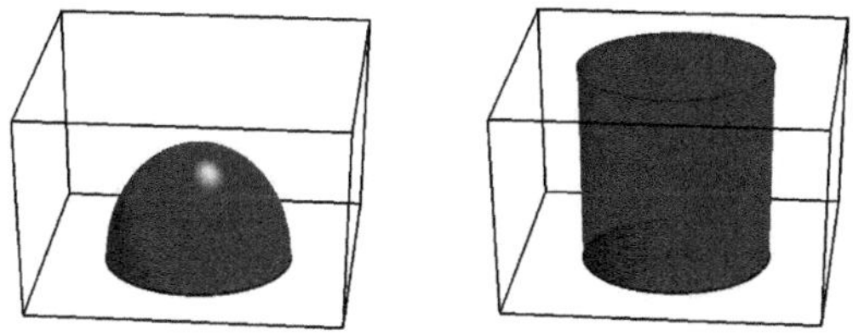

**Fig. 2.** The two cases considered in the analysis of the material concentration. On the *left* hand side we assume a hemi-spherical condensation of the material. On the *right* hand side the material is distributed evenly over the volume.

**Proposition 2.** *The relaxed solution can be projected to the set of binary functions in such a way that the resulting binary function preserves the user-specified volume $V_t$.*

*Proof.* It suffices to order the voxels $x$ by decreasing values $u(x)$. Subsequently, one sets the value of the first $V_t$ voxels to 1 and the value of the remaining voxels to 0.

Concerning an optimality bound the following holds:

**Proposition 3.** *Let $u^r_{opt}$ be the global optimal solution of the relaxed energy and $u_{opt}$ the global optimal solution of the binary problem. Then*

$$E(u_{bin}) - E(u_{opt}) \leq E(u_{bin}) - E(u^r_{opt}) \ . \tag{13}$$

## 3 Theoretical Analysis of Material Concentration

As we have seen above, the proposed convex relaxation technique does not guarantee global optimality of the binary solution. The thresholding theorem [14] – applicable in the unconstrained problem – no longer applies to the volume-constrained problem. While the relaxation naturally gives rise to aposteriori optimality bounds, one may take a closer look at the given problem and ask why the relaxed volume labeling $u$ should favor the emergence of solid objects rather than distribute the prescribed volume equally over all voxels.

In the following, we will prove analytically that the proposed functional has an energetic preference for material concentration. For simplicity, we will consider the case that the object silhouette in the image is a disk. And we will compare the two extreme cases of all volume being concentrated in a ball (a known solution of the Cheeger problem) compared to the case that the same volume is distributed equally over the feasible space (namely a cylinder) – see Figure 2. Note that in the following proof it suffices to consider the volume only on one side of the silhouette.

**Proposition 4.** *Let $u_{sphere}$ denote the binary solution which is 1 inside the sphere and 0 outside – Fig. 2, left side – and let $u_{cyl}$ denote the solution which*

**Fig. 3.** By simply increasing the target volume with the help of a slider, the reconstruction is intuitively inflated. Due to a highly parallelized implementation the result can be computed almost instantly. In this example the intial rendering of the volume with 175x135x80 voxels took 3.9 seconds. Starting from there each subsequent volume adaptation took only about 1 second.

*is uniformly distributed (i.e. constant) over the entire cylinder – Fig. 2, right side. Then we have*

$$E(u_{sphere}) < E(u_{cyl}), \tag{14}$$

*independent of the height of the cylinder.*

*Proof.* Let $R$ denote the radius of the disk. Then the energy of $u_{sphere}$ is simply given by the area of the half-sphere:

$$E(u_{sphere}) = \int |\nabla u_{sphere}| d^2x = 2\pi R^2. \tag{15}$$

If instead of concentrated to the half-sphere, the same volume, i.e. $V = \frac{2\pi}{3} R^3$, is distributed uniformly over the cylinder of height $h \in (0, \infty)$, we have

$$u_{cyl}(x) = \frac{V}{\pi R^2 h} = \frac{2\pi R^3}{3\pi R^2 h} = \frac{2}{3}\frac{R}{h}. \tag{16}$$

inside the entire cylinder, and $u_{cyl}(x) = 0$ outside the cylinder. The respective surface energy of $u_{cyl}$ is given by the area of the cylinder weighted by the respective jump size:

$$E(u_{cyl}) = \int |\nabla u_{cyl}| d^2x = \left(1 - \frac{2R}{3h}\right) \pi R^2 + \frac{2R}{3h}(\pi R^2 + 2\pi Rh) = \frac{7}{3}\pi R^2 > E(u_{sphere}). \tag{17}$$

## 4 Experimental Results

Having detailed the idea of variational implicit weighted surfaces and their fast computation, in this section we will study their properties and applicability within an interactive reconstruction environment. We will compare our approach to methods which resort to heuristic inflation techniques and finally show that appealing and realistic 3D models can be generated with minimal user input.

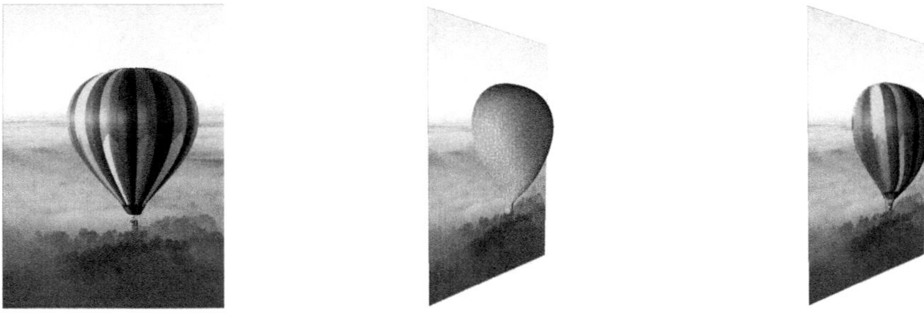

Input Image Reconstructed Geometry Textured Geometry

**Fig. 4.** The proposed Cheeger set approach favors minimal surfaces for a user-specified volume. Therefore the reconstruction algorithm is ideally suited to compute smooth, round reconstructions.

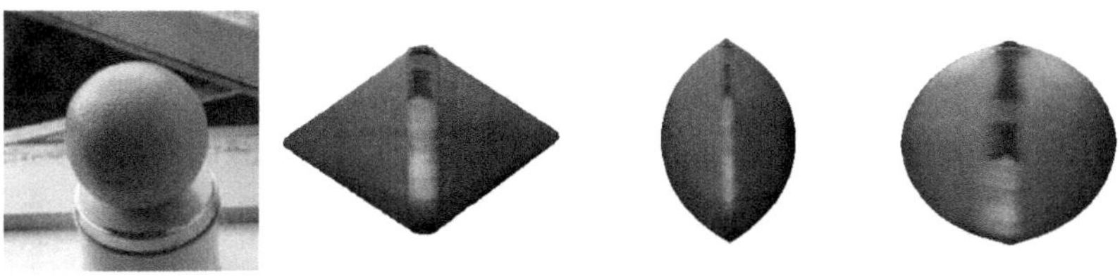

Input Image Data Term as Shape Prior Reconstruction with Data Term Our Method

**Fig. 5.** Using a silhouette distance transform as shape prior the relation between data term (*second from left*) and reconstruction (*third from left*) is not easy to assess for a user. With only one parameter our method delivers more intuitive and natural results.

### 4.1 Cheeger Sets and Single View Reconstruction

Solutions to (6) are Cheeger sets, i.e. minimal surfaces for a fixed volume. In the simplest case of a circle-shaped silhouette one therefore expects to get a ball. Fig. 4 demonstrates that in fact round silhouette boundaries (in the unweighted case) result in round shapes.

### 4.2 Fixed Volume vs. Shape Prior

Many approaches to volume reconstruction incorporate a shape prior in order to avoid surface collapsing. A common heuristic is to use a distance transform of the silhouette boundary for depth value estimation. We show that the fixed-volume approach solves several problems of such a heuristic.

Fig. 5 shows that it is hard to obtain ball-like surfaces with a silhouette distance transform as a shape prior. Another issue is the strong bias a shape prior inflicts on the reconstruction resulting in cone-like shapes (see Fig. 6) and inhibiting the flexibility of the model. The uniform fixed-volume approach fills both gaps while exhibiting the favorable properties of the distance transform (as seen in Fig. 8). With the results in Fig. 5 and 6 we directly compare our method to [10] and [9], in which the reconstruction volume is inflated artificially.

Input Image | Reconstruction with Data Term as Shape Prior | Our Method

**Fig. 6.** In contrast to the approach in [10] (*center*), the proposed method (*right*) does not favor a specific shape and generates more pleasing $3D$ models. Although in the center reconstruction the dominating shape prior can be mitigated by a higher smoothness, this ultimately leads to the vanishing of thin structures like the handle.

### 4.3 Varying the Volume

Apart from the weighting function of the TV-norm (see next section), the only parameter we have to determine for our reconstruction is the target volume $V_t$. The effect on the appearance of the surface can be witnessed in Fig. 3. One can see that changing the target volume has an intuitive effect on the resulting shape. This is important for a user driven reconstruction.

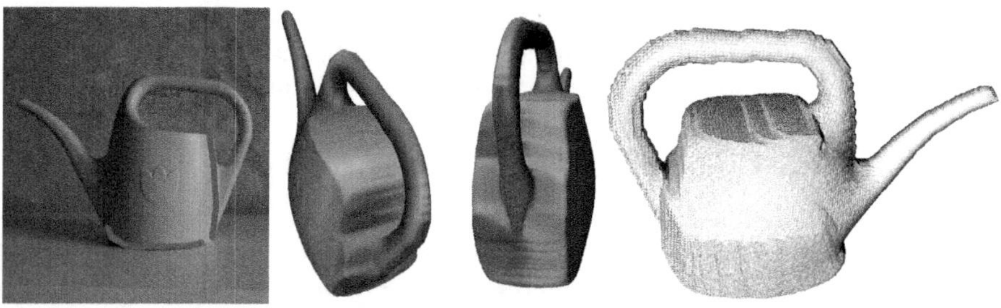

Image with User Input | Reconstructions | Geometry

**Fig. 7.** The proposed approach allows to generate $3D$ models with sharp edges, marked by the user as locations of low smoothness (see section 4.4). Along the red user strokes (*second from left*) the local smoothness weighting is decreased.

### 4.4 Weighted Minimal Surface Reconstruction

So far all presented reconstructions came along without further user input. The weight $g(x)$ of the TV-norm in (9) can be used to locally control the smoothness of the reconstruction: with a low $g(x)$, the smoothness condition on the surface is locally relaxed, allowing for creases and sharp edges to form. Conversely setting $g(x)$ to a high value locally enforces surface smoothness. For controlling the weighting function we employ a user scribble interface. The parameter associated to each scribble marks the local smoothness within the respective scribble area

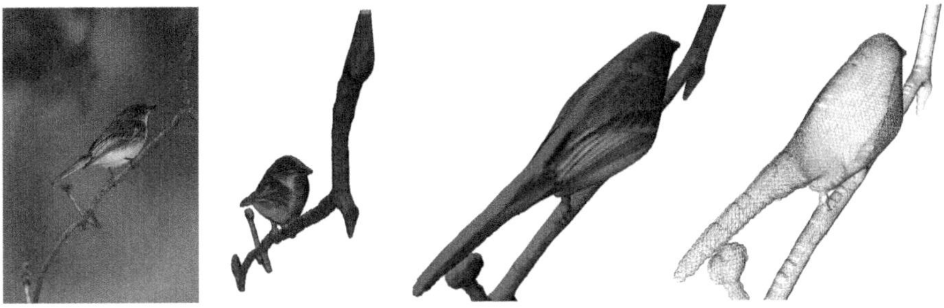


**Fig. 8.** Volume inflation dominates where the silhouette area is large (bird) whereas thin structures (twigs) are inflated less

and is propagated through the volume along projection direction. What we show in Fig. 7 is that with this tool not only round, but other very characteristic shapes can be modeled with minimal user interaction.

The air plane in Fig. 1 represents an example, where a parametric shape prior would fail to offer the necessary flexibility required for modeling protrusions. Since our fixed-volume approach does not impose points of inflation, user input can influence the reconstruction result in well-defined ways: Marking the wings as highly non-smooth (i.e. low $g(x)$) effectively allows them to form.

Note that apart from Fig. 1, 7 the adaption of the target volume was the only user input for all experiments.

## 5 Conclusion

We presented a novel framework for single view reconstruction which allows to compute $3D$ models from a single image in form of Cheeger sets, i.e. minimal surfaces for a fixed user-specified volume. The framework allows for appealing and realistic reconstructions of curved surfaces with minimal user input. The combinatorial problem of finding a silhouette-consistent surface with minimal area for a user defined volume is solved by reverting to an implicit surface representation and convex relaxation. The resulting convex energy is optimized globally using an efficient provably convergent primal-dual scheme. Parallel GPU implementation allows for computation times of a few seconds, allowing the user to interactively increase or decrease the volume. We proved that the computed surfaces are within a bound of the optimum and that they exactly fulfill the target volume. On a variety of challenging real world images, we showed that the proposed method compares favorably over existing implicit approaches, that volume variations lead to families of realistic reconstructions and that additional user scribbles allow to locally reduce smoothness so as to easily create protrusions.

**Acknowledgments.** We thank Andrew Fitzgibbon and Mukta Prasad for fruitful discussions on single view reconstruction.

## References


1. Cheeger, J.: A lower bound for the smallest eigenvalue of the laplacian. In: Problems in analysis, Princeton Univ. Press, Princeton (1970)
2. Liebowitz, D., Criminisi, A., Zisserman, A.: Creating architectural models from images. In: Proc. EuroGraphics, vol. 18, pp. 39–50 (1999)
3. Criminisi, A., Reid, I., Zisserman, A.: Single view metrology. Int. J. Comput. Vision 40, 123–148 (2000)
4. Horry, Y., Anjyo, K.I., Arai, K.: Tour into the picture: using a spidery mesh interface to make animation from a single image. In: SIGGRAPH 1997: Proceedings of the 24th Annual Conference on Computer Graphics and Interactive Techniques, pp. 225–232. ACM Press/Addison-Wesley Publishing Co, New York, NY, USA (1997)
5. Hoiem, D., Efros, A.A., Hebert, M.: Automatic photo pop-up. ACM Trans. Graph. 24, 577–584 (2005)
6. Igarashi, T., Matsuoka, S., Tanaka, H.: Teddy: a sketching interface for 3d freeform design. In: SIGGRAPH 1999, pp. 409–416. ACM Press/Addison-Wesley Publishing Co., New York, NY, USA (1999)
7. Nealen, A., Igarashi, T., Sorkine, O., Alexa, M.: Fibermesh: designing freeform surfaces with 3d curves. ACM Trans. Graph. 26, 41 (2007)
8. Zhang, L., Dugas-Phocion, G., Samson, J.S., Seitz, S.M.: Single view modeling of free-form scenes. In: Proc. of CVPR, pp. 990–997 (2001)
9. Prasad, M., Zisserman, A., Fitzgibbon, A.W.: Single view reconstruction of curved surfaces. In: CVPR, pp. 1345–1354 (2006)
10. Oswald, M.R., Toeppe, E., Kolev, K., Cremers, D.: Non-parametric single view reconstruction of curved objects using convex optimization. In: Pattern Recognition (Proc. DAGM), Jena, Germany (2009)
11. Cohen, L.D., Cohen, I.: Finite-element methods for active contour models and balloons for 2-d and 3-d images. IEEE Trans. on Patt. Anal. and Mach. Intell. 15, 1131–1147 (1993)
12. Rother, C., Kolmogorov, V., Blake, A.: GrabCut: interactive foreground extraction using iterated graph cuts. ACM Trans. Graph. 23, 309–314 (2004)
13. Ambrosio, L., Fusco, N., Pallara, D.: Functions of bounded variation and free discontinuity problems. In: Oxford Mathematical Monographs, The Clarendon Press/Oxford University Press, New York (2000)
14. Chan, T., Esedoḡlu, S., Nikolova, M.: Algorithms for finding global minimizers of image segmentation and denoising models. SIAM Journal on Applied Mathematics 66, 1632–1648 (2006)
15. Pock, T., Cremers, D., Bischof, H., Chambolle, A.: An algorithm for minimizing the piecewise smooth mumford-shah functional. In: IEEE Int. Conf. on Computer Vision, Kyoto, Japan (2009)
16. Boyle, J.P., Dykstra, R.L.: An method for finding projections onto the intersection of convex sets in Hilbert spaces, vol. 37, pp. 28–47 (1986)

# Network Connectivity via Inference over Curvature-Regularizing Line Graphs

Maxwell D. Collins[1,2], Vikas Singh[2,1], and Andrew L. Alexander[3]

[1] Department of Computer Sciences
[2] Department of Biostatistics and Medical Informatics
[3] Waisman Laboratory for Brain Imaging, Departments of Medical Physics and Psychiatry, University of Wisconsin-Madison, Madison, WI
mcollins@cs.wisc.edu, vsingh@biostat.wisc.edu, alalexander2@wisc.edu

**Abstract.** Diffusion Tensor Imaging (DTI) provides estimates of local directional information regarding paths of white matter tracts in the human brain. An important problem in DTI is to infer tract connectivity (and networks) from given image data. We propose a method that infers high-level network structures and connectivity information from Diffusion Tensor images. Our algorithm extends principles from perceptual contours to construct a weighted line-graph based on how well the tensors agree with a set of proposal curves (regularized by length and curvature). The problem of extracting high-level anatomical connectivity is then posed as an optimization problem over this curvature-regularizing graph – which gives subgraphs which comprise a representation of the tracts' network topology. We present experimental results and an open-source implementation of the algorithm.

## 1 Introduction

Diffusion-tensor imaging (DT-MR or DTI) is an imaging modality that measures the diffusion of water molecules in brain tissues [1]. DTI exploits the fact that bundles of neural tissues with a certain orientation *preferentially restrict* water diffusivity (especially perpendicular to the direction of the fibers), which is otherwise isotropic in an unrestricted medium [1,2]. The diffusion data is given as a $3 \times 3$ positive semidefinite matrix at each voxel [3,4], and provides an estimate of the microstructural organization in the brain. DT-MR images are important to quantify how the neural fiber organization varies with cognitive change, age, and diseases [5], and therefore are very promising in the context of many neuroscience questions.

Research in DTI has extensively focused on the design of tools to facilitate the process of obtaining (from raw DTI data) connectivity maps of the entire human brain; in other words, the strength of connectivity in axonal brain networks. One approach toward deriving such information is to calculate, as a first step, the "network pathways" (or fiber tracts) between regions. This procedure is referred to as *tractography* [6,7,8]. It is reasonable to expect that if the underlying diffusion signal is ideal, a simple streamline propagation process (along

R. Kimmel, R. Klette, and A. Sugimoto (Eds.): ACCV 2010, Part I, LNCS 6492, pp. 65–78, 2011.

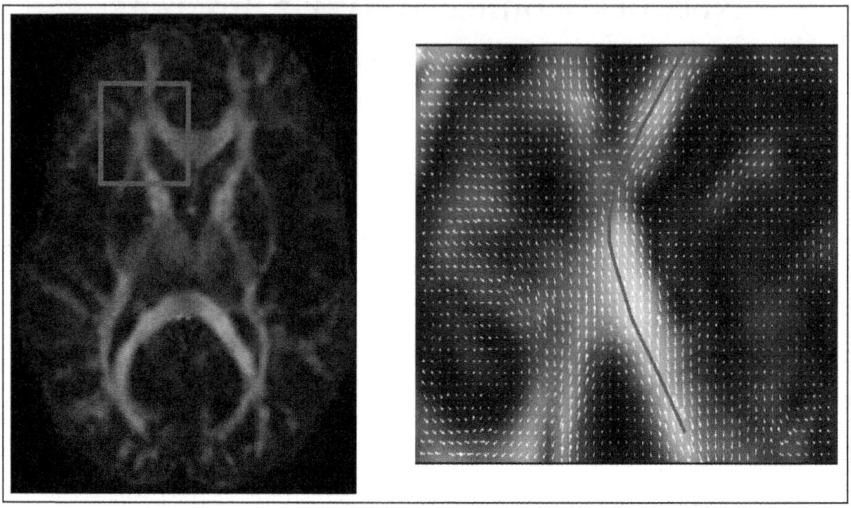

**Fig. 1.** A color-map of the orientation of White Matter in an image with a selected region shown in a red box (left); The vector field of the selected region and a possible pathway in red (right)

orientation specific diffusion) will lead to the desired solution. That is, one sequentially follows the principal eigenvector of the diffusion tensor at each voxel to reconstruct the underlying tract [9,10], see Fig. 1. Unfortunately, there is significant signal drop-off in areas where the diffusion is isotropic (common in regions containing "crossing" fibers). Noise in the estimation of the tensors (or in the acquisition itself) further exacerbates the problem of estimating the underlying pathway. While several local methods have been proposed for tractography, they occasionally make mistakes in the presence of noise and ambiguity (see [11] for a discussion). These errors accumulate as the tracking proceeds, and may also mislead the process into pursuing erroneous paths [12]. Further, the tracking may get lost when it passes through "uncertain" regions (i.e., where the magnitudes of the first and second eigenvectors are similar). Such limitations are common in most local tractography methods which is why recent work in this area suggests a preference toward strategies that lead to more *global* solutions [13]. While this idea is interesting and seems to be an appropriate solution to the problem, there is a significant associated cost. Many of the global methods proposed in the literature are very computationally demanding, and some take more than *one month* of processing time per image [13]. Our primary focus in this paper is to come up with efficient methods for this problem: *to infer reliable long range connections globally from (potentially erroneous or ambiguous) local orientation information.*

**Related Work.** Local methods track fibers through a series of small steps, where each step provides an estimate of the local fiber direction within a prespecified neighborhood. The simplest tractography methods, known as *streamline tracking* [14,15,16], employ path integration based on the diffusion direction.

Noise in the principal eigenvector (due to estimation or acquisition), however, may lead to significant inaccuracies in this streamline process. To address this issue, Tensor Deflection (TEND) [11] takes into account the shape of the diffusion tensor as well in each update step (i.e., not relying on only the principal eigenvector). A number of subsequent papers have demonstrated that introducing some degree of randomness/stochasticity in the local estimation process helps in reducing errors, and yields better results in general [17].

Local methods described above have been extensively used in Neuroscience studies using DTI data. However, in an effort to identify more subtle group-level differences in statistical evaluations (which clearly require more accurate tractography solutions), there is a great deal of interest in leveraging more global methods for this problem. To this end, some authors have made use of Bayesian methods [18], while others have proposed the incorporation of strong priors on the tracking by assuming probability density models of the fiber directions [19,20]. However, these strategies have certain limitations, especially in cases where the pre-specified model is far from the actual orientation distribution in the given image volume. A recently proposed technique for this problem [13], Gibbs Tracking, formulates the problem as an energy minimization, which is solved via a variant of Markov Chain Monte Carlo methods. The approach is quite interesting and seeks to consider the entire image at once in order to handle fiber crossings, while incorporating local agreement with the data and higher order connectedness properties. Unfortunately, such algorithms are known to require significant computational resources and turn out to be rather inefficient for practical applications. The efficiency/quality trade-off described above is quite significant in many cases. This has led a number of authors [21,22,23,24] to investigate the utility of concepts from graph theory – that is, by constructing a weighted neighborhood graph over the voxels. For example, in [24], the orientation distribution function at a given voxel is evaluated for the vector in the direction of the outgoing edge, to obtain a weight which corresponds to the likelihood that a tract connects the two voxels. Then, a simple shortest-path algorithm on this graph finds the most probable tract between any pair of locations. Observe that the success of such a method varies with the richness of the underlying graph representation (e.g., whether it is curvature regularized or not, and how/whether the higher order dependencies are modeled). It also raises the question whether more powerful graph algorithms can lead to tangible improvements. We seek to address both these issues in this paper. A recent work that is relevant to ours is the method in [25]. Here, the authors model tracts as helices between *triplets* of tensors. They introduce *co-helicity* to model the setting where the orientations at each point in the triplet can be joined by a helix. Then, a local search method is used to find orientations for each tensor that leads to the "most probable" helices that match the data. The choice of a helix representation here seems to be a distinct weakness. While our algorithm is similar to [25] in spirit, we present a different geometrically based scheme for generation of primitives (which offers some distinct advantages over [25]). Our most probable set of primitives is then optimized *globally* rather than via local search.

The **main contributions** of this paper are: **(i)** We propose construction of line graph primitives as basic building blocks of a tractography solution. Our strategy adapts ideas from Perceptual Contours and Tensor Voting for the generation of a set of "proposal splines" to geometrically describe the local context in an image volume. **(ii)** We propose a simple optimization model (equipped with connectivity and branching constraints) to select a subset of edges from a curvature regularized graph to infer the final solution, from the given data. **(iii)** We present results on a set of Diffusion Tensor images noting that the proposed solution is of independent interest in the context of general Vision problems dealing with connectivity inference (for example, see [26,27]).

## 2 Preliminaries

Certain ideas from the Computer Vision literature (called *perceptual contours*) can inform approaches to DTI connectivity and tractography; we briefly review the relevant details, and then move to the presentation of our formalization.

**Perceptual Contours/Perceptual Grouping.** The use of energy functions that prefer connections or curves minimal length and curvature has a long history in the perceptual contour literature [28]. Recent work has exploited such functions in a graph setting (with different types of regularization) for a variety of problems and applications [29,30,31]. For example, [31] presents a linear program over a graph of region and boundary segments to find a segmentation of the image – this minimizes an objective which includes curvature regularity on the boundary. *Tensor Voting* [32] is a type of perceptual grouping which takes as input a tensor field[1]. Each tensor field provides a distribution over *normals* rather than *tangents*. The tensor voting algorithm consists of iterations where each member of a tensor field casts a *vote* to its neighbors. The vote itself consists of a tensor based on the orientations of "proposal curves". In the case of the *stick vote*, which is the vote cast by a purely anisotropic stick tensor, we simply consider a proposal curve between the voting tensor and its neighbor with *known* orientation at all points. A stick vote is thus cast to the neighboring tensor in the same direction as the proposal curve's orientation at the neighbor. The magnitude of the stick vote is scaled by a curvature regularity function termed the *saliency decay function*:

$$\exp\left(\frac{s^2 + c\kappa^2}{\sigma}\right), \tag{1}$$

where $s$ and $\kappa$ are the length and curvature of the proposal arc, and $c$ and $\sigma$ are parameters expressing a neighborhood size. In the presence of uncertainty, the vote consists of an integration over multiple proposal curves. The magnitude of the resulting vote is then a function of how well the two tensors can be linked by simple and probable proposal curves. We will use this idea in our modeling.

[1] The term "tensor" here is used in a more general sense, but is similar in terms of interpretation to the usage in the context of DTI.

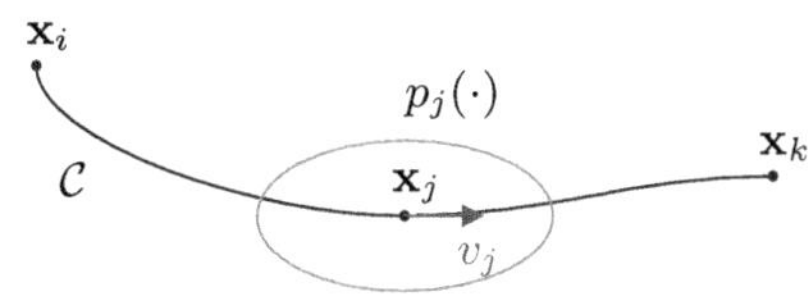


**Fig. 2.** Information in a triplet as described in Section 3.1. Probability distributions $p_i, p_k$ and unit tangents $v_i, v_k$ not shown.

# 3 Weighted Line Graph

Our strategy is to encode curvature regularity in the context of weights on a certain line graph over the tensor field. We will first outline a graph construction, and then discuss how a proposal set of curves can be calculated. Later, in Section 4, we will describe the optimization model to obtain our final solution.

## 3.1 Graph Construction

Consider a set of voxels $\mathcal{V} = \{\mathbf{x}_1, ..., \mathbf{x}_n\}$. Over these voxels an orientation distribution field is defined, so at $\mathbf{x}_i$ we have a distribution function $p_i(\hat{\mathbf{v}})$ corresponding to the local probability that a fiber *through* $\mathbf{x}_i$ has *tangent* along $\hat{\mathbf{v}}$ in that voxel. Adapting DTI data to such a distribution can be done by using $p_i(\hat{\mathbf{v}}) \propto \hat{\mathbf{v}}^T D_i \hat{\mathbf{v}}$, where $D_i$ is the symmetric PSD matrix representing the tensor at voxel $i$ [33]. Note that the formulation easily generalizes to other diffusion imaging modalities [34,35].

We now define a neighborhood graph $\mathcal{G} = (\mathcal{V}, \mathcal{E})$ over voxels in the input image. For simplicity, we can assume that $\mathcal{G}$ is the complete graph $\mathcal{K}_n$, though one may also use a threshold to limit the number of edges introduced (i.e. $\mathcal{E} = \{(ij) \mid \|\mathbf{x}_i - \mathbf{x}_j\| < \delta\}$ for some $\delta$). Let $(\mathcal{E}, \mathcal{L})$ denote the *line graph* of $\mathcal{G}$, the edge set $\mathcal{L}$ is the set of *pairs* of edges of $\mathcal{G}$ which share a common endpoint. It is convenient to view an edge $((ij),(jk)) \in \mathcal{L}$ as a triplet $(ijk) \in \mathcal{L}$ of vertices such that $(ij) \in \mathcal{E}$ and $(jk) \in \mathcal{E}$, which elides the repeat of $j$. With the above notation in place, we can represent groups of white matter tracts at a global level as subgraphs of $\mathcal{L}$. If $\mathcal{H} \subset \mathcal{L}$ is a tractography and $(ijk) \in \mathcal{H}$, this expresses the belief that there is a tract segment $\mathcal{C}$ which connects the voxels $\mathbf{x}_i, \mathbf{x}_j, \mathbf{x}_k$ (in order) and passes through *no other voxels* in between, see Fig. 2 for an illustration. Using the line graph enables us to explicitly model the topology of the tracts, as discussed in [36] and used in Section 4.

## 3.2 Minimum Energy Proposal Curves

To equip the graph, $\mathcal{G}$, with edge weights we use an energy function over curves $\mathcal{C}$ which consists of a weighted sum of total curvature and squared speed [28,37].

$$\mathrm{E}(\mathcal{C}) = \int_{-1}^{1} K \cdot \kappa_{\mathcal{C}}(t)^2 + \|\mathcal{C}'(t)\|^2 dt, \tag{2}$$

where $K$ is a user-defined weight which determines a neighborhood size. This energy function serves as a regularizer to prefer shorter and smooth curves.

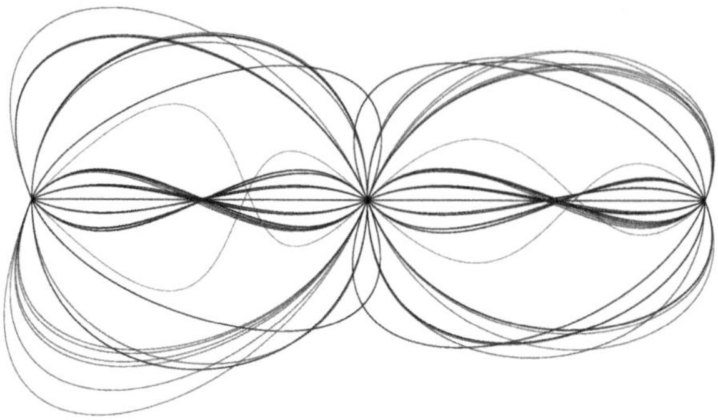

**Fig. 3.** Plot of the proposal spline function (3) when in two dimensions. We take a set of splines constructed and weighted as in section 3.2 passing through $(-1,0), (0,0), (1,0)$, with orientation constraints ranging over a discretization of $\left[0, \frac{\pi}{2}\right]$. Lower-energy curves are shaded darker.

Given known orientations at each point in the triplet $\hat{v}_i, \hat{v}_j, \hat{v}_k$, we can propose a most likely curve. This is equivalent to the proposal arc in tensor voting's stick vote [32] and is the basic building block in our weight construction. Briefly, given a family of curves parameterized over $[-1,1]$, we can generate a proposal curve by choosing $\mathcal{C}$ according to:

$$\begin{aligned} \underset{\mathcal{C}}{\operatorname{argmin}} \quad & E(\mathcal{C}) \\ \text{subject to} \quad & \mathcal{C}(-1) = \mathbf{x}_i, \mathcal{C}(0) = \mathbf{x}_j, \mathcal{C}(1) = \mathbf{x}_k \\ & \frac{\mathcal{C}'(-1)}{\|\mathcal{C}'(-1)\|} = \hat{v}_i, \quad \frac{\mathcal{C}'(0)}{\|\mathcal{C}'(0)\|} = \hat{v}_j, \quad \frac{\mathcal{C}'(1)}{\|\mathcal{C}'(1)\|} = \hat{v}_k, \end{aligned} \tag{3}$$

where $E(\cdot)$ denotes the energy. The proposal curves used here are cubic hermite splines with two segments. The knots have positions $\mathbf{x}_i, \mathbf{x}_j, \mathbf{x}_k$ and derivatives $m_i\hat{v}_i, m_j\hat{v}_j, m_k\hat{v}_k$ respectively. Note that we must distinguish between an orientation and a tangent. The tensors provide a distribution over orientations which we express as a unit vector $\hat{v}_.$, parallel (or antiparallel) to the tangent of any proposal curve. The tensors provide no information on the magnitudes $m_.$ of the tangents, so they are chosen to minimize the curve energy in (2) through gradient descent. We calculate the gradient for $m_.$ by approximating the integral (the longer version of the paper includes all relevant details):

$$\frac{\partial}{\partial m_.}(E(\mathcal{C})) = \int_{-1}^{1} \frac{\partial}{\partial m_.}\left(K\kappa_{\mathcal{C}}(t)^2 + \|\mathcal{C}'(t)\|^2\right) dt \tag{4}$$

Thus, given the positions and orientations at each point we can find a proposal spline and its corresponding energy. An illustration is provided in Fig. 3.

### 3.3 Expected Curve Energy

The local information from the tensors in a triplet and (4) can now be used to calculate weight for the corresponding line edge. If we call a solution to (3)

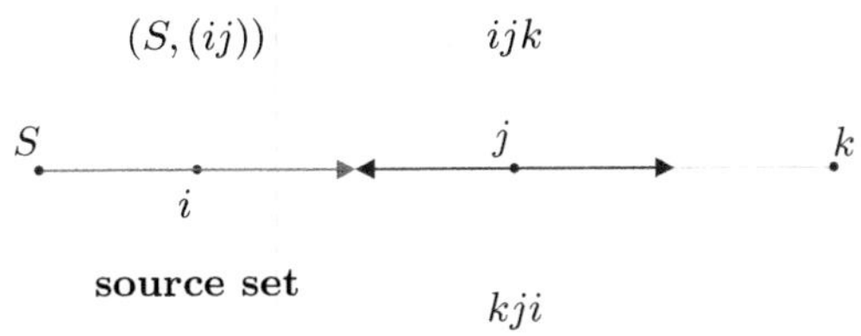


**Fig. 4.** Elements of the flow graph used in (7). Displayed is a source set of $\{i\}$ and a neighborhood graph $G$ with the edes $(ij)$, $(jk)$, inducing the directed line graph edges $(ijk), (kji)$. An edge $(S, (ij))$ is added to introduce flow from the source to $(ij)$.

Curve$(\mathbf{x}_{\{i,j,k\}}, \hat{v}_{\{i,j,k\}})$, then by taking the expectation

$$w_{ijk} = \mathop{\mathbb{E}}_{\hat{v}_{\{i,j,k\}} \sim p_{\{i,j,k\}}} \left[ E(\mathrm{Curve}(\mathbf{x}_{\{i,j,k\}}, \hat{v}_{\{i,j,k\}})) \right] \tag{5}$$

This yields weights for our line edges, and are analogous to edge weights calculated in other graph-based methods in tractography [21,23,24] and segmentation [30,31]. Observe that it is possible to quickly calculate the weights for a large connectivity graph over a regular grid (via a preprocessing step) by taking advantage of the fact that the energy function is invariant to rigid body transformations of the corresponding curve.

## 4 Inferring Connectivity

We can find the most probable (least-weight) tracts connecting an arbitrary pair of regions by solving an *augmented* min-cost flow problem over the digraph $(\mathcal{V}_f, \mathcal{E}_f)$. $\mathcal{V}_f = \mathcal{E} \cup \{S, T\}$, where $S$ and $T$ are nodes for the source and sink respectively. $\mathcal{E}_f$ is the union of the symmetric digraph equivalent to the line graph $\mathcal{L}_\pm$, and a set of edges $(S, (ij))$ or $((ij), T)$ for all $(ij) \in \mathcal{E}$ incident to voxels in the source or sink set respectively, see Figure 4. We set $d_v$ for $v \in \mathcal{V}_f$ to be the flow divergences:

$$d_e = \begin{cases} N & \text{if } e = S \\ -N & \text{if } e = T \\ 0 & \text{if } e = (ij) \in \mathcal{E} \end{cases} \tag{6}$$

where we wish to recover $N$ most-likely tracts.

We use $\alpha_{ijk}$ as indicator variables to give the presence of flow across voxel $j$ moving from $i$ to $k$. A penalty for branching or crossing is imposed by adding variables $\beta_j$ for each voxel $j = \{1, \cdots, n\}$. Each $\beta_j$ is the total amount of flow passing across that voxel. The model imposes a (user-specified) penalty of $\lambda$ for each unit of flow over 1. The effect is a hinge loss which produces topologically simple tractographies with large groups of parallel tracts, except where the data strongly suggests that the reduction in the weight term introduced by a crossing (or branching) is greater than $\lambda$. This hyperparameter encodes a prior based on the topology of the tracts, and can be tuned by a qualitative analysis on how much branching is expected in the true tracts.

$$\begin{aligned} \min_{\alpha,\beta} \quad & \sum_{(ijk)\in\mathcal{L}_\pm} w_{ijk}\alpha_{ijk} + \lambda \sum_{j\in\mathcal{V}} \beta_j \\ \text{subject to} \quad & \sum_{(kij)\in\mathcal{L}_\pm} \alpha_{kij} - \sum_{(ijk)\in\mathcal{L}_\pm} \alpha_{ijk} = d_{(ij)} \quad \forall (ij) \in \mathcal{E}_f \\ & \beta_j \geq \sum_{(ijk)\in\mathcal{L}_\pm} \alpha_{ijk} \quad \forall j \in \mathcal{V} \\ & \beta_j \geq 1 \quad \forall j \\ & \alpha_{ijk} \in \{0,1\} \quad \forall (ijk) \in \mathcal{L}_\pm \end{aligned} \tag{7}$$

We can solve the above IP using a number of solvers; from the solution the resulting tractography is interpreted as $\mathcal{H} = \{(ijk) \in \mathcal{L} \mid \alpha^*_{ijk} + \alpha^*_{kji} \geq 1\}$ where $\alpha^*$ is the solution to (7). Note that the minimum-cost flow problem is a special case of (7) for $\lambda = 0$, and can be solved exactly with the LP relaxation $\alpha_{ijk} \in [0,1]$.

We can derive a new model by modifying (7), relaxing the constraints on the endpoints of the tracts. In such a relaxation, the flow constraints will be replaced with *continuation* constraints similar to [31], see Fig. 5. The continuation constraints express the dichotomy that for a given edge pair and direction, there is another edge pair the continues that tract, otherwise selecting this edge pair introduces at least one "endpoint" to the tractography. We use variables $\gamma_{ijk}$ as indicator variables equal to 1 in such a situation (i.e., for an endpoint), and 0 otherwise. Each $\gamma$ incurs a penalty of $\mu$, optionally relaxed at a set of voxels $M$ considered likely endpoints (i.e. the GM-WM boundary). This hyperparameter allows for "soft" endpoint constraints, so that $M$ can be specified approximately. Note that as $\mu \to +\infty$, the optimal $\gamma^* = \mathbf{0}$ for all $(ijk) \in \mathcal{L}$, and all tracts found will either be cycles or end within $M$. The reverse of this is that if connecting a tract will incur a penalty in the other remaining terms of less than $\mu$, the corresponding connection *will* be made in the optimal subgraph. Our model is given as

$$\begin{aligned} \min_{\alpha,\beta,\gamma} \quad & -\sum_{(ijk)\in\mathcal{L}} w_{ijk}\alpha_{ijk} + \lambda \sum_{j\in\mathcal{V}} \beta_j^s + \mu \sum_{(ijk)\in\mathcal{L}\setminus\mathcal{L}(M)} \gamma_{ijk} \\ \text{subject to} \quad & \alpha_{ijk} \in \{0,1\} \\ & \beta_j \geq \sum_{(ijk)\in\mathcal{L}} \alpha_{ijk} \\ & \gamma_{ijk} \geq 0 \\ & \gamma_{ijk} \geq \alpha_{ijk} - \sum_l \alpha_{lij} \\ & \gamma_{ijk} \geq \alpha_{ijk} - \sum_l \alpha_{jkl}, \end{aligned} \tag{8}$$

where $\mathcal{L}(M) = \{(ijk) \in \mathcal{L} \mid i \text{ or } k \in M\}$, and $s \in \{1,2\}$ is a hyperparameter governing the kind of sparsity penalty. For instance, $s = 2$ penalizes high levels of branching and crossing. Further, the binary constraint on $\alpha_{ijk} \in \{0,1\}$ can

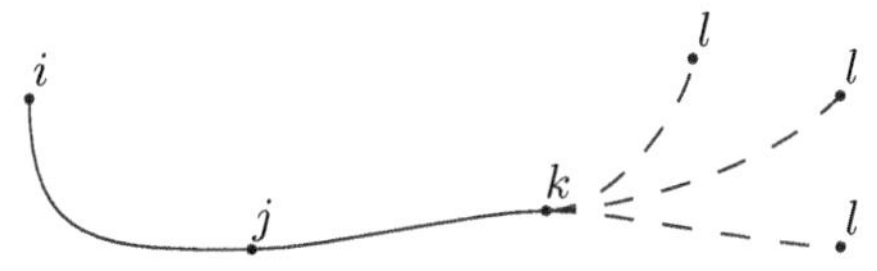


**Fig. 5.** Illustration of continuation constraints

be relaxed to $\alpha_{ijk} \in [0, 1]$ (in which case we obtain a LP for $s = 1$ and QP for $s = 2$).

## 5 Experiments

Our experiments were designed to evaluate where the proposed model can reliably (and efficiently) recover tract connections among different brain regions by incorporating local geometric context within a global optimization model. To this end, we first compared results from our algorithm relative to other streamline-based tracking methods on a number of synthetic tensor datasets. These experiments are useful to answer if the method can correctly resolve crossing fibers, as well as its applicability in general curve inference problems. We also evaluated our results on a set of DT-MR images. Since obtaining ground-truth data on such images is clearly impractical, our evaluations were mainly qualitative – by focusing on pairs of some important regions (e.g., Corpus Callosum) we can reliably assess consistency between our solution and known organizations of tracts in those regions. We present our experimental results next.

### 5.1 Simulated Data

Synthetic tensor fields were constructed from manually specified linear paths. At each voxel along the line, we add the stick tensor in the direction of the ground truth line and a random tensor of magnitude up to a given SNR (10:1 in our experiments). At voxels containing a crossing we take the average of all of these tensors for the crossing tracts. Finally, areas not occupied by a tract are filled with random noise.

In Fig. 6, we illustrate the necessity of a global approach to tractography by showing an example where two fiber cross at an angle. A source set was placed at the lower-left tract endpoints and a sink set was placed at the upper-right endpoints, the union of these two sets was used to seed TEND [11]. Notice that due to partial voluming, the principal diffusion direction in the region of the crossing is between the two tracts. Local methods will typically infer tracts which follow this average direction and infer higher-curvature tracts or leave the region occupied by the tracts entirely. This is compared relative to our global method which extracts the minimum-curvature tracts that match the ground truth rather well.

We also demonstrate the usefulness of the model in (8) on such a synthetic dataset in Fig. 7. With $M$ set to the tensors at the edges of the image, we infer the tracts with no user intervention beyond specifying the appropriate hyperparameters for these artificial tensor fields.

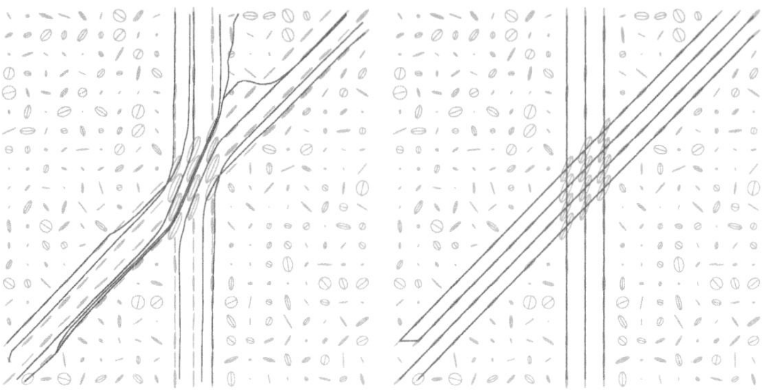

(a) (b)

**Fig. 6.** Comparison of local tractography (a) with ROI pair method from section 4 (b) on a basic problem with a crossing

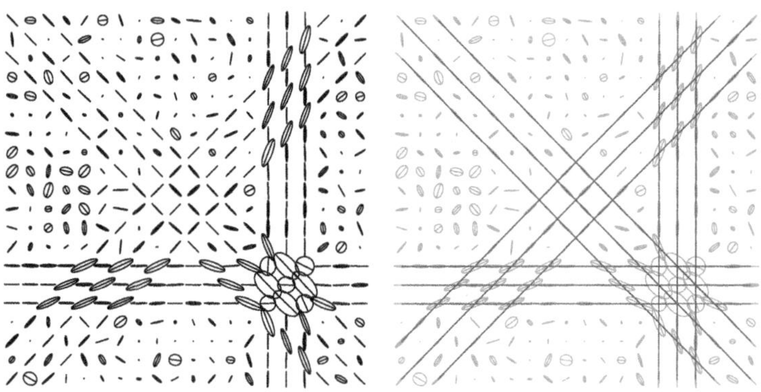

**Fig. 7.** Demonstration of (8)

### 5.2 Anatomical Structures

**Acquisition Setup.** DT-MR images used in our evaluations were acquired with a GE SIGNA 3-Tesla scanner. DW images were taken with 12 non-collinear diffusion directions and a diffusion weighting factor of $b = 1000\text{s/mm}^2$. Eddy current related distortion and head motion of each data set were corrected, using standard methods. Distortions from field inhomogeneities were corrected using field maps. From the raw data, tensor elements were estimated using methods available in Camino, and the data was registered to a common template [38]. The resulting DTI images were then resampled to $128 \times 128 \times 64$ voxels, each of size 1.5 mm $\times$ 1.75 mm $\times$ 2.25 mm. White matter was segmented using the FAST tool available as part of FSL [39]. The tract networks inferred by our algorithm were restricted to lie within this mask.

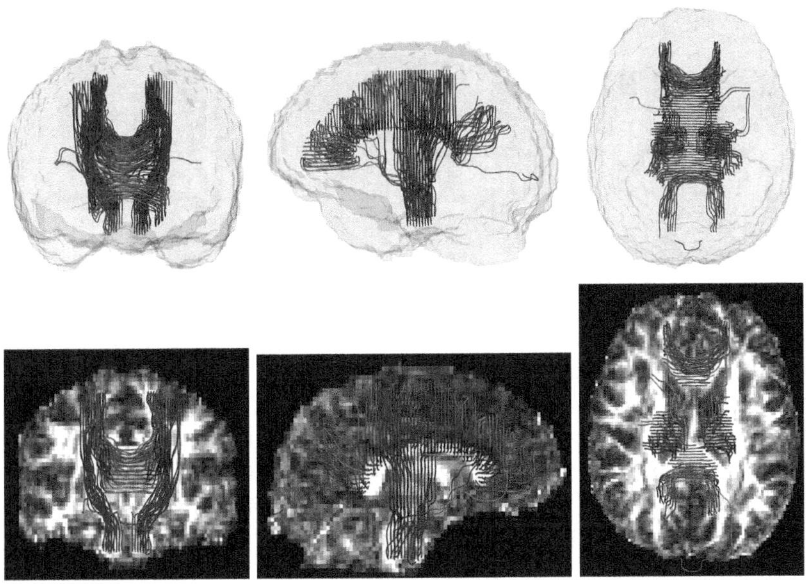

**Fig. 8.** A visualization of our tractography solution in six different views using Trackvis

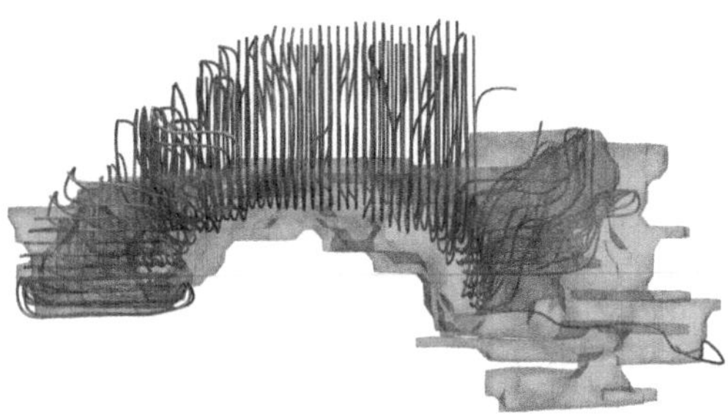

**Fig. 9.** Callosal fibers overlaid on automatic segmentation of the corpus callosum

We constructed a 6-neighborhood graph over the white matter region, and the graph weights were calculated using (5). Source and sink sets were specified for two prominent tract groups: the callosal and projective fibers [40]. We set the minimum flows to recover a total of 200 pairs of tract endpoints, with 120 in the callosal fibers. Tracts were extracted from the optimal line graph by finding the endpoint-to-endpoint paths between voxels within the subgraph, using ideas discussed in Section 4. We find the edge pair leading up to an endpoint as defined there, then repeatedly consider those edge pairs which share the next edge. This leads to paths among the voxels, and B-splines are then fit to the ordered sequence of their centers to obtain the final tracts. Representative results of our method are presented in Figure 8. In general, our results are consistent with known/expected connection pathways in these regions, as well as results obtained via other methods. Some local artifacts are seen due to the low angular

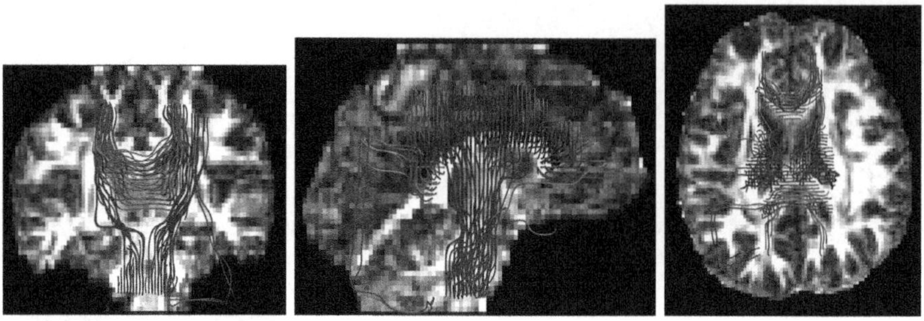

**Fig. 10.** Fibers from another subject using the same source and sink sets

resolution of a 6-neighborhood graph over the voxels. This can be addressed by considering a more computationally-intensive higher-degree graph or by increasing the smoothing when fitting B-spline streamlines during postprocessing.

We note, however, that the proposed method is global and is unaffected by voxels with crossing fibers (which occasionally leads to inaccuracies and errors in other methods). In addition, while streamlines serve as a visualization of inferred tracks, our core algorithm outputs a line graph from which a wide range of measures can be calculated. This includes overall connectivity between ROIs for use within group studies and identifying points of crossing and branches.

## 6 Conclusions

In this paper, we have presented an algorithm for inferring tract connectivity information from DT-MR images. Our algorithm constructs a line graph, whose edge weights are calculated based on a set of proposal splines. This helps equip our algorithm with local geometric context. Once such primitives are generated, a global optimization procedure gives the final tractography solution. Our global model is inspired by network flow algorithms but includes additional constraints that penalize extensive branching and encourage strong region-to-region connectivity. Such requirements are imposed to ensure that the resultant solution is consistent with the topology of white matter tract networks in the brain. We have presented experimental results on synthetic data as well as on brain images, where the proposed method performs well. While extensive further evaluations on a variety of datasets are still required to assess the advantages and limitations of this method, and we believe that the proposed model provides a framework for incorporation of local *and* global context for tractography. Our C++ implementation will be made publicly available concurrently with publication.

**Acknowledgements.** Funding was provided by NIH grants R21-AG034315 (Singh) and MH62015 (Alexander). Partial support for this research was provided by the University of Wisconsin-Madison UW ICTR (1UL1RR025011). Collins was also supported by an NLM training grant to the CIBM Training Program (NLM 5T15LM007359) and the Morgridge Institute for Discovery. Thanks to Nagesh Adluru for assistance with DTI data.

## References

1. Bihan, D.L.: Molecular diffusion nuclear magnetic resonance imaging. Magnetic Resonance Quarterly 7, 1–30 (1991)
2. Bammer, R.: Basic principles of diffusion-weighted imaging. European J. of Radiology 45, 169–184 (2003)
3. Bihan, D.L., Mangin, J., Poupon, C., et al.: Diffusion tensor imaging: Concepts and applications. J. of Magnetic Resonance Imaging 13, 534–546 (2001)
4. Basser, P.J., Jones, D.K.: Diffusion-tensor MRI: theory, experimental design and data analysis - a technical review. NMR Biomed. 15, 456–467 (2002)
5. Naggara, O., Oppenheim, C., Rieu, D., et al.: Diffusion tensor imaging in early Alzheimer's disease. Psychiatry Research: Neuro Imaging 146, 243–249 (2006)
6. Cook, P., Bai, Y., Nedjati-Gilani, S., Seunarine, K., Hall, M., Parker, G., Alexander, D.: Camino: Open-source diffusion-MRI reconstruction and processing. In: ISMRM, Seattle, WA, USA, p. 2759 (2006)
7. Basser, P.J., Mattiello, J., Bihan, D.L.: MR diffusion tensor spectroscopy and imaging. Biophys. J. 66, 259–267 (1994)
8. Stieltjes, B., Kaufmann, W.E., Zijl, P.V., et al.: Diffusion tensor imaging and axonal tracking in the human brainstem. NeuroImage 14, 723–735 (2001)
9. Bammer, R., Acar, B., Moseley, M.E.: In vivo MR tractography using diffusion imaging. European J. of Radiology 45, 223–234 (2003)
10. Roebroeck, A., Galuske, R., Formisano, E., et al.: High-resolution diffusion tensor imaging and tractography of the human optic chiasm at 9.4t. NeuroImage 39, 157–168 (2008)
11. Lazar, M., Weinstein, D.M., Tsuruda, J.S., et al.: White matter tractography using diffusion tensor deflection. Human Brain Mapping 18, 306–321 (2003)
12. Basser, P.J., Pajevic, S., Pierpaoli, C., et al.: In vivo fiber tractography using dt-mri data. Magnetic Resonance in Medicine 44, 625–632 (2000)
13. Kreher, B.W., Mader, I., Kiselev, V.G.: Gibbs Tracking: A novel approach for the reconstruction of neuronal pathways. Magnetic Resonance in Medicine 60, 953–963 (2008)
14. Basser, P.J., Pajevic, S., Pierpaoli, C., et al.: In vivo fiber tractography using DT-MRI data. Magnetic Resonance in Medicine 44, 625–632 (2000)
15. Conturo, T.E., Lori, N.F., Cull, T.S., et al.: Tracking neuronal fiber pathways in the living human brain. In: Proc. of the National Academy of Sciences, vol. 96 (1999)
16. Mori, S., Crain, B.J., Chacko, V.P., et al.: Three-dimensional tracking of axonal projections in the brain by magnetic resonance imaging. Annals of Neurology 45, 265–269 (1999)
17. Bjornemo, M., Brun, A., Kikinis, R., et al.: Regularized stochastic white matter tractography using diffusion tensor MRI. In: Dohi, T., Kikinis, R. (eds.) MICCAI 2002. LNCS, vol. 2488, pp. 435–442. Springer, Heidelberg (2002)
18. Jbabdi, S., Woolrich, M.W., Andersson, J.L.R., et al.: A Bayesian framework for global tractography. Neuroimage 37, 116–129 (2007)
19. Wedeen, V.J., Hagmann, P., Tseng, W.I., et al.: Mapping complex tissue architecture with diffusion spectrum magnetic resonance imaging. Magnetic Resonance in Medicine 54, 1377–1386 (2005)
20. Schmahmann, J.D., Pandya, D.N., Wang, R., et al.: Association fibre pathways of the brain: parallel observations from diffusion spectrum imaging and autoradiography. Brain 130, 630–653 (2007)

21. Iturria-Medina, Y., Sotero, R.C., Canales-Rodrguez, E.J., et al.: Studying the human brain anatomical network via diffusion-weighted MRI and graph theory. NeuroImage 40, 1064–1076 (2008)
22. Lifshits, S., Tamir, A., Assaf, Y.: Combinatorial fiber-tracking of the human brain. NeuroImage 48, 532–540 (2009)
23. Sotiropoulos, S.N., Bai, L., Morgan, P.S., et al.: Brain tractography using Q-ball imaging and graph theory: Improved connectivities through fibre crossings via a model-based approach. NeuroImage 49, 2444–2456 (2010)
24. Zalesky, A.: DT-MRI fiber tracking: A shortest paths approach. IEEE Trans. Medical Imaging 27, 1458–1471 (2008)
25. Savadjiev, P., Campbell, J., Pike, G., et al.: 3D curve inference for diffusion mri regularization and fibre tractography. Medical Image Analysis 10, 799–813 (2006)
26. Peng, T., Jermyn, I.H., Prinet, V., et al.: Incorporating generic and specific prior knowledge in a multi-scale phase field model for road extraction from VHR images. IEEE Trans. Geoscience and Remote Sensing 1, 139–146 (2008)
27. Rochery, M., Jermyn, I.H., Zerubia, J.: Higher-order active contour energies for gap closure. J. of Mathematical Imaging and Vision 29, 1–20 (2007)
28. Kass, M., Witkin, A., Terzopoulos, D.: Snakes: Active contour models. International J. of Computer Vision 1, 321–331 (1988)
29. Boykov, Y., Kolmogorov, V.: Computing geodesics and minimal surfaces via graph cuts. In: ICCV, vol. 1, pp. 26–33 (2003)
30. Schoenemann, T., Cremers, D.: Introducing curvature into globally optimal image segmentation: Minimum ratio cycles on product graphs. In: ICCV (2007)
31. Schoenemann, T., Kahl, F., Cremers, D.: Curvature regularity for region-based image segmentation and inpainting: A linear programming relaxation. In: ICCV (2009)
32. Mordohai, P., Medioni, G.: Tensor Voting: A Perceptual Organization Approach to Computer Vision and Machine Learning. Morgan & Claypool Publishers (2006)
33. Parker, G., Haroon, H., Wheeler-Kingshott, C.: A framework for a streamline-based probabilistic index of connectivity (PICo) using a structural interpretation of MRI diffusion measurements. J. of Magnetic Resonance Imaging 18, 242–254 (2003)
34. Barmpoutis, A., Hwang, M., Howland, D., et al.: Regularized positive-definite fourth order tensor field estimation from DW-MRI. NeuroImage 45, 153–162 (2009)
35. Tuch, D.S.: Q-ball imaging. Magnetic Resonance in Medicine 52, 1358–1372 (2004)
36. Savadjiev, P., Campbell, J.S., Descoteaux, M., Deriche, R., Pike, G.B., Siddiqi, K.: Labeling of ambiguous subvoxel fibre bundle configurations in high angular resolution diffusion MRI. NeuroImage 41, 58–68 (2008)
37. Guy, G., Medioni, G.: Inferring global perceptual contours from local features. International J. of Computer Vision 20, 113–133 (1996)
38. Zhang, H., Yushkevich, P., Alexander, D., et al.: Deformable registration of diffusion tensor MR images with explicit orientation optimization. Med. Image Analysis 10, 764–785 (2006)
39. Zhang, Y., Brady, M., Smith, S.: Segmentation of brain MR images through a hidden markov random field model and the expectation-maximization algorithm. IEEE Trans. Medical Imaging 20, 45–57 (2001)
40. Wakana, S., Jiang, H., Nagae-Poetscher, L.M., et al.: Fiber tract-based atlas of human white matter anatomy. Radiology 230, 77–87 (2004)

# Image and Video Decolorization by Fusion

Codruta O. Ancuti[1], Cosmin Ancuti[1,2], Chris Hermans[1], and Philippe Bekaert[1]

[1] Hasselt University - tUL -IBBT,
Expertise Center for Digital Media, Wetenschapspark 2, Diepenbeek, 3590, Belgium
[2] University Politehnica Timisoara, Piata Victoriei 2, 300006, Romania

**Abstract.** In this paper we present a novel decolorization strategy, based on image fusion principles. We show that by defining proper inputs and weight maps, our fusion-based strategy can yield accurate decolorized images, in which the original discriminability and appearance of the color images are well preserved. Aside from the independent $R$,$G$,$B$ channels, we also employ an additional input channel that conserves color contrast, based on the Helmholtz-Kohlrausch effect. We use three different weight maps in order to control saliency, exposure and saturation. In order to prevent potential artifacts that could be introduced by applying the weight maps in a per pixel fashion, our algorithm is designed as a multi-scale approach. The potential of the new operator has been tested on a large dataset of both natural and synthetic images. We demonstrate the effectiveness of our technique, based on an extensive evaluation against the state-of-the-art grayscale methods, and its ability to decolorize videos in a consistent manner.

## 1 Introduction

Although color plays an important role in images, applications such as compression, visualization of medical imaging, aesthetical stylization, and printings require reliable decolorized image versions. The widely-used standard color-to-grayscale conversion employs the luminance channel only, disregarding the important loss of color information. In many cases, a decolorized image obtained in this way will not fulfill our expectations, as the global appearance is not well preserved (illustrated in Figure 1). This limitation of the standard transformation is due to the fact that isoluminant regions are mapped onto the same output intensity.

In this paper we present a novel decolorization method, built on the principle of image fusion. This well-studied topic of computational imaging has found many useful applications, such as single image dehazing [1], interactive photomontage [2], image editing [3], image compositing [4,5] and HDR imaging [6,7]. The main idea is to combine several images into a single one, retaining only the most significant features.

The main difference between fusion methods that makes them application-specific, is the choice of inputs and weights. Our algorithm employs the three independent $RGB$ channels and an additional image that conserves the color

R. Kimmel, R. Klette, and A. Sugimoto (Eds.): ACCV 2010, Part I, LNCS 6492, pp. 79–92, 2011.

**Fig. 1.** In contrast with the standard color-grayscale approach, our method seeks to preserve the global appearance

contrast, based on Helmholtz-Kohlrausch effect, as image inputs. This fourth input better preserves the global appearance of the image, as it enforces a more consistent gray-shades ordering. The weights used by our algorithm are based on three different forms of local contrast: (a) a saliency map which helps us preserve the saliency of the original color image; (b) a second weight map that advantages well-exposed regions; and (c) a chromatic weight map which enhances the color contrast in addition to the effect of $H - K$ input. In order to minimize artifacts introduced by the weight maps, our approach is designed in a multi-scale fashion, using a Laplacian pyramid representation of the inputs combined with Gaussian pyramids of normalized weights.

To the best of our knowledge we are the first that introduce a fusion-based decolorization technique. Our method performs faster than existing color-to-gray methods since it does not employ color quantization [8], that tends to introduce artifacts, or cost function optimization, which commonly is computationally expensive (e.g. Gooch et al. [9] approach) and risks not converging to a global extremum. Our new operator has been tested on a large dataset of both natural and synthetic images. In addition, we demonstrate that our operator is able to decolorize videos. Our multi-scale fusion approach demonstrates consistency over varying palettes, and is able to maintain temporal coherence within videos. Furthermore, we have performed a comparative evaluation of the contrast enhancement qualities of the recent state-of-the-art color-to-grayscale techniques.

## 2 Related Work

Recently, grayscale image conversion has received an increasing amount of attention in the computer vision and graphics communities. There have been attempts to solve dimensionality reduction problem by both local [8,9,10,11] and global [12,13,14] mapping strategies, using different linear but also non-linear mapping techniques.

The gamut-mapping method of Bala et al. [10] combines the luminance channel with a high-pass filtered chroma channel. Gooch et al. [9] attempt to preserve the sensitivity of the human visual system by iteratively comparing each color pixel value with the average of its surrounding region. Since, in their experiments, small neighborhoods appear to introduce artifacts, the entire image has been used as neighbor region by default. Similarly, the method of Rasche et al. [8] computes the distribution of all the image colors previously quantized in a number of *landmark* points. The main drawback of these methods [8,9] is that they are computationally expensive, as their computation time depends heavily on the number of colors within the image.

Queiroz et al. [15] generate grayscale images that encode color information in the output image. The colors are mapped onto low-visibility high-frequency texture, which can be identified by a decoder, and finally recovered. Their work has proven to be a practical solution for office documents. Neumann et al. [12] compute the gradient field in both the *CIEL*a*b* and *Coloroid* [16] color space. Based on an extensive user-study, the indices of relative luminance are determined. The global mapping technique of Grundland and Dodgson [13] performs a dimensionality reduction using predominant component analysis, which is similar to principal component analysis (PCA). This technique reduces the processing cost substantially, compared to previous approaches. The multi-spectral method of Alsam and Drew [17] improves the approach of Socolinsky and Wolff [18] in terms of computation efficiency. Both approaches aim to preserve the *maximum local contrast*.

Smith et al. [11] developed a decolorization algorithm that exploits the Helmholtz-Kohlrausch effect, by applying the lightness measure of Fairchild and Pirrotta [19]. The algorithm uses a multi-scale chromatic filter to enhance the discriminability over the salient color features. Additionally, the authors have proven the applicability of their method to decolorizing videos. The recent technique of Kim et al. [14] optimizes a nonlinear global mapping function. The method is built on the Gooch at al. [9] approach, but is computationally more effective.

In contrast with the existing techniques, our decolorization algorithm employs a multi-scale fusion strategy. The method is straightforward to implement, and only uses classical concepts. By selecting the appropriate weights and inputs, our approach has demonstrated robustness and consistency in decolorizing both images and videos.

## 3 Fusion-Based Decolorization Approach

The standard grayscale transformation tends to reduce the amount of variations and sharpness within an image. Qualitatively, the dull appearance is due to the loss of contrast that is more visually noticeable on dimmed highlights and shadows. In order to obtain pleasing decolorized images, photographers might

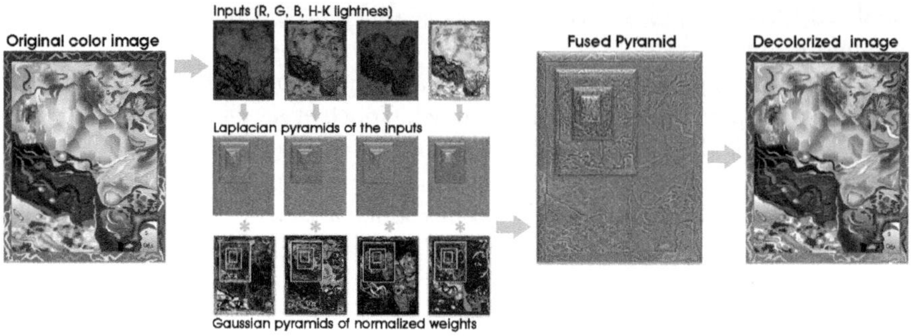


**Fig. 2.** An overview of our fusion-based approach. Based on the original input image, we derive four input images ($R$, $G$, $B$ and $H-K$ lightness) and three weight maps that blended by a multi-scale image fusion strategy yields the decolorized output.

compensate these limitations by tedious work in the darkroom, applying elaborate lighting techniques or using photo-editor programs to manually adjust the contrast, luminance or histogram distribution.

We argue that the image appearance in black-and-white is tightly connected with models of color appearance, and that measurable values like salient features and color contrast are difficult to integrate by simple per pixel blending, without introducing artifacts into the image structure. For this reason, we have opted for the multi-scale approach of image fusion, combining the Helmholtz-Kohlrausch lightness predictor [19] with a set of pixel weights depending on important image qualities. This will ensure that regions with superior gain are well depicted in the decolorized image. Practically, the resulted grayscale image is obtained by fusing four input images (a lightness image that incorporates the $H-K$ effect, and the $R$, $G$, $B$ color channels), weighted by normalized coefficients maps determined by saliency, pixel exposure, and chromatic weights. An overview of our approach is given in Figure 2.

### 3.1 $H-K$ Chromatic Adapted Lightness

Our algorithm requires four input images to be used in the fusion process. Besides the color channels $R$, $G$, $B$ , we define an additional input that preserves the global contrast, based on Helmholtz-Kohlrausch effect. As observed by Smith et al. [11], the $H-K$ effect can be used to resolve potential ambiguities regarding the difference between the isoluminant colors. Therefore, given two isoluminant patches, the most colorful one will be mapped onto a brighter output intensity. For this fusion input channel, we used Fairchild's chromatic lightness metric [19], which predicts the $H-K$ effect, defined in the $CIEL^*c^*h^*$ color space by the expression:

$$L_{H-K} = L^* + (2.5 - 0.025L^*)(0.116 \left| sin\left(\frac{h^* - 90}{2}\right)\right| + 0.085)c^* \quad (1)$$

This $L_{H-K}$ predictor has also been used in the work of Smith et al. [11], in which it was demonstrated to be more appropriate for the task of image decolorization than the chromatic lightness metric of Nayatani [20]. Nayatani [20] predictor often tends to map bright colors to white, which makes it harder to discriminate between images that contain bright isoluminant colors. However, as can be observed in the comparative results (see the image but as well video results of Smith et al. [11]), relying only to the Helmholtz-Kohlrausch effect the decolorized outputs might not preserve accurately the original saliency (this feature is mostly ensured in our operator by integrating the saliency weight map).

## 3.2 Weight Maps Assignment

In the following section, we present how the weight maps are defined in our fusion-based decolorization algorithm. Our approach is based on the principle that the output decolorized image needs be both visually pleasing and meet the application requirements. In the case of grayscale conversion, aside from the luminance which is the main contributor to the perceived lightness, there are also several other image qualities that guide our visual system during its analysis of the incoming light. Practically, the attention of an observer tends to be focused on the salient regions that stand-out within their neighborhood. In order to maintain this focus, it is desirable that these predominant regions are well preserved by the grayscale version. Therefore, in order to meet this requirement, we first introduce a *saliency weight map*. Furthermore, as commonly the over- and underexposed regions are advantaged by the saliency map, we also define an *exposedness weight map* that overcomes perception degradation in these regions. Finally, we assign a third weight map, the *chromatic weight map*, which has the main goal of balancing the influence of chromatic stimuli into the perception of lightness. Practically, by smoothly fusing the input channels weighted by these weight maps, the original consistency of the image is well preserved, while ghosting and haloing artifacts are reduced. Moreover, we believe that these weight maps are intuitive concepts for the users.

***Saliency weight map.*** ($\mathcal{W}_S$) reveals the degree of conspicuousness with respect to the neighborhood regions. For this measurement, our algorithm employs the recent saliency algorithm of Achanta et al. [21]. Their strategy is inspired by the biological concept of center-surround contrast. The saliency weight at pixel position $(x, y)$ of input $I^k$ is defined as:

$$\mathcal{W}_S(x, y) = \left\| I^k_{\mu} - I^k_{\omega_{hc}} \right\| \tag{2}$$

where $I^k_{\mu}$ represents the arithmetic mean pixel value of the input $I^k$ while $I^k_{\omega_{hc}}$ is the blurred version of the same input that aims to remove high frequency noise and textures. $I^k_{\omega_{hc}}$ is obtained by employing a small $5 \times 5$ ($\frac{1}{16}[1, 4, 6, 4, 1]$) separable binomial kernel with the high frequency cut-off value $\omega_{hc} = \pi/2.75$. For small kernels the binomial kernel is a good approximation of its Gaussian counterpart, but it can be computed more effectively. The approach of Achanta et al. [21] is very fast, and has the additional advantage of the extracted maps being characterized by well-defined boundaries and uniformly highlighted salient regions, even at high resolution scales. Based on extensive experiments, we found

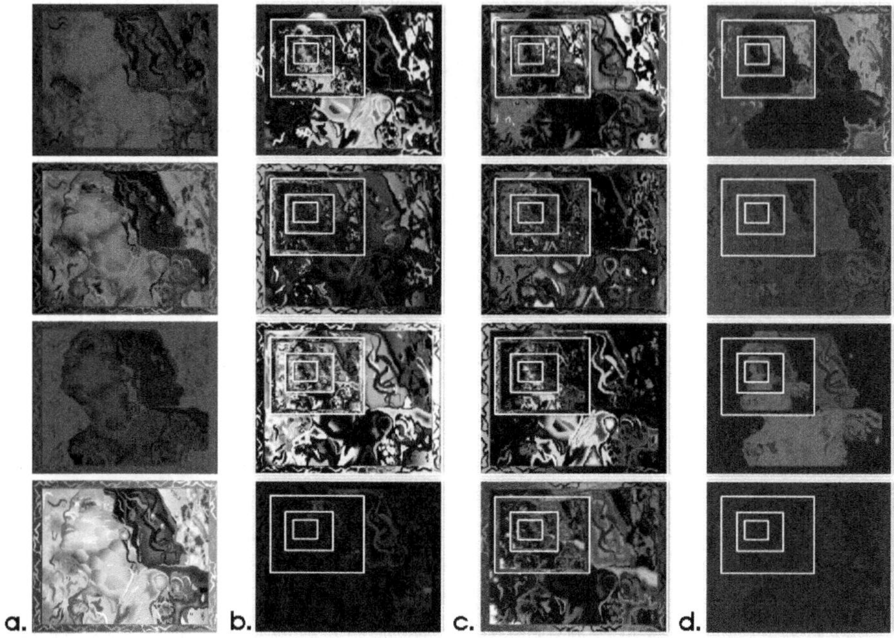


**Fig. 3.** a. The four image fusion inputs ($R$, $G$, $B$ and $H - KLightness$) and the corresponding Gaussian pyramids of the image fusion weights (b. saliency; c. exposedness; d. chromatic )

that this saliency map tends to favor highlighted areas. In order to increase the accuracy of results, we introduce the exposedness map to protect the mid tones that might be altered in some specific cases.

***Exposedness weight map.*** ($\mathcal{W}_E$) estimates the degree to which a pixel is exposed. The function of this weight map is to maintain a constant appearance of the local contrast, neither exaggerated nor understated. Practically, this weight avoids an over- or underexposed look by constraining the result to match the average luminance. Pixels are commonly better exposed when they have normalized values close to the average value of 0.5. Inspired by the approach of Mertens et al. [7], who employ a similar weight in the context of tone mapping, the exposedness weight map is expressed as a Gaussian-modeled distance to the average normalized range value (0.5):

$$\mathcal{W}_E(x, y) = exp\left(-\frac{(I^k(x, y) - 0.5)^2}{2\sigma^2}\right) \tag{3}$$

where $I^k(x, y)$ represents the value of the pixel location $(x, y)$ of the input image $I^k$, while the standard deviation is set to $\sigma = 0.25$. This mapping conserves those tones that are characterized by distance close to zero, while larger distance values are related with the over- and underexposed regions. As a result, the impact of over- and underexposed regions filtered by the saliency map is tempered, keeping the original image appearance well preserved.

***Chromatic weight map.*** ($\mathcal{W}_C$) controls the saturation contribution of the inputs in the decolorized image. This is expressed as the *standard deviation* between every input and the saturation $S$ (in $HSL$ color space) of the original image. Due to the fact that in general humans prefer increased saturation, it is desirable that more saturated areas are mapped onto brighter tones. This balances the chromatic contrast loss with the desired amount of enhancement. We have observed that the impact of this gain is reduced for the $H-K$ chromatic adapted lightness input.

In our framework these weight maps (saliency, exposedness and chroma) have the same contribution to the resulted decolorized images. As an example, Figure 3 shows the computed weights for the considered inputs.

### 3.3 Multi-scale Fusion of the Inputs

Having defined the inputs ($R$, $G$, $B$ color channels and $H-K$ chromatic adapted lightness) and the weight maps, in the following section we present how this information is blended by our fusion strategy. As previously mentioned, during the fusion process the inputs are weighted by specific maps in order to conserve the most significant features, and finally combined into a single output image:

$$\mathcal{F}(x,y) = \sum_{k=1}^{K} \bar{\mathcal{W}}^k(x,y)\mathcal{I}^k(x,y) \tag{4}$$

where the value of every pixel location $(x,y)$ of the fused result $\mathcal{F}$ is obtained by taking the sum of the corresponding locations of the inputs $\mathcal{I}^k$ ($k$ is the input index), weighted by the normalized weight maps $\bar{\mathcal{W}}^k$. The number of the inputs is counted by the index $k$ (in our case $K = 4$). The normalized weights $\bar{\mathcal{W}}$ are obtained by normalizing over the $M$ weight maps $\mathcal{W}$ ($M = 3$) in order that the value of each pixel $(x,y)$ weights to sum up to unity ($\sum \mathcal{W}^k = 1$ for each pixel location) (see Figure 4).

Unfortunately, applying eq. 4 directly sometimes introduces haloing artifacts, mainly in locations close to strong transitions between weight maps. In order to solve this problem, a more effective strategy needs to be devised. Generally, this task is solved by multi-scale decomposition strategies that use linear [22,23] or non-linear filters [24,25,26]. While the class of non-linear filters has shown to be better at preserving edges, the linear filters are computationally more effective. Even though more refined multi-scale solutions might be applied as well, we have opted for the classical multi-scale Laplacian pyramid decomposition [22]. In this linear decomposition, every input image is represented as a sum of patterns computed at different scales based on the Laplacian operator. The inputs are convolved by a Gaussian kernel, yielding a low pass filtered versions of the original. In order to control the cut-off frequency, the standard deviation is increased monotonically. To obtain the different levels of the pyramid, initially we need to

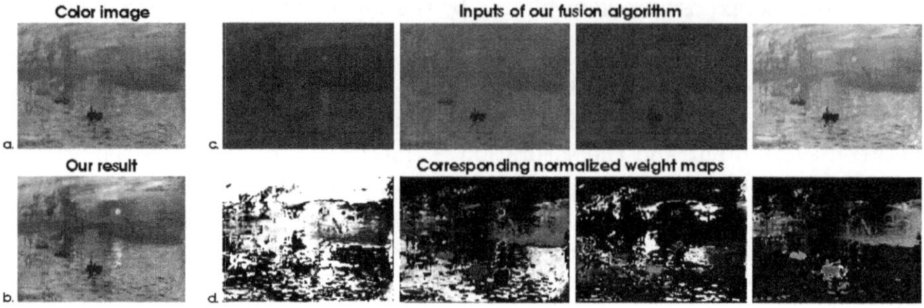


**Fig. 4.** From color to gray: from the original color image (a), we obtain our decolorized result (b) by applying an image fusion approach, using the four inputs (c), weighted by the corresponding normalized weight maps (d)

compute the difference between the original image and the low pass filtered image. From then on, the process is iterated by computing the difference between two adjacent levels of the Gaussian pyramid. The resulting representation, the Laplacian pyramid, is a set of quasi-bandpass versions of the image.

In our case, each input is decomposed into a pyramid by applying Laplacian operator to different scales. Similarly, for each normalized weight map $\bar{\mathcal{W}}$ a Gaussian pyramid is computed. Considering that both the Gaussian and Laplacian pyramids have the same number of levels, the mixing between the Laplacian inputs and Gaussian normalized weights is performed at each level independently yielding finally the fused pyramid:

$$\mathcal{F}^l(x,y) = \sum_{k=1}^{K} G^l\left\{\bar{\mathcal{W}}^k(x,y)\right\} L^l\left\{\mathcal{I}^k(x,y)\right\} \tag{5}$$

where $l$ represents the number of the pyramid levels (determined by the image dimensions), $L\{\mathcal{I}\}$ is the Laplacian version of the input $\mathcal{I}$, and $G\left\{\bar{\mathcal{W}}\right\}$ represents the Gaussian version of the normalized weight map $\bar{\mathcal{W}}$. This step is performed successively for each pyramid layer, in a bottom-up manner. Basically, this approach solves the *cut and paste* problem among the inputs with respect to the normalized masks. A similar adjustment of the Laplacian pyramid, but in the context of exposure fusion has been applied in other contexts [6,7]. The final decolorized image is obtained by summing the fused contribution of all inputs.

This linear multi-scale strategy performs relatively fast (takes approximately 1.4 seconds per image in our unoptimized MATLAB implementation) representing a good trade off between speed and accuracy. By employing independently a fusion process at every scale level the potential artifacts due to the sharp transitions of the weight maps are minimized. Multi-scale techniques are broadly used due to their efficiency in image compression, analysis and manipulation. This operation has the advantage that it respects the perceptual system of the human eye, which is known to be more sensitive to modifications into high frequencies than changes in low frequencies.

## 4 Results and Discussion

Our fusion-based approach addresses the preservation of several important image features: saliency, well-exposedness and chromatic contrast. One major benefit of fusing the inputs guided by weight maps is that this principle allows for a direct transfer of the important characteristics of the color image to the decolorized version. We believe that strong perceptual similarity between colorized and decolorized images can be obtained by algorithms that consider both global and local impressions. In our approach, the global appearance of the image is preserved by imposing a gray-shades order that respect the $H - K$ color appearance model. The weight maps contribute to the local preservation of the original relations between neighbor patches. A similar idea has been experimented with, using Poisson solvers [27] in a related approach of Gooch at al. [9]. However, their approach performs poorly for images with extended disconnected regions that represent isoluminant features. The main reason for this is that the Poisson solver ignores differences in gradients over distances larger than one pixels. Our fusion technique proves that by employing well defined quality measures and inputs, consistent results can be produced even for these difficult cases.

The new operator has been tested extensively for a large set of images. Figure 7 presents several comparative results against recent grayscale operators (for additional results the reader is referred to the supplementary material).

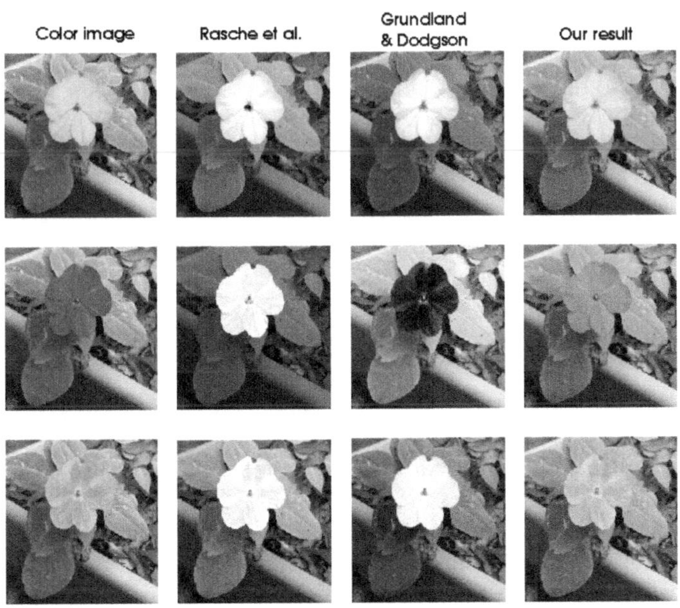


**Fig. 5.** Coherence of color-to-gray methods. Note how differently the methods map the background (e.g. leaves) and the flower. Compared with the methods of [8,13] our operator maps into the same grayscale level the leaves while the flower is converted into different grayscale levels.

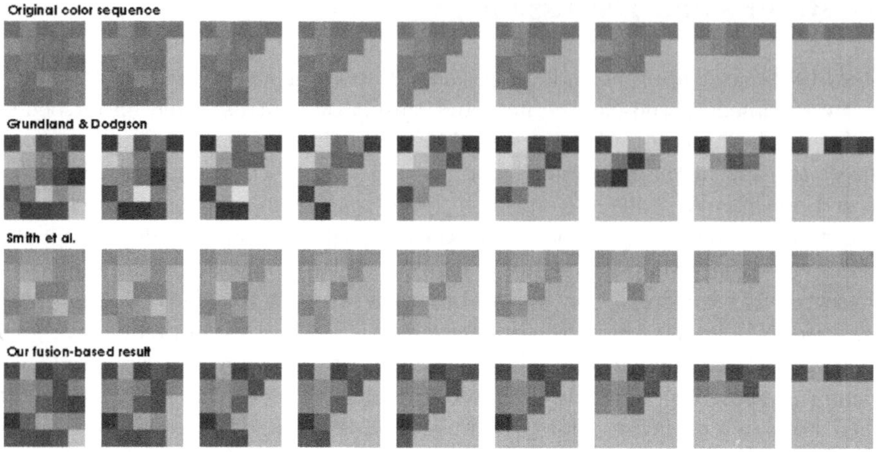


**Fig. 6.** Isoluminant video test. For a well consistency the same color patch needs to be converted into a similar gray level in all images. Notice the artifacts introduced by Smith et al. [11] approach but also the different grayscale mapping of the same colored patch yielded by Grundland and Dodgson method [13] (please refer to the supplementary material for the entire sequence).

### 4.1 Evaluation of Grayscale Operators

In order to measure the quality of the conversions, we performed a contrast-based evaluation of the recent state-of-the-art operators. For this task we adapted the recent technique of Aydin et al. [28], which is used to compare a pair of images with significantly different dynamic ranges. Instead of detecting only contrast changes, this metric is sensitive to three types of structural changes: *loss of visible contrast* (**green**) - a contrast that was visible in the reference image becomes invisible in the transformed version, *amplification of invisible contrast* (**blue**) - a contrast that was invisible in the reference image becomes visible in the transformed version and *reversal of visible contrast* (**red**) - a contrast is visible in both images, but has different polarity. They observed that the contrast loss (green) is related with blurring, while contrast amplification (blue) and reversal (red) with sharpening. An online implementation[1] of this metric is made available by the authors.

We tested several grayscale operators for a set of 24 images that have also been used in the perceptual evaluation of Cadik [29]. Besides the $CIEY$, Bala and Eschbach [10], Gooch et al. [9], Rasche et al. [8], Grundland and Dodgson [13], Coloroid [12], Smith et al. [11] methods, we also reviewed the recent technique of Kim et al. [14] and our fusion-based decolorization operator. The measure of Aydin et al. [28] is applied using the default parameter set of the authors, and considering the original color image as a reference. The results of applying the $IQA$ measure are shown in Figure 8. The graphics in the figure

[1] http://www.mpi-inf.mpg.de/resources/hdr/vis metric/

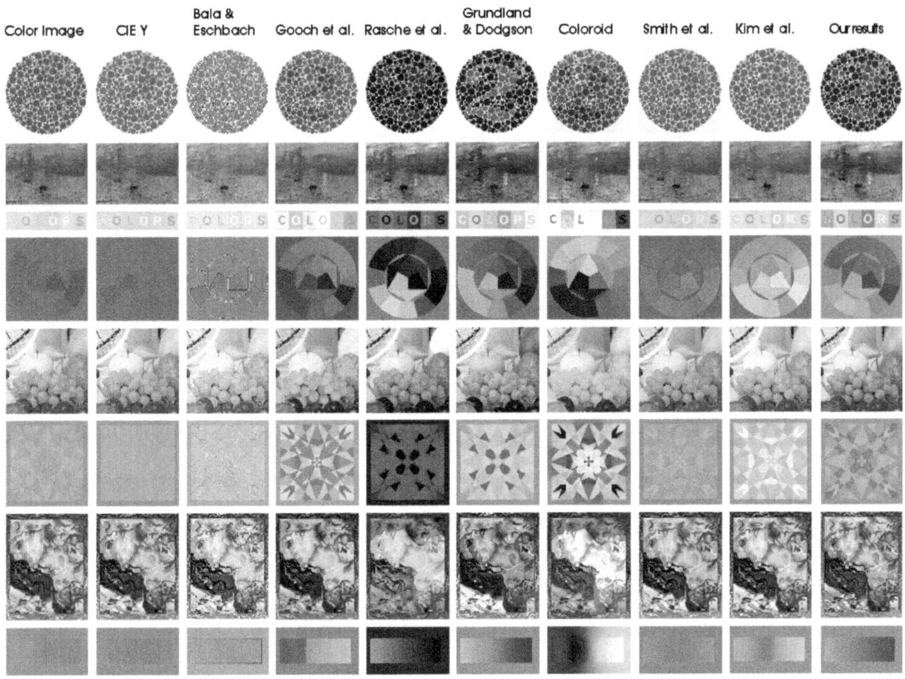


**Fig. 7.** Comparative results. From left to right the grayscale results obtained by applying $CIEY$, Bala and Eschbach [10], Gooch et al. [9], Rasche et al. [8] , Grundland and Dodgson [13], Coloroid [12], Smith et al. [11], Kim et al. [14] methods and our fusion-based operator.

display the average ratio over the 24 $IQA$ images of the pixels with the contrast changed after applying the corresponding transformation. Only the pixels with a probability higher than 70% have been counted.

Based on these graphics, we can observe that our operator, together with Smith et al. [11] and Kim et al. [14], shows the minimal amount of produced blurring artifacts (rendered with the *green* after applying IQA). Regarding the sharpening effects (*blue* and *red* pixels), in general all methods, except the one of Smith et al. [11] and $CIEY$, perform in a relative similar range of values.

## 4.2 Video Decolorization

Video decolorization adds an other dimension to the problem of image decolorization, as temporal coherence needs to be guaranteed for the entire video sequence. In order to speak of consistency, an algorithm has to map similar regions from the color input onto similar areas in the decolorized output. Recently Smith et al. [11] have shown that local approaches are suitable for this task. As our strategy retains both global and local characteristics, it is able to maintain consistency over varying palettes (see Figure 5), yielding temporal coherence for

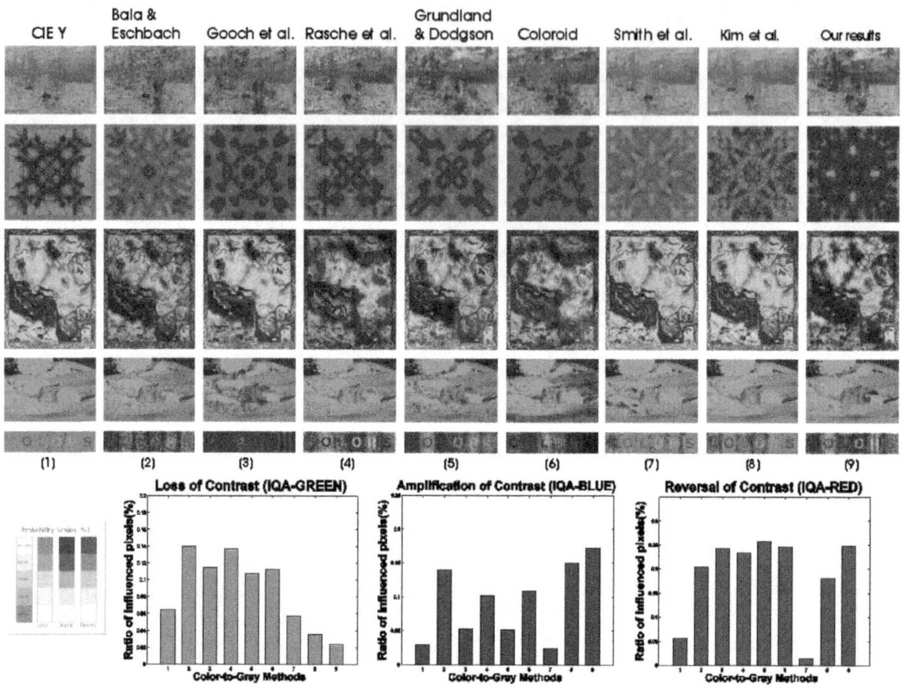


**Fig. 8.** IQA evaluation of operators. In the top are shown the results obtained by applying the contrast-based measure of Aydin et al. [28] between the original color and decolorized images. In the bottom part are displayed the graphics that plot the average ratio (over the complete set of images) of the pixels with the contrast changed after applying the corresponding transformation. Note that the *green* is related with blurring while *blue* and *red* are related with sharpening.

videos (see as well Figure 6). Figure 5 shows several versions of the same image, in which the flower is colored differently at each instance. Global pallet mapping techniques like the one of Grundland and Dodgson [13] generate dissimilar gray levels for the same region on different instances (note the leaves and the background mapping). Our operator and the method of Smith et al. [11] yield more consistent outputs.

A similar limitation can be observed by analyzing Figure 6 that displays the frames of a synthetically generated footage with isoluminant color patches. By a close inspection it might be seen that even thought the technique of Grundland and Dodgson [13] decolorizes each frame perceptually accurate, this technique is not able to preserve the same grayscale level corresponding to the same color patch along the entire sequence of the frames. On the other hand, the method of Smith et al. [11] introduces some non-homogeneity artifacts along edges (please refer to the supplementary material for the comparative videos).

In our extensive experiments, the fusion-based operator performed generally well. However, we observed that in some situation due to the inexact selection of the saliency map, our technique is unable to improve substantially the results yielded by standard conversion. In general, the chosen weights can yield proper results. As well, the algorithm shares a weight map related problem in common with other fusion algorithms. As noticed by previous methods [7], the exposedness map may generate an artificial appearance of the image when its gain is exaggerated.

Our algorithm is computationally effective (our unoptimized implementation takes approximately 2.5 seconds for a 800x600 image), having a processing time comparable to recent CPU approaches (e.g. Smith et al. [11] method takes 6.7 - 10.8 seconds for 570x593 image, Decolorize [13] -unoptimized code - 3.5 seconds for a 800x600 image and the (extremely) optimized code of Kim et al. [14] decolorizes a 800x600 image in 1-2 seconds). However, even relatively fast since it employs an effective nonlinear global mapping optimization, the method of Kim et al. [14] did not solve the rendering limitations of the related technique of Gooch et al [9], tending to diminish the global contrast and to loose the original saliency (please refer to their image results and the rose video in the additional material). In addition, we believe that an optimized CPU implementation would make our operator suitable for real-time applications.

## 5 Conclusion

In this paper we have introduced a new color-to-grayscale conversion strategy, in which we employ a multi-scale fusion algorithm. We have shown that by choosing the appropriate weight maps and inputs, an image fusion strategy can be used to effectively decolorize images. We performed an extensive evaluation against the recent decolorization operators. Moreover, our operator is able to transform color videos into a decolorized version that preserve the original saliency and appearance in a consistent manner.

To future work, we would like to experiment the potential of our operator for several other applications but also to perform a perceptual evaluation of the recent color-to-grayscale techniques.

## References

1. Ancuti, C.O., Ancuti, C., Bekaert, P.: Effective single image dehazing by fusion. In: IEEE Int. Conf. on Image Processing, ICIP (2010)
2. Agarwala, A., Dontcheva, M., Agrawala, M., Drucker, S.M., Colburn, A., Curless, B., Salesin, D., Cohen, M.F.: Interactive digital photomontage. ACM Trans. Graph, SIGGRAPH (2004)
3. Perez, P., Gangnet, M., Blake, A.: Poisson image editing. ACM Trans. Graph, SIGGRAPH (2003)
4. Brinkmann, R.: The art and science of digital compositing. Morgan Kaufmann, San Francisco (1999)
5. Grundland, M., Vohra, R., Williams, G.P., Dodgson, N.A.: Cross dissolve without cross fade: Preserving contrast, color and salience in image compositing. In: Computer Graphics Forum, EUROGRAPHICS (2006)

6. Burt, P.J., Hanna, K., Kolczynski, R.J.: Enhanced image capture through fusion. In: IEEE Int. Conf. on Computer Vision (1993)
7. Mertens, T., Kautz, J., Reeth, F.V.: Exposure fusion: A simple and practical alternative to high dynamic range photography. In: Computer Graphics Forum (2009)
8. Rasche, K., Geist, R., Westall, J.: Re-coloring images for gamuts of lower dimension. In: Computer Graphics Forum, EUROGRAPHICS (2005)
9. Gooch, A.A., Olsen, S.C., Tumblin, J., Gooch, B.: Color2gray: salience-preserving color removal. SIGGRAPH, ACM Trans. Graph. 24, 634–639 (2005)
10. Bala, R., Eschbach, R.: Spatial color-to-grayscale transform preserving chrominance edge information. In: Color Imaging Conf. (2004)
11. Smith, K., Landes, P.E., Thollot, J., Myszkowski, K.: Apparent greyscale: A simple and fast conversion to perceptually accurate images and video. In: Computer Graphics Forum (2008)
12. Neumann, L., Cadik, M., Nemcsics, A.: An efficient perception-based adaptive color to gray transformation. In: Proc. of Comput. Aesthetics, pp. 73–80 (2007)
13. Grundland, M., Dodgson, N.A.: Decolorize: Fast, contrast enhancing, color to grayscale conversion. Pattern Recognition 40 (2007)
14. Kim, Y., Jang, C., Demouth, J., Lee, S.: Robust color-to-gray via nonlinear global mapping. ACM Trans. Graph, SIGGRAPH ASIA (2009)
15. de Queiroz, R.L., Braun, K.: Color to gray and back: color embedding into textured gray images. IEEE Trans. on Image Processing 15, 1464–1470 (2006)
16. Nemcsis, A.: Color space of the coloroid color system. Color Research and Applications 12 (1987)
17. Alsam, A., Drew, M.S.: Fast multispectral2gray. Journal of Imaging Science and Technology (2009)
18. Socolinsky, D., Wolff, L.: Multispectral image visualization through first-order fusion. IEEE Transactions on Image Processing (2002)
19. Fairchild, M., Pirrotta, E.: Predicting the lightness of chromatic object colors using cielab. Color Research and Application (1991)
20. Nayatani, Y.: Relations between the two kinds of representation methods in the helmholtz-kohlrausch effect. Color Research and Application (1998)
21. Achanta, R., Hemami, S., Estrada, F., Süsstrunk, S.: Frequency-tuned salient region detection. In: IEEE CVPR (2009)
22. Burt, P., Adelson, T.: The laplacian pyramid as a compact image code. IEEE Transactions on Communication (1983)
23. Rahman, Z., Woodell, G.: A multi-scale retinex for bridging the gap between color images and the human observation of scenes. In: IEEE Trans. on Image Proc. (1997)
24. Durand, F., Dorsey, J.: Fast bilateral filtering for the display of high-dynamic-range images. ACM Trans. Graph, SIGGRAPH (2002)
25. Farbman, Z., Fattal, R., Lischinski, D., Szelinski, R.: Edge-preserving decompositions for multi-scale tone and detail manipulation. ACM Trans. Graph, SIGGRAPH (2008)
26. Subr, K., Soler, C., Durand, F.: Edge-preserving multiscale image decomposition based on local extrema. ACM Trans. Graph, SIGGRAPH ASIA (2009)
27. Fattal, R., Lischinski, D., Werman, M.: Gradient domain high dynamic range compression. ACM Trans. Graph, SIGGRAPH (2002)
28. Aydin, T., Mantiuk, R., Myszkowski, K., Seidel, H.S.: Dynamic range independent image quality assessment. ACM Trans. Graph, SIGGRAPH (2008)
29. Cadik, M.: Perceptual evaluation of color-to-grayscale image conversions. Computer Graphics Forum 27 (2008)

# Video Temporal Super-Resolution Based on Self-similarity

Mihoko Shimano[1,2], Takahiro Okabe[2], Imari Sato[3], and Yoichi Sato[2]

[1] PRESTO, Japan Science and Technology Agency
[2] The University of Tokyo
{miho,takahiro,ysato}@iis.u-tokyo.ac.jp
[3] National Institute of Informatics
imarik@nii.ac.jp

**Abstract.** We propose a method for making temporal super-resolution video from a single video by exploiting the self-similarity that exists in the spatio-temporal domain of videos. Temporal super-resolution is inherently ill-posed problem because there are an infinite number of high temporal resolution frames that can produce the same low temporal resolution frame. The key idea in this work to solve this ambiguity is exploiting self-similarity, *i.e.*, a self-similar appearance that represents integrated motion of objects during each exposure time of videos with different temporal resolutions. In contrast with other methods that try to generate plausible intermediate frames based on temporal interpolation, our method can increase the temporal resolution of a given video, for instance by resolving one frame to two frames. Based on the quantitative evaluation of experimental results, we demonstrate that our method can generate enhanced videos with increased temporal resolution thereby recovering appearances of dynamic scenes.

## 1 Introduction

There is a great demand for video enhancement to increase temporal resolution by generating a high-frame rate video from a given video captured at low frame rate. Several methods [1,2,3,4] based on temporal interpolation have been proposed for increasing the frame rate of a given video. These techniques produce a high-frame rate video with intermediate frames generated by interpolating input frames with pixel-wise correspondences (Fig. 1(a)).

The quality of video largely depends on temporal resolution, how much detailed events at different points in time can be resolved in relation to both the exposure time and frame rate. In a low-frame rate video recorded with *sub-exposure time*, *i.e.*, exposure time shorter than one frame period (Fig. 1(a)) discontinuous motion, called jerkiness, is often observed. On the other hand, motion blur may occur in a low-frame rate video recorded with *full-exposure time*, *i.e.*, exposure time equals to one frame period (Fig. 1(b)), due to camera shake or moving objects in a scene.

Videos often become underexposed if they are taken in sub-exposure time to avoid images getting too dark in scenes like dark cloudy scenes, night scenes,

R. Kimmel, R. Klette, and A. Sugimoto (Eds.): ACCV 2010, Part I, LNCS 6492, pp. 93–106, 2011.

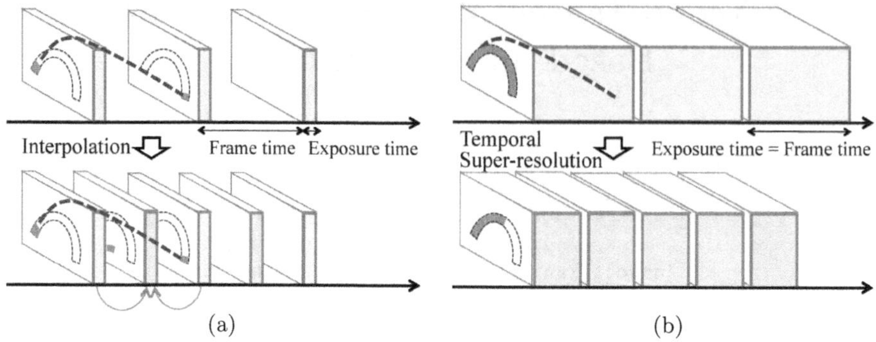


**Fig. 1.** Videos with different exposure times. (a) LTR video with sub-exposure time (upper), HTR video produced by interpolation (lower), (b) LTR video with full-exposure time (upper), and HTR video produced by temporal super-resolution (lower).

and indoor scenes. In that case, full-exposure time (Fig. 1(b)) should be selected automatically[1] or manually to capture the enough amount of light suppressing noise. For such videos recorded with full-exposure time, the above temporal interpolation methods assuming sub-exposure time are not applicable because they cannot resolve one frame into two frames.

In this paper, we propose a temporal super-resolution method for increasing the temporal resolution of a given video recorded with full-exposure time by resolving one frame to two frames as illustrated in Fig. 1(b). Temporal super-resolution is inherently ill-posed problem because there are an infinite number of high temporal resolution (HTR) frames that can produce the same low temporal resolution (LTR) frame. The key idea of this work is to exploit *self-similarity* in videos to resolve this ambiguity in creating HTR frames from LTR frames of the original video. Here, self-similarity means a self-similar structure in the spatio-temporal volume of videos with different temporal resolutions. This self-similar appearance in videos occurs because the appearance represents integrated motion of objects during each exposure time of videos with different temporal resolutions. Such self-similarity is often observed in various types of videos.

For simplicity, let us consider self-similarity in the case of 1D image sequences of a uniformly moving object (Fig. 2). Figure 2(a) shows a spatio-temporal slice of video (right) representing the dynamic scene (left). Figure 2(b) illustrates two image sequences, $V_0$ and $V_1$, with exposure time (=one frame period) $e/2$ and $e$. Consider, for example, a small green bordered 1D image patch of $V_0$ with exposure time $e/2$. In a blue-bordered 1D image patch captured with exposure time $e$ at the same position in x-t plane, the same object moves twice the distance. If the spatial size of the blue-bordered patch of $V_1$ is twice as that of the green-bordered patch of $V_0$, the green-bordered patch becomes similar to the corresponding blue-bordered patch because the object moves twice the distance in the image of $V_1$. This self-similar relationship can be extended to a 2D-image patch.

[1] We confirmed that several consumer cameras automatically selected full-exposure mode for underexposed outdoor and indoor scenes.

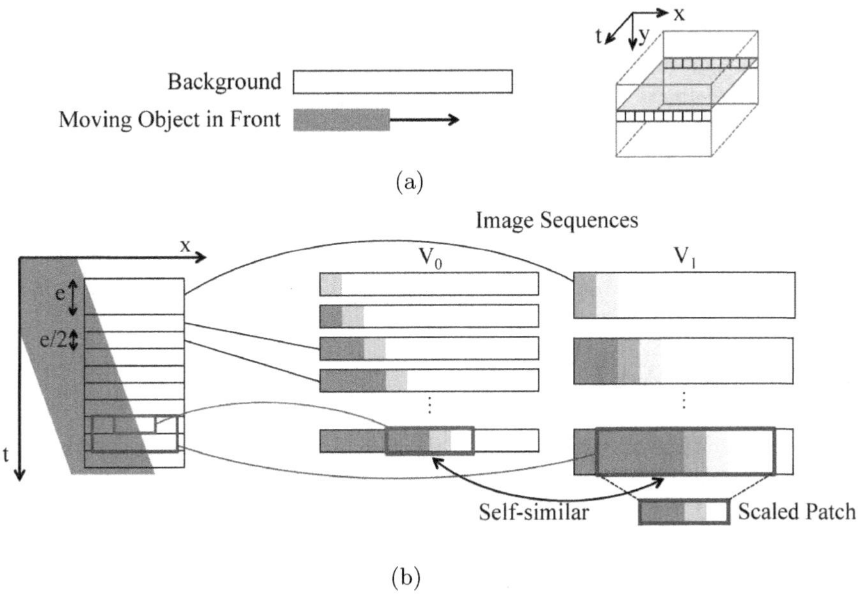


**Fig. 2.** Self-similarity. (a) Dynamic scene (left) and its continuous spatio-temporal volume of video (right), (b) x-t plane for single scan line, and 1-D image sequences with different temporal resolutions.

We exploit not only the self-similar relationship at the same position but also the relationship between different spatio-temporal positions in a video. A moving object in the scene can be recursively seen at many positions along its moving path in a spatio-temporal volume of video. It is also reported that patches in a natural still image tend to recur inside the image with different scales [5]. Due to these recurrences of the same and different objects, a lot of self-similar appearances between different positions can be found through a video. More thorough discussion on self-similarity will be given in Section 3.

On the basis of the self-similar appearances, we formulate the problem of temporal super-resolution as a Maximum a Posteri (MAP) estimate. Here, we use reconstruction constraints: a single frame of LTR video is equal to the average of the two corresponding frames of HTR video. Also, the following self-similar exemplar is utilized for computing a prior probability. We consider similarity at the different temporal resolutions. As shown in Fig. 3, consider patches in three videos of different temporal resolutions: HTR video $V_0$ which we try to create, the original LTR video $V_1$, and another video $V_2$ with even lower temporal resolution created by taking the average of two successive frames of $V_1$. If an image patch of $V_2$ is similar to a patch in $V_1$, a patch of $V_0$ is expected to be similar to the patch of $V_1$. In this way, we can create the prior probability of an HTR video $V_0$ from scaled self-similar exemplars of LTR video $V_1$.

The contribution of this work is that we propose a video temporal super-resolution method from a single image sequence by exploiting the self-similarity in the spatio-temporal domain. Such self-similarity helps us to make the unconstrained problem of temporal super-resolution tractable, while avoiding the explicit motion estimations required for the previous methods. The probabilistic model that reflects both reconstruction constraints and a self-similar prior can yield an HTR sequence, which is consistent with the input LTR sequence.

The rest of this paper is organized as follows. We briefly summarize related work in Section 2. We analyze self-similarity in videos with different temporal resolutions in Section 3. Our temporal super-resolution algorithm is introduced in Section 4. This method is validated in our experiments in comparison to the related work in Section 5, and concluding remarks are given in Section 6.

## 2 Related Work

One approach to increasing temporal resolution is frame rate conversion based on temporal interpolation from a single video. It is known that several methods have serious limitations; frame repetition suffers from jerky object motion as well as linear interpolation of successive frames enlarges blurred areas because of improperly mixed pixels of different objects. To cope with those problems, a number of methods based on motion compensation have been proposed [1,2,3,4]. For these methods based on motion compensation, motion vectors such as optical flows are commonly used to blend corresponding blocks or pixels. Recently, Mahajan *et al.* [6] proposed an interpolation method in a gradient domain to get rid of annoying effects such as ghost, blur, and block artifacts caused by errors in motion vectors. These methods achieve good results for videos with sub-exposure times as shown in Fig.1(a), but are not suitable for videos with full-exposure time as shown in Fig.1(b) because establishing dense correspondence is not always possible due to motion blur. Our method has a distinct advantage over those interpolation-based approaches because dense motion estimation is not required.

Another approach to increasing the temporal resolution or deblurring is to use multiple image sequences. Ben-Ezra and Nayar [7] and Tai *et al.* [8] proposed the use of a hybrid camera system to capture a sequence of low spatial resolution images at high frame rate and a sequence of high spatial resolution images at low frame rate. They estimate motion blur kernels from the former sequence and deblur the latter one. Watanabe *et al.* [9] proposed a method for combining two captured image sequences in the wavelet domain. Some methods used multiple image sequences that were simultaneously captured by co-located cameras with overlapping exposure times for both spatial and temporal super-resolution [10,11]. The use of multiple sequences can significantly enhance the resolution of a video. In contrast to methods that use multiple input image sequences captured simultaneously or by using special hardware, our method uses only a single input image sequence for temporal super-resolution by exploiting a self-similar appearance in video. In the context of spatial super-resolution, Glasner *et al.* [5] reported that natural images have self-similar textures, *i.e.* patches in a natural image tend to redundantly recur many times inside the image, both within the same scale as well as across different scales. On the basis of this observation, they propose a method for spatial super-resolution from a single image. Unlike this spatial super-resolution method using self-similarity in a still image, we focus on self-similarity in video for the first time to increase temporal resolution and can handle the dynamics of the scene.

In summary, we propose a method for temporal super-resolution from a single input image sequence with full-exposure time by introducing the self-similarity observed in the spatio-temporal domain. Unlike the frame rate conversion approach based on temporal interpolation, our method does not need dense motion estimation. Furthermore, unlike the approach using multiple image sequences, our method does not require multiple image sequences captured simultaneously or the use of special hardware.

## 3 Self-similarity

Here, we explain self-similarity in video in more detail, including the necessary conditions under which self-similarity is observed.

### 3.1 Definition

Let us consider an input video $V_1$ captured with an exposure time $e$ and construct a video $V_2$ with an exposure time $2e$ by averaging two adjacent image frames of $V_1$[2]. In the same manner, we recursively construct a video $V_k$ with an exposure time $2^{k-1}e$ from a video $V_{k-1}$. Self-similarity is a similarity structure observed among videos with different exposure times constructed from a single input video $V_1$.

For the sake of simplicity, let us begin with a video $V_1$ capturing a scene where an edge with an infinite length is in uniform linear motion as shown in Fig. 3. We denote a small image patch in $V_1$ as $f(\boldsymbol{p}, \boldsymbol{a}, e)$, where $\boldsymbol{p} = (x, y, t)^T$ is the

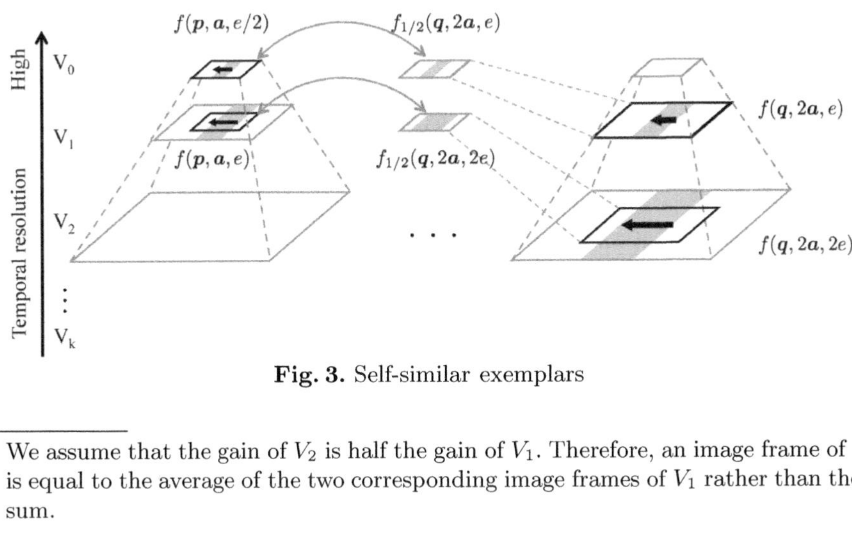


**Fig. 3.** Self-similar exemplars

[2] We assume that the gain of $V_2$ is half the gain of $V_1$. Therefore, an image frame of $V_2$ is equal to the average of the two corresponding image frames of $V_1$ rather than their sum.

space-time coordinates of the center, $\boldsymbol{a} = (a_x, a_y)$ is the spatial size, and $e$ is the exposure time. Similarly, we denote a small image patch in $V_2$ as $f(\boldsymbol{p}, \boldsymbol{a}, 2e)$, and so on. Since the exposure time of $V_2$ is twice as long as that of $V_1$, the edge moves twice the distance in a single image frame of $V_2$. Therefore, the image patch $f(\boldsymbol{p}, \boldsymbol{a}, e)$ in $V_1$ and the image patch $f(\boldsymbol{p}, 2\boldsymbol{a}, 2e)$ in $V_2$ are similar. In other words, $f(\boldsymbol{p}, \boldsymbol{a}, e)$ is equal to $f_{1/2}(\boldsymbol{p}, 2\boldsymbol{a}, 2e)$ which can be synthesized from $f(\boldsymbol{p}, 2\boldsymbol{a}, 2e)$ by scaling the spatial size by 1/2. In the same manner, $f(\boldsymbol{p}, \boldsymbol{a}, e)$ is equal to $f_{1/4}(\boldsymbol{p}, 4\boldsymbol{a}, 4e)$, and so on.

Generally, image patches $f(\boldsymbol{p}, \boldsymbol{a}, e)$ and $f(\boldsymbol{q}, c_s\boldsymbol{a}, c_t e)$ are self-similar, if there is a pair of scalars $c_s$ and $c_t$ such that $f(\boldsymbol{p}, \boldsymbol{a}, e) = f_{1/c_s}(\boldsymbol{q}, c_s\boldsymbol{a}, c_t e)$. Note that the self-similarity could be observed between image patches at different space-time coordinates.

### 3.2 Necessary Conditions

As defined in the previous section, self-similarity among videos with different exposure times requires similarities in both the spatial and temporal domains. For the spatial domain, self-similarity requires that small image patches appear similar when one changes the spatial scale according to the given temporal scale. For some families of images, as Glasner *et al.* [5] reported, patches in a natural image tend to redundantly recur many times inside the image, both within the same scale as well as across different scales. In addition, since we deal with small patches instead of the entire image, a small patch containing an edge longer than the spatial size of the patch, for example, is considered to be invariant to the scale change.

For the temporal domain, self-similarity requires that the motions observed in small patches are similar when one changes the temporal scale in accordance with the given spatial scale. This is exactly true for objects in uniform linear motion (*e.g.* $c_s = 2$ and $c_t = 2$) and in uniformly-accelerated motion (*e.g.* $c_s = 4$ and $c_t = 2$). In addition, since we deal with images captured during the exposure time, *i.e.* a short period of time, the velocity of an object of interest is often considered to be constant, and as a result, the behavior of the object can often be approximated by uniform linear motion.

## 4 Temporal Super-Resolution Algorithm

### 4.1 Overview

We propose an algorithm for increasing the temporal resolution of an input video twice[3]. That is, we reconstruct a video $V_0$ with an exposure time $e/2$ from a single input video $V_1$ with an exposure time $e$ and a set of videos $V_k (k = 2, 3, 4, ...)$ with an exposure time $2^{k-1}e$ constructed from $V_1$ by averaging adjacent image frames recursively.

Since we assume that the exposure time of the input video is almost equal to the frame time, a single image frame of $V_1$ is equal to the average of the two

[3] The algorithm for increasing the temporal resolution more than twice can be formulated straightforwardly.

corresponding image frames of $V_0$. We impose this constraint termed a *reconstruction constraint*[4] in order to decompose a single image frame of $V_1$ into two image frames of $V_0$, as described in Section 4.2. However, this decomposition is an ill-posed problem because an infinite number of combinations of image frames can satisfy the constraint.

To cope with this ill-posedness, we incorporate another constraint based on self-similarity. We take advantage of self-similar exemplars to make the ill-posed problem tractable(Section 4.3).

To this end, we formulate temporal super-resolution as a MAP estimation. Let us denote the image frame of $V_k$ at time $t$ as $F_k(t)$, or $F_k$ for short. According to Bayesf law, the posterior probability $P(F_0|F_1)$ of the random variable $F_0$ given an observation $F_1$ is

$$P(F_0|F_1) = \frac{P(F_1|F_0)P(F_0)}{P(F_1)}. \tag{1}$$

Here, $P(F_1|F_0)$ and $P(F_0)$ are the likelihood of $F_1$ given $F_0$ and the prior probability of $F_0$ respectively. Since the probability $P(F_1)$ is constant, the MAP estimation of $F_0$ results in minimization of the negative logarithm of $P(F_1|F_0)P(F_0)$:

$$F_0^{\mathrm{MAP}} = \arg\min_{F_0} \left[ - \ln P(F_1|F_0) - \ln P(F_0) \right]. \tag{2}$$

In the next two sections, we describe how the likelihood and the prior probability are computed from the reconstruction constraint and self-similar exemplars.

## 4.2 Likelihood Model

We compute the likelihood from the reconstruction constraint. Assuming that the pixel values of image frame $F_1(t)$ are contaminated by additive noise, the constraint is described as

$$F_1(x,y,t) = \frac{F_0(x,y,t) + F_0(x,y,t+e/2)}{2} + \eta(x,y,t). \tag{3}$$

Here, $F_k(x,y,t)$ is a pixel value of the image frame $F_k(t)$ at the image coordinates $(x,y)$, and $\eta(x,y,t)$ stands for additive noise.

We assume that the additive noise obeys an i.i.d. Gaussian distribution with standard deviation $\sigma_{\mathrm{like}}$. Accordingly, the first term of Eq.(2) becomes

$$- \ln P(F_1|F_0) = \frac{1}{2\sigma_{\mathrm{like}}^2} \sum_{\Omega} \left[ F_1(x,y,t) - \frac{F_0(x,y,t) + F_0(x,y,t+e/2)}{2} \right]^2, \tag{4}$$

where $\Omega$ is the image plane over which the summation is taken, and we have omitted constant terms with respect to $F_0$.

[4] Strictly speaking, this constraint assumes a linear camera response function (CRF). However, because adjacent image frames usually have similar pixel values, it is approximately satisfied when the CRF is approximated by a piecewise-linear function. In addition, one could calibrate the CRF and convert the pixel values in advance.

### 4.3 Prior Model

We compute the prior probability from self-similar exemplars in a similar manner to image hallucination [12,13]. The self-similar exemplars are exploited on the following assumption: if an image patch $f(\boldsymbol{p}, \boldsymbol{a}, e)$ in $V_1$ and a scaled patch $f_{1/2}(\boldsymbol{q}, 2\boldsymbol{a}, 2e)$ of $V_2$ are similar, the corresponding patch $f(\boldsymbol{p}, \boldsymbol{a}, e/2)$ in $V_0$ and the corresponding scaled patch $f_{1/2}(\boldsymbol{q}, 2\boldsymbol{a}, e)$ of $V_1$ are also similar as shown in Fig. 3.

Specifically, we find an image patch $f_{1/2}(\boldsymbol{q}, 2\boldsymbol{a}, 2e)$ of $V_2$ similar to $f(\boldsymbol{p}, \boldsymbol{a}, e)$ in $V_1$, and consider $f_{1/2}(\boldsymbol{q}, 2\boldsymbol{a}, e)$ of $V_1$ as a candidate for $f(\boldsymbol{p}, \boldsymbol{a}, e/2)$ in $V_0$. The candidate image patch $\tilde{f}(\boldsymbol{p}, \boldsymbol{a}, e/2)$ is given by

$$\tilde{f}(\boldsymbol{p}, \boldsymbol{a}, e/2) = f_{1/2}(\boldsymbol{q}, 2\boldsymbol{a}, e) \quad \text{where} \quad \boldsymbol{q} = \arg\min_{\boldsymbol{p}'} |f(\boldsymbol{p}, \boldsymbol{a}, e) - f_{1/2}(\boldsymbol{p}', 2\boldsymbol{a}, 2e)|. \tag{5}$$

The difference between image patches is defined as the sum of the squared differences (SSD) of the pixel values.

Next, the candidate image patches $\tilde{f}(\boldsymbol{p}, \boldsymbol{a}, e/2)$ are concatenated into the candidate image frame $\tilde{F}_0(t)$. Here, we assume that the deviation of $F_0$ from the candidate $\tilde{F}_0$ obeys an i.i.d. Gaussian distribution with standard deviation $\sigma_{\text{pri}}$. In addition, we do not use the image frames but the difference between image frames for modeling the prior probability because it performs better experimentally. Hence, the second term of Eq.(2) is given by

$$-\ln P(F_0) = \frac{1}{2\sigma_{\text{pri}}^2} \sum_{\Omega, t'} \left[\Delta F_0(x, y, t') - \Delta \tilde{F}_0(x, y, t')\right]^2, \tag{6}$$

where $t' = t, t + e/2$, and constant terms have been omitted.

Note that we can use more temporal-scale levels for determining a candidate image patch in Eq.(5).The current implementation computes the SSD between $f(\boldsymbol{p}, \boldsymbol{a}, 2e)$ and $f_{1/2}(\boldsymbol{p}', 2\boldsymbol{a}, 4e)$ in addition to the SSD between $f(\boldsymbol{p}, \boldsymbol{a}, e)$ and $f_{1/2}(\boldsymbol{p}', 2\boldsymbol{a}, 2e)$. That is, we use the videos $V_k$ constructed from the input video $V_1$ up to $k = 3$.

### 4.4 Optimization

To reduce spatial noise and temporal flicker, we add a smoothness term $S$ to Eq.(2). Substituting Eq.(4) and Eq.(6) into Eq.(2), the MAP estimation of $F_0$ results in

$$\begin{aligned} F_0^{\text{MAP}} &= \arg\min_{F_0} \left[-\ln P(F_1|F_0) - \ln P(F_0) + S\right] \\ &= \arg\min_{F_0} \left\{ \frac{1}{2\sigma_{\text{like}}^2} \sum_{\Omega} \left[F_1(x, y, t) - \frac{F_0(x, y, t) + F_0(x, y, t + e/2)}{2}\right]^2 \right. \\ &\quad + \frac{1}{2\sigma_{\text{pri}}^2} \sum_{\Omega, t'} \left[\Delta F_0(x, y, t') - \Delta \tilde{F}_0(x, y, t')\right]^2 \\ &\quad \left. + \lambda \sum_{\Omega, t'} \left[\sum_{\Lambda} w(x, y, t', m, n, q) L(m, n, q) F_0(x{+}m, y{+}n, t'{+}q)\right]^2 \right\}, \end{aligned}$$

where $\Delta F_0(x, y, t') = F_0(x, y, t' + e/2) - F_0(x, y, t')$.

Our smoothness term is based on the Laplacian filter L(m,n,q). To preserve spatiotemporal edges, we incorporate a pixel-wise weight $w(x, y, t', m, n, q)$. It is defined as $w(x, y, t', m, n, q) = w_d(x, y, t', m, n, q) / \sum_{m',n',q'} w_d(x, y, t', m', n', q')$, where $w_d(x, y, t', m, n, q) = \exp[-(F(x, y, t) - F(x+m, y+n, t'+q))^2/2\sigma_s^2]$.

The optimization problem of Eq.(7) is iteratively solved according to the following steps.

1. Ignore the smoothness term in Eq.(7), and compute the initial estimation $F_0^{r=0}$ by using the least-squares method.
2. Compute the weight $w(x, y, t', m, n, q)$ from $F_0^r$, and update the estimation as $F_0^{r+1}$ by minimizing Eq.(2) via least squares.
3. Repeat step 2 until the difference $|F^{r+1} - F^r|$ converges.

Usually, only a few iterations are needed for convergence.

## 5 Experimental Results

We evaluated various aspects of our approach in experiments using real videos. In Section 5.1, we compare the performance of our method with that of previously proposed temporal interpolation methods.We then clarified the importance of each component of our algorithm in Section 5.2.

We used CANON HF-20 and CASIO EX-F1 video cameras. The exposure time of the camera was set almost equal to one frame time. Eight kinds of videos were collected as $V_0$, by capturing dynamic scenes for about a few hundred frames: three videos capturing indoor scenes and five videos captured outside. The video of textured paper 1 and 2 were captured as scenes of uniform circular motion. The video of a small merry-go-round was captured as scenes with uniform linear motion as well as various other motions. Textured paper 1 included repetitions of the same texture, and textured paper 2 included several textures. The videos captured outside included various scenes such as walking people, cars, and trees. The frame images of the captured videos were converted to grayscale in all experiments. The captured videos were considered to be the ground truth high-temporal-resolution video $V_0$. We reconstructed $V_0$ from its low-temporal-resolution videos $V_1$, $V_2$, and $V_3$ that were provided from $V_0$, as mentioned in Section 3. Our algorithm used 5×5 pixel patches. We empirically set $\sigma_{like} = 1, \sigma_{pri} = 1.25, \lambda = 1.6$, and $\sigma_s$ was set from 10 to 100 to adjust to the brightness of the frames. The produced grayscale videos were converted to a YUV color space as follows. We obtained the Y channel as the produced grayscale video. For the U and V channels, we used those channels of the original low temporal resolution frame of $V_1$ for two successive high temporal-resolution frames.

### 5.1 Comparison with Other Methods

We compared the results of our technique with those of three methods: simple frame repetition, linear interpolation, and interpolation based on optical flows [4,14]. Linear interpolation and optical flow based interpolation generate

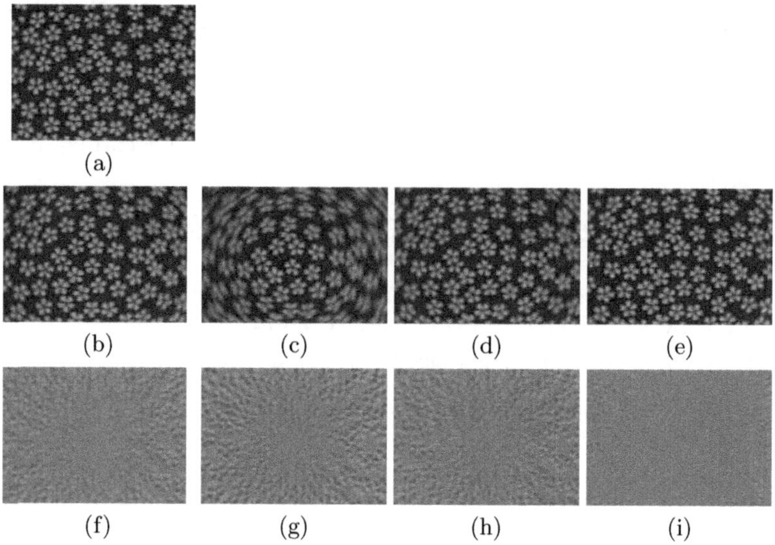

**Fig. 4.** Comparison of our method with other interpolation methods applied video in which textured paper1 rotates uniformly: (a) ground truth frame, frames with increased temporal resolution (b) frame repetition, (c) linear interpolation, (d) optical flow based interpolation, and (e) our method. (f)(g)(h)(i) The difference (image values are added offset 128 to the difference.) between each produced frame ((b),(c),(d), or (e)) and the ground truth (a).

same number of intermediate frames from input frames. The eight videos were used to examine the performance of each method on various motions.

We made a qualitative comparison of our method and the other interpolation methods. In Fig. 4 and Fig. 5, (a) shows the ground truth and (f)-(i) show the differences between the ground truth and the reconstructed frame (b)-(e) respectively by various methods ((b) frame repetition, (c) linear interpolation, (d) optical flow based interpolation, and (e) our method). As we see in (Fig. 4 (e) and Fig. 5 (e)), the visual quality of the frames recovered using our method is much better than those of the other interpolation methods. In the results obtained by optical flow based interpolation (d), the motion blur region remained or was enlarged in the interpolated frame. Motion blur likewise remained in the videos produced by frame repetition and was enlarged in the videos produced by linear interpolation. In Fig. 6, we show our results, zoom-ups in interesting areas, to compare to those of the input and the ground truth. In the scene of a crowded intersection with diagonal crosswalks, the moving leg and two heads moving apart in the frames produced by our method matched very closely with the ground truth (Fig. 6 (a)). Our method could also be used to generate frames to represent correctly the changing movie displayed on the big TV screen (Fig. 6 (b)) and the shadow hovering beyond the vein of the leaf (Fig. 6 (c)). Note that the information of the input frame was properly resolved to two frames with high temporal resolution produced by our method.

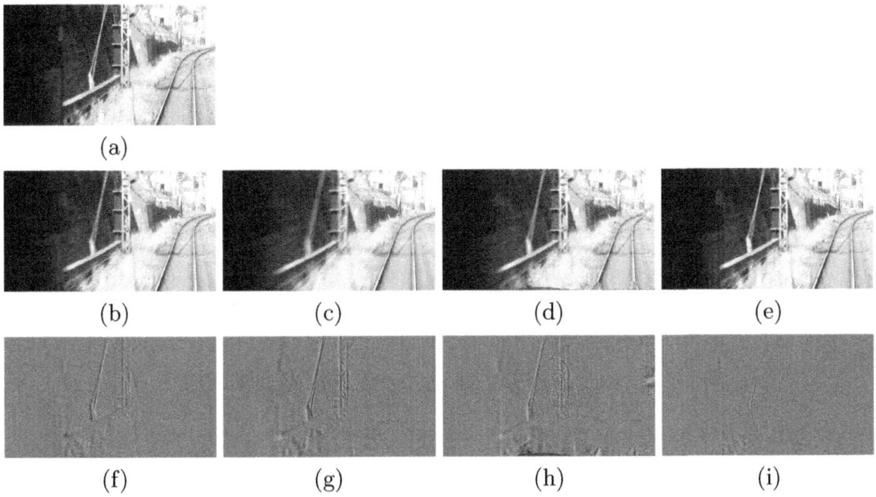


**Fig. 5.** Comparison of our method with other interpolation methods applied to video of outside1. (a) Ground truth frame, frames with increased temporal resolution (b) frame repetition, (c) linear interpolation, (d) optical flow based interpolation, and (e) our method. (f)(g)(h)(i) Difference (image values are added offset 128 to the difference.) between each produced frame ((b),(c), (d), or (e)) and the ground truth (a).

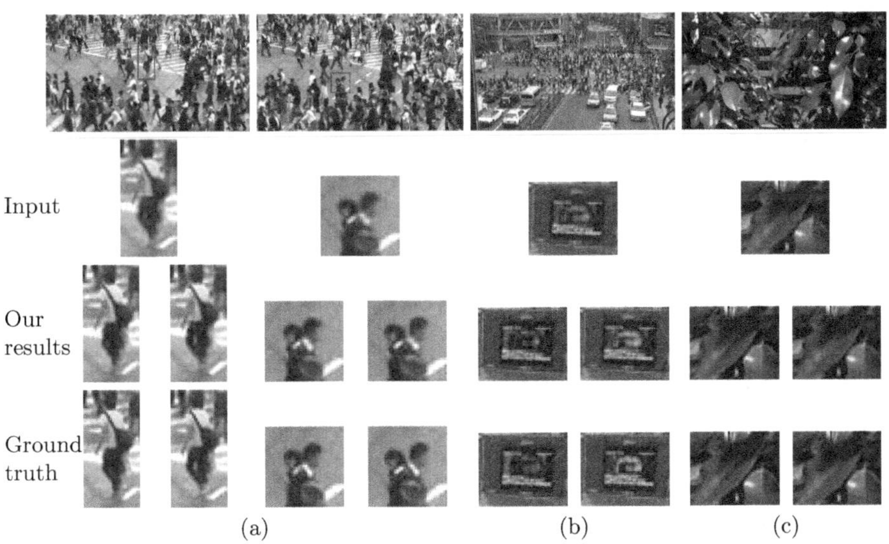


**Fig. 6.** Zoom-ups of high temporal resolution frames produced by our method applied to video of (a) outside2, (b) outside3, and (c) outside4. The example input frames (first row), zoom-ups of the marked region in the input frames (second row), zoom-ups of the marked region in the two frames with increased temporal resolution produced by our method (third row), and zoom-ups of the marked region in the two ground truth frames (fourth row).

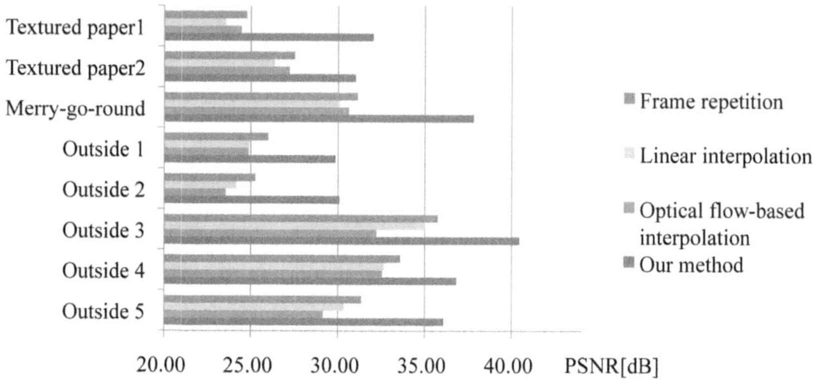


**Fig. 7.** Comparison of our method with other interpolation methods

To make a quantitative comparison, we also evaluated the peak signal to noise ratio (PSNR) between the frames produced by all four temporal super-resolution methods and the ground truth frames. The average PSNRs of all four methods are shown in Fig. 7. We clearly see that our method is superior to the other methods for all videos. The optical flow based interpolation have sometimes failed to estimate the optical flows because of the input blurred frames, and these errors may have led to more ghosting and blurring artifacts than in the frame repetition or the linear interpolation video. The qualitative and quantitative comparisons indicated that our method can be effectively used for compensating circular, linear, and various other motions.

## 5.2 Evaluation of Components

We conducted experiments to evaluate the contribution of each component in our method such as reconstruction constraints, the smoothness term, the use of temporal-scale levels up to $V_k$, and features describing self-similar structures (images or their differences) by using 30 frames each of three captured videos, as follows.

First, we examined the effectiveness of considering multiple temporal-scale levels $k$ of the videos. As shown in Table 1, cases using temporal-scale levels up to $k = 3, 4$ superior to the case using temporal-scale levels up to $k = 2$ and achieved a better PSNR. This shows that examining motion at different exposure times ($V_1$, $V_2$, and so on) is effective for revealing motion at higher frame rate.

To examine the contribution of each component, we conducted experiments by adding components incrementally. The PSNR of each case is shown in Table 2. Here, the quality of the produced videos is improved, as components are added. In other words, all the components play essential roles for improving the resulting image quality.

**Table 1.** Comparison of different temporal-scale levels. Our algorithmfs results using videos $V_k$ constructed from the input video $V_1$ up to $k = 4$ are compared.

| PSNR [dB] | | | |
|---|---|---|---|
| Sequence | k=2 | k=3 | k=4 |
| Textured paper1 | 15.68 | 24.06 | 24.01 |
| Textured paper2 | 18.60 | 27.45 | 27.40 |
| Merry-go-round | 21.77 | 34.34 | 33.19 |

**Table 2.** Evaluation of a variety of components. Note that use of several temporal-scale levels, image differences, reconstruction constraints, and the smoothness term are all important in our method.

| | | | | | PSNR (dB) | | |
|---|---|---|---|---|---|---|---|
| Case | k=3 | Image differences | Reconstruction constraints | Smoothness term | Textured paper1 | Textured paper2 | Merry-go-round |
| 1 | – | – | – | – | 15.55 | 19.06 | 17.96 |
| 2 | √ | – | – | – | 22.90 | 26.26 | 32.78 |
| 3 | √ | √ | – | – | 24.06 | 27.45 | 34.34 |
| 4 | √ | √ | √ | – | 25.99 | 28.82 | 35.49 |
| 5 | √ | √ | √ | √ | 26.19 | 28.72 | 35.58 |

## 6 Conclusions

We proposed an approach for making temporal super resolution video from a single input image sequence. To make this ill-posed problem tractable, we focused on *self-similarity* in video which is a self-similar appearances with different temporal resolutions. Our method increases the temporal resolution by using a MAP estimation that incorporates both a prior probability computed from self-similar appearances and reconstruction constraints *i.e.*, a single frame of LTR video should be equal to the average of the two corresponding frames of HTR video. The experimental results showed that our method could recover a high-quality video with increased temporal resolution, which keeps consistent with the input LTR video. We demonstrated that our approach produce results that are sharper and clearer than those of other techniques based on temporal interpolation. An interesting direction of future work is to increase the temporal and spatial resolutions at the same time.

## References


1. Choi, B.T., Lee, S.H., Ko, S.J.: New frame rate up-conversion using bi-directional motion estimation. IEEE Trans. CE 46, 603–609 (2000)
2. Ha, T., Lee, S., Kim, J.: Motion compensated frame interpolation by new block-based motion estimation algorithm. IEEE Trans. CE 50, 752–759 (2004)

3. ying Kuo, T., Kim, J., Jay Kuo, C.: Motion-compensated frame interpolation scheme for h.263 codec. In: Proc. ISCAS 1999, pp. 491–494 (1999)
4. Baker, S., Scharstein, D., Lewis, J., Roth, S., Black, M., Szeliski, R.: A database and evaluation methodology for optical flow. In: Proc. ICCV 2007, pp. 1–8 (2007)
5. Glasner, D., Bagon, S., Irani, M.: Super-resolution from a single image. In: Proc. ICCV 2009, pp. 349–356 (2009)
6. Mahajan, D., Huang, F.C., Matusik, W., Ramamoorthi, R., Belhumeur, P.: Moving gradients: a path-based method for plausible image interpolation. In: Proc. SIGGRAPH 2009, pp. 1–11 (2009)
7. Ben-Ezra, M., Nayar, S.K.: Motion deblurring using hybrid imaging. In: Proc. CVPR 2003, pp. I–657–I–664 (2003)
8. Tai, Y.W., Du, H., Brown, M.S., Lin, S.: Image/video deblurring using a hybrid camera. In: Proc. CVPR 2008, pp. 1–8 (2008)
9. Watanabe, K., Iwai, Y., Nagahara, H., Yachida, M., Suzuki, T.: Video synthesis with high spatio-temporal resolution using motion compensation and image fusion in wavelet domain. In: Narayanan, P.J., Nayar, S.K., Shum, H.-Y. (eds.) ACCV 2006. LNCS, vol. 3851, pp. 480–489. Springer, Heidelberg (2006)
10. Shechtman, E., Caspi, Y., Irani, M.: Space-time super-resolution. IEEE Trans. PAMI 27, 531–545 (2005)
11. Agrawal, A., Gupta, M., Veeraraghavan, A., Narasimhan, S.: Optimal coded sampling for temporal super-resolution. In: Proc. CVPR 2010, pp. 599–606 (2010)
12. Baker, S., Kanade, T.: Limits on super-resolution and how to break them. IEEE Trans. PAMI 24, 1167–1183 (2002)
13. Freeman, W., Jones, T., Pasztor, E.: Example-based super-resolution. IEEE Computer Graphics and Applications 22, 56–65 (2002)
14. Brox, T., Bruhn, A., Papenberg, N., Weickert, J.: High accuracy optical flow estimation based on a theory for warping. In: Pajdla, T., Matas, J(G.) (eds.) ECCV 2004. LNCS, vol. 3024, pp. 25–36. Springer, Heidelberg (2004)

# Temporal Super Resolution from a Single Quasi-periodic Image Sequence Based on Phase Registration

Yasushi Makihara, Atsushi Mori, and Yasushi Yagi

Osaka University, 8-1 Mihogaoka, Ibaraki, Osaka, 567-0047, Japan

**Abstract.** This paper describes a method for temporal super resolution from a single quasi-periodic image sequence. A so-called reconstruction-based method is applied to construct a one period image sequence with high frame-rate based on phase registration data in sub-frame order among multiple periods of the image sequence. First, the periodic image sequence to be reconstructed is expressed as a manifold in the parametric eigenspace of the phase. Given an input image sequence, phase registration and manifold reconstruction are alternately executed iteratively within an energy minimization framework that considers data fitness and the smoothness of both the manifold and the phase evolution. The energy minimization problem is solved through three-step coarse-to-fine procedures to avoid local minima. The experiments using both simulated and real data confirm the realization of temporal super resolution from a single image sequence.

## 1 Introduction

Image super resolution[1] is one of the fundamental image processing techniques for increasing spatial resolution. Methods for super resolution fall mainly into two categories: (1) reconstruction-based methods (RBM) using multiple images registered at sub-pixelorder[2][3][4], and (2) example-based methods using correspondence between low and high resolution image patches from training sets[5][6][7]. Furthermore, Glasner et al.[8] proposed a sophisticated framework, combining the two approaches by using the patch recurrence within and across scales of a single image. Most of the existing methods, however, are applicable to static scenes only.

Several methods deal with dynamic scenes in the context of near-real-time super resolution[9], taking advantage of hot-air optical turbulence[10] and image super resolution in the presence of motion debluring[11][12]. Nevertheless, these methods still focus on spatial super resolution, and fail to address temporal super resolution.

Contrary to the above methods, Shechtman et al.[13] proposed a space-time super resolution by combining information from multiple low-resolution video sequences of the same dynamic scene. This method falls into the so-called RBMs and hence, requires multiple video cameras to obtain multiple image sequences.

R. Kimmel, R. Klette, and A. Sugimoto (Eds.): ACCV 2010, Part I, LNCS 6492, pp. 107–120, 2011.

Inspired by the effective use of patch recurrence[8], it is noticeable that temporal recurrence in a single sequence can be used for temporal super resolution. In particular, given a periodic image sequence, multiple periods of subsequences with sub-frame order *phase displacement* can be used for the temporal super resolution. If a sequence can be segmented into period-based subsequences composed of more than a few frames, a spatio-temporal sequence registration[14] method is applicable. On the contrary, in cases where each subsequence contains only a few frames or a single frame, period segmentation and the spatio-temporal sequence registration may be error prone, since the method discards useful cues of inter-period phase continuity. Moreover, if the period fluctuates due to a non-uniform sampling rate or to target motion fluctuation appearing in human periodic actions (e.g., gait), phase registration becomes much more difficult.

Therefore, we propose a method to reconstruct a one period image sequence at a high frame-rate from a single low frame-rate quasi-periodic image sequence. Instead of segmenting the whole sequence into multiple subsequences, we solve the *phase assignment* problem for the whole sequence of frames to take advantage of the inter-period phase continuity. Then, a one period image sequence is expressed as a *manifold in phase-parametric eigenspace* and the manifold is reconstructed based on the registered phase information. These two processes are repeated in an energy minimization framework by taking into consideration the following three aspects: (1) data fitness, (2) smoothness of the manifold, and (3) smoothness of the phase evolution.

## 2 Related Work

**Temporal interpolation:** In addition to spatial interpolation, temporal interpolation has been developed for video editing (e.g., retiming[15]) and video compression based on motion compensation methods[16]. In particular, morphingapproaches[17][18] based on optical flow[19] are regarded as some of the most promising methods for temporal interpolation, with these ideas having been extended to view and time interpolation[20] and spatio-temporal interpolation from two sequences, one with high resolution and low frame rate and the other with low resolution and high frame rate[21]. Temporal interpolation does not, however, work well in cases where motion between two frames is relatively large, in other words, where the sampling interval is relatively long in fairly low frame-rate video.

**Geometric fitting:** Our manifold reconstruction step is similar to the fitting of geometric primitives, such as lines, ellipses, or conics[22], [23], in terms of parameter estimation from a series of sampling points. In geometric fitting, global parameters are used for the primitive representation (e.g., 5 parameters for an ellipse) and sampling points are treated independently. On the other hand, piecewise local parameters are used for manifold representation and sampling points are dependent on each other due to the phase evolution smoothness as described in the following sections.

# 3 Problem Setting

## 3.1 Assumptions

Several assumptions are used in this paper to allow us to focus only on temporal super resolution based on a phase registration framework. The first assumption is that an image sequence is spatially registered in advance. This assumption is basically true in several scenes, such as those with periodic sign language such as a waving hand, periodic actions at the same position such as jumping jacks or walking on a treadmill, or a rotating object such as a fan. Even in other scenes with periodic actions such as walking, skipping, and running on the ground, spatial registration is possible once the target object has been accurately tracked to some extent.

A further assumption is that motion blurs are negligible. Although this assumption may not be true in scenes with fast moving objects, there are still several possible situations in which low frame-rate image sequences with less motion blur are stored. For example, consider the situation where a CCTV camera captures a person walking on the street. In this case, the image sequence is often stored at a low frame-rate due to limited communication band width or storage size, and besides motion blur is not significant since walking motion is relatively slow compared with normal shutter speeds. Therefore, gait recognition in low frame-rate video is one of the typical applications of the proposed method.

## 3.2 Quasi-periodic Image Sequence

Next, we define a quasi-periodic image sequence. An image drawn from the periodic image sequence at time $t$ is denoted by the vector-form $\boldsymbol{x}(t)$, which satisfies

$$\boldsymbol{x}(t+P) = \boldsymbol{x}(t) \ \ \forall t, \tag{1}$$

where $P$ is a period. Then, two non-dimensional time parameters, *phase* $s$ and *relative phase* $\tilde{s}$, are introduced as

$$s = s_P(t) = \frac{t}{P} \tag{2}$$

$$\tilde{s} = s - \lfloor s \rfloor, \tag{3}$$

where $s_P$ is a phase evolution function and $\lfloor \rfloor$ is a floor function. Now, the periodic image sequence is represented in the phase domain as

$$\boldsymbol{x}_s(s) = \boldsymbol{x}(s_P^{-1}(s)). \tag{4}$$

Note that the periodic image sequence constructs a manifold with respect to the relative phase $\tilde{s} \in [0, 1]$ and which satisfies $\boldsymbol{x}_s(1) = \boldsymbol{x}_s(0)$.

On the other hand, an input image sequence is composed of $N^{in}$ discretely observed images $\boldsymbol{X}^{in} = \{\boldsymbol{x}_i^{in}\}(i = 0, \ldots, N^{in} - 1)$. In cases where the scene is completely periodic and where the frame-rate $f$ is completely constant during the input image sequence, the input image sequence $\boldsymbol{X}^{in}$, time sequence $\boldsymbol{t} = \{t_i\}$, and phase sequence $\boldsymbol{s}_P = \{s_{P,i}\}$ are denoted as

$$\boldsymbol{x}_i^{in} = \boldsymbol{x}(t_{P,i}) = \boldsymbol{x}_s(s_{P,i}) \tag{5}$$

$$t_i = t_0 + \frac{i}{f} \tag{6}$$

$$s_{P,i} = s_P(t_i) = s_0 + \frac{1}{fP}. \tag{7}$$

This assumption is, however, often violated due to fluctuations in the frame-rate or in the timing of the periodic motion. Hence, the input image sequence is degraded to a *quasi-periodic image sequence* $\boldsymbol{X}_Q^{in} = \{\boldsymbol{x}_{Q,i}^{in}\}$ defined as

$$\boldsymbol{x}_{Q,i}^{in} = \boldsymbol{x}_s(s_{Q,i}) \tag{8}$$

$$s_{Q,i} = s_i + \Delta s_i, \tag{9}$$

where $\boldsymbol{s}_Q = \{s_{Q,i}\}$ is a quasi-periodic phase sequence.

In summary, the problem setting can be stated as a simultaneous estimation problem of a periodic manifold $\boldsymbol{x}_s$ and a phase sequence $\boldsymbol{s}_Q$ from an input quasi-periodic image sequence $\boldsymbol{X}_Q^{in}$. Analogous to the spatial super resolution case, note that the phase sequence estimation process and periodic manifold estimation correspond to an image registration process and super-resolution process, respectively. In addition, the problem setting falls into the so-called reconstruction-based methods using multiple observations, although a single sequence is used in this problem setting. This is because *multiple periods* in the single sequence serve as multiple observations.

### 3.3 Manifold Representation

The periodic manifold $\boldsymbol{x}_s$ is represented by a parametric eigenspace method[24]. This is motivated by the property of the parametric eigenspace method that a phase for an input image can be estimated by projection to the manifold, which plays quite a significant role in the phase registration stage. More specifically, we use a cubic N-spline function parameterized by the phase in the eigenspace. Let us consider $N^{cp}$ control points $\{\boldsymbol{y}_j^{cp}\}$ in the $M$ dimensional eigenspace accompanied by corresponding phases $\{s_j^{cp}(= j/N^{cp})\}$, $(j = 0, \ldots, N^{cp} - 1)$. Next, a spline parameter vector for a $k$ power-term coefficient at the $j$th interval $[s_j^{cp}, s_{j+1}^{cp}]$ is denoted as $\boldsymbol{a}_{j,k}^{sp} \in \mathbb{R}^M$ and subsequently, a submatrix $A_j^{sp}$ at the $j$th interval and a total spline matrix $A^{sp}$ are defined as $A_j^{sp} = [\boldsymbol{a}_{j,0}^{sp}, \ldots, \boldsymbol{a}_{j,3}^{sp}]^T \in \mathbb{R}^{4 \times M}$ and $A^{sp} = [A_0^{sp\,T}, \ldots, A_{N^{cp}-1}^{sp}{}^T]^T \in \mathbb{R}^{4N^{cp} \times M}$, respectively. Then, an interpolated point $\hat{\boldsymbol{y}}(\tilde{s})$ in the eigenspace for a relative phase $\tilde{s}$ at the $j$th interval is expressed as

$$\hat{\boldsymbol{y}}(\tilde{s}) = A^{sp^T}\boldsymbol{w}(\tilde{s}) \tag{10}$$

$$\boldsymbol{w}(\tilde{s}) = [0,\ldots,0,1,w,w^2,w^3,0,\ldots,0]^T \tag{11}$$

$$w = \frac{\tilde{s}-s_j^{cp}}{s_{j+1}^{cp}-s_j^{cp}},\ (s_j^{cp} \leq \tilde{s} \leq s_{j+1}^{cp}) \tag{12}$$

where $\boldsymbol{w}(\tilde{s})$ is an interpolation coefficient vector whose components from $4j$ to $(4j+3)$ are $[1,w,w^2,w^3]$, and $w$ is the interpolation ratio between the control points.

On the other hand, the relation between a control points matrix $Y^{cp} = [\boldsymbol{y}_0^{cp},\ldots,\boldsymbol{y}_{N^{cp}-1}^{cp}]^T$ and a spline parameter matrix $A^{sp}$ is derived from the C2-continuous boundary conditions as

$$CA^{sp} = DY^{cp} \tag{13}$$

$$C = \begin{bmatrix} C_1 & C_2 & O & \cdots \\ \ddots & \ddots & \ddots & \ddots \\ C_2 & O & \cdots & C_1 \end{bmatrix},\ C_1 = \begin{bmatrix} 1&0&0&0 \\ 1&1&1&1 \\ 0&1&2&3 \\ 0&0&2&6 \end{bmatrix},\ C_2 = \begin{bmatrix} 0&0&0&0 \\ 0&0&0&0 \\ 0&-1&0&0 \\ 0&0&-2&0 \end{bmatrix} \tag{14}$$

$$D = \begin{bmatrix} D_1 & D_2 & O & \cdots \\ \ddots & \ddots & \ddots & \ddots \\ D_2 & O & \cdots & D_1 \end{bmatrix},\ D_1 = \begin{bmatrix} 1 \\ 0 \\ 0 \\ 0 \end{bmatrix},\ D_2 = \begin{bmatrix} 0 \\ 1 \\ 0 \\ 0 \end{bmatrix} \tag{15}$$

Hence, the spline parameter matrix $A^{sp}$ is linearly solved as $A^{sp} = C^{-1}DY^{cp}$ given the control points $Y^{cp}$. This also indicates that interpolation $\hat{\boldsymbol{y}}(\tilde{s})$ for a relative phase $\tilde{s}$ is obtained by Eqs. (10) and (15) once the control points $Y^{cp}$ have been given. Then, the reconstruction problem of the periodic manifold $\boldsymbol{x}_s(s;Y^{cp})$ can be replaced by an estimation problem of the control points $Y^{cp}$ as discussed in the following sections.

## 4 Energy Minimization Framework

### 4.1 Energy Function

Similar to the spatial super resolution case, we adopt an energy minimization approach. First, suppose that an input quasi-periodic image sequence in the eigenspace is expressed as $\boldsymbol{Y}_Q^{in} = \{\boldsymbol{y}_{Q,i}^{in}\}$ and recall that the accompanying phase sequence $\boldsymbol{s}_Q = \{s_{Q,i}\}$ is unknown. Subsequently the interpolation coefficient vector for the $i$th phase $\boldsymbol{s}_{Q,i}$ is defined as $\boldsymbol{w}(s_{Q,i})$, in the same way as Eq. (11), and then projection onto the periodic manifold by the $i$th phase $s_{Q,i}$ is

$$\hat{\boldsymbol{Y}}(Y^{cp}, s_{Q,i}) = A^{sp^T}\boldsymbol{w}(s_{Q,i}) = Y^{cp^T}(C^{-1}D)^T\boldsymbol{w}(s_{Q,i}). \tag{16}$$

The energy function is constructed by considering the following three aspects: (1) data fitness between the interpolation $\hat{\boldsymbol{y}}(Y^{cp}, s_{Q,i})$ and the input $\boldsymbol{y}^{in}_{Q,i}$, (2) smoothness of the periodic manifold $\boldsymbol{y}_s(s; Y^{cp})$ in the eigenspace, and (3) smoothness of the phase evolution $\boldsymbol{s}_Q$. The actual form of the function is

$$
\begin{aligned}
E(Y^{cp}, \boldsymbol{s}_Q) = & \frac{1}{N^{in}} \sum_{i=0}^{N^{in}-1} ||Y^{cpT}(C^{-1}D)^T \boldsymbol{w}(s_{Q,i}) - \boldsymbol{y}^{in}_{Q,i}||^2 \\
& + \lambda_m \frac{1}{N^{cp}} \int_0^1 \left|\left| \frac{d^2 \boldsymbol{y}_s(s; Y^{cp})}{ds^2} \right|\right|^2 ds \\
& + \lambda_s \frac{1}{N^{in}} \sum_{i=1}^{N^{in}-1} \left( s_{Q,i+1} - s_{Q,i} - \frac{1}{P'} \right)^2 ,
\end{aligned} \tag{17}
$$

where the first, second, and third terms are the data term, and the smoothness terms for the periodic manifold and the phase evolution, respectively, and $P'$ is a global period in the frame domain, which is defined as the product of the frame-rate $f$ and period $P$ in the time domain, that is, $P' = fP$. Since the integration in the second term is calculated in advance with respect to the domain of the interpolation ratio $w$ for each interval, it is rearranged as a quadratic form of $Y^{cp}$ as

$$
\int_0^1 \left|\left| \frac{d^2 \boldsymbol{y}_s(s; Y^{cp})}{ds^2} \right|\right|^2 ds = Y^{cpT}(C^{-1}D)^T B (C^{-1}D) Y^{cp} \tag{18}
$$

$$
B = \begin{bmatrix} B_{sub} & \cdots & O \\ \vdots & \ddots & \vdots \\ O & \cdots & B_{sub} \end{bmatrix}, \quad B_{sub} = \begin{bmatrix} 0 & 0 & 0 & 0 \\ 0 & 0 & 0 & 0 \\ 0 & 0 & 4 & 6 \\ 0 & 0 & 6 & 12 \end{bmatrix} \tag{19}
$$

We can see that the objective function $E(Y^{cp}, \boldsymbol{s}_Q)$ is a quadratic form with respect to the manifold control points $Y^{cp}$ and therefore, that a linear solution of the manifold control points $Y^{cp}$ is provided under the fixed phase $\boldsymbol{s}_Q$. On the other hand, the phase $\boldsymbol{s}_Q$ is a complex form, since the spline curves are switched piecewise based on phase $\boldsymbol{s}_Q$ and the interpolation ratio $w$ appears as a sixth-order polynomial in the data term. To solve this highly nonlinear optimization problem, three-step coarse-to-fine solutions are provided in the following sections.

## 4.2 Solution by Linear Approximation

In the first step, the phase sequence $\boldsymbol{s}_Q$ is limited to a completely periodic domain, that is, linear phase evolution $\boldsymbol{s}_P$. Because the initial phase $s_0$ is not significant in this problem setting, it is set to zero without loss of generality. Hence, the objective function in the first step is rewritten as

$$
E_{init}(P) = E(Y^{cp}_P, \boldsymbol{s}_P) \tag{20}
$$

$$
Y^{cp}_P = \arg\min_{Y^{cp}} E(Y^{cp}, \boldsymbol{s}_P) \tag{21}
$$

$$
s_{P,i} = \frac{i}{fP}. \tag{22}
$$

Note that manifold control points $Y_P^{cp}$ for each hypothesis of period $P$ are linearly solved as described in 4.1. On the other hand, the objective function $E_{init}(P)$ has many local minima, so a gradient descent method from multiple initial solutions is applied. Optimal solutions of the period, phase and manifold control points in the periodic-domain are denoted as $P^*$, $\boldsymbol{s}_{P^*}$, and $Y_{P^*}^{cp}$, respectively.

### 4.3 Dynamic Programming Solution

In the second step, in order to extend from a periodic domain to a quasi-periodic domain, continuous Dynamic Programming (DP)[25] is applied within a so-called corridor, that is, the neighboring search area from the periodic-domain phase solution $\boldsymbol{s}_{P^*}$ as $R_i^{cdr} = \{s | s_{P^*,i} - s^{cdr} \leq s \leq s_{P^*,i} + s^{cdr}\}$ under fixed manifold control points $Y_{P^*}^{cp}$. First, the phase space is quantized by the same interval as that of the manifold control points $(1/N^{cp})$. This means that the $j$th-phase interpolation coincides with the $j$th control point $\boldsymbol{y}_{P^*,j}^{cp}$ and that the phase difference between the $j$th and $k$th phases is equal to $(j-k)/N^{cp}$. Next, a cumulative cost and the optimal transition from a previous step at the $i$th input frame and $j$th phase are denoted as $c(i, j)$ and $p(i, j)$, respectively, and the optimal phase is provided by the following DP steps.

1. Initialize cost matrix

$$c(0, j) = ||\boldsymbol{y}_{P^*,j}^{cp} - \boldsymbol{y}_0^{in}||^2, \ \ \forall j \in R_0^{cdr} \tag{23}$$

2. Update cumulative cost and transition path

$$p(i, j) = \arg \min_{k \in R_{i-1}^{cdr}} \{c(i-1, k) + \lambda g_{P^*}(j, k)\} \tag{24}$$

$$\begin{aligned} c(i, j) = & \, c(i-1, p(i, j)) + \lambda g_{P^*}(j, p(i, j)) \\ & + ||\boldsymbol{y}_{P^*,j}^{cp} - \boldsymbol{y}_i^{in}||^2, \forall j \in R_i^{cdr} \end{aligned} \tag{25}$$

$$g_{P^*}(j, k) = \left( \frac{\min\{|j-k|, |j-k+N^{cp}|\}}{N^{cp}} - \frac{1}{fP^*} \right)^2 \tag{26}$$

3. Optimize the terminal phase

$$p^*(N^{in} - 1) = \arg \min_j c(N^{in} - 1, j), \ \ j \in R_{N^{in}-1}^{cdr} \tag{27}$$

4. Back track

$$p^*(i-1) = p(i, p^*(i)), \ \ \forall \, 1 \leq i \leq N^{cp} - 1 \tag{28}$$

Finally, the optimal phase $\boldsymbol{s}_{DP}^*$ is set based on the optimal path $\{p^*(i)\}$ and the manifold control points are updated as

$$Y_{DP}^{cp} = \arg \min_{Y^{cp}} E(Y^{cp}, \boldsymbol{s}_{DP}^*) \tag{29}$$

Strictly speaking, one can bypass the DP procedure and proceed to the following step where there is a sufficiently small fluctuation in phase from the periodic solution. On the other hand, in cases with a substantial fluctuation in phase, the following step may result in local optima only and therefore, the DP procedure is still essential for finding the global optimum.

### 4.4 Iterative Solution by Quadratic Approximation

A crucial procedure in the third step is quadratic approximation of the data term with respect to the phase within a narrow search range from the DP solution. First, the DP solution, $\boldsymbol{s}^*_{DP}$ and $Y^{cp}_{DP}$, is set as the initial solution for this step, $\boldsymbol{s}^0_Q$ and $Y^{cp,0}$, respectively. Then, the narrow search region of the phase at the $r$th iteration is set as $R^r_i = \{s | s^{r-1}_{Q,i} - s^{tol} \leq s \leq s^{r-1}_{Q,i} + s^{tol}\}$. The data term minimum for the $i$th phase $s_{Q,i}$ is found by the Newton method within the intervals that include at least a part of the search region $R^r_i$. After the phase minimizing the data term has been obtained as $s^{data,r^*}_{Q,i}$, the $i$th data term in Eq. (17) is approximated by a Taylor expansion up to the second-order terms as

$$\hat{E}^{data,r}_i(s_{Q,i}) = E^{data,r}_i(s^{data,r^*}_{Q,i}) + \left.\frac{dE^{data,r}_i}{ds_{Q,i}}\right|_{s_{Q,i}=s^{data,r*}_{Q,i}} (s_{Q,i} - s^{data,r^*}_{Q,i})$$
$$+ \frac{1}{2}\left.\frac{d^2E^{data,r}_i}{ds^2_{Q,i}}\right|_{s_{Q,i}=s^{data,r*}_{Q,i}} (s_{Q,i} - s^{data,r^*}_{Q,i})^2. \tag{30}$$

Now, the total energy function is a quadratic form with respect to the phase $\boldsymbol{s}_Q$ and thus the optimal phase $\boldsymbol{s}_Q$ is given as

$$\boldsymbol{s}^{r\,*}_Q = \arg\min_{\boldsymbol{s}_Q} \left\{ \frac{1}{N^{in}} \sum_{i=0}^{N^{in}-1} \hat{E}^{data,r}_i(s_{Q,i}) + \lambda_s \frac{1}{N^{in}} \sum_{i=1}^{N^{in}-1} \left( s_{Q,i+1} - s_{Q,i} - \frac{1}{fP^*} \right)^2 \right\} \tag{31}$$

$$\text{s.t.} \quad s^{r-1}_{Q,i} - s^{tol} \leq s_{Q,i} \leq s^{r-1}_{Q,i} + s^{tol} \tag{32}$$
$$s_{Q,i+1} \geq s_{Q,i}, \tag{33}$$

where Eqs. (32) and (33) are the lower and upper limit constraints and the monotonically increasing constraints, respectively. As a result, the problem is formulated as a convex quadratic programming one and is solved by the active set method. Then the manifold control points at the $r$th iteration are updated in the same way as in the previous steps.

$$Y^{cp,r} = \arg\min_{Y^{cp}} E(Y^{cp}, \boldsymbol{s}^{r\,*}_Q) \tag{34}$$

These procedures are iterated until convergence by gradually relaxing the manifold smoothness constraint so that the manifold fits the data.

## 5 Experiments

### 5.1 Experimental Setup

In these experiments, the proposed temporal super resolution method is applied to low frame-rate quasi-periodic image sequences. We used two different image sequences: the first is CG data of a conical pendulum viewed from an oblique direction (Fig. 1(top)), while the other is real data in the form of a silhouette

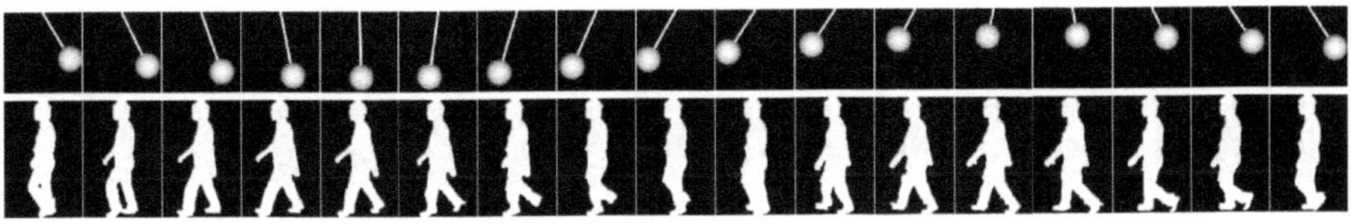

**Fig. 1.** Periodic image sequences of conical pendulum (top) and gait (bottom) used in the experiments

**Table 1.** Image sequence properties

| Image sequence | Image size [pixel] | Original frame-rate [fps] | Period [sec] |
|---|---|---|---|
| Conical pendulum | 64 × 64 | - (Arbitrary) | 1.17 |
| Gait | 88 × 128 | 60 | 1.17 |

sequence of a person walking on a treadmill (Fig. 1(bottom)). Regarding the real data, size normalization and spatial registration are done in advance. Moreover, images comprising one period are manually extracted as a subsequence and a completely periodic image sequence is constructed by repeating this subsequence. The image sizes, original frame rates, and periods of the two image sequences are listed in Table 1.

Then, low frame-rate periodic image sequences are down sampled from the original image sequences. In addition, quasi-periodic image sequences are produced by randomly changing the sampling interval up to a predefined phase noise level from the linear phase evolution. Figure 2 shows examples of quasi-periodic image sequences of the conical pendulum at 6 fps and 1 fps, respectively.

Finally, the parameters for the energy minimization framework are determined experimentally as follows. PCA is exploited for eigenspace projection and the information loss rate is set to 1%, in other words, the cumulative contribution ratios of the eigen values is 99%. The number of manifold control points is 100, and the search range for the dynamic programming $s^{cdr}$ and quadratic approximation $s^{tol}$ are 0.25 and 0.02, respectively. The smoothness term coefficients $\lambda_z$ and $\lambda_s$ are set to 50.0 and 1.0, respectively, with $\lambda_z$ reduced by half, down to a minimum of 1.0, in the iterative process of quadratic approximation.

### 5.2 Refinement Process

In this subsection, to explain the refinement process throughout the three steps, i.e., Linear Approximation (LA), Dynamic Programming (DP), and Quadratic Approximation (QA), we focus on a specific example of the conical pendulum image sequence, with frame-rate, number of frames, and phase noise set to 3 fps, 67 frames, and 20%, respectively.

First, phase estimation errors in the Ground Truth (GT) are shown in Fig. 3(a). While the estimated phases in LA deviate widely, since LA cannot absorb non-linear phase noise, those in DP and QA converge within a small range and their deviation is approximately periodic. Although phase error biases

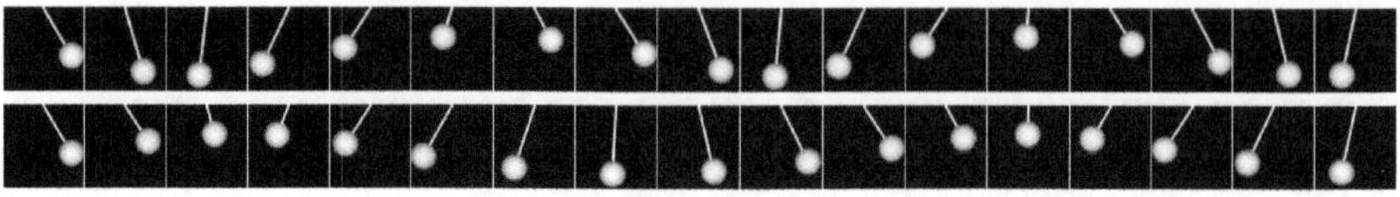

**Fig. 2.** Examples of input quasi-periodic conical pendulum image sequences with 10% phase noise at 6 fps (top) and 1 fps (bottom)

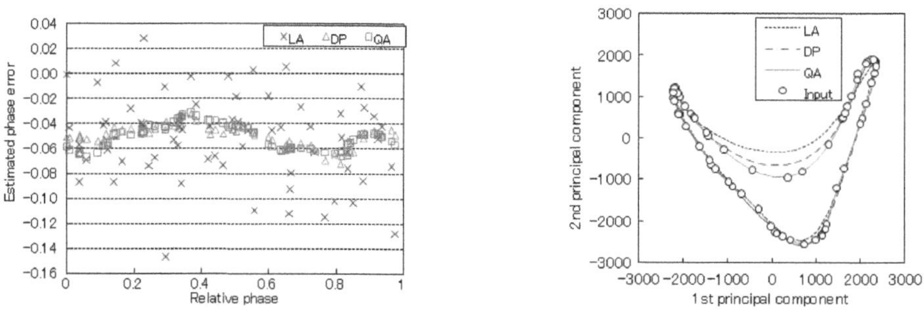


**Fig. 3.** Phase estimation error on relative phase (left) and manifold in two principal components of eigenspace (right)

from GT (approx. -0.05) are observed in DP and QA, they are not necessarily significant since phase registration up to a relative relation among all the input frames is sufficient for temporal super resolution. In other words, only the standard deviation of the phase estimation errors should be evaluated.

Next, to demonstrate the impact of the phase estimation errors in a more visual way, image sequences sorted by estimated relative phases are shown in Fig. 4. Based on these results, we can see several significant "phase backs" in LA and hence, a sufficiently large manifold smoothness coefficient $\lambda_z$ is essential for the LA step, otherwise a manifold disturbed by the phase backs is reconstructed. On the other hand, relative phase orders of DP and QA are almost consistent with that of the GT and hence, the smoothness coefficient $\lambda_z$ can be relaxed.

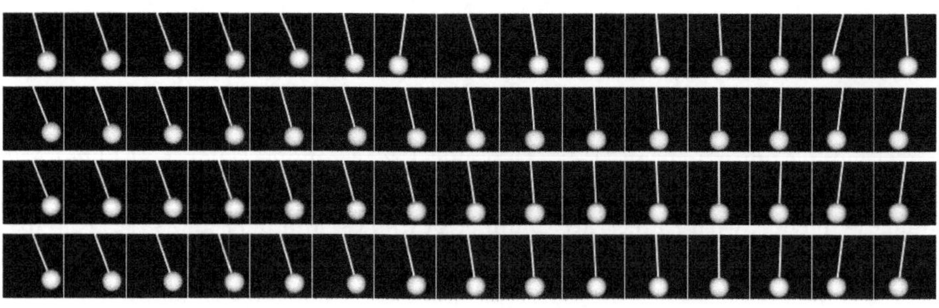

**Fig. 4.** A section of the conical pendulum image sequences sorted by estimated relative phase. Rows from the top to the bottom indicate LA, DP, QA, and GT, respectively.

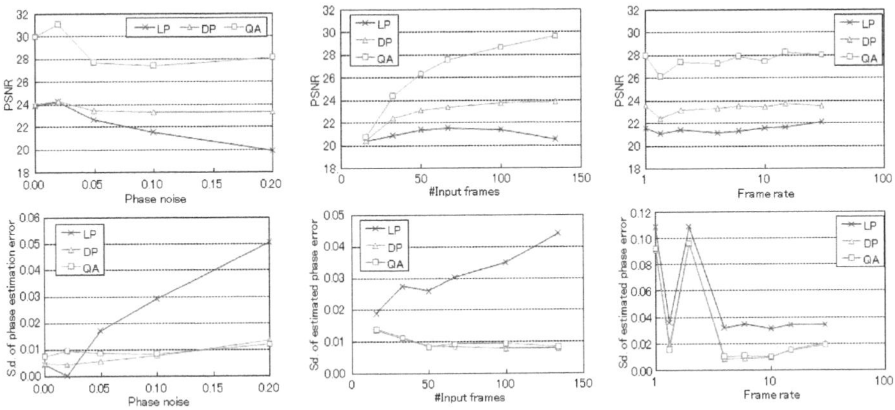


(a) Phase noise (b) # Frames used (c) Frame-rate (log scale)

**Fig. 5.** Performance evaluation of manifold PSNR (top row) and phase error SD (bottom row) using gait image sequence. The horizontal axis denotes each factor: (a) phase error, (b) number of frames used, and (c) frame-rate. The vertical axes in the top and bottom rows indicate the PSNR between the reconstructed manifold and ground truth and the SD of the phase estimation error, respectively.

Finally, manifolds in the two principal components of the eigenspace are shown in Fig. 3(b). We can see that the manifoldsf fitness to the input points improves throughout the three-step phase estimation refinement. In particular, the fitness of QA is the best, since the smoothness coefficient $\lambda_z$ is relaxed in the iteration process.

### 5.3 Evaluation

In this section, manifold reconstruction errors and phase estimation errors are evaluated. First, as for manifold reconstruction, a single period of ground truth images are initially projected to a manifold reconstructed from a low frame-rate quasi-periodic input image sequence and then they are back projected to the image domain to calculate Peak Signal to Noise Ratio (PSNR) between the ground truth images and reprojected images. As for phase estimation, errors between the ground truth phase and estimated phases of the input image sequence are calculated and then their Standard Deviation (SD) is evaluated, since their biases are not significant as discussed in the previous section.

The above two items are evaluated with respect to three factors: (1) phase noise, (2) number of frames used, and (3) frame-rate as shown in Fig. 5. The baseline settings for the three factors are 20%, 67 frames, and 3 fps, respectively.

As for the phase noise (Fig. 5(a)), although LA achieves the lowest phase error SD in the case of lower phase noise, both the manifold PSNR and the phase error SD in LA deteriorate in proportion to an increase in phase noise. On the contrary, such degradations in DP and QA are suppressed within a certain range.

Regarding the number of frames used (Fig. 5(b)), the performance naturally improves as the number of frames increases in DP and QA. On the other hand, the performance in LP does not improve in proportion to the frames used. This is because LP does not solve non-linear phase noise, and besides, an increase in the phase-error input points in the eigenspace makes the manifold reconstruction worse.

Regarding frame-rate (Fig. 5(c)), the manifold reconstruction PSNRs improve slightly as the frame-rate increases. On the other hand, critical phase errors are observed in an extremely low frame-rate (1 fps and 2 fps) and this is discussed further in the next section.

## 6 Discussion

In this section, prerequisites and limitations of the proposed temporal super resolution are discussed. In cases where a periodic image sequence is observed at a coarse sampling interval compared with its period, we need to consider two main issues: (1) the *stroboscopic effect* (*temporal aliasing*), and (2) the *wagon wheel effect* as reported in [13].

The *stroboscopic effect* typically occurs when the sampling interval coincides with the period of a moving object. In such cases, the observed image sequence appears to be standing still because the observed images are always the same even though the object is actually moving periodically.

From this observation, we can introduce a theoretical upper bound of the temporal super resolution from a single period image sequence. Intuitively speaking, temporal super resolution is impossible when each phase in one period is exactly the same as the corresponding phase in the other periods, that is, the period in the frame domain is not a sub-frame order, but merely a frame order defined as an integer. Let us consider a low frame-rate periodic image sequence and denote its period in the frame domain $P'$ [frame]. Then, assume that the period $P'$ is expressed as a fraction by relative prime numbers, $m$ and $n$,

$$P' = \frac{m}{n} \in \mathbb{Q}, \; m, n \in \mathbb{N}. \tag{35}$$

Now, if $n = 1$, the period $P'$ is degraded from a fraction to an integer, that is, the period is just a frame order and the temporal super resolution is impossible. On the other hand, if $n > 1$, the period $P'$ is a sub-frame order and hence different phase images can be observed among multiple periods and the temporal super resolution is possible. In summary, the low frame-rate image sequence can be up-converted to a $n$-times frame-rate, if $m$ input frames are given.

The *wagon wheel effect* typically occurs when the sampling interval is slightly smaller than the period of a moving object as shown in the bottom row of Fig. 2, where the sampling interval and period are 1.0 sec and 1.17 sec, respectively. In this case, the false minimum by backward play competes with the true minimum by forward play, and then the false minimum is often adopted in the LA step as shown in Fig. 6. This is the reason that the critical phase errors occur in the extremely low frame-rate in Fig. 5(c).

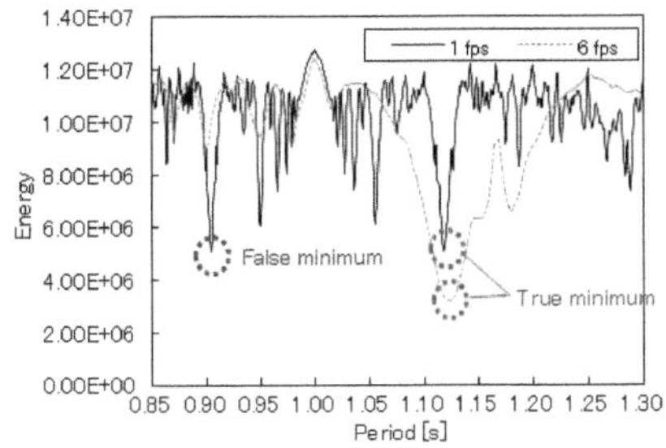


**Fig. 6.** False period detection by "Wagon wheel effects" in the LA step. The horizontal and vertical axes denote period and energy, respectively. False minimum by backward play (blue broken circle) competes with true minimum by forward play (red broken circle) at 1 fps, while true minimum is obviously the global minimum at 6 fps.

## 7 Conclusion

This paper described a method for temporal super resolution from a single quasi-periodic image sequence. The temporal super resolution was formulated as simultaneous phase registration and reconstruction of a manifold of the periodic image sequence in a phase parametric eigenspace. An energy minimization framework considering data fitness, and the smoothness of both the manifold and the phase evolution, was introduced and solved through three-step coarse-to-fine procedures to avoid local minima. Experiments using synthesized conical pendulum and real gait silhouette scenes were conducted to evaluate the effects of phase noise, the number of frames used, and frame-rate on the PSNR of the manifold reconstruction and the standard deviation of the phase estimation errors.

In this paper, we focused mainly on phase registration while ignoring blur effects, and hence, it is necessary, in future, to include the blur effects in the manifold reconstruction framework. Moreover, not only phase fluctuation, but also inter-period image deformation (e.g., view or speed transition in gait scenes) should be considered for real applications.

**Acknowledgement.** This work was supported by Grant-in-Aid for Scientific Research(S) 21220003.

## References

1. van Ouwerkerk, J.: Image super-resolution survey. Image and Vision Computing 24, 1039–1052 (2006)
2. Borman, S., Stevenson, R.: Spatial resolution enhancement of low-resolution image sequences: A comprehensive review with directions for future research. Technical report, University of Notre Dame (1998)
3. Irani, M., Peleg, S.: Improving resolution by image registration. Computer Vision, Graphics, and Image Processing 53, 231–239 (1991)
4. Tanaka, M., Okutomi, M.: A fast map-based super-resolution algorithm for general motion. In: Proc. of SPIE-IS& T Electronic Imaging 2006, Computational Imaging IV, vol. 6065, pp. 1–12 (2006)

5. Freeman, W., Jones, T., Pasztor, E.: Example-based super-resolution. IEEE Trans. on Computer Graphics and Applications 22, 56–65 (2002)
6. Liu, C., Shum, H., Zhang, C.: A two-step approach to hallucinating faces: Global parametric model and local non-parametric model. In: IEEE Int. Conf. on Computer Vision and Pattern Recognition, pp. 192–198 (2001)
7. Baker, S., Kanade, T.: Limits on super-resolution and how to break them. IEEE Trans. Pattern Anal. and Machine Intelligent 24, 1167–1183 (2002)
8. Glasner, D., Bagon, S., Irani, M.: Super-resolution from a single image. In: Proc. of the 12th Int. Conf. on Computer Vision (2009)
9. Tanaka, M., Okutomi, M.: Near-real-time video-to-video super-resolution. In: Yagi, Y., Kang, S.B., Kweon, I.S., Zha, H. (eds.) ACCV 2007, Part I. LNCS, vol. 4843, Springer, Heidelberg (2007)
10. Shimizu, M., Yoshimura, S., Tanaka, M., Okutomi, M.: Super-resolution from image sequence under influence of hot-air optical turbulence. In: Proc. of IEEE Computer Society Conference on Computer Vision and Pattern Recognition (CVPR), pp. 1–8 (2008)
11. Blake, A., Bascle, B., Zisserman, A.: Motion deblurring and super-resolution from an image sequence. In: Proc. European Conf. Computer Vision, pp. 312–320 (1996)
12. Sezan, M., Patti, A., Tekalp, A.: Superresolution video reconstruction with arbitrary sampling lattices and nonzero aperture time. IEEE Trans. Image Processing 6, 1064–1076 (1997)
13. Shechtman, E., Caspi, Y., Irani, M.: Space-time super-resolution. IEEE Transactions on Pattern Analysis and Machine Intelligence 27, 531–545 (2005)
14. Caspi, Y., Irani, M.: Spatio-temporal aignment of sequences. IEEE Trans. Pattern Analysis and Machine Intelligence 24, 1409–1425 (2002)
15. REALVIZ2: Retimer (2000), `http://www.realviz.com/products/rt`
16. Wang, Y., Ostermann, J., Zhang, Y.Q.: Video Processing and Communications. Prentice Hall, Englewood Cliffs (2002)
17. Beymer, D., Poggio, T.: Image representations for visual learning. Science 272, 1905–1909 (1996)
18. Stich, T., Magnor, M.: Image morphing for space-time interpolation. In: SIGGRAPH 2007: ACM SIGGRAPH 2007 sketches, vol. 87, ACM, New York (2007)
19. Lucas, B., Kanade, T.: An iterative image registration technique with an application to stereo vision. In: Proc. Seventh International Joint Conf. Artificial Intelligence, pp. 674–679 (1981)
20. Stich, T., Linz, C., Albuquerque, G., Magnor, M.: View and time interpolation in image space. Computer Graphics Forum (Proc. Pacific Graphics) 27 (2008)
21. Watanabe, K., Iwai, Y., Nagahara, H., Yachida, M., Suzuki, T.: Video synthesis with high spatio-temporal resolution using spectral fusion. In: Gunsel, B., Jain, A.K., Tekalp, A.M., Sankur, B. (eds.) MRCS 2006. LNCS, vol. 4105, pp. 683–690. Springer, Heidelberg (2006)
22. Taubin, G.: Estimation of planar curves, surfaces, and nonplanar space curves defined by implicit equations with applications to edge and range image segmentation. IEEE Trans. Pattern Anal. Mach. Intell. 13, 1115–1138 (1991)
23. Kanatani, K.: Ellipse fitting with hyperaccuracy. IEICE Transactions on Information and Systems E89-D, 2653–2660 (2006)
24. Murase, H., Nayar, S.K.: Parametric eigenspace representation for visual learning and recognition. In: Proc. of SPIE, vol. 2031 (1993)
25. Oka, R.: Spotting method for classification of real world data. Computer Journal 41, 559–565 (1998)

# Solving MRFs with Higher-Order Smoothness Priors Using Hierarchical Gradient Nodes

Dongjin Kwon[1], Kyong Joon Lee[1], Il Dong Yun[2], and Sang Uk Lee[1]

[1] School of EECS, ASRI, Seoul Nat'l Univ., Seoul, 151-742, Korea
[2] School of EIE, Hankuk Univ. of F. S., Yongin, 449-791, Korea
{djk,kjoon}@cvl.snu.ac.kr, yun@hufs.ac.kr, sanguk@ipl.snu.ac.kr

**Abstract.** In this paper, we propose a new method for solving the *Markov random field* (MRF) energies with higher-order smoothness priors. The main idea of the proposed method is a graph conversion which decomposes higher-order cliques as hierarchical auxiliary nodes. For a special class of smoothness priors which can be formulated as gradient-based potentials, we introduce an efficient representation of an auxiliary node called a gradient node. We denote a graph converted using gradient nodes as a *hierarchical gradient node* (HGN) graph. Given a label set $\mathcal{L}$, the computational complexity of message passings of HGN graphs are reduced to $\mathcal{O}(|\mathcal{L}|^2)$ from exponential complexity of a conventional factor graph representation. Moreover, as the HGN graph can integrate multiple orders of the smoothness priors inside its hierarchical structure, this method provides a way to combine different smoothness orders naturally in MRF frameworks. For optimizing HGN graphs, we apply the tree-reweighted (TRW) message passing which outperforms the belief propagation. In experiments, we show the efficiency of the proposed method on the 1D signal reconstructions and demonstrate the performance of the proposed method in three applications: image denoising, sub-pixel stereo matching and nonrigid image registration.

## 1 Introduction

Currently, it is popular to model computer vision problems as discrete *Markov random field* (MRF) energy optimization schemes whose solution space is defined on discrete label sets. The MRF energy is optimized using discrete optimization methods such as the belief propagation (BP) [1,2] and the graph cut (GC) [3]. This approach has been successfully applied to many important vision problems including image restoration, image segmentation, stereo matching, optical flow, image registration, etc [3,4,5,6,7,8].

However, most MRF energies for problems mentioned above use pairwise potentials which regularize first-order smoothness priors. As first-order smoothness priors are often insufficient to represent the full statistics of problems, these approaches showed limited performances. For vision problems using 1D label sets including image restoration, image segmentation and stereo matching, solutions obtained by MRFs with first order smoothness priors have fronto-parallel bias. For problems using 2D label sets including optical flow and image registration,

R. Kimmel, R. Klette, and A. Sugimoto (Eds.): ACCV 2010, Part I, LNCS 6492, pp. 121–134, 2011.

MRFs with first order smoothness priors often produce irregular displacements between neighboring nodes when object is transformed by rotation and scaling.

By these limitations of MRFs with first-order smoothness priors, incorporating higher-order priors into the MRF frameworks recently has come into the spotlight. Firstly, we introduce the GC [3] based approaches. Kohli *et al.* [9,10] proposed efficient optimization methods for a specific class of higher-order potentials: the $P^n$ Potts model [9] and the Robust $P^n$ model [10]. Rother *et al.* [11] introduced a compact representation of higher-order MRFs obtained by transforming arbitrary higher-order potentials into quadratic ones. Their method is quite suitable for specific vision problems whose energy models imply the *sparseness*. Ishikawa [12] introduced a technique reducing higher-order MRFs with binary labels to first-order ones. For the problems with multiple labels, the fusion move algorithm is applied, where the solution depends on the quality of proposals. Next, the BP [1,2] based approaches are introduced. Liao *et al.* [13] proposed a *computation template* that specifies how to propagate messages through the clique. They applied this idea for the efficient summation over large numbers of nodes[1]. Potetz [14] proposed an efficient belief propagation method using linear constraint nodes. Using this method, computational complexity of message passings increased linearly with respect to clique size[2]. Komodakis and Paragios [15] proposed an efficient higher-order MRF optimization framework on the basis of their dual decomposition method [16]. This method works more efficiently when higher-order potentials have sparse patterns.

In this paper, we focus on solving MRFs with higher-order smoothness priors. Firstly, in Section 2, preliminaries are introduced and a method composing higher-order cliques using auxiliary nodes follows in Section 3. In Section 4, a *gradient node* which is the efficient representation of an auxiliary node is described. The gradient node is applied to a special class of smoothness priors which can be formulated as gradient-based potentials. We denote a graph converted using gradient nodes as a *hierarchical gradient node* (HGN) graph. In Section 5, an efficient optimization method for the HGN graphs based on the TRW algorithm is addressed. In Section 6, we verify the efficiency of the HGN graph using signal reconstruction tests and we show the performance of the HGN graph with the TRW method in three applications: image denoising, sub-pixel stereo matching and nonrigid image registration.

## 2 Preliminaries

**Notations:** Let $\mathcal{G}_{\mathcal{E}} = (\mathcal{V}, \mathcal{E})$ be an undirected graph where $\mathcal{V}$ and $\mathcal{E}$ are a set of nodes and edges, respectively, and $\mathcal{G}_{\mathcal{F}} = (\mathcal{V}, \mathcal{F})$ be a factor graph [17] with a set

[1] The proposed method is designed for applying various optimization schemes while [13] uses the BP only. The HGN graph is specialized for higher-order smoothness priors while the purpose of [13] is summation. Also, an auxiliary node can have a unary potential in each order, so the effects of multiple-order priors can be naturally mixed as we showed in experiments.

[2] The HGN graph is a more efficient representation than factor graphs used in [14] and various optimization schemes can be applied. Also, it can integrate multiple-order smoothness priors in its hierarchical structures.

of factors $\mathcal{F}$. We denote $\mathcal{E}_{\mathcal{F}}$ is a set of factor to node edges. For each $s \in \mathcal{V}$, let $x_s$ be a label taking value in some discrete set $\mathcal{L}_s$. If we define a function $d : \mathcal{L} \rightarrow \mathbb{R}$ for mapping labels to the values of the solution space, each label $x_s$ corresponds to $d(x_s)$. For convenience, we assume the granularity of $d(x_s)$ is fixed: $d(l_a)$ - $d(l_b) = d(l_b)$ - $d(l_c)$ if $\{l_a, l_b, l_c\} \in \mathcal{L}_s$ are three consecutive labels. In next section, we will describe a factor graphs to auxiliary node graphs conversion method. In this procedure, we define an auxiliary node as a node created to replace a factor. Let $\mathcal{G}_{\mathcal{A}} = (\mathcal{V} \cup \mathcal{V}_{\mathcal{A}}, \mathcal{E}_{\mathcal{V}} \cup \mathcal{E}_{\mathcal{A}})$ be an auxiliary node graph where $\mathcal{V}_{\mathcal{A}}$ is a set of auxiliary nodes, $\mathcal{E}_{\mathcal{V}}$ is a set of edges between $\mathcal{V}$ and $\mathcal{V}_{\mathcal{A}}$, and $\mathcal{E}_{\mathcal{A}}$ is a set of edges between $\mathcal{V}_{\mathcal{A}}$ and $\mathcal{V}_{\mathcal{A}}$. Let $\overline{a}$ (a variable with over-line) be a factor node, $\underline{a}$ be an auxiliary node (a variable with under-line), and $\mathbf{x}_{\overline{a}}$ (or $\mathbf{x}_{\underline{a}}$) be a vector of labels for nodes connected with a factor $\overline{a}$. Similarly, we define $\mathcal{V}_{\overline{a}}$ as a set of nodes connected to $\overline{a}$ and $\mathcal{V}_{\underline{a}}$ as a set of child nodes of $\underline{a}$.

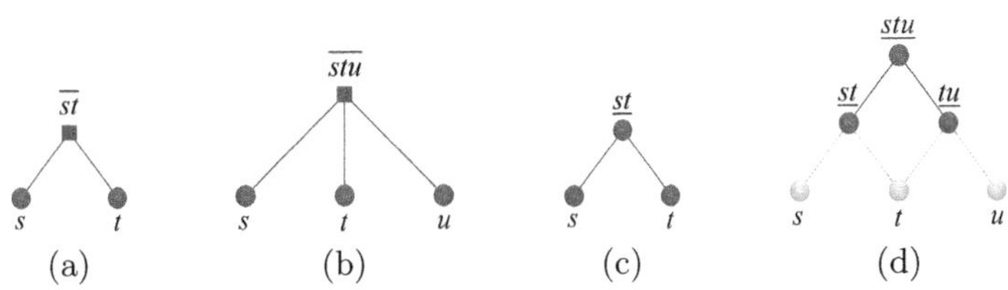


**Fig. 1.** Examples of factor graphs (a) (pairwise) and (b) (ternary) and their corresponding hierarchical auxiliary node graphs (c) and (d), respectively

**MRF Energy Model:** In this paper, we assume a set of nodes $\mathcal{V}$ of a graph $\mathcal{G}_{\mathcal{F}}$ or $\mathcal{G}_{\mathcal{A}}$ is placed on a grid. For image denoising or stereo matching, the grid corresponds to pixels, and for image registration, the grid is defined as a mesh with a constant spacing. Given a graph, following MRF energy model is applied[3]:

$$E(x|\theta) = \sum\nolimits_{s \in \mathcal{V}} \theta_s(x_s) + \sum\nolimits_{\overline{a} \in \mathcal{F}} \theta_{\overline{a}}(\mathbf{x}_{\overline{a}}) \; . \tag{1}$$

In this model, a data cost $\theta_s$ and a smoothness prior $\theta_{\overline{a}}$ are defined according to applications.

## 3 Composing Higher-Order Cliques Using Hierarchical Auxiliary Nodes

In this section, we introduce how to compose higher-order cliques using high-dimensional auxiliary nodes. Firstly, we describe a pairwise clique case and generalize the concept to higher-order cliques in the next. Let us consider a factor $\overline{st}$ which has a pairwise potential $\theta_{\overline{st}}(x_s, x_t)$. The conversion is simply done by replacing a factor node $\overline{st}$ with an auxiliary variable node $\underline{st}$. (See Fig.1(a) and 1(c).) We associate a new label $z^{\underline{st}} \in \mathcal{Z}_{\underline{st}}$ with the auxiliary node $\underline{st}$ which replaces the factor $\overline{st}$ where $\mathcal{Z}_{\underline{st}}$ is Cartesian product of label spaces of connected

[3] We separate a factor node connected with a single node from $\mathcal{F}$, and treat it explicitly as a data cost.

two nodes $s$ and $t$ ($\mathcal{Z}_{\underline{st}} = \mathcal{L}_s \times \mathcal{L}_t$). Then unary and pairwise potentials of this converted graph are described as follows [1,18]

$$\psi_{\underline{st}}(z^{\underline{st}}) = \theta_{\overline{st}}(z_s^{\underline{st}}, z_t^{\underline{st}}) \;, \quad \psi_s(x_s) = \psi_t(x_t) = 0 \;, \tag{2}$$

$$\psi_{\underline{st}i}(z^{\underline{st}}, x_i) = \begin{cases} 0 & \text{if } z_i^{\underline{st}} = x_i \\ \infty & \text{otherwise} \end{cases} \quad \forall i \in \{s,t\} \tag{3}$$

where any possible value $z^{\underline{st}}$ has one-to-one correspondence with a pair $(z_s^{\underline{st}}, z_t^{\underline{st}})$ if $z_s^{\underline{st}} \in \mathcal{L}_s$ and $z_t^{\underline{st}} \in \mathcal{L}_t$. Then the pairwise potential is converted to the sum of these pairwise interactions:

$$\psi_{\underline{st}}(z^{\underline{st}}) + \sum\nolimits_{i\in\{s,t\}} \left(\psi_i(x_i) + \psi_{\underline{st}i}(z^{\underline{st}}, x_i)\right) \,. \tag{4}$$

This augmented distribution is marginalized down to $\theta_{\overline{st}}(x_s, x_t)$.

The factors connecting more than 2 nodes are converted using auxiliary nodes hierarchically. We construct a hierarchical auxiliary node graph by gathering two neighboring nodes recursively. If we denote $\underline{a}$ is a parent auxiliary node of $\underline{b}$ and $\underline{c}$, relations as below are held in this hierarchy:

$$\mathcal{V}_{\underline{st}} = \{s,t\} \;, \quad \mathcal{V}_{\underline{a}} = \mathcal{V}_{\underline{b}} \cup \mathcal{V}_{\underline{c}} \;. \tag{5}$$

We associate a new label $z^{\underline{a}} \in \mathcal{Z}_{\underline{a}}$ with an auxiliary node $\underline{a}$ where $\mathcal{Z}_{\underline{a}}$ is Cartesian product of label spaces for the node set $\mathcal{V}_{\underline{a}}$. The labels $z^{\underline{b}}$ and $z^{\underline{c}}$ are defined as well. Then unary and pairwise potentials of converted graph are described as follows

$$\psi_{\underline{i}}(z^{\underline{i}}) = \begin{cases} \theta_{\overline{i}}(\mathbf{z}^{\underline{i}}) & \text{if } \underline{i} \text{ has no parent} \\ 0 & \text{otherwise} \end{cases} \quad \forall \underline{i} \in \{\underline{a}, \underline{b}, \underline{c}\} \;, \tag{6}$$

$$\psi_{\underline{ai}}(z^{\underline{a}}, z^{\underline{i}}) = \begin{cases} 0 & \text{if } z_k^{\underline{a}} = z_k^{\underline{i}} \quad \forall k \in \mathcal{V}_{\underline{a}} \cap \mathcal{V}_{\underline{i}} \\ \infty & \text{otherwise} \end{cases} \quad \forall \underline{i} \in \{\underline{b}, \underline{c}\} \tag{7}$$

where $\theta_{\overline{i}}$ is a potential assigned to a factor $\overline{i}$ and $\mathbf{z}^{\underline{i}}$ is a $|\mathcal{V}_{\underline{i}}|$-dimensional vector satisfying $\mathbf{z}_k^{\underline{i}} = z_k^{\underline{i}} \; \forall k \in \mathcal{V}_{\underline{i}}$.

For example, we describe a conversion procedure for triple cliques. Let us consider a factor $\overline{stu}$ which represents a ternary potential $\theta_{\overline{stu}}(x_s, x_t, x_u)$. For the conversion, we replace a factor node $\overline{stu}$ with an auxiliary variable node $\underline{stu}$ connected with pairwise auxiliary nodes $\underline{st}$ and $\underline{tu}$. (See Fig.1(b) and 1(d).) The unary and pairwise potentials of converted graph are defined as (6) and (7) with following conditions:

$$\underline{a} = \underline{stu}, \quad \underline{b} = \underline{st}, \quad \underline{c} = \underline{tu} \;, \quad \psi_{\underline{a}}(z^{\underline{a}}) = \theta_{\overline{stu}}(z_s^{\underline{a}}, z_t^{\underline{a}}, z_u^{\underline{a}}) \;. \tag{8}$$

Then the ternary potential is converted to the sum of these pairwise interactions:

$$\psi_{\underline{a}}(z^{\underline{a}}) + \sum_{\underline{i}\in\{\underline{b},\underline{c}\}} \left\{ \psi_{\underline{i}}(z^{\underline{i}}) + \psi_{\underline{ai}}(z^{\underline{a}}, z^{\underline{i}}) + \sum_{k\in\mathcal{V}_{\underline{i}}} \left(\psi_k(x_k) + \psi_{\underline{i}k}(z^{\underline{i}}, x_k)\right) \right\} . \tag{9}$$

This augmented distribution is marginalized down to $\theta_{\overline{stu}}(x_s, x_t, x_u)$.

# 4 Gradient Nodes: Reducing Complexity of Auxiliary Nodes

On general clique potentials, the conversion procedure of Section 3 does not have much benefit. Thus, we focus on special types of clique potentials which reduce the complexity of auxiliary nodes: gradient-based potentials. We define gradient-based potentials as a set of functions whose variables can be reduced as a $n^{th}$-order derivative forms. We describe gradient-based potentials up to $3^{rd}$-order derivatives as follows:

$$\theta_{\overline{st}}(x_s, x_t) = \lambda_{\overline{st}} g(d(x_s) - d(x_t)) , \tag{10}$$
$$\theta_{\overline{stu}}(x_s, x_t, x_u) = \lambda_{\overline{stu}} g(d(x_s) - 2d(x_t) + d(x_u)) , \tag{11}$$
$$\theta_{\overline{stuv}}(x_s, x_t, x_u, x_v) = \lambda_{\overline{stuv}} g(d(x_s) - 3d(x_t) + 3d(x_u) - d(x_v)) , \tag{12}$$
$$\theta_{\overline{pqrstu}}(x_p, x_q, x_r, x_s, x_t, x_u) = \lambda_{\overline{pqrstu}} g(d(x_p) - 2d(x_q) + d(x_r) - d(x_s) + 2d(x_t) - d(x_u)) \tag{13}$$

where $\lambda_{\overline{a}}$ is the regularization constant and $g(\cdot)$ is a prior distribution function. We assume $s$, $t$, $u$, $v$ or $p$, $q$, $r$ are consecutive nodes on the x or y direction of a grid. For example, nodes $\{s, t, u\}$ can be located $\{(x, y), (x+1, y), (x+2, y)\}$ or $\{(x, y), (x, y+1), (x, y+2)\}$ coordinates of the grid. Among gradient-based potentials mentioned above, (10) is used for first-order smoothness priors, (11) is used for second-order smoothness priors, and (12) and (13) are used for third-order smoothness priors.

When clique potentials are gradient-based forms, the size of the label set of auxiliary nodes can be reduced significantly. To demonstrate that, we associate a new label $y$ with an auxiliary node as follows:

$$d(y^{\underline{st}}) = d(x_s) - d(x_t) , \quad d(y^{\underline{a}}) = d(y^{\underline{b}}) - d(y^{\underline{c}}) \tag{14}$$

where $\underline{a}$ is a parent auxiliary node of $\underline{b}$ and $\underline{c}$ as before, and $y^{\underline{st}} \in \mathcal{Y}^{\underline{st}}$ and $y^{\underline{a}} \in \mathcal{Y}^{\underline{a}}$ correspond to one distinct value of a set $\{d(x_s) - d(x_t) | x_s \in \mathcal{L}_s \wedge x_t \in \mathcal{L}_t\}$ and $\{d(y^{\underline{b}}) - d(y^{\underline{c}}) | y^{\underline{b}} \in \mathcal{Y}^{\underline{b}} \wedge y^{\underline{c}} \in \mathcal{Y}^{\underline{c}}\}$, respectively. For example, if both $d(y^{\underline{b}})$ and $d(y^{\underline{c}})$ are elements in the same set $\{-L, -L+1, \dots, L-1, L\}$, then $d(y^{\underline{a}})$ is an element in a set $\{-2L, -2L+1, \dots, 2L-1, 2L\}$.

We call an auxiliary node with a label $y$ as a *gradient node* because $y$ represents first-order derivative of labels of its two child nodes. And we denote a converted graph using gradient nodes as a *hierarchical gradient node* (HGN) graph. The size of label sets of gradient nodes which are defined on (14) are follows:

$$|\mathcal{Y}^{\underline{st}}| = |\mathcal{L}_s| + |\mathcal{L}_t| - 1 , \quad |\mathcal{Y}^{\underline{a}}| = |\mathcal{Y}^{\underline{b}}| + |\mathcal{Y}^{\underline{c}}| - 1 . \tag{15}$$

Compared to the size of label sets of auxiliary nodes ($|\mathcal{Z}^{\underline{st}}| = |\mathcal{L}_s| \cdot |\mathcal{L}_t|$ and $|\mathcal{Z}^{\underline{a}}| = |\mathcal{Z}^{\underline{b}}| \cdot |\mathcal{Z}^{\underline{c}}|$), the complexity of gradient node is reduced from $\mathcal{O}(|\mathcal{L}|^{|\mathcal{V}_{\underline{a}}|})$ to $\mathcal{O}(|\mathcal{L}|)$.

However, if we reduce a label space of parent nodes, whenever we fix labels of two child nodes, many feasible labels of the parent node can be selected though we want an unique label of the parent node as specified in (14). It is not

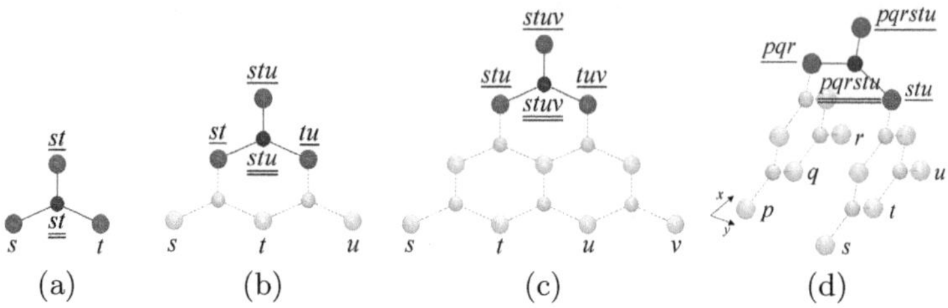


**Fig. 2.** Examples of HGN graphs. (a) and (b) are correspond to hierarchical auxiliary node graphs (c) and (d) in Fig. 1, respectively. (c) and (d) are HGN graphs for third-order priors.

possible to design proper pairwise potentials between parent and child nodes in this case. To eliminate this problem, we add an intermediate node $\underline{\underline{a}}$ between the parent node $\underline{a}$ and child nodes $\underline{b}$ and $\underline{c}$. We associate a label $z^{\underline{\underline{a}}} \in \mathcal{Z}^{\underline{\underline{a}}}$ with an intermediate node $\underline{\underline{a}}$ where $\mathcal{Z}^{\underline{\underline{a}}}$ is Cartesian product of label spaces of child nodes $\underline{b}$ and $\underline{c}$ ($\mathcal{Z}^{\underline{\underline{a}}} = \mathcal{Y}^{\underline{b}} \times \mathcal{Y}^{\underline{c}}$). In $\mathcal{Z}^{\underline{\underline{a}}}$, any possible value $z^{\underline{\underline{a}}}$ has one-to-one correspondence with a pair $(z_{\underline{b}}^{\underline{\underline{a}}}, z_{\underline{c}}^{\underline{\underline{a}}})$ if $z_{\underline{b}}^{\underline{\underline{a}}} \in \mathcal{Y}^{\underline{b}}$ and $z_{\underline{c}}^{\underline{\underline{a}}} \in \mathcal{Y}^{\underline{c}}$. In Fig.2, HGN graphs constructed with intermediate nodes are shown. In the figure, (a), (b) and (c)-(d) correspond to HGN graphs for first, second and third-order priors, respectively.

For first-order clique potentials, we define unary and pairwise potentials of the converted graph as follows:

$$\psi_{\underline{st}}(y^{\underline{st}}) = \lambda_{\overline{st}} g(d(y^{\underline{st}})), \quad \psi_s(x_s) = \psi_t(x_t) = 0 , \quad \psi_{\underline{\underline{st}}}(z^{\underline{\underline{st}}}) = 0 , \tag{16}$$

$$\psi_{\underline{\underline{st}}i}(z^{\underline{\underline{st}}}, x_i) = \begin{cases} 0 & \text{if } z_i^{\underline{\underline{st}}} = x_i \\ \infty & \text{otherwise} \end{cases} \quad \forall i \in \{s, t\} , \tag{17}$$

$$\psi_{\underline{st}\underline{\underline{st}}}(y^{\underline{st}}, z^{\underline{\underline{st}}}) = \begin{cases} 0 & \text{if } d(y^{\underline{st}}) = d(z_s^{\underline{\underline{st}}}) - d(z_t^{\underline{\underline{st}}}) \\ \infty & \text{otherwise} \end{cases} . \tag{18}$$

For higher-order cliques potentials, we define following unary and pairwise potentials for the converted graph as follows:

$$\psi_{\underline{i}}(y^{\underline{i}}) = \begin{cases} \lambda_{\overline{i}} g(d(y^{\underline{i}})) & \text{if } \underline{i} \text{ has no parent} \\ 0 & \text{otherwise} \end{cases} \quad \forall \underline{i} \in \{\underline{a}, \underline{b}, \underline{c}\} , \quad \psi_{\underline{\underline{a}}}(z^{\underline{\underline{a}}}) = 0 , \tag{19}$$

$$\psi_{\underline{\underline{a}}\underline{i}}(z^{\underline{\underline{a}}}, y^{\underline{i}}) = \begin{cases} 0 & \text{if } z_{\underline{i}}^{\underline{\underline{a}}} = y^{\underline{i}} \\ \infty & \text{otherwise} \end{cases} \quad \forall \underline{i} \in \{\underline{b}, \underline{c}\} , \tag{20}$$

$$\psi_{\underline{a}\underline{\underline{a}}}(y^{\underline{a}}, z^{\underline{\underline{a}}}) = \begin{cases} 0 & \text{if } d(y^{\underline{a}}) = d(z_{\underline{b}}^{\underline{\underline{a}}}) - d(z_{\underline{c}}^{\underline{\underline{a}}}) \\ \infty & \text{otherwise} \end{cases} . \tag{21}$$

To describe the entire energy model, we denote $\mathcal{V}_{\mathcal{I}}$ as a set of intermediate nodes, $\mathcal{E}_{\mathcal{IA}}$ as a set of edges between intermediate nodes and their child auxiliary nodes, and $\mathcal{E}_{\mathcal{AI}}$ as a set of edges between intermediate nodes and their parent nodes. Then, using conversion formulas mentioned above, the energy model (1) is converted as follows:

$$E(x,y,z|\theta,\psi) = \sum_{s\in\mathcal{V}} \theta_s(x_s) + \sum_{\underline{a}\in\mathcal{V}_\mathcal{A}} \psi_{\underline{a}}(y^{\underline{a}}) + \sum_{\underline{\underline{a}}\in\mathcal{V}_\mathcal{I}} \psi_{\underline{\underline{a}}}(z^{\underline{\underline{a}}}) + \sum_{(\underline{\underline{st}},i)\in\mathcal{E}_\mathcal{V}} \psi_{\underline{\underline{st}}i}(z^{\underline{\underline{st}}}, x_i)$$
$$+ \sum_{(\underline{\underline{a}},\underline{i})\in\mathcal{E}_{\mathcal{IA}}} \psi_{\underline{\underline{a}}\underline{i}}(z^{\underline{\underline{a}}}, y^{\underline{i}}) + \sum_{(\underline{a},\underline{\underline{a}})\in\mathcal{E}_{\mathcal{AI}}} \psi_{\underline{a}\underline{\underline{a}}}(y^{\underline{a}}, z^{\underline{\underline{a}}}) \,. \quad (22)$$

This augmented distribution is marginalized down to (1).

**Mixed-Order Smoothness Priors:** In the HGN graph structures, multiple orders of the smoothness priors can be integrated into the hierarchy. To this purpose, (19) is replaced by following node definition for the converted graph

$$\psi_{\underline{i}}(y^{\underline{i}}) = \begin{cases} \lambda_{\overline{i}}\, g(d(y^{\underline{i}})) & \exists \overline{i} \\ 0 & \text{otherwise} \end{cases} \quad \forall \underline{i} \in \{\underline{a}, \underline{b}, \underline{c}\} \,. \quad (23)$$

In the factor graph structures, we have to add factor-to-node edges whenever factors are added into the graph. However, in the gradient node graph structures, we do not need to add edges for adding lower-order smoothness priors when the graph is constructed for the higher-order smoothness priors. Moreover, HGN graphs have structural efficiency as intermediate edges are shared to represent neighboring smoothness priors.

# 5 Optimization

## 5.1 Max-product BP on Factor Graphs and HGN Graphs

To optimize energy models described in this paper, the max-product belief propagation (BP) can be used. The message update equations for (1) are as follows [2]

$$m_{\overline{a}\to s}(x_s) = \min_{\mathcal{V}_{\overline{a}}\setminus s} \Big\{ \theta_{\overline{a}}(\mathbf{x}_{\overline{a}}) + \sum_{t\in N(\overline{a})\setminus s} n_{t\to\overline{a}}(x_t) \Big\}, \quad n_{s\to\overline{a}}(x_s) = \theta_s(x_s) + \sum_{\overline{c}\in N(s)\setminus\overline{a}} m_{\overline{c}\to s}(x_s) \quad (24)$$

where $N(s)$ is a set of neighboring factors for a node $s$ and $N(\overline{a})$ is a set of neighboring nodes for a factor $\overline{a}$. The complexity of calculating $m_{\overline{a}\to s}$ is $\mathcal{O}(|\mathcal{L}|^{|\mathcal{V}_{\overline{a}}|})$. The complexity is increased exponentially with a number of nodes $\mathcal{V}_{\overline{a}}$ connected with a factor $\overline{a}$.

The message update equations for (22) are as follows

$$m_{i\to\underline{\underline{st}}}(z^{\underline{\underline{st}}}) = \min_{x_i} \Big\{ \theta_i(x_i) + \sum\nolimits_{\underline{\underline{j}}\in N(i)\setminus\underline{\underline{st}}} m_{\underline{\underline{j}}\to i}(x_i) + \psi_{\underline{\underline{st}}i}(z^{\underline{\underline{st}}}, x_i) \Big\} \quad \forall i \in \{s,t\} \,, \quad (25)$$

$$m_{\underline{\underline{st}}\to i}(x_i) = \min_{z^{\underline{\underline{st}}}} \Big\{ \psi_{\underline{\underline{st}}}(z^{\underline{\underline{st}}}) + \sum\nolimits_{j\in N(\underline{\underline{st}})\setminus i} m_{j\to\underline{\underline{st}}}(z^{\underline{\underline{st}}}) + \psi_{\underline{\underline{st}}i}(z^{\underline{\underline{st}}}, x_i) \Big\} \quad \forall i \in \{s,t\} \,, \quad (26)$$

$$m_{\underline{\underline{a}}\to\underline{a}}(y^{\underline{a}}) = \min_{z^{\underline{\underline{a}}}} \Big\{ \psi_{\underline{\underline{a}}}(z^{\underline{\underline{a}}}) + \sum\nolimits_{\underline{j}\in N(\underline{\underline{a}})\setminus\underline{a}} m_{\underline{j}\to\underline{\underline{a}}}(z^{\underline{\underline{a}}}) + \psi_{\underline{a}\underline{\underline{a}}}(y^{\underline{a}}, z^{\underline{\underline{a}}}) \Big\} \,, \quad (27)$$

$$m_{\underline{a}\to\underline{\underline{a}}}(z^{\underline{\underline{a}}}) = \min_{y^{\underline{a}}} \Big\{ \psi_{\underline{a}}(y^{\underline{a}}) + \sum\nolimits_{\underline{\underline{j}}\in N(\underline{a})\setminus\underline{\underline{a}}} m_{\underline{\underline{j}}\to\underline{a}}(y^{\underline{a}}) + \psi_{\underline{a}\underline{\underline{a}}}(y^{\underline{a}}, z^{\underline{\underline{a}}}) \Big\} \,, \quad (28)$$

$$m_{\underline{i}\to\underline{\underline{a}}}(z^{\underline{\underline{a}}}) = \min_{y^{\underline{i}}} \Big\{ \psi_{\underline{i}}(y^{\underline{i}}) + \sum\nolimits_{\underline{\underline{j}}\in N(\underline{i})\setminus\underline{\underline{a}}} m_{\underline{\underline{j}}\to\underline{i}}(y^{\underline{i}}) + \psi_{\underline{\underline{a}}\underline{i}}(z^{\underline{\underline{a}}}, y^{\underline{i}}) \Big\} \quad \forall \underline{i} \in \{\underline{b}, \underline{c}\} \,, \quad (29)$$

$$m_{\underline{\underline{a}}\to\underline{i}}(y^{\underline{i}}) = \min_{z^{\underline{\underline{a}}}} \Big\{ \psi_{\underline{\underline{a}}}(z^{\underline{\underline{a}}}) + \sum\nolimits_{\underline{j}\in N(\underline{\underline{a}})\setminus\underline{i}} m_{\underline{j}\to\underline{\underline{a}}}(z^{\underline{\underline{a}}}) + \psi_{\underline{\underline{a}}\underline{i}}(z^{\underline{\underline{a}}}, y^{\underline{i}}) \Big\} \quad \forall \underline{i} \in \{\underline{b}, \underline{c}\} \quad (30)$$

where $N(i)$ is a set of neighboring nodes for a node $i$ and $\underline{a}$ is a parent auxiliary node of $\underline{b}$ and $\underline{c}$ as before. As pairwise potentials (17), (18), (20) and (21) add

**Table 1.** Complexity of Message Passings for Factor Graph and HGN Graph

| Smoothness Prior | Factor Graph | | HGN Graph | | | |
|---|---|---|---|---|---|---|
| | $\lvert\mathcal{E}_{\mathcal{F}}\rvert$ | $\lvert m_{\mathcal{F}\to\mathcal{V}}\rvert$ | $\lvert\mathcal{E}_{\mathcal{V}}\rvert+\lvert\mathcal{E}_{\mathcal{IA}}\rvert$ | $\lvert m_{\mathcal{V}_{\mathcal{I}}\leftrightarrow\mathcal{V}}\rvert, \lvert m_{\mathcal{V}_{\mathcal{I}}\leftrightarrow\mathcal{V}_{\mathcal{A}}}\rvert$ | $\lvert\mathcal{E}_{\mathcal{AI}}\rvert$ | $\lvert m_{\mathcal{V}_{\mathcal{A}}\leftrightarrow\mathcal{V}_{\mathcal{I}}}\rvert$ |
| First-Order | $4\lvert\mathcal{V}\rvert$ | $\mathcal{O}(\lvert\mathcal{L}\rvert^2)$ | $4\lvert\mathcal{V}\rvert+0$ | $\mathcal{O}(\lvert\mathcal{L}\rvert^2)$ | $2\lvert\mathcal{V}\rvert$ | $\mathcal{O}(\lvert\mathcal{L}\rvert^2)$ |
| Second-Order | $6\lvert\mathcal{V}\rvert$ | $\mathcal{O}(\lvert\mathcal{L}\rvert^3)$ | $4\lvert\mathcal{V}\rvert+4\lvert\mathcal{V}\rvert$ | $\mathcal{O}(\lvert\mathcal{L}\rvert^2)$ | $4\lvert\mathcal{V}\rvert$ | $\mathcal{O}(\lvert\mathcal{L}\rvert^2)$ |
| Third-Order | $20\lvert\mathcal{V}\rvert$ | $\mathcal{O}(\lvert\mathcal{L}\rvert^4)+\mathcal{O}(\lvert\mathcal{L}\rvert^6)$ | $4\lvert\mathcal{V}\rvert+12\lvert\mathcal{V}\rvert$ | $\mathcal{O}(\lvert\mathcal{L}\rvert^2)$ | $10\lvert\mathcal{V}\rvert$ | $\mathcal{O}(\lvert\mathcal{L}\rvert^2)$ |

nothing to the messages, we can logically discard all possible combinations of $(z, y)$ when $\psi = \infty$ for $(z, y)$. Using this information, the efficient representations of (25)-(30) are constructed as follows

$$m_{i\to\underline{\underline{st}}}(z^{\underline{\underline{st}}}) = \theta_i(z_i^{\underline{\underline{st}}}) + \sum\nolimits_{\underline{\underline{j}}\in N(i)\setminus\underline{\underline{st}}} m_{\underline{\underline{j}}\to i}(z_i^{\underline{\underline{st}}}) \quad \forall i \in \{s,t\}\ , \tag{31}$$

$$m_{\underline{\underline{st}}\to i}(x_i) = \min_{\{z^{\underline{\underline{st}}} \mid z_i^{\underline{\underline{st}}} = x_i\}} \left\{\sum\nolimits_{j\in N(\underline{\underline{st}})\setminus i} m_{j\to\underline{\underline{st}}}(z^{\underline{\underline{st}}})\right\} \quad \forall i \in \{s,t\}\ , \tag{32}$$

$$m_{\underline{\underline{a}}\to\underline{a}}(y^{\underline{a}}) = \min_{\{z^{\underline{\underline{a}}} \mid d(z_{\underline{b}}^{\underline{\underline{a}}}) - d(z_{\underline{c}}^{\underline{\underline{a}}}) = d(y^{\underline{a}})\}} \left\{m_{\underline{b}\to\underline{\underline{a}}}(z^{\underline{\underline{a}}}) + m_{\underline{c}\to\underline{\underline{a}}}(z^{\underline{\underline{a}}})\right\}\ , \tag{33}$$

$$m_{\underline{a}\to\underline{\underline{a}}}(z^{\underline{\underline{a}}}) = \psi_{\underline{a}}(y^{\underline{a}}) + \sum\nolimits_{\underline{\underline{j}}\in N(\underline{a})\setminus\underline{\underline{a}}} m_{\underline{\underline{j}}\to\underline{a}}(y^{\underline{a}}) \quad \text{where} \quad d(y^{\underline{a}}) = d(z_{\underline{b}}^{\underline{\underline{a}}}) - d(z_{\underline{c}}^{\underline{\underline{a}}})\ , \tag{34}$$

$$m_{\underline{i}\to\underline{\underline{a}}}(z^{\underline{\underline{a}}}) = \psi_{\underline{i}}(z_{\underline{i}}^{\underline{\underline{a}}}) + \sum\nolimits_{\underline{\underline{j}}\in N(\underline{i})\setminus\underline{\underline{a}}} m_{\underline{\underline{j}}\to\underline{i}}(z_{\underline{i}}^{\underline{\underline{a}}}) \quad \forall \underline{i} \in \{\underline{b}, \underline{c}\}\ , \tag{35}$$

$$m_{\underline{\underline{a}}\to\underline{i}}(y^{\underline{i}}) = \min_{\{z^{\underline{\underline{a}}} \mid z_{\underline{i}}^{\underline{\underline{a}}} = y^{\underline{i}}\}} \left\{\sum\nolimits_{\underline{j}\in N(\underline{\underline{a}})\setminus\underline{i}} m_{\underline{j}\to\underline{\underline{a}}}(z^{\underline{\underline{a}}})\right\} \quad \forall \underline{i} \in \{\underline{b}, \underline{c}\}\ . \tag{36}$$

Considering the number of labels of $y$ is $\mathcal{O}(|\mathcal{L}|)$, the complexity of calculating messages (31)-(36) are all $\mathcal{O}(|\mathcal{L}|^2)$. We summarize the message passing complexity for 2D grid graphs in Table 1. In the table, $|\mathcal{E}|$ is a total number of edges $\mathcal{E}$, $m_{\mathcal{A}\to\mathcal{B}}$ represents messages from $\mathcal{A}$ to $\mathcal{B}$, $m_{\mathcal{A}\leftrightarrow\mathcal{B}}$ represents messages between $\mathcal{A}$ and $\mathcal{B}$, and $|m|$ is the time complexity of computing messages $m$.

### 5.2 TRW Message Passing on HGN Graphs

The major problems of BP on the higher-order factor graphs are that message update scheduling is not easy and the convergence is not guaranteed. It is more difficult to obtain the convergence on the HGN energy (22) using BP as HGN adds many tight loops after the conversion procedure.

To achieve better optimization performances, we apply tree-reweighted (TRW) message passing [18,19] which is used for minimizing the MRF energy (22) of HGN graphs. In a recent comparative study, TRW shows the state-of-the-art performances among various discrete optimization methods [20]. The TRW method decomposes a graph into trees in which an exact MAP can be calculated by max-product belief propagation (BP) on each tree. In this scheme, a lower bound for the energy is calculated from the energies of each tree. This lower bound can then be used as a measure of convergence. In particular, the sequential TRW (TRW-S) [19] adjusts the message updating schedule and yields a lower bound guaranteed not to decrease. Thus, this method is preferable to conventional BP methods, in which scheduling is heuristic and convergence is unknown.

The algorithm works by message passings which are similar to BP. The message update equations for the TRW which correspond to (25)-(28) are described as follows (we dropped some arguments for simplicity.)

$$m_{i\rightarrow \underline{\underline{st}}}(z^{\underline{\underline{st}}}) = \min_{x_i}\Big\{ \gamma_i\Big(\theta_i + \sum\nolimits_{\underline{\underline{j}}\in N(i)} m_{\underline{\underline{j}}\rightarrow i}\Big) - m_{\underline{\underline{st}}\rightarrow i} + \psi_{\underline{\underline{st}}i}(z^{\underline{\underline{st}}}, x_i)\Big\} , \tag{37}$$

$$m_{\underline{\underline{st}}\rightarrow i}(x_i) = \min_{z^{\underline{\underline{st}}}}\Big\{ \gamma_{\underline{\underline{st}}}\Big(\psi_{\underline{\underline{st}}} + \sum\nolimits_{j\in N(\underline{\underline{st}})} m_{j\rightarrow \underline{\underline{st}}}\Big) - m_{i\rightarrow \underline{\underline{st}}} + \psi_{\underline{\underline{st}}i}(z^{\underline{\underline{st}}}, x_i)\Big\} , \tag{38}$$

$$m_{\underline{\underline{a}}\rightarrow \underline{a}}(y^{\underline{a}}) = \min_{z^{\underline{\underline{a}}}}\Big\{ \gamma_{\underline{\underline{a}}}\Big(\psi_{\underline{\underline{a}}} + \sum\nolimits_{\underline{j}\in N(\underline{\underline{a}})} m_{\underline{j}\rightarrow \underline{\underline{a}}}\Big) - m_{\underline{a}\rightarrow \underline{\underline{a}}} + \psi_{\underline{a}\underline{\underline{a}}}(y^{\underline{a}}, z^{\underline{\underline{a}}})\Big\} , \tag{39}$$

$$m_{\underline{a}\rightarrow \underline{\underline{a}}}(z^{\underline{\underline{a}}}) = \min_{y^{\underline{a}}}\Big\{ \gamma_{\underline{a}}\Big(\psi_{\underline{a}} + \sum\nolimits_{\underline{\underline{j}}\in N(\underline{a})} m_{\underline{\underline{j}}\rightarrow \underline{a}}\Big) - m_{\underline{\underline{a}}\rightarrow \underline{a}} + \psi_{\underline{a}\underline{\underline{a}}}(y^{\underline{a}}, z^{\underline{\underline{a}}})\Big\} , \tag{40}$$

$$m_{\underline{i}\rightarrow \underline{\underline{a}}}(z^{\underline{\underline{a}}}) = \min_{y^{\underline{i}}}\Big\{ \gamma_{\underline{i}}\Big(\psi_{\underline{i}} + \sum\nolimits_{\underline{\underline{j}}\in N(\underline{i})} m_{\underline{\underline{j}}\rightarrow \underline{i}}\Big) - m_{\underline{\underline{a}}\rightarrow \underline{i}} + \psi_{\underline{\underline{a}}\underline{i}}(z^{\underline{\underline{a}}}, y^{\underline{i}})\Big\} , \tag{41}$$

$$m_{\underline{\underline{a}}\rightarrow \underline{i}}(y^{\underline{i}}) = \min_{z^{\underline{\underline{a}}}}\Big\{ \gamma_{\underline{\underline{a}}}\Big(\psi_{\underline{\underline{a}}} + \sum_{\underline{j}\in N(\underline{\underline{a}})} m_{\underline{j}\rightarrow \underline{\underline{a}}}\Big) - m_{\underline{i}\rightarrow \underline{\underline{a}}} + \psi_{\underline{\underline{a}}\underline{i}}(z^{\underline{\underline{a}}}, y^{\underline{i}})\Big\} \tag{42}$$

where $\gamma_{\underline{a}} = 1/n_{\underline{a}}$ and $n_{\underline{a}}$ is the number of trees containing node $\underline{a}$. Using the same method applied in (31)-(36), efficient message passings of (37)-(42) are possible and their computational complexities are all $\mathcal{O}(|\mathcal{L}|^2)$.

# 6 Experimental Results

## 6.1 Signal Reconstruction

To verify the efficiency of the HGN graph, we conduct simple 1D signal reconstruction tests. For test signals, we use Sine, Line, Sawtooth, Circle as showed in Fig. 4. The width of signals is 80 and signal values range in $[-40, 40]$. The input signals are generated by adding Gaussian noise with $\sigma = 5$. The amount of noise is cropped in $[-10, 10]$ as we want to make $|\mathcal{L}|$ is not too large. For $\mathcal{V}$, we create a node for each $x$-coordinate. For unary potentials, we use $\theta_s(x_s) = |y_s - d(x_s)|$ where $y_s$ is a value at $s$ of the input signal and $d(x_s) = d_s^0 + d^1(x_s)$ where $d^1(x_s) \in \{-10, \ldots, 10\}$. We set $d_s^0 = y_s$ which means we find the solution $d(\hat{x}_s)$ in the range of $[y_s - 10, y_s + 10]$. For smoothness potentials, we use a truncated linear function $g(x) = \min(|x|, T)$ where we use $T = 10$ for first-order, $T = 50$ for second and third-order smoothness priors. For regularization parameters, we use $\lambda_{\overline{st}} = 10$ for first-order, $\lambda_{\overline{st}} = 0$, $\lambda_{\overline{stu}} = 10$ for second-order, $\lambda_{\overline{st}} = \lambda_{\overline{stu}} = 0$, $\lambda_{\overline{stuv}} = 10$ for third-order, $\lambda_{\overline{st}} = 1$ and $\lambda_{\overline{stu}} = \lambda_{\overline{stuv}} = 5$ for mixed-order smoothness priors (denoted as $1^{st}$+$2^{nd}$+$3^{rd}$). The parameters are empirically chosen.

Using these parameters, we construct factor graphs (denoted as FG) and HGN graphs. The MRF energies for these graphs are optimized using BP and TRW described in Section 5. We show the time-energy curves for combinations of {FG,HGN} and {BP,TRW} in Fig. 3. In the figure, HGN+TRW shows best optimization performances while HGN+BP is not converged in second and third-order cases. The running time[4] of one message passing iteration for each method is shown in Table 2. As expected in Table 1, running time of FG+BP or FG+TRW is increasing exponentially as higher smoothness order is used while that of HGN+BP or HGN+TRW is increasing linearly. The signal reconstruction results using HGN+TRW is shown in Fig. 4. In the first-order case, the reconstructed signals have many stair case errors. The second-order case shows

[4] The running time is measured on an Intel Xeon, 2.33GHz Machine.

**Table 2.** Running Time of One Message Passing Iteration

| Method | First-Order | Second-Order | Third-Order |
|---|---|---|---|
| FG+BP | 0.012 sec | 0.946 sec | 115.41 sec |
| FG+TRW | 0.014 sec | 1.000 sec | 81.36 sec |
| HGN+BP | 0.001 sec | 0.004 sec | 0.011 sec |
| HGN+TRW | 0.001 sec | 0.004 sec | 0.011 sec |

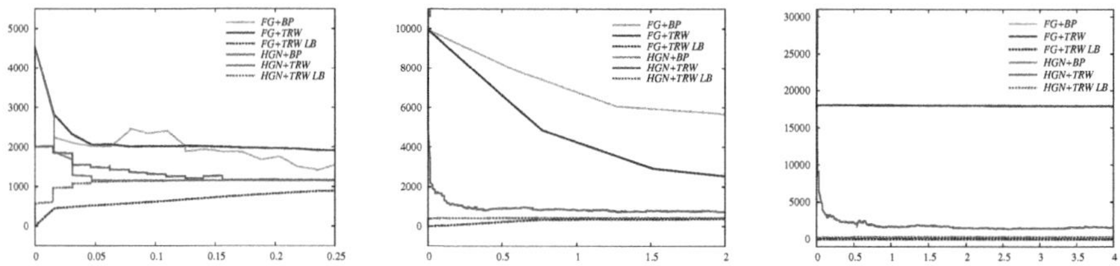


(a) First-Order (b) Second-Order (c) Third-Order

**Fig. 3.** The time-energy curve of signal reconstruction tests for first, second and third-order smoothness priors. In the figures, $x$-axis is the time (second), $y$-axis is the energy and TRW LB means the lower bound for the energy.

overall good reconstruction results, while the third-order case focuses on the fine structures of the noise. What is remarkable is the mixed-order prior produces better results than those of other priors. The results seem to collect advantages of each prior.

### 6.2 Image Denoising

Image denoising is the process of recovering original images from input images corrupted by noise. For test images, we use conventional ones named Boat, House, Lena, and Peppers. The noisy image is generated by adding Gaussian noise with $\sigma = 10$ while the amount of noise is cropped in $[-15, 15]$. The definition of potentials are same with those in Section 6.1 except that a node is created for each pixel and $d^1(x_s)$ is in $\{-15, \ldots, 15\}$. For regularization parameters, we use $\lambda_{\overline{st}} = 2$ for first-order, $\lambda_{\overline{st}} = 0$, $\lambda_{\overline{stu}} = 0.5$ for second-order, $\lambda_{\overline{st}} = 0.5$ and $\lambda_{\overline{stu}} = \lambda_{\overline{stuv}} = \lambda_{\overline{pqrstu}} = 0.1$ for mixed-order smoothness priors (denoted as $1^{st}+2^{nd}+3^{rd}$). The parameters are empirically chosen. For optimization, HGN+TRW is used. We also use HGN+TRW in the following two applications.

The results of image denoising are shown in Table 3 and Fig. 5. The similarity of denoising results with the original image is measured using the peek signal-to-noise ratio (PNSR)[5]. The first-order prior produces many fronto-parallel planes and the second-order priors over-smoothes the details. The field of experts (FoE) model [7] also tends to over-smooth the details of the image. The PSNR of mixed-order smoothness prior is comparable with the FoE model. It is impressive because the unary potential $\theta_s$ does not involve learned data as in the FoE model.

[5] The PSNR is defined as $20 \cdot \log_{10}\{255/\sum_s [y_s^0 - (y_s + d(\hat{x}_s))]\}$ where $y_s^0$ is a pixel value at $s$ of the original image.

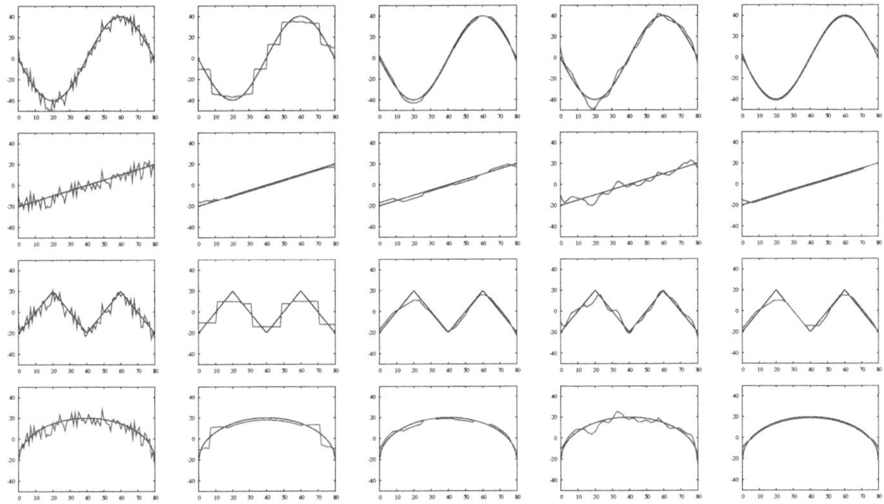

(a) Original (b) First-Order (c) Second-Order (d) Third-Order (e) $1^{st}+2^{nd}+3^{rd}$

**Fig. 4.** Signal reconstruction results using HGN graphs: The first to fourth rows show Sine, Line, Sawtooth and Circle signals, respectively. (a) shows original (blue) and input (red) signals. (b)-(e) show original (blue) and reconstructed signals (red).

**Table 3.** Results for Image Denoising ($\sigma = 10$)

| Smoothness Prior | PSNR (dB) Boat | House | Lena | Peppers | Mean |
|---|---|---|---|---|---|
| First-Order | 30.69 | 33.65 | 30.93 | 31.88 | 31.79 |
| Second-Order | 32.27 | 34.34 | 33.01 | 33.89 | 33.38 |
| $1^{st}+2^{nd}+3^{rd}$ | 32.95 | 35.50 | 33.47 | 34.45 | 34.09 |
| FoE [7] | 33.10 | 35.46 | 33.61 | 34.74 | 34.23 |

## 6.3 Sub-pixel Stereo Matching

Given a stereo image pair, the integer disparity $d^0 \in \mathbb{N}$ is computed using the conventional pairwise MRF energy with a first-order smoothness prior [3]. Given $d^0$, our problem is to find the sub-pixel disparity $d(x_s) = d_s^0 + d^1(x_s)$ for each node $s$ where the range for $d^1(x_s) \in \mathbb{R}$ is defined as $\{-w_d \cdot s_d, -(w_d - 1) \cdot s_d, \ldots, (w_d - 1) \cdot s_d, w_d \cdot s_d\}$. We use [21] for unary potentials. For smoothness potentials, a truncated linear function $g(x) = \min(|x|, T)$ is used where $T = 10$ for first and $T = 50$ for second and third order. For regularization parameters, we use $\lambda_{\overline{st}} = 25$ for first-order, $\lambda_{\overline{st}} = 25$, $\lambda_{\overline{stu}} = 5$ for second-order with first-order (denoted as $1^{st}+2^{nd}$), $\lambda_{\overline{st}} = 25$, $\lambda_{\overline{stu}} = \lambda_{\overline{stuv}} = \lambda_{\overline{pqrstu}} = 5$ for mixed-order smoothness priors (denoted as $1^{st}+2^{nd}+3^{rd}$). For label spaces, $w_d = 10$, $s_d = 0.25$ are used. For test images, we use Cloth1, Cloth2, Cloth3 and Cloth4 of the Middlebury stereo dataset [22]. The results of sub-pixel stereo matching are shown in Table 4 and Fig. 6. The errors are measured in the non occluded region (nonocc). In results, $1^{st}+2^{nd}+3^{rd}$ performs best among the first-order, $1^{st}+2^{nd}$ and depth enhancement method [8] which over-smoothes the detail structures.

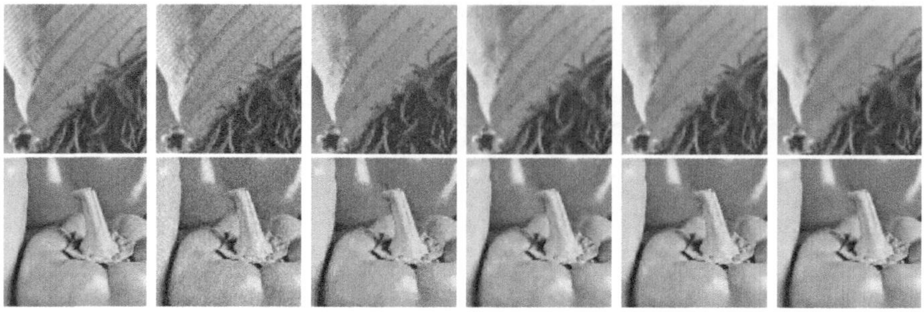


**Fig. 5.** Selected results of image denoising: Lena (top) and Peppers (bottom). To visualize the details, we magnify the interest regions having 100x100 size.

**Table 4.** Results for Sub-Pixel Stereo Matching

| Smoothness Prior | nonocc (error threshold = 0.5) | | | |
|---|---|---|---|---|
| | Cloth1 | Cloth2 | Cloth3 | Cloth4 |
| Integer Disparity $d^0$ | 13.60 | 31.02 | 20.07 | 24.16 |
| First-Order | 2.68 | 20.97 | 8.33 | 15.66 |
| $1^{st}+2^{nd}$ | 2.08 | 19.35 | 7.88 | 14.77 |
| $1^{st}+2^{nd}+3^{rd}$ | 1.47 | 17.38 | 7.25 | 13.62 |
| Depth Enhancement [8] | 4.74 | 27.01 | 14.90 | 19.85 |

### 6.4 Nonrigid Image Registration

Nonrigid image registration is the process of determining the geometric transformation between two images which are not related to simple rigid or affine transforms. For test images, we generate a synthetically deformed data set (10 images for each $\sigma$) given base images $I_1$ and $I_2$. The deformed image generated by applying random rotation, translation, scaling, and warping with control points perturbed with random variation $[-\sigma, \sigma]$. More detailed procedure for generating synthetic images and testing registrations in this section are referred to [6].

For an input image, we construct a set $\mathcal{V}$ which consists of nodes $s$ placed in a grid with a spacing $\delta = 8$. A label $x_s$ of a node $s$ represents a 2D displacement vector $(d_x(x_s), d_y(x_s))$. We assume $d_x$ and $d_y$ have same discrete set of values $[-8, 8]$. For unary potentials, we use normalized cross correlation (NCC) measure between two input images. For smoothness potentials, we use a truncated linear function $g(x) = \min(|x|, 10)$ for first-order, a quadratic function $g(x) = x^2$ for second and third-order smoothness priors. For regularization parameters, we use $\lambda_{\overline{st}} = 10^{-4}$ for first-order, $\lambda_{\overline{st}} = 0$, $\lambda_{\overline{stu}} = 10^{-4}$ for second-order, $\lambda_{\overline{st}} = 0$, $\lambda_{\overline{stu}} = 10^{-4}$, $\lambda_{\overline{stuv}} = \lambda_{\overline{pqrstu}} = 2 \cdot 10^{-5}$ for mixed-order smoothness priors (denoted as $1^{st}$+$2^{nd}$+$3^{rd}$). We show average displacements errors (measured as root mean square error (RMSE)) in Table 5. The result shows second-order [6] and mixed-order smoothness priors perform better than the first-order one [5]. The performance difference between second-order and mixed-order is not significant.

(a) Cloth2 (b) Ground Truth (c) Integer Disparity $d^0$ (d) $1^{st}+2^{nd}+3^{rd}$ (e) Depth Enhancement [8]

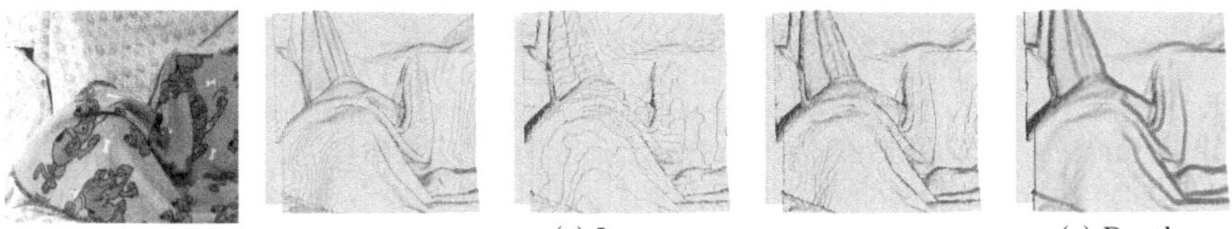

(a) Cloth3 (b) Ground Truth (c) Integer Disparity $d^0$ (d) $1^{st}+2^{nd}+3^{rd}$ (e) Depth Enhancement [8]

**Fig. 6.** Selected results of sub-pixel stereo matching. $1^{st}+2^{nd}+3^{rd}$ (d) restores the fine details of ground truths (b) from integer disparity $d^0$ (c).

**Table 5.** Results for Nonrigid Image Registration

| Smoothness Prior | Data Set $I_1$ (RMSE) | | | Data Set $I_2$ (RMSE) | | |
|---|---|---|---|---|---|---|
| | $\sigma$=3 | $\sigma$=6 | $\sigma$=9 | $\sigma$=3 | $\sigma$=6 | $\sigma$=9 |
| First-Order [5] | 0.72 | 1.43 | 4.06 | 0.82 | 1.13 | 2.86 |
| Second-Order [6] | 0.65 | 1.16 | 3.55 | 0.72 | 1.03 | 2.64 |
| $1^{st}+2^{nd}+3^{rd}$ | 0.66 | 1.15 | 3.65 | 0.74 | 0.98 | 2.58 |

## 7 Conclusion

We proposed a new method for solving the MRF energies with higher-order smoothness priors. It converts a factor graph to a hierarchical auxiliary node graph, and reduces the complexity of auxiliary nodes using gradient nodes when smoothness potentials are formulated as $n^{th}$-order derivative forms. In HGN graphs, the computational complexity message passing is reduced to $\mathcal{O}(\mathcal{L}^2)$ from exponential complexity of a factor graph representation. In addition, HGN graphs can integrate multiple orders of smoothness priors into its node hierarchy. We introduced an efficient optimization method using TRW message passings for HGN graphs, which works better than the belief propagation for difficult graphs. In experiments, we verified the efficient optimization is possible on HGN graphs compared to factor graphs and showed the MRFs with higher-order smoothness priors produce better results than first-order ones in various applications including image denoising, sub-pixel stereo matching and nonrigid image registration. We also showed that, among higher-order priors, mixing various smoothness orders performs better than single-order ones in many cases.

**Acknowledgement.** This work was supported by the National Research Foundation of Korea (NRF) Grant funded by the Korean Government (MEST) (2009-0083815).

## References


1. Weiss, Y., Freeman, W.T.: On the Optimality of Solutions of the Max-Product Belief-Propagation Algorithm in Arbitrary Graphs. IEEE Trans. Inf. Theory 47, 736–744 (2001)
2. Yedidia, J.S., Freeman, W.T., Weiss, Y.: Constructing Free-Energy Approximations and Generalized Belief Propagation Algorithms. IEEE Trans. Inf. Theory 51, 2282–2312 (2005)
3. Boykov, Y., Veksler, O., Zabih, R.: Fast Approximate Energy Minimization via Graph Cuts. IEEE Trans. Pattern Anal. Mach. Intell. 23, 1222–1239 (2001)
4. Boykov, Y., Jolly, M.P.: Interactive Graph Cuts for Optimal Boundary and Region Segmentation of Objects in N-D Images. In: ICCV (2001)
5. Glocker, B., Komodakis, N., Paragios, N., Tziritas, G., Navab, N.: Inter and Intramodal Deformable Registration: Continuous Deformations Meet Efficient Optimal Linear Programming. In: IPMI (2007)
6. Kwon, D., Lee, K.J., Yun, I.D., Lee, S.U.: Nonrigid Image Registration Using Dynamic Higher-Order MRF Model. In: Forsyth, D., Torr, P., Zisserman, A. (eds.) ECCV 2008, Part I. LNCS, vol. 5302, pp. 373–386. Springer, Heidelberg (2008)
7. Roth, S., Black, M.J.: Fields of Experts. Int. J. Comput. Vision 82, 205–229 (2009)
8. Yang, Q., Wang, L., Yang, R., Stewénius, H., Nistér, D.: Stereo Matching with Color-Weighted Correlation, Hierarchical Belief Propagation, and Occlusion Handling. IEEE Trans. Pattern Anal. Mach. Intell. 31, 492–504 (2009)
9. Kohli, P., Kumar, M.P., Torr, P.H.S.: $P^3$ & Beyond: Solving Energies with Higher Order Cliques. In: CVPR (2007)
10. Kohli, P., Ladický, L., Torr, P.H.S.: Robust Higher Order Potentials for Enforcing Label Consistency. In: CVPR (2008)
11. Rother, C., Kohli, P., Feng, W., Jia, J.: Minimizing Sparse Higher Order Energy Functions of Discrete Variables. In: CVPR (2009)
12. Ishikawa, H.: Higher-Order Clique Reduction in Binary Graph Cut. In: CVPR (2009)
13. Liao, L., Fox, D., Kautz, H.: Location-based Activity Recognition. In: NIPS (2005)
14. Potetz, B.: Efficient Belief Propagation for Vision Using Linear Constraint Nodes. In: CVPR (2007)
15. Komodakis, N., Paragios, N.: Beyond Pairwise Energies: Efficient Optimization for Higher-order MRFs. In: CVPR (2009)
16. Komodakis, N., Paragios, N., Tziritas, G.: MRF Optimization via Dual Decomposition: Message-Passing Revisited. In: ICCV (2007)
17. Kschischang, F.R., Frey, B.J., Loeliger, H.A.: Factor Graphs and the Sum-Product Algorithm. IEEE Trans. Inf. Theory 47, 498–519 (2001)
18. Wainwright, M.J., Jaakkola, T., Willsky, A.S.: MAP Estimation Via Agreement on Trees: Message-Passing and Linear Programming. IEEE Trans. Inf. Theory 51, 3697–3717 (2005)
19. Kolmogorov, V.: Convergent Tree-Reweighted Message Passing for Energy Minimization. IEEE Trans. Pattern Anal. Mach. Intell. 28, 1568–1583 (2006)
20. Szeliski, R., Zabih, R., Scharstein, D., Veksler, O., Kolmogorov, V., Agarwala, A., Tappen, M., Rother, C.: A Comparative Study of Energy Minimization Methods for Markov Random Fields. In: Leonardis, A., Bischof, H., Pinz, A. (eds.) ECCV 2006. LNCS, vol. 3952, pp. 16–29. Springer, Heidelberg (2006)
21. Birchfield, S., Tomasi, C.: A Pixel Dissimilarity Measure That Is Insensitive to Image Sampling. IEEE Trans. Pattern Anal. Mach. Intell. 20, 401–406 (1998)
22. Middlebury stereo datasets, `http://vision.middlebury.edu/stereo/data`

# An Efficient RANSAC for 3D Object Recognition in Noisy and Occluded Scenes

Chavdar Papazov and Darius Burschka

Technische Universität München (TUM), Germany
{papazov,burschka}@in.tum.de

**Abstract.** In this paper, we present an efficient algorithm for 3D object recognition in presence of clutter and occlusions in noisy, sparse and unsegmented range data. The method uses a robust geometric descriptor, a hashing technique and an efficient RANSAC-like sampling strategy. We assume that each object is represented by a model consisting of a set of points with corresponding surface normals. Our method recognizes multiple model instances and estimates their position and orientation in the scene. The algorithm scales well with the number of models and its main procedure runs in linear time in the number of scene points. Moreover, the approach is conceptually simple and easy to implement. Tests on a variety of real data sets show that the proposed method performs well on noisy and cluttered scenes in which only small parts of the objects are visible.

## 1 Introduction

Object recognition is one of the most fundamental problems of computer vision. In recent years, advances in 3D geometry acquisition technology have led to a growing interest in object recognition techniques which work with three-dimensional data. Referring to [1], the 3D object recognition problem can be stated as follows. Given a set $\mathcal{M} = \{\mathbf{M}_1, \ldots, \mathbf{M}_q\}$ of models and a scene $\mathbf{S}$ are there transformed subsets of some models which match a subset of the scene? The output of an object recognition algorithm is a set $\{(\mathbf{M}_{k_1}, T_1), \ldots, (\mathbf{M}_{k_r}, T_r)\}$ where $\mathbf{M}_{k_j} \in \mathcal{M}$ is a recognized model instance and $T_j$ is a transform which aligns $\mathbf{M}_{k_j}$ to the scene $\mathbf{S}$. In this paper, we discuss a special instance of this problem which is given by the following assumptions.

(i) Each model $\mathbf{M}_i$ is a finite set of oriented points, i.e., $\mathbf{M}_i = \{(\mathbf{p}, \mathbf{n}) : \mathbf{p} \in \mathbb{R}^3, \mathbf{n} \text{ is the normal at } \mathbf{p}\}$.
(ii) Each model is representing a non-transparent object.
(iii) The scene $\mathbf{S} = \{\mathbf{p}_1, \ldots, \mathbf{p}_s\} \subset \mathbb{R}^3$ is a range image.
(iv) The transform $T_j$ which aligns $\mathbf{M}_{k_j}$ to $\mathbf{S}$ is a rigid transform.

Even under these assumptions the problem remains hard because of several reasons: it is a priori not known which objects are in the scene and how they are oriented; the scene points are typically corrupted by noise and outliers; the

R. Kimmel, R. Klette, and A. Sugimoto (Eds.): ACCV 2010, Part I, LNCS 6492, pp. 135–148, 2011.

**Fig. 1.** Three views of a typical recognition result obtained with our method. The scene is shown as a blue mesh and the four recognized model instances are rendered as yellow point clouds and superimposed over the scene mesh (see Section 4 for details).

objects are only partially visible due to scene clutter, occlusions and scan device limitations.

**Contributions and Overview.** In this paper, we introduce an efficient algorithm for 3D object recognition in noisy, sparse and unsegmented range data. We make the following contributions: (i) We use a hash table for rapid retrieval of pairs of oriented model points which are similar to a sampled pair of oriented scene points. (ii) A new efficient RANSAC-like sampling strategy for fast generation of object hypotheses is introduced. (iii) We provide a complexity analysis of our sampling strategy and derive the number of iterations needed to recognize model instances with a predefined success probability. (iv) A new measure for the quality of an object hypothesis is presented. (v) We use a non-maximum suppression to remove false positives and to achieve a consistent scene explanation by the given models.

The rest of the paper is organized as follows. After reviewing previous work in Section 2, we describe our algorithm in Section 3. Section 4 presents experimental results. Conclusions are drawn in the final Section 5 of the paper.

## 2 Related Work

Object recognition should not be confused with object classification/shape retrieval. The latter methods only measure the similarity between a given input shape and shapes stored in a model library. They do not estimate a transform which maps the input to the recognized model. Moreover, the input shape is assumed to be a subset of some of the library shapes. In our case, however, the input contains points originating from multiple objects and scene clutter.

There are two major classes of 3D object recognition methods. One class consists of the so-called voting methods. Well-known representatives are the generalized Hough transform [2] and geometric hashing [1]. The generalized Hough transform has a favorable space and time complexity of $O(nk^3)$, where $n$ is the number of scene points and $k$ is the number of bins for each dimension of the discretized rotation space. Unfortunately, the method scales bad with the number of models since one has to match sequentially each one of them to the scene. The geometric hashing approach [1] allows for a simultaneous recognition of all models without the need of sequential matching. However, it tends to be very costly since its space complexity is $O(m^3)$ and its worse case time complexity is $O(n^4)$, where $m$ and $n$ are the number of model and scene points, respectively. A more recent voting approach is the tensor matching algorithm [3]. It performs well on complex scenes but the authors did not present tests on noisy and sparse data sets.

The correspondence based methods belong to the second class of object recognition approaches. First, correspondences between the models and the scene are established usually using local geometric descriptors. In the second step, the aligning rigid transform is calculated based on the established correspondences. There is a vast variety of descriptors which can be used in a correspondence based object recognition framework. A list includes, without being nearly exhaustive, spin images [4], local feature histograms [5], 3D shape context, harmonic shape context [6] and integral invariants [7]. In [8], classic 2D image descriptors were extended to the domain of 2-manifolds embedded in $\mathbb{R}^3$ and applied to rigid and non-rigid matching of meshes. Intrinsic isometry invariant descriptors were developed in [9] and shown to be effective for the matching of articulated shapes. All correspondence based algorithms rely heavily on the assumption that the models to be recognized have distinctive feature points, i.e., points with rare descriptors. In many cases, however, this assumption does not hold. A cylinder, for example, will have too many points with similar descriptors. This results in many ambiguous correspondences between the model and the scene and the recognition method degenerates to a brute force search.

In our recognition approach, we combine a robust descriptor, a hashing technique and an efficient RANSAC variant. A similar strategy was proposed in [10]. In contrast to [10], where a hash table is used only for fast indexing into a large collection of geometry descriptors of *single* model points, we use a hash table to store descriptors of *pairs* of oriented model points (called doublets). This not only enables us to efficiently determine the model doublets which are similar to a sampled scene doublet but also allows for a very easy computation of the aligning rigid transform since it is uniquely defined by two corresponding doublets. Furthermore, in [10], a complex scene preprocessing is performed before running the actual object recognition: (i) multiple views of the scene are registered in order to build a more complete scene description and (ii) a scene segmentation is executed to separate the object from the background. In contrast to this, our method copes with a single view of the scene and does not require any segmentation. Moreover, the scenes used in all tests presented in [10] contain a single object and some background clutter. In this paper, we deal with the more challenging problem of object recognition and pose estimation in scenes which contain multiple object instances plus background clutter.

Before we describe our algorithm in detail, we briefly review the surface registration technique presented in [11] and include a short discussion on RANSAC [12] since both are of special relevance to our work.

**Fast Surface Registration.** [11] To put it briefly, the task of rigid surface registration is to find a rigid transform which aligns two given surfaces. Let $\mathbf{S}$ be a surface given as a set of oriented points. For a pair of oriented points $(\mathbf{u}, \mathbf{v}) = ((\mathbf{p}_u, \mathbf{n}_u), (\mathbf{p}_v, \mathbf{n}_v)) \in \mathbf{S} \times \mathbf{S}$, a descriptor $f : \mathbf{S} \times \mathbf{S} \to \mathbb{R}^4$ is defined by

$$f(\mathbf{u}, \mathbf{v}) = \begin{pmatrix} f_1(\mathbf{u}, \mathbf{v}) \\ f_2(\mathbf{u}, \mathbf{v}) \\ f_3(\mathbf{u}, \mathbf{v}) \\ f_4(\mathbf{u}, \mathbf{v}) \end{pmatrix} = \begin{pmatrix} \|\mathbf{p}_u - \mathbf{p}_v\| \\ \angle(\mathbf{n}_u, \mathbf{n}_v) \\ \angle(\mathbf{n}_u, \mathbf{p}_v - \mathbf{p}_u) \\ \angle(\mathbf{n}_v, \mathbf{p}_u - \mathbf{p}_v) \end{pmatrix}, \tag{1}$$

where $\angle(\mathbf{a}, \mathbf{b})$ denotes the angle between $\mathbf{a}$ and $\mathbf{b}$. In order to register two surfaces $\mathbf{S}_1$ and $\mathbf{S}_2$, oriented point pairs $(\mathbf{u}, \mathbf{v}) \in \mathbf{S}_1 \times \mathbf{S}_1$ and $(\tilde{\mathbf{u}}, \tilde{\mathbf{v}}) \in \mathbf{S}_2 \times \mathbf{S}_2$ are sampled uniformly and the corresponding descriptors $f(\mathbf{u}, \mathbf{v})$ and $f(\tilde{\mathbf{u}}, \tilde{\mathbf{v}})$ are computed and stored in a four-dimensional hash table. The hash table is continuously filled in this way until a collision occurs, i.e., until a descriptor of a pair from $\mathbf{S}_1 \times \mathbf{S}_1$ and a descriptor of a pair from $\mathbf{S}_2 \times \mathbf{S}_2$ end up in the same hash table cell. Computing the rigid transform which best aligns (in least square sense) the colliding pairs gives a transform hypothesis for the surfaces.

According to [11], this process is repeated until a hypothesis is good enough, a predefined time limit is reached or all combinations are tested. Non of these stopping criteria is well-grounded: the first two are ad hoc and the last one is computationally infeasible. In contrast to this, we compute the number of iterations required to recognize model instances with a user-defined success probability. Furthermore, a direct application of the above described registration technique to 3D object recognition will have an unfavorable computational complexity since it will require a sequential registration of each model to the scene.

**RANSAC.** [12] can be seen as a general approach for model recognition. It works by uniformly drawing minimal point sets from the scene and computing a transform which aligns the model with the minimal point set[1]. The score of the resulting hypothesis is computed by counting the number of transformed model points which lie within a certain $\epsilon$-band of the scene. After a given number of trials, the model is considered to be recognized at the locations defined by the hypotheses which achieved a score higher than a predefined threshold. In order to recognize the model with a probability $P_S$ we need to perform

$$N = \frac{\ln(1 - P_S)}{\ln(1 - P_M)}, \tag{2}$$

trials, where $P_M$ is the probability of recognizing the model in a single iteration.

The RANSAC approach has the advantages of being conceptually simple, very general and robust against outliers. Unfortunately, its direct application to the 3D object recognition problem is computationally very expensive. In order to compute an aligning rigid transform, we need at least three pairs of corresponding model $\leftrightarrow$ scene points. Under the simplifying assumption that the model is

[1] A minimal point set is the smallest set of points required to uniquely determine a given type of transform.

completely contained in the scene, the probability of drawing three such pairs in a single trial is $P_M(n) = \frac{3!}{(n-2)(n-1)n}$, where $n$ is the number of scene points. Since $P_M(n)$ is a small number we can approximate the denominator in (2) by its Taylor series $\ln(1 - P_M(n)) = -P_M(n) + O(P_M(n)^2)$ and get for the number of trials as a function of the number of scene points:

$$N(n) \approx \frac{-\ln(1 - P_S)}{P_M(n)} = O(n^3). \tag{3}$$

Assuming $q$ models in the library the complexity of RANSAC is $O(qn^3)$.

There are many modifications of the classic RANSAC scheme. Some recently proposed methods like ASSC [13] and ASKC [14] significantly improve outlier robustness by using a different score function. However, these variants are not designed to enhance the performance of RANSAC. In [15], an efficient RANSAC-like registration algorithm was proposed. However, it is not advisable to directly apply the method to 3D object recognition since it will require a sequential matching of each model to the scene. In [16], another efficient RANSAC variant for primitive shape detection was introduced. The method is related to ours since the authors also used a localized minimal point set sampling. Their method, however, is limited to the detection of planes, spheres, cylinders, cones and tori.

## 3 Method Description

Like most object recognition methods, ours consists of two phases. The first phase — the model preprocessing — is done offline. It is executed only once for each model and does not depend on the scenes in which the model instances have to be recognized. The second phase is the online recognition which is executed on the scene using the model representation computed in the offline phase.

### 3.1 Model Preprocessing Phase

For a given object model $\mathbf{M}$, we sample all pairs of oriented points $(\mathbf{u}, \mathbf{v}) = ((\mathbf{p}_u, \mathbf{n}_u), (\mathbf{p}_v, \mathbf{n}_v)) \in \mathbf{M} \times \mathbf{M}$ for which $\mathbf{p}_u$ and $\mathbf{p}_v$ are approximately at a distance $d$ from each other. For each pair, the descriptor $f(\mathbf{u}, \mathbf{v}) = (f_2(\mathbf{u}, \mathbf{v}), f_3(\mathbf{u}, \mathbf{v}), f_4(\mathbf{u}, \mathbf{v}))$ is computed as defined in (1) and stored in a three-dimensional hash table. Note that since $d$ is fixed we do not use $f_1$ as part of the descriptor. Furthermore, in contrast to [11], we do *not* consider all pairs of oriented points, but only those which fulfill $\|\mathbf{p}_u - \mathbf{p}_v\| \in [d - \delta_d, d + \delta_d]$, for a given tolerance value $\delta_d$. This has several advantages. The space complexity is reduced from $O(m^2)$ to $O(m)$, where $m$ is the number of points in $\mathbf{M}$ (this is an empirical measurement further discussed in [17]). For large $d$, the pairs we consider are wide-pairs which allow a much more stable computation of the aligning rigid transform than narrow-pairs do [17]. A further advantage of wide-pairs is due to the fact that the larger the distance the less pairs we have. Thus, computing and storing descriptors of wide-pairs leads to less populated hash table cells which means that we will have to test less transform hypotheses in the online recognition phase and will save computation time.

Note, however, that the pair width $d$ can not be arbitrary large due to occlusions in real world scenes. For a typical value for $d$, there are still a lot of pairs with similar descriptors, i.e., there are hash table cells with too many entries. To avoid this overpopulation, we remove as many of the most populated cells as needed to keep only a fraction $K$ of the pairs in the hash table (in our implementation $K = 0.1$). This strategy leads to some information loss about the object shape. We take this into account in the online phase of our algorithm.

The final representation of all models $\mathbf{M}_1, \dots, \mathbf{M}_q$ is computed by processing each $\mathbf{M}_i$ in the way described above using *the same* hash table. In order not to confuse the correspondence between pairs and models, each cell contains a list for each model which has pairs stored in the cell. In this way, new models can be added to the hash table without recomputing it.

### 3.2 Online Recognition Phase

The online recognition phase can be outlined as follows:

1. Initialization
   (a) Compute an octree for the scene $\mathbf{S}$ to produce a modified scene $\mathbf{S}^*$.
   (b) $\mathcal{T} \leftarrow \emptyset$ (an empty solution list).
2. Compute a number of iterations $N$ needed to achieve a probability for successful recognition higher than a predefined value $P_S$.

**[repeat** $N$ **times]**

3. Sampling
   (a) Sample a point $\mathbf{p}_u$ uniformly from $\mathbf{S}^*$.
   (b) Sample $\mathbf{p}_v \in \mathbf{S}^*$ uniformly from all points at a distance $d \pm \delta_d$ from $\mathbf{p}_u$.
4. Estimate normals $\mathbf{n}_u$ and $\mathbf{n}_v$ at $\mathbf{p}_u$ and $\mathbf{p}_v$, respectively, to get an oriented scene point pair $(\mathbf{u}, \mathbf{v}) = ((\mathbf{p}_u, \mathbf{n}_u), (\mathbf{p}_v, \mathbf{n}_v))$.
5. Compute the descriptor $f_{\mathbf{uv}} = (f_2(\mathbf{u}, \mathbf{v}), f_3(\mathbf{u}, \mathbf{v}), f_4(\mathbf{u}, \mathbf{v}))$ (see (1)).
6. Use $f_{\mathbf{uv}}$ as a key to the model hash table to retrieve the oriented model point pairs $(\mathbf{u}_j, \mathbf{v}_j)$ similar to $(\mathbf{u}, \mathbf{v})$.
   **[repeat for each** $(\mathbf{u}_j, \mathbf{v}_j)$**]**
   (a) Get the model $\mathbf{M}$ of $(\mathbf{u}_j, \mathbf{v}_j)$.
   (b) Compute the rigid transform $T$ that best aligns $(\mathbf{u}_j, \mathbf{v}_j)$ to $(\mathbf{u}, \mathbf{v})$.
   (c) Set $\mathcal{T} \leftarrow \mathcal{T} \cup (\mathbf{M}, T)$ if $(\mathbf{M}, T)$ is accepted by an acceptance function $\mu$.
   **[end repeat]**

**[end repeat]**

7. Filter conflicting hypotheses from $\mathcal{T}$.

For our algorithm to be fast, we need to search efficiently for closest points (in step 4) and for points lying on a sphere around a given point (in step 3b). These operations are greatly facilitated if a neighborhood structure is available for the point set. Note that the 2D range image grid defines such a structure which, however, is not well suited for the above mentioned geometric operations. This is due to the fact that points which are neighbors on the gird are not necessarily close to each other in $\mathbb{R}^3$ because of perspective effects and scene depth discontinuities. A very efficient way to establish spatial proximity between points in $\mathbb{R}^3$ is to use an octree.

**Step 1, Initialization.** In step 1a of the algorithm, we construct an octree with a fixed leaf size $L$ (the edge length of a leaf). The full octree leaves (the ones which contain at least one point) can be seen as voxels ordered in a regular axis-aligned 3D grid. Thus, each full leaf has unique integer coordinates $(i, j, k)$. We say that two full leaves are neighbors if the absolute difference between their corresponding integer coordinates is $\leq 1$. Next, we down-sample $\mathbf{S}$ by setting the new scene points in $\mathbf{S}^*$ to be the centers of mass of the full leaves. The center of mass of a full leaf is defined to be the average of the points it contains. In this way, a one-to-one correspondence between the points in $\mathbf{S}^*$ and the full octree leaves is established. Two points in $\mathbf{S}^*$ are neighbors if the corresponding full leaves are neighbors.

**Step 2, Number of Iterations.** This step is explained in Section 3.3.

**Step 3, Sampling.** In the sampling stage, we make extensive use of the scene octree. As in the classic RANSAC, we sample minimal sets from the scene. In our case, a minimal set consists of two oriented points. However, in contrast to RANSAC, they are *not* sampled uniformly. Only the first point, $\mathbf{p}_u$, is drawn uniformly from $\mathbf{S}^*$. In order to draw the second point, $\mathbf{p}_v$, we first retrieve the set $\mathbf{L}$ of all full leaves which are intersected by the sphere with center $\mathbf{p}_u$ and radius $d$, where $d$ is the pair width used in the offline phase (see Section 3.1). This operation can be implemented very efficiently due to the hierarchical structure of the octree. Finally, a leaf is drawn uniformly from $\mathbf{L}$ and $\mathbf{p}_v$ is set to be its center of mass.

**Step 4, Normal Estimation.** The normals $\mathbf{n}_u$ and $\mathbf{n}_v$ are estimated by performing a Principal Component Analysis. $\mathbf{n}_u$ and $\mathbf{n}_v$ are set to be the eigenvectors corresponding to the smallest eigenvalues of the covariance matrix of the points in the neighborhood of $\mathbf{p}_u$ and $\mathbf{p}_v$, respectively. The result is the oriented scene point pair $(\mathbf{u}, \mathbf{v}) = ((\mathbf{p}_u, \mathbf{n}_u), (\mathbf{p}_v, \mathbf{n}_v))$.

**Steps 5 and 6, Hypotheses Generation and Testing.** Step 5 involves the computation of the descriptor $f_{\mathbf{uv}} = (f_2(\mathbf{u}, \mathbf{v}), f_3(\mathbf{u}, \mathbf{v}), f_4(\mathbf{u}, \mathbf{v}))$ (see (1)). In step 6, $f_{\mathbf{uv}}$ is used as a key to the model hash table to retrieve all model pairs $\mathbf{u}_j, \mathbf{v}_j)$ which are similar to $(\mathbf{u}, \mathbf{v})$. For each $(\mathbf{u}_j, \mathbf{v}_j)$, the model $\mathbf{M}$ corresponding to $(\mathbf{u}_j, \mathbf{v}_j)$ is retrieved (step 6a) and the rigid transform $T$ which best aligns (in least squares sense) $(\mathbf{u}_j, \mathbf{v}_j)$ to $(\mathbf{u}, \mathbf{v})$ is computed (step 6b). The result of these two sub-steps is the hypothesis that the model $\mathbf{M}$ is in the scene at the location defined by $T$. In order to save the hypothesis in the solution list, it has to be accepted by the acceptance function $\mu$.

**Step 6c, The Acceptance Function.** $\mu$ measures the quality of a hypothesis $(\mathbf{M}, T)$ and consists of a support term and a penalty term. As in RANSAC, the support term, $\mu_S$, is proportional to the number $m_S$ of transformed model points (i.e., points from $T(\mathbf{M})$) which fall within a certain $\epsilon$-band of the scene. More precisely, $\mu_S(\mathbf{M}, T) = m_S/m$, where $m$ is the number of model points. To compute $m_S$, we back project $T(\mathbf{M})$ in the scene range image and count the number of points which have a z-coordinate in the interval $[z - \epsilon, z + \epsilon]$, where $z$ is the z-coordinate of the corresponding range image pixel.

In contrast to RANSAC, our algorithm contains a penalty term, $\mu_P$, which is proportional to the size of the transformed model parts which occlude the scene. It is clear that in a scene viewed by a camera a correctly recognized (non-transparent) object can not occlude scene points reconstructed from the same viewpoint. We penalize hypotheses which violate this condition. We compute the penalty term by counting the number $m_P$ of transformed model points which are between the projection center of the range image and a valid range image pixel and thus are "occluding" reconstructed scene points. We set $\mu_P(\mathbf{M}, T) = m_P/m$, where $m$ is the number of model points.

For $(\mathbf{M}, T)$ to be accepted as a valid hypothesis it has to have a support higher than a predefined $S \in [0, 1]$ and a penalty lower than a predefined $P \in [0, 1]$.

**Step 7, Filtering Conflicting Hypotheses.** We say that an accepted hypothesis $(\mathbf{M}, T)$ explains a set $\mathbf{P} \subset \mathbf{S}^*$ of scene points if for each $\mathbf{p} \in \mathbf{P}$ there is a point from $T(\mathbf{M})$ which lies within the octree leaf corresponding to $\mathbf{p}$. Note that the points from $\mathbf{P}$ explained by $(\mathbf{M}, T)$ are *not* removed from $\mathbf{S}^*$ because there could be a better hypothesis, i.e., one which explains a superset of $\mathbf{P}$. Two hypotheses are conflicting if the intersection of the point sets they explain is non-empty. At the end of step 6, many conflicting hypotheses are saved in the list $\mathcal{T}$. To filter the weak ones, we construct a so called conflict graph. Its nodes are the hypotheses in $\mathcal{T}$ and an edge is connecting two nodes if the hypotheses they represent are conflicting ones. To produce the final output, the solution list is filtered by performing a non-maximum suppression on the conflict graph: a node is removed if it has a better neighboring node.

### 3.3 Time Complexity

The complexity of the proposed algorithm is dominated by three major factors: (i) the number of iterations (the loop after step 2), (ii) the number of pairs per hash table cell (the loop in step 6) and (iii) the cost of evaluating the acceptance function for each object hypothesis (step 6c). In the following, we discuss each one in detail.

**(i)** Consider the scene $\mathbf{S}^*$ consisting of $|\mathbf{S}^*| = n$ points and a model instance $\mathbf{M}$ therein consisting of $|\mathbf{M}| = m$ points. We already saw in the discussion on RANSAC at the end of Section 2 that we need

$$N = \frac{\ln(1 - P_S)}{\ln(1 - P_M)} \tag{4}$$

iterations to recognize $\mathbf{M}$ with a predefined success probability $P_S$, where $P_M$ is the probability of recognizing $\mathbf{M}$ in a single iteration. Recall from Section 2 that in the classic RANSAC applied to 3D object recognition we have $P_M \approx 1/n^3$.

Our sampling strategy and the use of the model hash table lead to a significant increase of $P_M$ and thus to a reduction of the complexity. In the following, we estimate $P_M$.

Let $P(\mathbf{p}_u \in \mathbf{M}, \mathbf{p}_v \in \mathbf{M})$ denote the probability that both points are sampled from $\mathbf{M}$ (see step 3 in Section 3.2). Thus, the probability of recognizing $\mathbf{M}$ in a single iteration is

$$P_M = KP(\mathbf{p}_u \in \mathbf{M}, \mathbf{p}_v \in \mathbf{M}), \tag{5}$$

where $K$ is the fraction of oriented point pairs for which the descriptors are saved in the model hash table (see Section 3.1). Using conditional probability and the fact that $P(\mathbf{p}_u \in \mathbf{M}) = m/n$ we can rewrite (5) to get

$$P_M = (m/n)KP(\mathbf{p}_v \in \mathbf{M}|\mathbf{p}_u \in \mathbf{M}). \tag{6}$$

$P(\mathbf{p}_v \in \mathbf{M}|\mathbf{p}_u \in \mathbf{M})$ is the probability to sample $\mathbf{p}_v$ from $\mathbf{M}$ given that $\mathbf{p}_u \in \mathbf{M}$. Recall from Section 3.2 that $\mathbf{p}_v$ is not independent of $\mathbf{p}_u$ because it is sampled uniformly from the set $\mathbf{L}$ consisting of the scene points which lie on the sphere with center $\mathbf{p}_u$ and radius $d$, where $d$ is the pair width used in the offline phase. Under the assumptions that the visible object part has an extent larger than $2d$ and that the reconstruction is not too sparse, $\mathbf{L}$ contains points from $\mathbf{M}$. Thus, $P(\mathbf{p}_v \in \mathbf{M}|\mathbf{p}_u \in \mathbf{M}) = |\mathbf{L} \cap \mathbf{M}|/|\mathbf{L}|$ is well-defined and greater than zero. $|\mathbf{L} \cap \mathbf{M}|/|\mathbf{L}|$ depends on the scene, i.e., it depends on the extent and the shape of the visible object part. Estimating $C = |\mathbf{L} \cap \mathbf{M}|/|\mathbf{L}|$ by, e.g., 1/4 (this is what we use in our implementation) accounts for up to 75% outliers and scene clutter. Thus, we get for $P_M$ as a function of $n$ (the number of scene points)

$$P_M(n) = (m/n)KC. \tag{7}$$

Again, approximating the denominator in (4) by its Taylor series $\ln(1 - P_M(n)) = -P_M(n) + O(P_M(n)^2)$ we get for the number of iterations

$$N(n) \approx \frac{-\ln(1 - P_S)}{P_M(n)} = \frac{-n\ln(1 - P_S)}{mKC} = O(n). \tag{8}$$

This proves that the number of iterations depends linearly on the number of scene points. Furthermore, it is guaranteed that the model instances will be recognized with the desired probability $P_S$.

**(ii)** The number of pairs per hash table cell depends on the number of models as well on the number of points of each model. An algorithm is considered to scale well with the number of models if its runtime is less than the sum of the runtime needed for the recognition of each model separately [1,10]. In other words, an algorithm should need less time than it is needed for a sequential matching of each model to the scene. The use of the model hash table ensures this in the case of our method. For almost all real world objects it holds that a hash table cell does not store pairs from all models. Furthermore, not all pairs originating from a model end up in the same hash table cell.

**(iii)** The acceptance function $\mu$ runs in $O(l)$ time, where $l$ is the number of model points. Note that $\mu$ does not depend on the number of scene points since back projecting a model point in the range image is performed in constant time.

# 4 Experimental Results

**Comparison with spin images [4] and tensor matching [3].** In the first test scenario, we compare the recognition rate of our algorithm with the spin images [4] and the tensor matching [3] approaches on occluded real scenes. We test our method on the same 50 data sets which are used in [3]. This allows for a precise comparison without the need of re-implementing neither of the two algorithms. The models of the four toys to be recognized are shown in the upper row of Fig. 2. Each test scene contains the toys (not necessary all four of them) in different positions and orientations. Each scene is digitized with a laser range finder from a single viewpoint which means that the back parts of the objects are not visible. Furthermore, in most scenes the toys are placed such that some of them occlude others which makes the visible object parts even smaller. The lower row of Fig. 2 shows exemplary four (out of 50) test scenes with the corresponding recognition results obtained with our algorithm. Since our algorithm is a probabilistic one we run 100 recognition trials on each scene

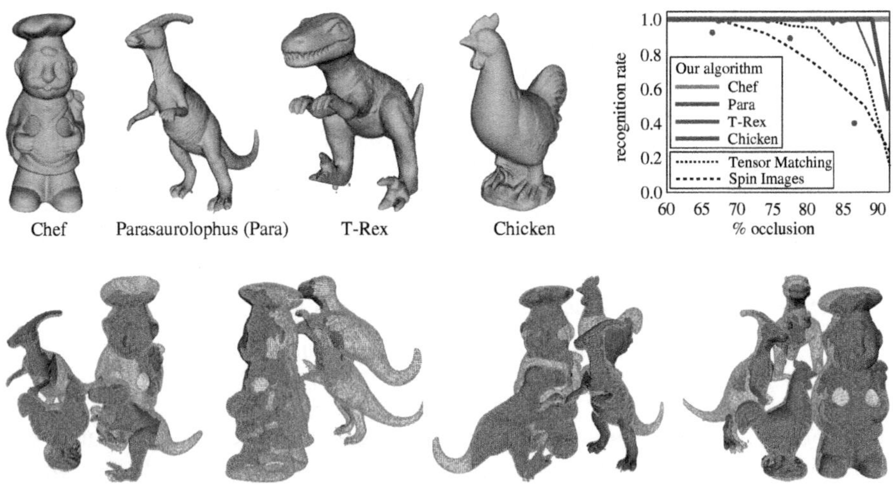


**Fig. 2.** (Upper left) The models used in the comparison test case. (Upper right) The continuous lines indicate the recognition rate of our algorithm for each object as a function of its occlusion. The dashed lines give the recognition rate of the spin images and the tensor matching approaches on the same scenes as reported in [3]. Note that our method outperforms both algorithms. The chef is recognized in all trials, even in the case of occlusion over 91%. The blue dots represent the recognition rate in the three chicken test scenes in which our method performs worse than the other algorithms. This is due to the fact that in these scenes only the chicken's back part is visible which contains strongly varying normals which makes it difficult to compute a stable aligning transform. (Lower row) Four (out of 50) test scenes and the corresponding recognition results. The recognized models are rendered as yellow point clouds and superimposed over the scenes which are rendered as blue meshes. These are challenging examples since only small parts of the objects are visible.

and compute the recognition rate for each object represented in the scene in the following way. We visually inspect the result of each of the 100 trials. If object **A** was recognized $n$ times $(0 \leq n \leq 100)$ then the recognition rate for **A** is $n/100$. Since the occlusion of every object in each scene is known we report the recognition rate for each object as a function of its occlusion. According to [4], the occlusion for an object model is given by $1 - \frac{\text{area of visible model surface}}{\text{total area of model surface}}$. The results of the tests and the comparison with the spin images [4] and the tensor matching [3] approaches are summarized in the upper right part of Fig. 2.

**Noisy and Sparse Scenes.** In the second scenario, we run tests under varying noisy conditions. The models to be recognized are the same as in the last test case and the scene is the third one in the lower row of Fig. 2. Next, several versions of the scene are computed by degrading it by zero-mean Gaussian noise with different variance values $\sigma$. Again, we perform 100 recognition trials for each noisy scene and compute the recognition rate, the mean number of false positives and the mean RMS error as functions of $\sigma$. For a point set $\mathbf{P}$, a (rigidly) transformed copy $\mathbf{Q}$ and a (rigid) transform $T$ the RMS error measures how close each point $\mathbf{p}_i \in \mathbf{P}$ comes to its corresponding point $\mathbf{q}_i \in \mathbf{Q}$ after transforming $\mathbf{Q}$ by $T$. Thus RMS measures the quality of $T$. It is given by

$$\mathrm{RMS}(T) = \sqrt{\frac{1}{N}\sum_{i=1}^{N} \|\mathbf{p}_i - T(\mathbf{q}_i)\|^2}, \tag{9}$$

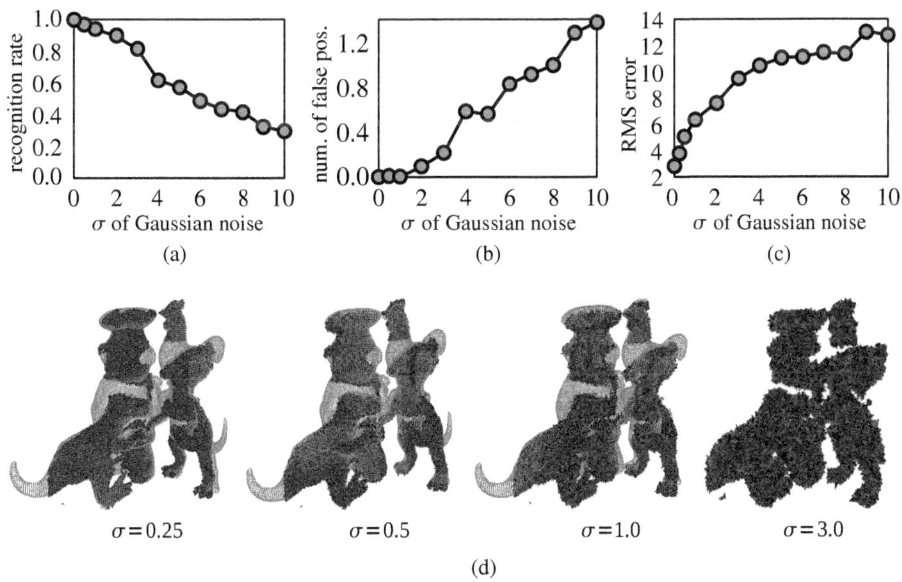


**Fig. 3.** (a) - (c) Recognition rate, mean number of false positives and mean RMS error as functions of the $\sigma$ of Gaussian noise. One $\sigma$ unit equals 1% of the bounding box diagonal length of the scene. The RMS units are in millimeters. (d) Typical recognition results for noise degraded data sets.

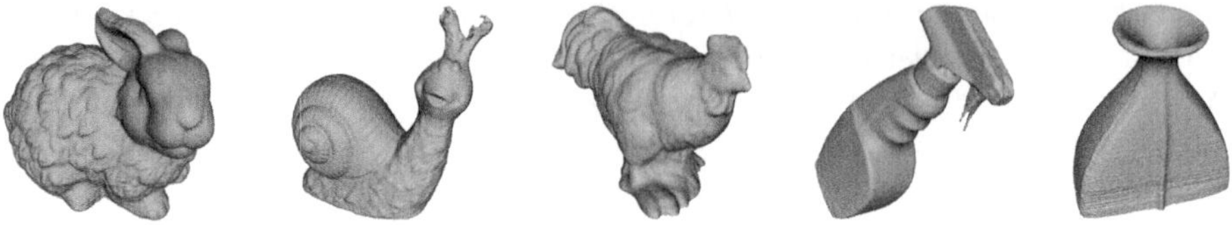


**Fig. 4.** The models used for object recognition in scenes reconstructed with a low-cost light intersection based device

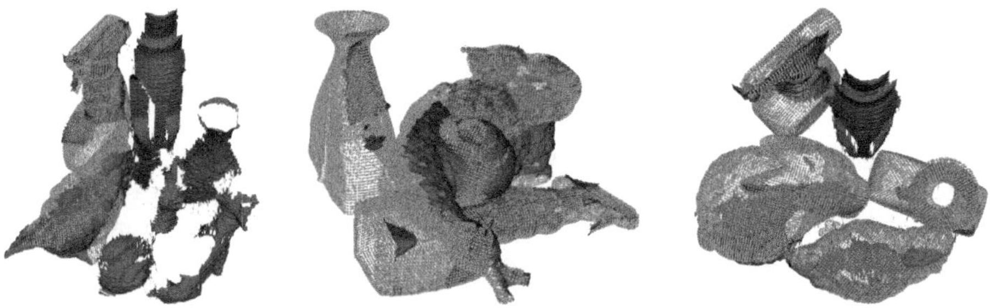

**Fig. 5.** Typical recognition results obtained with our method for three test scenes. The scenes are shown as blue meshes and the recognized model instances are rendered as yellow point clouds and superimposed over the meshes. Some of the scenes contain unknown objects (the left and the right one). Note that the scene reconstruction contains only small portions of the objects.

where $N$ is the number of points in $\mathbf{P}$. Since we know the ground truth location of each model in the test scene the RMS error of the rigid transform computed by our method can be easily calculated[2]. The results of all noise tests are summarized in Fig. 3(a) – (c). Typical recognition results and four of the noisy scenes are shown in Fig. 3(d).

Next, we demonstrate the ability of our method to deal with data sets corrupted by noise which is not artificially generated but originates in scan device imprecision. Note that the scenes used in [4] and [3] are dense and have a relatively good quality. We use a low-cost light section based scanner which gives sparser and noisier data sets. The models used in this test scenario are shown in Fig. 4. Typical recognition results of our method are shown in Fig. 1 and Fig. 5.

**Runtime.** In the last test scenario, we experimentally verify the two main claims regarding the time complexity of our algorithm, namely that it needs less time than it is required for a sequential matching of each model to the scene and that it has a linear complexity in the number of scene points.

First, we measure the runtime dependency on the number of models. The models used in this test case are the ones shown in Fig. 2 and Fig. 4 and the

[2] The ground truth rigid transform for the models for each scene is available on the webpage of the authors of [3].

| model | comp. time (sec) |
|---|---|
| Chef | 0.568 |
| Para | 0.533 |
| T-Rex | 0.5 |
| Chicken | 0.522 |
| Rabbit | 0.536 |
| Snail | 0.546 |
| Chicken 2 | 0.551 |
| Bottle | 0.577 |
| Vase | 0.566 |

(a)

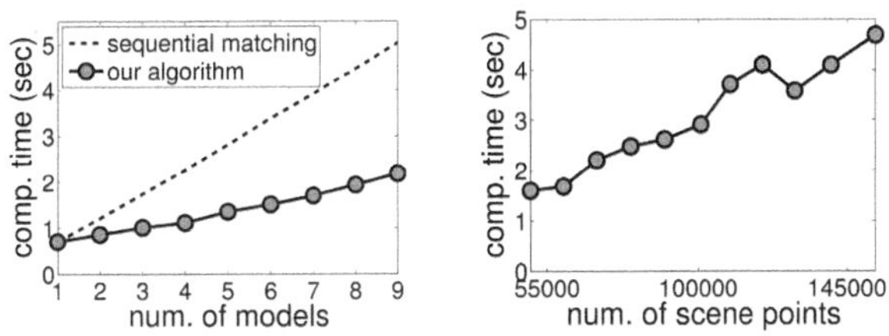


(b) (c)

**Fig. 6.** (a) Recognition time for each model. (b) Computation time for a simultaneous recognition of multiple objects (solid line) compared to a sequential matching of each model to the scene (dashed line). The runtime in the case of the sequential matching is the sum of the times reported in (a) for each model. (c) Linear time complexity in the number of scene points for the simultaneous recognition of 9 models.

scene is the leftmost one in Fig. 5. The recognition time for each object (when it is the only one loaded in the hash table) is reported in Fig. 6(a). In Fig. 6(b), the computation time of our algorithm as a function of the number of models loaded in the hash table is compared with the time needed for a sequential matching of each model to the scene. The difference in the performance is obvious.

Second, we measure how the runtime depends on the number of scene points. There are eleven different data sets involved in this test case — a subset from the scenes used in the comparison test case. It is important to note that we do *not* take a single data set and down/up-sample it to get the desired number of points. Instead we choose eleven *different* scenes with varying scene extent, number of points and number of objects. This suggests that the results will hold for arbitrary scenes. We report the results of this test in Fig. 6(c).

The algorithm presented in this paper is implemented in C++ and all tests were performed on a laptop with an Intel Core 2 Duo 3GHz CPU and 4GB RAM.

## 5 Conclusions

In this paper, we introduced a new algorithm for multiple 3D object recognition in noisy, sparsely reconstructed and unsegmented range data. The method combines a robust descriptor, a hashing technique and an efficient RANSAC-like sampling strategy. We provided a complexity analysis of the algorithm and derived the number of iterations required to recognize the model instances with a given probability. In the experimental part of the paper, it was verified that the proposed algorithm scales well with the number of models and that it has a linear time complexity in the number of scene points. Furthermore, we showed that our method performs well on noisy, sparse and unsegmented scenes in which only small parts of the objects are visible. A comparison showed that our method outperforms the spin images [4] and the tensor matching [3] approaches in terms of recognition rate.

## References


1. Lamdan, Y., Wolfson, H.: Geometric Hashing: A General And Efficient Model-based Recognition Scheme. In: ICCV, pp. 238–249 (1988)
2. Ballard, D.H.: Generalizing the Hough Transform to Detect Arbitrary Shapes. Pattern Recognition 13, 111–122 (1981)
3. Mian, A.S., Bennamoun, M., Owens, R.A.: Three-Dimensional Model-Based Object Recognition and Segmentation in Cluttered Scenes. IEEE TPAMI 28, 1584–1601 (2006)
4. Johnson, A., Hebert, M.: Using Spin Images for Efficient Object Recognition in Cluttered 3D Scenes. IEEE TPAMI 21, 433–449 (1999)
5. Hetzel, G., Leibe, B., Levi, P., Schiele, B.: 3D Object Recognition from Range Images Using Local Feature Histograms. In: CVPR, pp. 394–399 (2001)
6. Frome, A., Huber, D., Kolluri, R., Bülow, T., Malik, J.: Recognizing Objects in Range Data Using Regional Point Descriptors. In: Pajdla, T., Matas, J(G.) (eds.) ECCV 2004. LNCS, vol. 3023, pp. 224–237. Springer, Heidelberg (2004)
7. Gelfand, N., Mitra, N., Guibas, L., Pottmann, H.: Robust Global Registration. In: Eurographics Symposium on Geometry Processing, pp. 197–206 (2005)
8. Zaharescu, A., Boyer, E., Varanasi, K., Horaud, R.: Surface Feature Detection and Description with Applications to Mesh Matching. In: CVPR, pp. 373–380 (2009)
9. Sun, J., Ovsjanikov, M., Guibas, L.J.: A Concise and Provably Informative Multi-Scale Signature Based on Heat Diffusion. Comput. Graph. Forum 28, 1383–1392 (2009)
10. Matei, B., Shan, Y., Sawhney, H.S., Tan, Y., Kumar, R., Huber, D.F., Hebert, M.: Rapid Object Indexing Using Locality Sensitive Hashing and Joint 3D-Signature Space Estimation. IEEE TPAMI 28, 1111–1126 (2006)
11. Winkelbach, S., Molkenstruck, S., Wahl, F.M.: Low-Cost Laser Range Scanner and Fast Surface Registration Approach. In: Proceedings of 28th DAGM Symposium Pattern Recognition, pp. 718–728 (2006)
12. Fischler, M.A., Bolles, R.C.: Random sample consensus: a paradigm for model fitting with applications to image analysis and automated cartography. Commun. ACM 24, 381–395 (1981)
13. Wang, H., Suter, D.: Robust Adaptive-Scale Parametric Model Estimation for Computer Vision. IEEE TPAMI 26, 1459–1474 (2004)
14. Wang, H., Mirota, D., Hager, G.D.: A Generalized Kernel Consensus-Based Robust Estimator. IEEE TPAMI 32, 178–184 (2010)
15. Chen, C.S., Hung, Y.P., Cheng, J.B.: RANSAC-Based DARCES: A New Approach to Fast Automatic Registration of Partially Overlapping Range Images. IEEE TPAMI 21, 1229–1234 (1999)
16. Schnabel, R., Wahl, R., Klein, R.: Efficient RANSAC for Point-Cloud Shape Detection. Comput. Graph. Forum 26, 214–226 (2007)
17. Aiger, D., Mitra, N.J., Cohen-Or, D.: 4-points Congruent Sets for Robust Pairwise Surface Registration. ACM Trans. Graph. 27 (2008)
18. Shan, Y., Matei, B., Sawhney, H.S., Kumar, R., Huber, D.F., Hebert, M.: Linear Model Hashing and Batch RANSAC for Rapid and Accurate Object Recognition. In: CVPR, pp. 121–128 (2004)

# Change Detection for Temporal Texture in the Fourier Domain

Alexia Briassouli and Ioannis Kompatsiaris

Informatics and Telematics Institute
Centre for Research and Technology, Hellas
6th km, Charilaou-Thermis
Thermi, 57001

**Abstract.** Research on temporal textures has concerned mainly modeling, synthesis and detection, but not finding changes between different temporal textures. Shot change detection, based on appearance, has received much research attention, but detection of changes between temporal textures has not been addressed sufficiently. Successive temporal textures in a video often have a similar appearance but different motion, a change that shot change detection cannot discern. In this paper, changes between temporal textures are captured by deriving a non-parametric statistical model for the motions via a novel approach, based on properties of the Fourier transform. Motion statistics are used in a sequential change detection test to find changes in the motion distributions, and consequently the temporal textures. Experiments use a wide range of videos of temporal textures, groups of people, traffic. The proposed approach leads to correct change detection, at a low computational cost.

## 1 Introduction

The analysis of motion in video is fundamental for characterizing its contents, motion segmentation, activity recognition, and various other tasks. Temporal textures, also known as dynamic textures, are videos of textures which evolve over time. Numerous methods have been developed for the modeling and subsequent analysis of temporal textures, based on appropriately designed features of the video, or a model of the temporal texture process [1]. Initial work on temporal textures modeled them based on a combination of spatial and flow features [2]. Model based techniques have been developed for the recognition and synthesis of dynamic textures. The Spatiotemporal Auto-Regressive model (STAR) has been used extensively [3], while improvements have been made on it to account for rotational, non-translational motions [4]. In [5], temporal texture motions are modeled as a random process by using adaptive, predictive models. In [6], they are modeled by a linear dynamical system, and then spatially segmented by generalized PCA. A similar problem, that of spatially segmenting multiple temporal textures in each video frame, is addressed in [7], where motion co-occurrence statistics are used, as they provide a correspondence between the features of single and multiple textures.

R. Kimmel, R. Klette, and A. Sugimoto (Eds.): ACCV 2010, Part I, LNCS 6492, pp. 149–160, 2011.

**Proposed approach:** This work presents a novel method for the detection of changes in videos of successive temporal textures. We refer to each sequence of frames that corresponds to a temporal texture as a "temporal texture subsequence". The videos considered contain one temporal texture over a series of video frames, followed by a different temporal texture over another series of frames. The proposed method detects the frame at which a change occurs from one temporal texture subsequence to the next. For example, the frame at which the rippling of leaves changes intensity or density, the frame at which a crowd's motion changes, are found. These changes are difficult to detect by traditional shot change detection methods, which are most often based on appearance, because in many practical applications, such as detecting the change in the flow of traffic in a highway, there is a very small differentiation in the appearance of successive temporal textures. Traditional motion estimation methods are not suited for such videos either, as the motions of temporal textures are highly non-rigid [8]. This paper addresses the problem of separating temporal textures based on their (different) motion characteristics, rather than appearance features. The motion is modeled as following a random distribution based on properties of the FT, as detailed in Sec. 2, 3. The resulting statistical model is non-parametric and offers a complete description of the motion characteristics of each temporal texture. It is used to detect when a change takes place in the video from one temporal texture to another via sequential change detection techniques, which are designed precisely for the problem of detecting a change from one distribution to the next, by using each sample (video frame) as it arrives.

**Comparison with existing methods:** In [5], change detection refers to the separation of a foreground undergoing rigid object motion from non-rigid background motion via an adaptive predictive model for the temporal textures. Temporal texture motion field estimation takes place in [9] based on the STAR model, giving an accurate flow field, but at a high computational cost. In this work, the motion field is approximated as a random vector following a non-parametric probability distribution that is extracted at a low cost from the Fourier Transform (FT) of the video (Sec. 2, 3). This provides sufficient information for the separation of different temporal texture subsequences, without the computational burden of estimating a precise motion field. In [10], the change point between successive temporal textures is detected by modeling the video by a linear dynamic system whose parameters are determined after training with the EM algorithm. Our work differs from that of [10], as we derive a non-parametric description of the temporal texture motion, derived empirically from the data. Thus, we avoid issues such as tuning the EM algorithm parameters to ensure convergence, or determining the optimal clustering technique for segmenting the video into similar temporal textures, that is required in [10], [11]. The non-parametric model derived in this work provides a better description of the motion statistics than a simple histogram of the motion vectors, as the latter are not easy to calculate with accuracy and speed in temporal texture videos. Both approaches present a solution to the challenging problem of separating temporal textures, which cannot be dealt with the conventional optical flow or appearance based video segmentation methods, due to the stochasticity of the motion in temporal textures, its high non-rigidity and the large number of moving entities present in them (e.g. tree leaves, people in a crowd etc.).

This paper is organized as follows. In Sec. 2, a statistical model for a video containing multiple random motions or, equivalently, temporal textures, is presented. Sec. 3 provides the theoretical description of the method proposed for approximating the random motion distribution, based on the video FT. Sequential change detection for finding the moments of change between temporal texture subsequences is presented in Sec. 4. Experimental results for videos with various temporal texture subsequences, crowds of people walking, highway traffic, are shown in Sec. 5, and conclusions are provided in Sec. 6.

## 2 Statistical Model for Multiple Random Motions, Temporal Textures

We consider the case of a video containing successive temporal texture subsequences. Each subsequence contains a single temporal texture over all frame pixels, i.e. we do not examine the case where there are more than one temporal textures present in a frame (as in [7]). The pixel intensity of frame $k$ of the video sequence at pixel $\bar{r}$ is represented by $a(\bar{r}, k)$, and is considered to consist of $M$ moving "objects" $s_i(\bar{r})$, $1 \leq i \leq M$. The areas of the video that do not contain a moving object, e.g. the sky behind fluttering tree leaves, belong to background pixels, whose illumination is denoted as $s_b(\bar{r})$. Then, frame 1 is given by:

$$a(\bar{r}, 1) = s_b(\bar{r}) + s_1(\bar{r}) + ... + s_M(\bar{r}). \tag{1}$$

In videos of a group of people or of traffic, the moving objects $s_i(\bar{r})$ are the people or cars. In videos of temporal textures like flowing water or leaves moving in the wind, the $M$ objects represent each moving pixel or small group of pixels (e.g. the pixels forming a leaf), since those motions are highly non-rigid. In reality, the relation between the video frame illumination and that of the background and the objects' illumination is not additive, since the object pixels actually mask background pixels. However, in many temporal texture videos, there is no background at all (all pixels are moving e.g. in a video of flowing water). Even in cases of temporal textures with a background, the random movements are small, so the resulting occlusion is insignificant. This is verified by the experimental results with various types of backgrounds, where the change detection provides accurate results. The Fourier transform (FT) of the first video frame is given by:

$$A(\bar{\omega}, 1) = S_b(\bar{\omega}) + S_1(\bar{\omega}) + ... + S_M(\bar{\omega}), \tag{2}$$

where $\bar{\omega}$ is the two-dimensional frequency of the FT. At frame $k$, every object (group of pixels corresponding to a group of people, cars, or a single leaf, flower on a tree) has been displaced by random $\bar{r}_i$, $1 \leq i \leq M$, so frame $k$ is given by:

$$a(\bar{r}, k) = s_b(\bar{r}) + s_1(\bar{r}, k) + ... + s_M(\bar{r}, k) = s_b(\bar{r}) + s_1(\bar{r} - \bar{r}_1) + ... + s_M(\bar{r} - \bar{r}_M), \tag{3}$$

and since $S_i(\bar{\omega},k) = S_i(\bar{\omega})e^{-j\bar{\omega}^T\bar{r}_i}$, its FT becomes:

$$A(\bar{\omega},1) = S_b(\bar{\omega}) + S_1(\bar{\omega})e^{-j\bar{\omega}^T\bar{r}_1} + ... + S_M(\bar{\omega})e^{-j\bar{\omega}^T\bar{r}_M}. \quad (4)$$

It is known from probability theory that the characteristic function of a probability density function is equal to the complex conjugate of its FT [12], i.e.

$$\Phi(\bar{\omega}) = \mathfrak{F}^*[f(\bar{r})] = \int_{-\infty}^{+\infty} f(\bar{r})e^{j\bar{\omega}^T\bar{r}}d\bar{r} = E[e^{j\bar{\omega}^T\bar{r}}]. \quad (5)$$

For videos of temporal textures, each displacement $\bar{r}_i$ is considered to follow a random distribution $f_i(\bar{r})$.The expected value of eq. (4) is then given by:

$$\begin{aligned} E[A(\bar{\omega},k)] &= E[S_b(\bar{\omega})] + E[S_1(\bar{\omega})e^{-j\bar{\omega}^T\bar{r}_1}] + ... + E[S_M(\bar{\omega})e^{-j\bar{\omega}^T\bar{r}_M}] \\ &= S_b(\bar{\omega}) + S_1(\bar{\omega})E[e^{-j\bar{\omega}^T\bar{r}_1}] + ... + S_M(\bar{\omega})E[e^{-j\bar{\omega}^T\bar{r}_M}] \\ &= S_b(\bar{\omega}) + S_1(\bar{\omega})\Phi_1^*(\bar{\omega}) + ... + S_M(\bar{\omega})\Phi_M^*(\bar{\omega}), \end{aligned} \quad (6)$$

where the expected value operator $E[\cdot]$ only affects the displacements since the other quantities are not random, and where we have used the definition of eq. (5) to include the characteristic function in eq. (6).

The characteristic function provides the most complete statistical description of a random variable, as that random variable's pdf and all its existing moments can be derived from the characteristic function. The probability density function $f_i(\bar{r})$ corresponding to each characteristic function $\Phi_i(\bar{\omega})$ is given by:

$$f_i(\bar{r}) = \frac{1}{2\pi}\int_{-\infty}^{+\infty} \Phi_i(\bar{\omega})e^{-j\bar{\omega}^T\bar{r}}d\bar{\omega}. \quad (7)$$

Thus, the expression of each video frame's FT as a function of its random motion's characteristic function provides a comprehensive description of the activity in it, motivating us to use it for detecting changes between temporal texture subsequences.

## 3 Statistical Model for One Type of Random Motion, Temporal Textures

In this work we focus on videos containing one kind of random motion, or one kind of temporal texture, in the frames of a temporal texture subsequence. This encompasses a wide range of temporal texture videos found in practical applications, such as videos of trees, water flowing, groups of people walking, and traffic. Since one kind of random motion takes place, we have $\bar{r}_i \sim f_0(\bar{r})$ with characteristic function $\Phi_0(\bar{\omega})$, and the model of eq. (6) can be re-written as:

$$E[A(\bar{\omega},k)] = S_b(\bar{\omega}) + S_1(\bar{\omega})\Phi_0^*(\bar{\omega}) + ... + S_M(\bar{\omega})\Phi_0^*(\bar{\omega}) = S_b(\bar{\omega}) + \Phi_0^*(\bar{\omega})\sum_{i=1}^{M} S_i(\bar{\omega}). \quad (8)$$

For videos with a static background or a background that undergoes small changes, $S_b(\bar{\omega})$ can be removed via one of the numerous background removal techniques that are available [13], [14]. In practice, very many videos of temporal textures contain no background, or a very small static background area, so its removal is either very simple or not necessary at all. Consequently, the background can be removed or ignored (if it is very small or non-existent) and eq. (8) can be written as:

$$E[A(\bar{\omega}, k)] = \Phi_0^*(\bar{\omega}) \sum_{i=1}^{M} S_i(\bar{\omega}). \tag{9}$$

The FT of the first frame of such a video is given by:

$$E[A(\bar{\omega}, 1)] = \sum_{i=1}^{M} S_i(\bar{\omega}). \tag{10}$$

By combining eqs. (9) and (10), we have:

$$\frac{E[A(\bar{\omega}, k)]}{E[A(\bar{\omega}, 1)]} = \frac{\Phi_0^*(\bar{\omega}) \sum_{i=1}^{M} S_i(\bar{\omega})}{\sum_{i=1}^{M} S_i(\bar{\omega})} = \Phi_0^*(\bar{\omega}). \tag{11}$$

Thus, if the expected values of the frames' FTs $E[A(\bar{\omega}, 1)]$, $E[A(\bar{\omega}, k)]$ are known, we can obtain a complete description of the random motion in the temporal texture video. The video frames are available and their FT in Eq. (11) can easily be computed. In order to estimate the expected value of these FTs, several instantiations of each temporal texture video need to be available under the ergodicity assumption. However, this is not practically feasible, as usually only one instance of each video is available. In order to overcome this issue, we make the observation that neighboring video frames are characterized by similar motions, so a subsequence of $w_0$ frames can be considered to consist of instantiations of a random motion that follows the distribution $f_0$. Thus, the expected value of $A(\bar{\omega}, 1)$ and $A(\bar{\omega}, k)$ can be approximated as follows:

$$E[A(\bar{\omega}, 1)] = \frac{1}{w_0} \sum_{i=1}^{w_0} A(\bar{\omega}, i), \quad E[A(\bar{\omega}, k)] = \frac{1}{w_0} \sum_{i=k}^{k+w_0-1} A(\bar{\omega}, i), \tag{12}$$

leading to the approximation of the motion characteristic function near frame $k$:

$$\Phi_0(\bar{\omega}) = \frac{\sum_{i=k}^{k+w_0-1} A^*(\bar{\omega}, i)}{\sum_{i=1}^{w_0} A^*(\bar{\omega}, i)}. \tag{13}$$

As mentioned in Sec. 2, and known from probability theory [12], the probability density function can be obtained from the characteristic function, so by estimating $\Phi_0(\bar{\omega})$, we also know the pdf $f_0(\bar{r})$. The change detection algorithm presented in the section that follows is based on knowledge of this pdf. The pdf of Eq. (13) is extracted from the ratio of FTs, so only the information in the FT phase is retained: all appearance information is eliminated, and the proposed method uses only on the extracted *motion* distribution information.

## 4 Change Detection for Temporal Textures

As mentioned in Sec. 1, this work focuses on the detection of the moments of change in videos of temporal textures. Its aim is to detect at which frame the random motion changes, i.e. the motion follows a different pdf. For example, the change between water flowing slowly and then fast, traffic changing from heavy to light, is to be detected. It should be noted that the problem addressed here is particularly challenging, as not only the moment of change $M$ is unknown, but the motion distributions before and after the change are also not known. Sequential change detection methods are well suited to this task, since they can detect changes in a sequence of random variables as they arrive. For the data sequence $\bar{x} = [x_1, x_2, ..., x_N]$, where the first $L$ variables follow a pdf $f_0$, while the rest follow pdf $f_1$, the following hypothesis test applies:

$$\begin{aligned} H_0 &: x_n \sim f_0 \\ H_1 &: x_n \sim f_1. \end{aligned} \tag{14}$$

Sequential change detection examines the data for a change by estimating the log-likelihood ratio (LLRT) of the input data, as proposed by Page in [15], in order to detect if a change has happened at each frame $k$. For the data from frames 1 to $k$, the LLRT is given by:

$$T_{1,k} = \ln \frac{f_1(\bar{x}_{1,k})}{f_0(\bar{x}_{1,k})}, \tag{15}$$

where $\bar{x}_{1,k} = [x_1, ..., x_k]$ are all the samples from frames 1 to $k$. If these samples are independent and identically distributed (i.i.d.), $T_{1,k}$ becomes:

$$T_{1,k} = \ln \prod_{n=1}^{k} \frac{f_1(x_n)}{f_0(x_n)} = \sum_{n=1}^{k} \ln \frac{f_1(x_n)}{f_0(x_n)}. \tag{16}$$

The i.i.d. assumption is reasonable when neighboring pixels in a video frame move independently from each other, as is the case in temporal textures videos, which usually contain highly non-rigid activity. Additionally, the i.i.d. assumption is often necessary in practice, as the joint pdf of the motion over all data samples can be very cumbersome and impractical to estimate. For the case of i.i.d. samples, the test statistic of Eq. (16) is expressed in a computationally efficient iterative form [15] as a cumulative sum, giving the CUSUM test:

$$T_k = \max\left(0, T_{k-1} + \ln \frac{f_1(x_n)}{f_0(x_n)}\right), \tag{17}$$

where $T_k = T_{1,k}$ and $T_0 = 0$. This test detects a change when the distribution changes from $f_0$ to $f_1$, making the test statistic $T_k$ higher than a pre-defined threshold. Currently, there is no generally applicable way to obtain a closed-form expression for the threshold that will lead to the highest change detection rates, with the smallest amount of false alarms. A method for learning the threshold sequentially has been presented in [16], where the threshold is the solution to a diffusion equation. However, that approach is limited to computing a threshold for only a few simple cases, and also requires training with samples that follow both $H_0$ and $H_1$, which is often impractical. In [17], another method of threshold learning for the CUSUM test is proposed, based on parametric models for $f_1 = f(\theta_1), f_0 = f(\theta_0)$, but requires a non-parametric empirical approximation of the probability distributions before and after a change. Thus, the thresholds for the most general case are determined empirically, by experimental tuning with training data [18]. Here, after training with several videos, it is found that the threshold at frame $k$ can be obtained by the formula:

$$\eta = \mu_k + c \cdot \sigma_k, \tag{18}$$

where $\mu_k$, $\sigma_k$ are the mean and standard deviation, respectively, of the test statistic from frames 1 to $k-1$. When there is a change, $T_k$ becomes significantly higher than its previous values, and consequently the threshold $\eta$, which takes them into account. Experiments with this formula applied to videos in the categories included in our experiments show that correct detection but few false alarms are obtained for $c = 2$ to $c = 3$.

In order to implement the test of eq. (17), the pdfs $f_0$ and $f_1$ need to be known. They can be estimated from their characteristic functions that are derived as described in Section 3. In order to calculate $f_0$, or equivalently $\Phi_0$, we make the assumption that the first $w_0$ frames of the video under examination correspond to the initial motion, with "baseline" distribution $f_0$. In order to approximate $f_1$, we assume that frame $k$ and its neighboring $w_0$ frames follow $f_1$, and use this data in eq. (13). As mentioned in Sec. 3, it is reasonable to assume that in neighboring frames similar motion is taking place. This approach is common in similar problems [19], where the online estimation of distributions is necessary because of the complete lack of knowledge regarding them. When a change occurs, this assumption is violated temporarily, until all data being used follows the new distribution.

If the data from frames $k - w_0 + 1$ to $k$ actually follows $f_0$, it will produce a "current" distribution approximation $f_1$ which is close to $f_0$, leading to a low value of the LLRT in eq. (17). When a change has occurred, the pdf approximation for $f_1$ will deviate from that for $f_0$, leading to a higher value for the LLRT, which will indicate that a change took place. It should be noted that the CUSUM test has been proven to provide the fastest detection of change, so there will not be a significant delay between the changepoint and the instant at which the test statistic $T_k$ becomes higher than the detection threshold.

## 5 Experiments

Many different kinds of video sequences are examined to test the proposed system, and compare it with shot change detection. In the experiments the shot change detection fails to detect changes, whereas our approach can find meaningful changes. Table 1 contains the detected changes for the proposed approach and shot change detection. The proposed method runs faster than shot change detection, taking a fourth of the time to run, although it is implemented in Matlab, while the shot detection is implemented in C++. This can be attributed to the simpler nature of the algorithm.

### 5.1 Temporal Textures

A series of videos containing subsequences of temporal textures that undergo changes is examined. These videos can be found in the supplementary material provided with this paper, where the changes in texture motion are evident. In the video of flowers fluttering (Fig. 1(a)), in the first 200 frames the flowers are moving very fast, and in the rest they move more slowly. The proposed method correctly detects a change at frame 215. There is an ambiguity of a few frames because the change is detected using $w_0 = 10$ frames around each time instant to approximate the current pdf at frame $k$, as described in Sec. 3. This ambiguity is negligible, since a difference of about 10 frames is not visible in practice. A video of candles flickering is examined, where the speed of their flickering increases significantly at frame 400. The proposed algorithm detects a change at frame 428. The motion in a video of flamingoes walking becomes much slower at frame 190, and our method finds a change at frame 194. A video of seawater that rain is falling on, first very rapidly, and then more slowly, is examined. The change of the speed of the rain is correctly found at frame 67, as the actual change occurs at frame 70. In a sequence of a toilet flushing, the frame where the water stops running at frame 100 is detected at frame 96, which is very close. Fig. 1(e), (f) show frames 95 and 110, before and after the change, respectively. In the second video, changes are detected at frames 35 when the water starts flowing, and frame 160 when it stops running. Fig. 1(g)-(j) shows frames before and after these changes. In this case false alarms are also found at frames 200 and 220, caused by minor changes in the motion of the trickling water. A video of a barbeque is also examined, where the smoke changes direction and intensity. The change of the smoke's direction is correctly found at frame 82, as seen in Fig. 1(k), (l). The shot change detection of [20] is applied to these videos and, in all cases, is unable to detect the changes, as shown the top half of Table 1.

### 5.2 Random Motions in Groups

In this section, we examine videos of groups of entities moving together, such as people in crowds, cars and trucks in traffic. Here, the change takes place over a few frames rather than at one frame, e.g. a crowd does not enter a scene instantaneously, but over several frames. The ground truth for the correct moment of change is derived by observing the video and is considered to be the central frame of the frame subsequence during which the change takes place.

**Highway sequence: varying traffic density:** A video of traffic on a highway, whose density changes in the middle of the video, is examined in this experiment. Frame 52 before the change and frame 55 after it are shown in Figs. 2 (a), (b), where the difference in the traffic density is apparent to the observer. The

**Fig. 1.** Temporal textures. (a)-(d): video frames. (e)-(l): before/after a change.

CUSUM test statistic computed by the proposed method is plotted in Fig. 2 (c), where it is clear that the moment of change in the motion distribution has been captured. This results in the detection of a change at frame 60, i.e. with a small delay of 5 frames.

**Traffic sequence: trucks followed by cars:** In this video of a highway, initially there are mostly trucks and buses in the scene, followed by traffic consisting mostly of cars. In frame 50 before the change there are indeed mostly cars, and after the change at frame 140, e.g. in frame 180 there are mostly cars (Fig. 2(d), (e)). The CUSUM test statistic computed by the proposed method are plotted in Fig. 2, where it is clear that the moment of change in the motion distribution has been captured almost exactly, at frame 142.

**Crowd crossing street sequence:** A crowd of people crossing a street is modeled as a random texture. The people approach each other at frame 130 (frame 140 is shown in Fig. 2(g)) and the crowds merge until frame 290, after which they are separated (frame 320 is shown in Fig. 2(h)). The CUSUM of Fig. 2(i) show that changes are indeed correctly detected at frames 140 and 300, which are close to the true change points.

**Pedestrians crossing street:** A more challenging video showing pedestrians crossing a street with traffic is examined: the changes in the pedestrians' motion are difficult to capture as many people are walking in opposite directions at the same time. The proposed method only detects changes after frame 200 (Fig. 2(l)), which corresponds to a change between many pedestrians to almost none crossing the street. More subtle changes, like the crowds of pedestrians moving in opposite directions merging at frames 70, 150 are not detected, as the motion information extracted from them is not sufficient (Fig. 2(j), (k)). The failure of the algorithm for this sequence is evident from the way the values of the CUSUM test statistic change with time, as shown in Fig. 2(m): these statistics change very little around frame 70, making that change difficult to detect. However, a clear change occurs and is detected after frame 200, after which there are very pedestrians left crossing the street.

**Walking sequence:** A video of small groups of people appearing, walking in front of a building and exiting is examined. Fig. 2(n)-(p) show the frames where changes were detected, with the groups of people that appear in this video. The CUSUM values of Fig. 2(q) lead to the detection of change at frames 18, when the first group is leaving the scene, at frame 37 when the second group is entering the scene, and frame 59 when they exit the scene. Thus, the proposed method produces correct results for this kind of video as well.

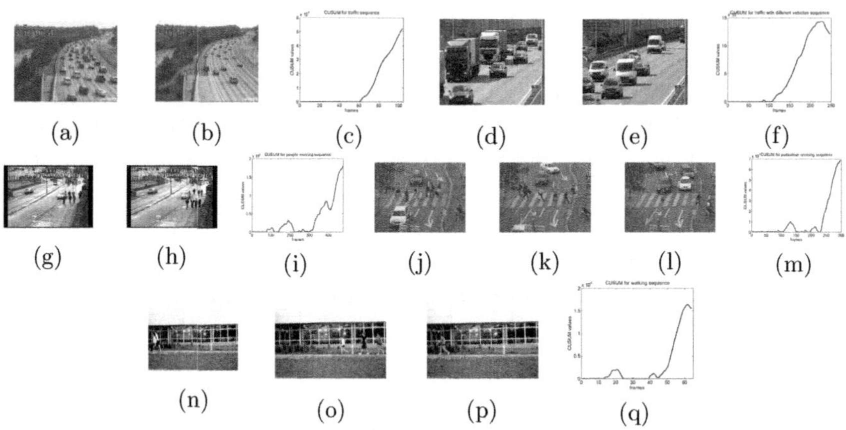


**Fig. 2.** Random motions in groups. Frames before and after a change in temporal texture, CUSUM Test statistic.

The shot change detection method of [20] is applied to the videos examined here and the results are shown in Table 1 below. The proposed method finds the changes in all cases, except the last video, where only one change is found. However, shot change detection is only able to find a change in the first traffic video, as there is a small change in the scene appearance.

**Table 1.** Change detection results for Temporal Texture videos

| Videos | Real ch. | Det. ch. | Shot ch. det | Motion before ch. | Motion after ch. |
|---|---|---|---|---|---|
| Flowers | 200 | 215 | 3 | Fast motion | Slow motion |
| Candles | 400 | 428 | 3 | Flicker | Fast flicker |
| Flamingoes | 190 | 194 | 3 | Fast walk | Normal walk |
| Water | 70 | 67 | 3 | Fast rain on sea | Normal rain on sea |
| Toilet 1 | 100 | 96 | 3 | No water | Water flushes |
| Toilet 2 | 34, 150 | 35, 160 | 3 | No water, water flushes | Water flushes, stop |
| BBQ | 90 | 82 | 3 | BBQ flame, left | flame, right |
| Traffic 1 | 53 | 54 | 3 | Light traffic | Heavy traffic |
| Traffic 2 | 140 | 142 | 3 | Trucks | Cars |
| Walking | 20, 35, 57 | 18, 37, 59 | 3 | Exit, enter, exit | Walk, |
| Crowd | 130, 290 | 140, 300 | 476 | Meets, part | Walk separately |
| Pedestrians | 70, 210 | 200 | 297 | Peds. meet, stop crossing | No peds. |

## 6 Conclusions

In this work, a novel approach to the segmentation of a video consisting of temporal texture subsequences is presented. The proposed approach provides an approximation to the temporal texture's motion distribution based on properties of the Fourier Transform, leading to a non-parametric model for the motion. Statistical sequential testing, namely the CUSUM test, is then applied to the resulting distributions, in order to detect changes between successive temporal texture subsequences. The need for fitting an appropriate statistical model to the data is avoided, as well as tuning its parameters appropriately. It is necessary to perform empirical testing with training data beforehand, in order to determine a general threshold formula for the categories of videos examined. Experiments show that the proposed method detects the moment of change between temporal texture subsequences with accuracy, based only on their motion characteristics, i.e. without using any appearance information. Comparisons with traditional shot change detection methods show that the CUSUM based approach, with online non-parametric distribution modeling, provides the same or better result, at a much lower computational cost. Shot change detection methods detect changes at the moment of change in sequences where a minor change in appearance has occurred, however they fail in the more challenging videos of crowds of people walking, or of traffic. Future work includes the application of the proposed approach in more complicated videos, where changes are more complex, as well as to videos containing more than one temporal texture in the scene.

**Acknowledgements.** The research leading to these results has received funding from the European Community's Seventh Framework Programme FP7/2007-2013 under grant agreements: FP7-214306 - JUMAS.

## References

1. Rahman, A., Murshed, M.: Temporal texture characterization: A review. In: Computational Intelligence in Multimedia Processing: Recent Advances, pp. 291–316 (2008)
2. Polana, R., Nelson, R.: Recognition of motion from temporal texture. In: Proc. IEEE Conference on Computer Vision and Pattern Recognition, pp. 129–134 (1992)
3. Szummer, M., Picard, R.: Temporal texture modeling. In: IEEE International Conference on Image Processing (ICIP), pp. 823–826 (1996)
4. Doretto, G., Chiuso, A., Soatto, S., Wu, Y.: Dynamic textures. International Journal of Computer Vision (IJCV) 51, 91–109 (2003)
5. Mittal, A., Monnet, A., Paragios, N.: Scene modeling and change detection in dynamic scenes: A subspace approach. Comput. Vis. Image Underst. 113, 63–79 (2009)
6. Vidal, R., Ravichandran, A.: Optical flow estimation and segmentation of multiple moving dynamic textures. In: CVPR 2005: Proceedings of the 2005 IEEE Computer Society Conference on Computer Vision and Pattern Recognition (CVPR 2005), vol. 2, pp. 516–521 (2005)
7. Rahman, A., Murshed, M.: Detection of multiple dynamic textures using feature space mapping. IEEE Transactions on Circuits and Systems for Video Technology 19(5), 766–771 (2009)
8. Jacobson, L., Wechsler, H.: Spatio-temporal image processing: Theory and scientific applications. In: Jähne, B. (ed.) Spatio-Temporal Image Processing. LNCS, vol. 751. Springer, Heidelberg (1993)
9. Edwards, D., Chang, J.T., Shi, L., Yu, Y.: Motion field estimation for temporal textures. In: Digital Image Computing: Techniques and Applications DICTA, pp. 389–398 (2003)
10. Chan, A.B., Vasconcelos, N.: Modeling, clustering, and segmenting video with mixtures of dynamic textures. IEEE Transactions on Pattern Analysis and Machine Intelligence 30(5), 909–926 (2008)
11. Chan, A.B., Vasconcelos, N.: Mixtures of dynamic textures. In: Proceedings of IEEE International Conference on Computer Vision (2005)
12. Papoulis, A.: Probability, Random Variables, and Stochastic Processes, 2nd edn. McGraw-Hill, New York (1987)
13. Stauffer, C., Grimson, W.: Adaptive background mixture models for real-time tracking. In: Proceedings of IEEE Computer Society Conference on Computer Vision and Pattern Recognition, CVPR 1999, vol. 2, pp. 2246–252 (1999)
14. Zivkovic, Z., van der Heijden, F.: Efficient adaptive density estimation per image pixel for the task of background subtraction. Pattern Recogn. Lett. 27, 773–780 (2006)
15. Page, E.S.: Continuous inspection scheme. Biometrika 41, 100–115 (1954)
16. Bershad, N.J., Sklansky, J.: Threshold learning and brownian motion. IEEE Transactions on Information Theory 17, 350–352 (1971)
17. Hory, C., Kokaram, A., Christmas, W.: Threshold learning from samples drawn from the null hypothesis for the generalized likelihood ratio CUSUM test. In: IEEE Workshop on Machine Learning for Signal Processing, pp. 111–116 (2005)
18. Basseville, M., Nikiforov, I.: Detection of Abrupt Changes: Theory and Application. Prentice-Hall, Inc., Englewood Cliffs (1993)
19. Muthukrishnan, S., van den Berg, E., Wu, Y.: Sequential Change Detection on Data Streams. In: Seventh IEEE International Conference on Data Mining Workshops (ICDMW 2007), pp. 551–550 (2007)
20. Chavez, G.C., Cord, M., Foliguet, S.P., Precioso, F., de, A., Araujo, A.: Robust scene cut detection by supervised learning. In: EUPISCO (2006)

# Stream-Based Active Unusual Event Detection

Chen Change Loy, Tao Xiang, and Shaogang Gong

School of EECS, Queen Mary University of London, United Kingdom
{ccloy,txiang,sgg}@eecs.qmul.ac.uk

**Abstract.** We present a new active learning approach to incorporate human feedback for on-line unusual event detection. In contrast to most existing unsupervised methods that perform passive mining for unusual events, our approach automatically requests supervision for critical points to resolve ambiguities of interest, leading to more robust and accurate detection on subtle unusual events. The active learning strategy is formulated as a stream-based solution, i.e. it makes decision on-the-fly on whether to query for labels. It adaptively combines multiple active learning criteria to achieve (i) quick discovery of unknown event classes and (ii) refinement of classification boundary. Experimental results on busy public space videos show that with minimal human supervision, our approach outperforms existing supervised and unsupervised learning strategies in identifying unusual events. In addition, better performance is achieved by using adaptive multi-criteria approach compared to existing single criterion and multi-criteria active learning strategies.

## 1 Introduction

Video surveillance data is typically characterised by highly imbalanced class distribution, i.e. most of the samples corresponding to normal event classes whilst the remaining unusual event (rare or abnormal events that should be examined further) classes only constituent a small percentage of the entire dataset. In addition, normal patterns are often known *a-priori*, whilst the unusual events are unforeseeable. Consequently, most unusual event detection methods [1,2,3,4] employ outlier detection strategy, in which a model is trained using normal events through unsupervised one-class learning and events that deviate statistically from the resulting normal profile are deemed unusual. This strategy offers a practical way of bypassing the problems of imbalanced class distribution and inadequate unusual event training samples. However, the unsupervised nature of this outlier detection methods is subject to a few inextricable limitations:

1. Difficulty in detecting unusual events whose distributions are partially overlapped with normal events. Specifically, in a busy public scene, unusual events are visually similar to a large number of normally behaving objects co-existing in a scene (see Fig. 1 for example). Without human supervision, it is hard to spot these subtle unusual events.
2. No subsequent exploitation of flagged unusual events. Outlier detection approach is therefore less effective in distinguishing the true unusual events from noise.

R. Kimmel, R. Klette, and A. Sugimoto (Eds.): ACCV 2010, Part I, LNCS 6492, pp. 161–175, 2011.

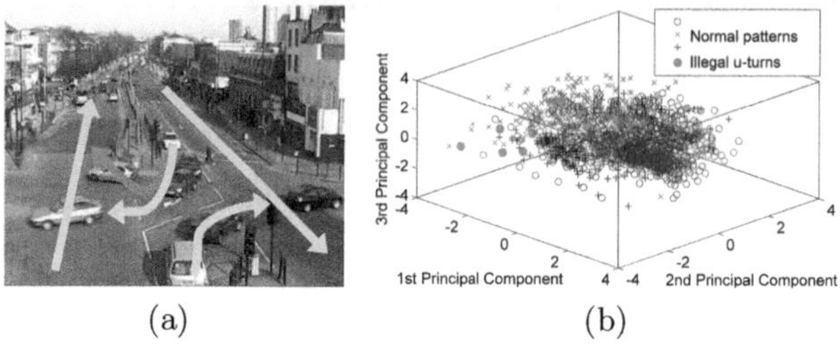


**Fig. 1.** An example of illegal u-turn event (Fig. 1a). It is subtle due to its visual similarity with large number of co-occurring normal patterns in a scene. This can observed from a plot in a principal component analysis space (Fig .1b), where similar u-turn cases (plotted as green dots) are partially overlapped with other normal patterns.

3. Large amount of uninteresting outliers causing false alarms. Normal behaviour patterns in a public scene are complicated and highly diverse. Hence, preparation of well-defined and complete normal data for off-line learning becomes unfeasible. Training a model using incomplete normal patterns is likely to result in a large amount of uninteresting outliers, since some outlying regions of the normal class may be consistently and wrongly flagged as unusual event.

In most video surveillance tasks, human knowledge is readily available in practise to remedy the aforementioned issues. Although it is unfeasible to label every single instances at hand, it is still desirable to make use of occasional human inputs for guiding the creation of an activity model. In particular, human inputs may exist in the form of feedback, i.e. indicating the exact event classes or whether a particular detection is right/wrong, when the activity model encounters difficulty in distinguishing an equivocal event or subtle unusual activity. The feedback would be extremely useful to resolve ambiguities of interest and to strengthen the decision boundary of activity classes on what is normal/abnormal, leading to more robust detection on inconspicuous and unknown unusual events.

Active learning strategy emerges as an compelling alternative to conventional supervised and unsupervised unusual event detection methods, since it is capable of seeking human feedback automatically on critical instances to improve event classification performance according to some predefined query criteria [5]. Note that it differs from supervised or semi-supervised strategies, which perform random labelling that treats all samples equally, whilst in essence, not all samples are critical for learning the correct decision boundary. In this study, we formulate a novel *stream-based active learning* strategy with several key features outlined as follows: (1) The method is formulated as a stream-based approach to ensure real-time response, i.e. the model makes immediate decision on whether to query for labels as new video data are streamed in. (2) Multiple criteria are employed for joint exploration and exploitation. In particular, some classes, especially unusual event classes have to be discovered (exploration) since they are not available in the early stage of training. At the same time, it is necessary to gradually improve the model by refining the decision boundary (exploitation). Thus, different criteria are needed to achieve these goals. (3) Query criterion is adaptively selected from multiple criteria. This is important because good active learning criteria are *dataset dependent* [5]. Importantly, we typically do not know the best suited criterion for a specific dataset at different phases of learning.

In our approach, the first query criterion is a *likelihood criterion*, which favours samples that have low likelihood w.r.t. the current model. Consequently, unknown classes or unexplored regions of existing classes can be discovered. Note that our method does not assume availability of predefined classes, i.e. once a new class is discovered, the model will expand itself automatically. The second criterion is an *uncertainty criterion* based on a modified Query-by-Committee (QBC) algorithm [6,7,8]. It is used to refine the decision boundary by selecting controversial samples in uncertain regions that give rise to the most disagreement among classes, with more emphasis given to the regions surrounding unusual event classes to address the problem of imbalance class distribution. The two query criteria are dynamically re-weighted based on the Kullback-Leibler (KL) divergence [9] measured on the model before and after it is trained using a queried sample. The premise behind this adaptive weighting scheme is to favour the criterion that is more likely to return a queried sample that brings most influence to the current model.

Comparative experiments are carried out on busy public space surveillance videos. We show that by exploiting a small cost of human supervision through active learning, more robust and accurate detection of subtle unusual events is achieved compared to conventional supervised and unsupervised learning methods. In addition, the results also suggest that our adaptive multi-criteria approach outperforms single criterion and multi-criteria methods we evaluated.

## 2 Related Work

Most existing unusual event detection methods follow *unsupervised one-class learning strategy* by employing different models such as topic models [1,2,3] and Markov random field [4]. On the other hand, there are several studies that perform event classification [10,11] based on *supervised strategy*. Our method differs significantly from these methods in that our approach is capable of discovering unknown classes and resolve inter-class ambiguities by exploiting human feedback. It is thus more suitable for on-line mining of unusual events. It is worth pointing out that Sillito and Fisher [12] attempt to incorporate human feedback using a one-class semi-supervised model. However, their method is limited to learning the normal event class. Therefore, it is still facing the same problems encountered by unsupervised approaches (see Sec. 1).

In the active learning perspective, most studies to date assume pool-based setting [13,14,15,16], which requires access to a fixed pool of unlabelled data for searching the most informative instance for querying. For surveillance task since activity patterns are dynamic and unusual events are often unpredictable, preparing a pool of unlabelled data that encompasses complete event classes is impractical. Moreover, performing exhaustive search in the pool is expensive therefore unsuitable for surveillance task that demands real-time performance. Stream-based setting is preferred in this context as it is capable of making immediate query decision without the need of accessing a data pool.

Most existing stream-based approaches are based on single query criterion [6,8,17], which are obviously not sufficient for exploration and exploitation that pursue different goals in nature. Even though there are attempts in combining multi-criteria for active learning, they are either not adaptive [13,14] or limited to pool-based setting [16,18]. Non-adaptive methods (e.g. iterate over different criteria with constant weights) cannot apply the right criteria at different phases of learning, e.g. the active learner may waste effort refining the boundary before discovering the right classes, or vice versa. Methods proposed by Baram et al. [16] and Cebron and Berthold [18], though adjusting weights of different criteria online, they require access to a pool of unlabelled data, which are often unavailable to stream-based environments.

Our uncertainty criterion is based the QBC algorithm [6,8], in which an ensemble of committee members are maintained. Query will be triggered if class label of a sample is controversial among the members. Various measures of disagreement have been proposed [6,7,19]. These measures, however, only return the disagreement score among members without identifying conflicting classes, i.e. the classes closest to the uncertain point. We formulate a new QBC scoring method to identify conflicting classes, thereby incorporate a prior constraint to favour uncertain samples surrounding unusual event classes, leading to a more balanced sample selection for class imbalanced data.

In summary, the main novelties and contributions of this study are:

1. We propose a new active learning approach to incorporate crucial human supervision to resolve ambiguities for more robust and accurate unusual event detection over conventional unsupervised and supervised approaches. To the best of our knowledge, this problem has not been addressed before.
2. We introduce a new adaptive weighting scheme suitable for combining multiple query criteria in a stream-based setting. This method does not need to access a fixed pool of unlabelled data.
3. To have a more balanced sample selection, we introduce prior to constrain uncertainty criterion to favour unusual classes during decision boundary formation. For this purpose, a new QBC scoring method is formulated to identify conflicting classes.

## 3 Active Unusual Event Detection

We consider active learning in a stream-based setting in which an unlabelled sample $\mathbf{x}_t$ is observed at each time step $t$ from an input data stream $\mathcal{X} = (\mathbf{x}_1, \ldots, \mathbf{x}_t, \ldots)$. Consequently, a classifier $\mathcal{C}_t$ is required to determine on-the-fly whether or not to query for label $y_t$ or discard $\mathbf{x}_t$. Our goal is to select critical samples from $\mathcal{X}$ for annotation to achieve two tasks simultaneously: (1) to discover unusual event classes or unknown region of existing classes in the input feature space and (2) to refine the classification boundary with higher priority being given to regions surrounding the unusual classes so as to improve the detection accuracy of unusual events.

### 3.1 Activity Representation

We wish to represent activity patterns using location-specific motion information over a temporal window without relying on object segmentation and tracking. This is achieved through the following steps: (1) Given an input video, we extract optical flow in each pair of consecutive frames using [20]. (2) A method similar to that in [21] is employed to automatically decompose a complex scene into $D$ regions, $\mathbf{r} = \{r_i | i = 1, \ldots, D\}$ according to the spatial-temporal distribution of motion patterns observed (Fig. 2(b,d)). (3) Motion direction of each moving pixel in each region are quantised into four directions and put into bins. (4) A histogram $\text{hist}_{f,r_i}$ with a size of four bins is constructed for each region $r_i$ in each frame $f$. We uniformly divide the whole video sequence into non-overlapping clips, each having 50 frames in length. We then sum up individual bins of a regional histogram within each clip $t$ as $\text{hist}_{t,r_i} = \sum_{f \in \text{clip } t} \text{hist}_{f,r_i}$. (5) Non-dominant motion directions are then removed as they are more likely to be caused by error in optical flow computation[1]. (6) Finally, the histogram is discretised to construct a codebook with 16 words $\omega_j$, $j \in \{1, 2, ..., 16\}$, representing the dominant motion directions of each region. For example, word $\omega_1$ represents motionless region, word $\omega_2$ means only direction bin 1 is observed, and word $\omega_4$ indicates both occurrence of direction bins 1 and 2, etc. Consequently, the $i$th region of the $t$th clip is represented as a variable $x_{i,t}$ of 16 possible discrete values $x_{ij}$ according to its word label and the clip is denoted as $\mathbf{x}_t = (x_{1,t}, \ldots, x_{D,t})$.

### 3.2 Bayesian Classification

We wish to classify the $D$-dimensional observed vector $\mathbf{x} = (x_1, \ldots, x_D)$ into one of the $K$ classes, where a class variable is represented by $y = k \in \{1, \ldots, K\}$. We approach the classification task as Bayesian classification. To facilitate efficient incremental learning, we employ a naïve Bayesian classifier with Bayesian learning by assuming conditional independence among the distributions of input attributes $x_1, \ldots, x_D$ given the class label. The classifier is quantified by a parameter set $\boldsymbol{\theta}$ specifying the conditional probability distributions (CPDs). We assume separate multinomial distribution $p(x_i|y)$ on each $x_i$ for each class label. Consequently, we use $\boldsymbol{\theta}_{x_i|y}$ to represent a vector of parameters $\theta_{x_{ij}|y}$ for multinomial $p(x_i|y)$. Given the multinomial CPDs, the conditional probability $p(\mathbf{x}|y = k)$ for an observed vector given class $y = k$ is given as $p(\mathbf{x}|y = k) = \prod_{i=1}^{D} p(x_i|y = k)$. Given $p(\mathbf{x}|y)$ and $p(y)$, posterior conditional distribution $p(y|\mathbf{x})$ can be computed via Bayes rule. A class $y^*$ that best explains $\mathbf{x}$ can then be obtained as follows:

---

[1] The four direction bins are ranked in a descending order based on their values. The dominant motion directions is identified from the first few bins in the rank that account for a given fraction $P \in [0, 1]$ of total bin values ($P = 0.8$ in this study). Motion directions in the remainder of the bins are considered as non-dominant.

$$y^* = \underset{k \in \{1,\ldots,K\}}{\operatorname{argmax}} \; p(y = k|\mathbf{x}) = \underset{k \in \{1,\ldots,K\}}{\operatorname{argmax}} \; p(y = k)p(\mathbf{x}|y = k). \tag{1}$$

**Incremental learning** - Efficient incremental learning is required for stream-based active learning. Since we have fully observed data, we use conjugate prior to facilitate efficient Bayesian learning. The conjugate prior of a multinomial distribution with parameters $\boldsymbol{\theta}_{x_i|y}$ is the Dirichlet distribution, which is given as:

$$\mathrm{Dir}(\boldsymbol{\theta}_{x_i|y} \mid \boldsymbol{\alpha}_{x_i|y}) \propto \prod_j [\theta_{x_{ij}|y}]^{\alpha_{x_{ij}|y}-1}, \tag{2}$$

where $\alpha_{x_{ij}|y} \in \mathbb{R}^+$ are hyper-parameters of the distribution.

### 3.3 Query Criteria

In stream-based setting, the query decision is typically determined by a query score $p^{\text{query}}$ derived from a query criterion $Q$. The query score will be compared against a threshold Th. Specifically, if $p^{\text{query}} \geq \text{Th}$, query is made; otherwise $\mathbf{x}_t$ is discarded. In this study, we propose to employ two widely used criteria with clear complementary nature, namely likelihood criterion and uncertainty criterion for joint unknown event discovery and classification boundary refinement. Next we formulate methods to compute the respective query scores based on these criteria.

**Likelihood criterion** - Using this criterion a point is selected by comparing its likelihood against current distribution modelled by the classifier. In particular, given a sample $\mathbf{x}$, we first find a class $y^*$ that best explains the sample according to Eqn. (1). Secondly, for each feature node, we compute the normalised probability score of $x_i$ given $y^*$:

$$\hat{p}(x_i|y^*) = \frac{p(x_i|y^*) - \mathrm{E}\left[p(x_{ij}|y^*)\right]}{\sqrt{\mathrm{E}\left[p(x_{ij}|y^*) - \mathrm{E}\left[p(x_{ij}|y^*)\right]\right]}}.$$

The normalised probability score $\hat{p}(x_i|y^*)$ is bounded to ensure $-0.5 \leq \hat{p}(x_i|y^*) \leq 0.5$. Finally, the likelihood score at time step $t$ is calculated as:

$$p_t^l = 1 - \left( \frac{1}{2} + \frac{1}{D} \sum_{i=1}^{D} \hat{p}(x_i|y^*) \right). \tag{3}$$

The likelihood score lies within [0,1]. If $p_t^l$ of a sample is closer to 1, it is more likely to be queried.

**Uncertainty criterion** - Our uncertainty criterion is re-formulated from the existing QBC algorithm [7,6], with additional consideration on conflicting classes for yielding a more balanced sample selection.

(1) *Generating a committee* - Given a classifier $\mathcal{C}_t$ and training data $\mathcal{S}_t$, we generate $M$ committee members corresponding to hypotheses $\mathbf{h} = \{h_i\}$ of the

hypotheses space $\mathcal{H}_t$, where each hypothesis is consistent with the training data seen so far [8], i.e., $h_i \in \mathcal{H}_t | \forall (\mathbf{x}, y) \in \mathcal{S}_t, h_i(\mathbf{x}) = y$.

In a naïve Bayes classifier with multinomial CPDs, this can be done by sampling new parameters from the posterior Dirichlet distribution of classifier [7,6]. It has been proven that parameters of a Dirichlet distribution can be generated from a Gamma distribution (Chapter XI, Theorem 4.1 in [22]). Consequently, we sample $\hat{\boldsymbol{\theta}}_{x_i|y}$ from its posterior Dirichlet distribution $\text{Dir}(\hat{\boldsymbol{\theta}}_{x_i|y}|\boldsymbol{\alpha}_{x_i|y})$, by drawing new weights $\hat{\alpha}_{x_{ij}|y}$ from the Gamma distribution, i.e. $\hat{\alpha}_{x_{ij}|y} \sim \text{Gam}\left(\alpha_{x_{ij}|y}\right)$. The parameter of a committee member is then estimated as:

$$\hat{\theta}_{x_{ij}|y} = \frac{\hat{\alpha}_{x_{ij}|y} + \lambda}{\sum_j \left(\hat{\alpha}_{x_{ij}|y} + \lambda\right)}. \tag{4}$$

where $\lambda$ is a weight added to compensate data sparseness, i.e. to prevent zero probabilities for infrequently occurring values $x_{ij}$.

(2) *Measure of disagreement* - As discussed in Sec. 2, existing approaches of measuring member disagreement are not able to return the corresponding classes that cause the most disagreement. In this study, we formulate a new uncertainty score as follows: first, a class disagreement score is computed over all possible class labels:

$$s_{y=k,t} = \left\{ \max_{h_i \in \mathcal{H}_t, h_j \in \mathcal{H}_t} \left[ p_i(y = k|\mathbf{x}_t) - p_j(y = k|\mathbf{x}_t) \right] \right\}, \tag{5}$$

where $i \neq j$. Consequently, the top two classes that return the highest $s_{y=k,t}$ are identified as $c_1$ and $c_2$. The final uncertainty score is computed as:

$$p_t^u = \frac{1}{2} \cdot \gamma_u \cdot \left[ s_{y=c_1,t} + s_{y=c_2,t} \right], \tag{6}$$

where $\gamma_u$ is the prior introduced to favour the learning of classification boundary for unusual classes. Specifically, $\gamma_u$ is set to a low value if $c_1$ and $c_2$ are both normal event class and a high value if any one of $c_1$ and $c_2$ is unusual event class. If $p_t^u$ of a sample is closer to 1, it is more likely to be queried.

### 3.4 Adaptive Selection of Multiple Query Criteria

As explained in Sec. 1, adaptive selection of multiple criteria is necessary for joint unknown event discovery and classification boundary refinement. In particular, different criteria can be more suitable for different datasets as well as different learning stages. Since we usually do not know the right choice *a priori*, selecting different criteria adaptively has the potential to provide a more reliable and even more optimal solution than using any single criterion alone.

To this end, we formulate an adaptive approach in selecting different query criteria for stream-based active learning. Specifically, given multiple query criteria $\mathcal{Q} \in \{Q_1, \ldots, Q_a, \ldots, Q_A\}$, a weight $w_{a,t}$ is assigned to each query criterion $Q_a$ at time step $t$. A criterion is then chosen by sampling from a multinomial distribution, $a \sim \text{Mult}(\mathbf{w}_t)$, where $\mathbf{w}_t \in \{w_{1,t}, \ldots, w_{a,t}, \ldots, w_{A,t}\}$.

The weights $\mathbf{w}$ are guided by the change in distribution modelled by our naïve Bayes classifier before and after it is updated using a newly queried sample. Intuitively, a criterion is preferred, therefore being assigned higher weight if it asks for samples that give greater impact to the existing distribution modelled by the classifier. To measure the distance between two distributions $p_{\boldsymbol{\theta}}(\mathbf{x})$ and $p_{\tilde{\boldsymbol{\theta}}}(\mathbf{x})$, we employ the KL-divergence, which is given as $\mathcal{KL}(\boldsymbol{\theta} \parallel \tilde{\boldsymbol{\theta}}) = \sum_{\mathbf{x}} p_{\boldsymbol{\theta}}(\mathbf{x}) \ln \frac{p_{\boldsymbol{\theta}}(\mathbf{x})}{p_{\tilde{\boldsymbol{\theta}}}(\mathbf{x})}$.

**Algorithm 1.** Stream-based active unusual event detection.

```
Input: Data stream X = (x_1, ..., x_t, ...), an initial classifier C_0 trained
       with a small set of labelled samples from known classes
Output: A set of labelled samples S and a classifier C trained with S
Set S_0 = a small set of labelled samples from known classes ;
for t from 1, 2, ... until the data stream runs out do
    Receive x_t;
    Compute p_t^l (Eqn. (3)) ;
    Compute p_t^u (Eqn. (6)) ;
    Select query criterion by sampling a ~ Mult(w), assign p_t^query based
    on the selected criterion ;
    if p_t^query >= Th then
        Request y_t and set S_t = S_{t-1} ∪ {(x_t, y_t)} ;
        Obtain classifier C_{t+1} by updating classifier C_t with {(x_t, y_t)} ;
        Update query criteria weights w (Eqn. (9)) ;
    else
        S_t = S_{t-1};
Unusual event is detected if p(y = unusual|x) is higher than Th_unusual;
```

In particular, given a naïve Bayes classifier $\mathcal{C}_t$ and an updated classifier $\mathcal{C}_{t+1}$ trained using $\mathcal{S}_t \bigcup \{(\mathbf{x}_t, y_t)\}$, the KL-divergence between their distributions can be decomposed as follows [23]:

$$\mathcal{KL}(\boldsymbol{\theta} \parallel \tilde{\boldsymbol{\theta}}) = \sum_{i=1}^{D} \mathcal{KL}\left(p_{\boldsymbol{\theta}}(x_i|y) \parallel p_{\tilde{\boldsymbol{\theta}}}(x_i|y)\right)$$
$$= \sum_{i=1}^{D} \sum_{k=1}^{K} p(y=k) \mathcal{KL}\left(p_{\boldsymbol{\theta}}(x_i|y=k) \parallel p_{\tilde{\boldsymbol{\theta}}}(x_i|y=k)\right). \quad (7)$$

where $\boldsymbol{\theta}$ and $\tilde{\boldsymbol{\theta}}$ represent sets of parameters of classifiers $\mathcal{C}_t$ and $\mathcal{C}_{t+1}$ respectively. A symmetric KL-divergence $\overline{\mathcal{KL}}(\boldsymbol{\theta} \parallel \tilde{\boldsymbol{\theta}})$ is computed as follows:

$$\overline{\mathcal{KL}}(\boldsymbol{\theta} \parallel \tilde{\boldsymbol{\theta}}) = \frac{1}{2} \cdot \left[\mathcal{KL}(\boldsymbol{\theta} \parallel \tilde{\boldsymbol{\theta}}) + \mathcal{KL}(\tilde{\boldsymbol{\theta}} \parallel \boldsymbol{\theta})\right]. \quad (8)$$

A weight $w_{a,t}$ at time step $t$ associated to query $Q_a$ is defined as:

$$w_{a,t} = \beta w_{a,t-1} + (1-\beta) \frac{\overline{\mathcal{KL}}_a(\boldsymbol{\theta} \parallel \tilde{\boldsymbol{\theta}})}{\sum_{a=1}^{A} \overline{\mathcal{KL}}_a(\boldsymbol{\theta} \parallel \tilde{\boldsymbol{\theta}})}, \quad (9)$$

where $\overline{\mathcal{KL}}_a(\boldsymbol{\theta} \parallel \tilde{\boldsymbol{\theta}})$ (see Eqn. (8)) represents the symmetric KL-divergence yielded by a query criterion $Q_a$ when it last triggered a query. Parameter $\beta$ is an update coefficient that controls the updating rate of weights. Algorithm 1 summaries the proposed approach.

# 4 Experiments

## 4.1 Datasets and Settings

Two public video datasets[2] captured at busy public scenes are employed in our experiments.

**MIT traffic dataset** [1] - This dataset with an approximate length of 1.5 hours (168822 frames), is recorded at 30 fps and scaled to a frame size of $360 \times 240$. The traffic is controlled with traffic lights and dominated by five different traffic flows (Fig. 2a). The scene decomposition result is given in Fig. 2b, showing the fourteen regions discovered.

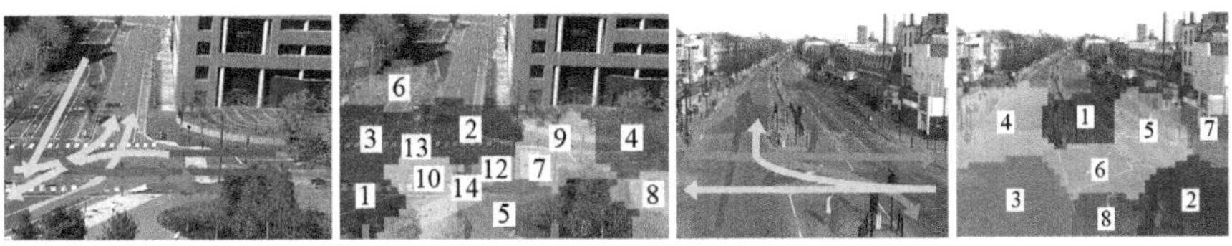


(a) (b) (c) (d)

**Fig. 2.** Dominant traffic flows observed in MIT traffic dataset (a) and Junction dataset (c) are treated as normal event classes. The scene decomposition results of both datasets according to the spatial distribution of activity patterns are shown in (b) and (d), respectively.

**Junction dataset** - The length of the video is approximately 60 minutes (89999 frames) captured with $360 \times 288$ frame size at 25 fps. The traffic is regulated by traffic lights and dominated with three traffic flows as shown in Fig. 2c. The scene decomposition result is depicted in Fig. 2d, showing the eight regions discovered.

Both datasets feature complex activities exhibited by multiple objects. In particular, behaviours and the correlations among vehicles are determined by not only the traffic light cycles, but also traffic volume and driving habits of drivers. For instance, vehicles waiting in region 6 of Junction dataset can perform horizontal turning whenever there is a gap in vertical flow. This type of activity is more frequent in MIT dataset, in which vehicles are allowed to do turning between gaps of traffic flows. As a consequence, the traffic phases of MIT traffic dataset are less distinctive visually and become harder to model compared to Junction dataset.

[2] Processed data with ground truth are available for download at: `http://www.eecs.qmul.ac.uk/~ccloy/files/accv_2010_dataset.zip`

**Table 1.** Ground truth

| Class | No. of clips (% from total) | Description |
|---|---|---|
| | | MIT Traffic Dataset |
| 1 | 874 (25.89) | Horizontal traffic flow (red arrows in Fig. 2a) |
| 2 | 1249 (37.00) | Vertical traffic flow (yellow arrows in Fig. 2a) |
| 3 | 376 (11.14) | Right-turn from zone 1 toward zone 4 (green arrow in Fig. 2a) |
| 4 | 185 (5.48) | Left-turn from zone 3 toward zone 4 (magenta arrow in Fig. 2a) |
| 5 | 517 (15.31) | Turning from left-exit toward zone 2, turning from zone 9 to zone 1 (cyan arrows in Fig. 2a) |
| 6 | 75 (2.22) | [Unusual] Left-turn from zone 1 to left-exit |
| 7 | 79 (2.34) | [Unusual] Turning right from zone 7 to zone 2 |
| 8 | 21 (0.62) | [Unusual] U-turn at zone 7 |
| | | Junction Dataset |
| 1 | 1078 (59.89) | Vertical traffic flow (red arrows in Fig. 2c) |
| 2 | 323 (17.94) | Rightward traffic flow (yellow arrows in Fig. 2c) |
| 3 | 355 (19.72) | Leftward traffic flow (green arrows in Fig. 2c) |
| 4 | 29 (1.61) | [Unusual] Illegal u-turns from zone 1 to zone 4 via zone 6 |
| 5 | 3 (0.17) | [Unusual] Emergency vehicles using an improper lane of traffic |
| 6 | 12 (0.67) | [Unusual] Traffic interruptions by fire engines |

**Ground truth** - The videos were segmented into non-overlapping clips of 50 frames long each, resulting 1800 clips and 3376 clips for Junction dataset and MIT traffic dataset respectively. Each clip was manually labelled into different event classes as listed in Table 1. The ground truth is used as feedback returned to a classifier when it requests for labels during active learning process[3]. It is also employed for comparison during testing phase.

**Settings** - The clips (see Table 1) were randomly partitioned into training/test sets with equal size. Different partitions were used in different runs in the experiments. In this study, all experimental results were averaged over 30 runs.

We followed similar experimental setting reported in [17]. In particular, if a model did not request for any labels after observing a sufficiently large number of samples (100 in this study), the query threshold Th (preset to 0.5) was reduced to Th$'$ where Th$'$ was the largest $p^{\text{query}}$ computed since the last query. A budget constraint, i.e. the number of samples a classifier can request on the data stream was specified as 250. There are three free parameters in our active learning approach, namely $\lambda$ (Eqn. (4)), uncertainty weights $\gamma_u$ (Eqn. (6)), and update coefficient $\beta$ (Eqn. (9)). We used coarse values in parameters setting without optimisation: $\lambda = 0.1$ for a weak prior, $\gamma_u = 0.9$ among normal classes, $\gamma_u = 10$ among normal-unusual classes, and $\beta = 0.9$ for slow adaptation rate. The number of committee members for all QBC approaches was set to three. It is reported in [7] that a committee size of three is sufficient and varying the size has little effect. Initially, the classifier was given a sample from a random normal event class to start the learning process. In the QBC approaches, two random samples from different classes were needed.

[3] In reality, these labels are assumed to be provided by human operators.

## 4.2 Active Learning vs. Unsupervised Learning

We first compared the proposed method with unsupervised learning approach. To build an unsupervised model, a random set of 250 normal samples was selected and normal classes were automatically determined through Gaussian clustering with automatic model selection based on Bayesian Information Criterion score [24]. The samples together with the predicted cluster labels were then employed to train a model described in Sec. 3.2. Note that the unsupervised learning strategy employed here is similar in spirit to that in [1,2,3,4]. For fair comparison, we used the same feature representation and model for both active learning and unsupervised learning strategies. As can be seen from Fig. 3a, our method outperformed the fully unsupervised method given as little as 90 samples annotated through active learning. Figure 3b suggests that the performance of an unsupervised model was still inferior than the proposed approach even if the number of unlabelled samples used for unsupervised learning was increased to 800. It is evident from the results that without exploiting human feedback, unsupervised learning method was unable to learn the correct decision boundary even if a large amount of unlabelled samples were employed.

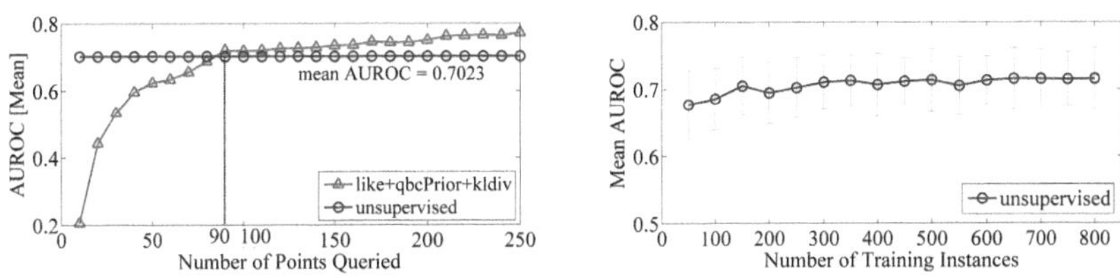


(a) (b)

**Fig. 3.** Comparison with unsupervised approach. Results were averaged over 30 runs.

## 4.3 Active Learning vs. Passive Supervised Learning and Other Active Learning Strategies

Passive supervised learning (random sampling strategy) and different active learning strategies evaluated in this experiment are summarised as follow:

1. *rand* - passive supervised learning (random sampling strategy), i.e. samples are randomly chosen from the data stream
2. *like* - likelihood criterion as explained in Sec. 3.3
3. *qbcEntropy* - QBC approach with vote entropy measure [6]
4. *qbcPrior*[4] - QBC approach with the proposed measure as described in Sec. 3.3
5. *like+qbcPrior+interleave* - combine both *like* and *qbcPrior* using interleave strategy, i.e. iterating different criteria during learning. This method is similar to the multi-criteria strategy proposed in [14]

[4] Matlab codes are available for download at:
`http://www.eecs.qmul.ac.uk/~ccloy/files/qbcPrior.zip`

6. *like+qbcPrior+KLdiv* - combine both *like* and *qbcPrior* using the KL-divergence-based strategy as described in Sec. 3.4

We evaluated different active learning strategies according to: (1) how fast they can discover unknown classes (including normal and unusual event classes) and (2) how accurately the learned classifier can detect unusual events. The former case was measured based on the number of classes discovered vs. number of samples queried. The latter case was evaluated using Area Under Receiver Operating Characteristic curve (AUROC) computed in each active learning iteration against the number of queried samples. ROC was obtained by varying $\text{Th}_{\text{unusual}}$.

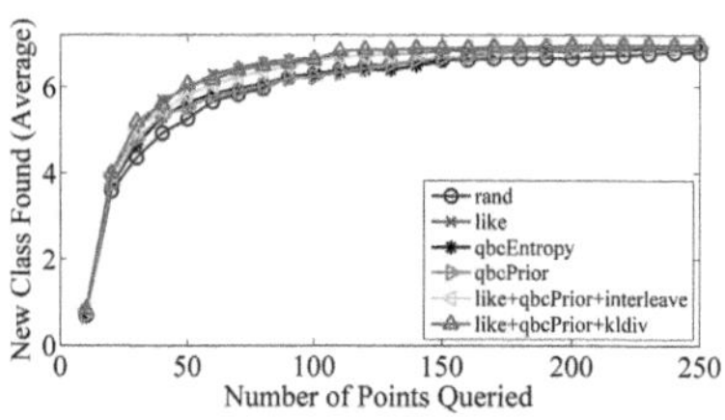


(a) MIT Traffic dataset

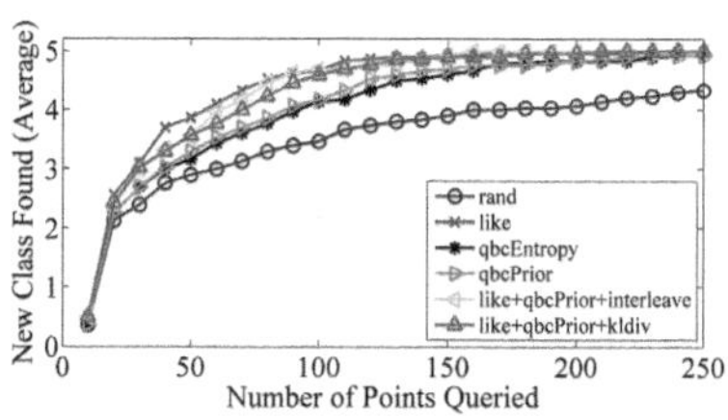


(b) Junction dataset

**Fig. 4.** Class discovery performance

**Discover unknown event classes** - As can be seen from Figure 4, *like* showed the best performance in discovering unknown event classes in both datasets. The QBC approaches (*qbcEntropy* and *qbcPrior*) yielded slightly inferior result compared to *like* but still performed better than random sampling strategy. Specifically, with the introduction of prior (see Eqn. (6)) for dealing with the imbalanced data problem, the performance of our *qbcPrior* is better to that of *qbcEntropy* (see Fig. 4b). Both the proposed *like+qbcPrior+KLdiv* method and the alternative multi-criteria method *like+qbcPrior+interleave* showed comparable results and performed better than *qbcPrior* and *qbcEntropy* after combining with *like*.

**Unusual event detection** - Figure 5 shows the performance of different active learning strategies in detecting unusual events, measured as averaged AUROC over 30 runs. Overall, it can be seen that the detection performance of all active learning methods monotonically increase as more data is queried and importantly, all methods significantly outperformed random sampling (passive supervised learning). In particular, we observed that by incorporating prior constraint into uncertainty criterion (*qbcPrior*) yielded slightly better result compared to method without prior constraint (*qbcEntropy*). This is because that, without the prior constraint, *qbcEntropy* wasted some effort in refining boundary between normal classes, whilst *qbcEntropy* focussed on the uncertainty regions surrounding the unusual event classes.

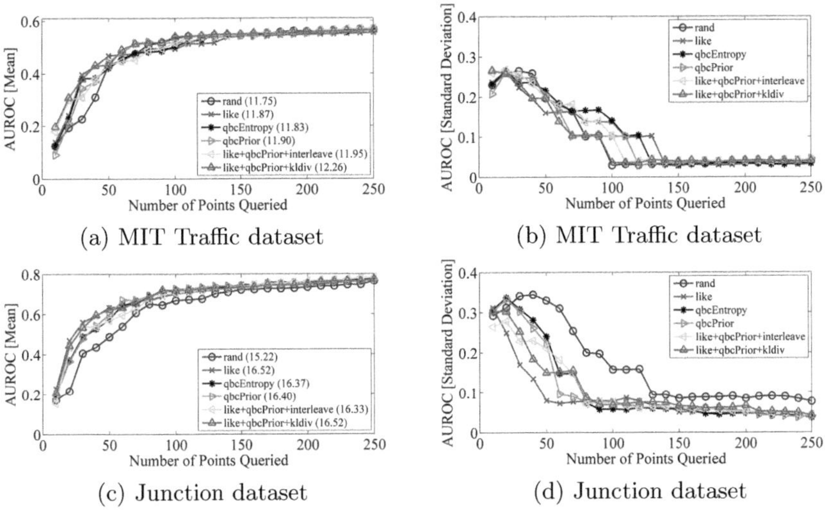


(a) MIT Traffic dataset (b) MIT Traffic dataset

(c) Junction dataset (d) Junction dataset

**Fig. 5.** Unusual events detection performance. Numbers shown in the brackets within graph legend are area under the mean AUROC of different approaches.

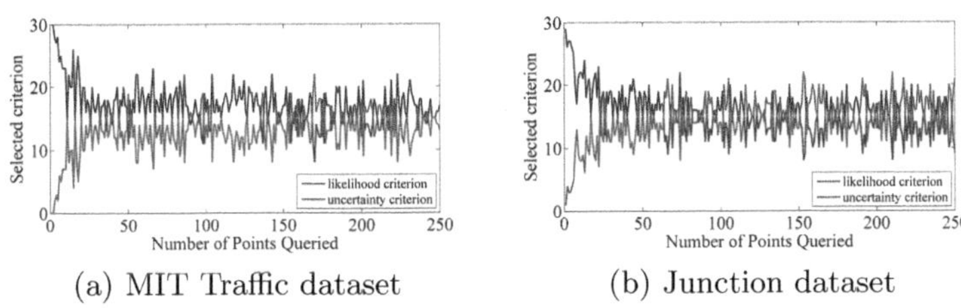


(a) MIT Traffic dataset (b) Junction dataset

**Fig. 6.** Selected criterion over 30 runs

In both datasets, as can be seen from Fig. 5a and 5c, the proposed method *like+qbcPrior+KLdiv* yields the best performance. Adaptive selection of multiple criteria leads to reliable and good performance with our *like+qbcPrior+KLdiv* outperforming the alternative *like+qbcPrior+interleave*. The reliability of the multi-criteria methods is also reflected by the smaller variance across multiple trials shown in Fig. 5b and 5d.

**Adaptive selection of criteria** - In contrast to iterative strategy reported in [14], the proposed strategy assigns weights adaptively to different criterion at different stage of active learning based on the KL-divergence of a model (see Sec.3.4). For the MIT traffic dataset (Fig. 6a), likelihood criterion was more frequently selected than uncertainty criterion before the number of queried samples reached 150. This observation suggests that when the visual distinctiveness between event classes were less obvious (see Sec. 4.1), our method was capable of selecting the right criterion and avoid uncertainty criterion that may keep querying uncertain points located at highly overlapped area of class boundary, which are less useful for improving the detection performance. On the other hand, in the Junction dataset (Fig. 6b), likelihood criterion dominated at the beginning, since it discovered unknown events that caused greater change in parameter values to the classifier compared to uncertainty criterion. The model eventually switched from likelihood criterion to uncertainty criterion (after 80 samples were queried) to refine the classification boundary when the exploratory learning was no longer fruitful (see Fig. 4b, approximately 90% of total event classes were discovered after 80 samples).

## 5 Conclusion

To our best knowledge, this study is the first investigation into the use of active learning to exploit human feedback for on-line unusual event detection. Importantly, the proposed approach yielded more robust and accurate detection on subtle unusual events in public space as compared to conventional supervised and unsupervised learning strategies, by exploiting a small cost of human supervision through active learning. Experimental results also showed that the proposed stream-based multi-criteria approach is capable of balancing different query criteria for joint unknown event discovery and decision boundary refinement. It therefore results in a more reliable detection performance than using single criterion alone, and it outperforms an existing multi-criteria strategy [14] applied in a stream-based manner. In addition, by introducing a prior to deal with imbalanced data, our re-formulated QBC criterion improves the performance.

## References

1. Wang, X., Ma, X., Grimson, W.E.L.: Unsupervised activity perception in crowded and complicated scenes using hierarchical Bayesian models. TPAMI 31, 539–555 (2009)
2. Hospedales, T., Gong, S., Xiang, T.: A Markov clustering topic model for mining behaviour in video. In: ICCV (2009)
3. Mehran, R., Oyama, A., Shah, M.: Abnormal crowd behaviour detection using social force model. In: CVPR, pp. 935–942 (2009)
4. Kim, J., Grauman, K.: Observe locally, infer globally: a space-time MRF for detecting abnormal activities with incremental updates. In: CVPR, pp. 2921–2928 (2009)
5. Settles, B.: Active learning literature survey. Technical report, University of Wisconsin Madison (2010)
6. Argamon-Engelson, S., Dagan, I.: Committee-based sample selection for probabilistic classifiers. JAIR 11, 335–360 (1999)
7. McCallum, A.K., Nigam, K.: Employing EM in pool-based active learning for text classification. In: ICML, pp. 350–358 (1998)
8. Seung, H.S., Opper, M., Sompolinsky, H.: Query by committee. In: COLT, pp. 287–294 (1992)

9. Cover, T., Thomas, J.: Information Theory. Wiley, Chichester (1991)
10. Remagnino, P., Jones, G.: Classifying surveillance events from attributes and behaviour. In: BMVC (2001)
11. Hu, Y., Cao, L., Lv, F., Yan, S., Gong, Y., Huang, T.S.: Action detection in complex scenes with spatial and temporal ambiguities. In: ICCV (2009)
12. Sillito, R., Fisher, R.: Semi-supervised learning for anomalous trajectory detection. In: BMVC (2008)
13. Pelleg, D., Moore, A.: Active learning for anomaly and rare-category detection. In: NIPS (2004)
14. Stokes, J.W., Platt, J.C., Kravis, J., Shilman, M.: ALADIN: Active learning of anomalies to detect intrusions. Technical report, Microsoft Research (2008)
15. Kapoor, A., Grauman, K., Urtasun, R., Darrell, T.: Gaussian processes for object categorization. IJCV (2009)
16. Baram, Y., El-Yaniv, R., Luz, K.: Online choice of active learning algorithms. JMLR 5, 255–291 (2004)
17. Ho, S.S., Wechsler, H.: Query by transduction. TPAMI 30, 1557–1571 (2008)
18. Cebron, N., Berthold, M.R.: Active learning for object classification: from exploration to exploitation. Data Min. Knowl. Disc. 18, 283–299 (2008)
19. Dagan, I., Engelson, S.: Committee-based sampling for training probabilistic classifiers. In: ICML, pp. 150–157 (1995)
20. Lucas, B.D., Kanade, T.: An iterative image registration technique with an application to stereo vision. In: Imaging Understanding Workshop, pp. 121–130 (1981)
21. Loy, C.C., Xiang, T., Gong, S.: Multi-camera activity correlation analysis. In: CVPR, pp. 1988–1995 (2009)
22. Devroye, L.: Non-Uniform Random Variate Generation. Springer, Heidelberg (1986)
23. Tong, S., Koller, D.: Active learning for parameter estimation in Bayesian networks. In: NIPS, pp. 647–653 (2000)
24. Xiang, T., Gong, S.: Video behaviour profiling for anomaly detection. TPAMI 30, 893–908 (2008)

# Asymmetric Totally-Corrective Boosting for Real-Time Object Detection

Peng Wang[1], Chunhua Shen[2,3], Nick Barnes[2,3], Hong Zheng[1], and Zhang Ren[1]

[1] Beihang University, Beijing 100191, China
[2] NICTA, Canberra Research Laboratory, Canberra, ACT 2601, Australia
[3] Australian National University, Canberra, ACT 0200, Australia

**Abstract.** Real-time object detection is one of the core problems in computer vision. The cascade boosting framework proposed by Viola and Jones has become the standard for this problem. In this framework, the learning goal for each node is asymmetric, which is required to achieve a high detection rate and a moderate false positive rate. We develop new boosting algorithms to address this asymmetric learning problem. We show that our methods explicitly optimize asymmetric loss objectives in a totally corrective fashion. The methods are totally corrective in the sense that the coefficients of all selected weak classifiers are updated at each iteration. In contract, conventional boosting like AdaBoost is stage-wise in that only the current weak classifier's coefficient is updated. At the heart of the totally corrective boosting is the column generation technique. Experiments on face detection show that our methods outperform the state-of-the-art asymmetric boosting methods.

## 1 Introduction

Due to its important applications in video surveillance, interactive human-machine interface *etc*, real-time object detection has attracted extensive research recently. Although it was introduced a decade ago, the boosted cascade classifier framework of Viola and Jones [1] is still considered as the most promising approach for object detection, and many papers have extended this framework.

One difficulty in object detection is the problem is highly asymmetric. A common method to detect objects in an image is to exhaustively search all sub-windows at all possible scales and positions in the image, and use a trained model to detect target objects. Typically, there are only a few targets in millions of searched sub-windows. The cascade classifier framework partially solves the asymmetry problem by splitting the detection process into several nodes. Only those sub-windows passing through all nodes are classified as true targets. At each node, we want to train a classifier with a very high detection rate (*e.g.* , 99.5%) and a moderate false positive rate (*e.g.*, around 50%). The learning goal of each node should be asymmetric in order to achieve optimal detection performance. A drawback of standard boosting like AdaBoost in the context of the cascade framework is that it is designed to minimize the overall false rate.

R. Kimmel, R. Klette, and A. Sugimoto (Eds.): ACCV 2010, Part I, LNCS 6492, pp. 176–188, 2011.

The losses are equal for misclassifying a positive example and a negative example, making it unfit to build an optimal classifier for the asymmetric learning goal.

Many incremental works try to improve the performance of object detectors by introducing asymmetric loss functions to boosting algorithms. Viola and Jones proposed asymmetric AdaBoost [2], which applies an asymmetric multiplier to one of the classes. However, this asymmetry is absorbed immediately by the first weak classifier because AdaBoost's optimization strategy is greedy. In practice, they manually apply the $n$-th root of the multiplier on each iteration to keep the asymmetric effect throughout the entire training process. Here $n$ is the number of weak classifiers. This heuristic cannot guarantee the solution to be optimal and the number of weak classifiers need to be specified before training. AdaCost presented by Fan *et al.* [3] adds a cost adjustment function on the weight updating strategy of AdaBoost. They also pointed out that the weight updating rule should consider the cost not only on the initial weights but also at each iteration. Li and Zhang [4] proposed FloatBoost to reduce the redundancy of greedy search by incorporating floating search with AdaBoost. In FloatBoost, the poor weak classifiers are deleted when adding the new weak classifier. Xiao *et al.* [5] improved the backtrack technique in [4] and exploited the historical information of preceding nodes into successive node learning. Hou *et al.* [6] used varying asymmetric factors for training different weak classifiers. However, because the asymmetric factor changes during training, the loss function remains unclear. Pham *et al.* [7] presented a method which trains the asymmetric AdaBoost [2] classifiers under a new cascade structure, namely multi-exit cascade. Like soft cascade [8], boosting chain [5] and dynamic cascade [9], multi-exit cascade is a cascade structure which takes the historical information into consideration. In multi-exit cascade, the $n$-th node "inherits" weak classifiers selected at the preceding $n-1$ nodes. Wu *et al.* [10] stated that feature selection and ensemble classifier learning can be decoupled. They designed a linear asymmetric classifier (LAC) to adjust the linear coefficients of the selected weak classifiers. Kullback-Leibler Boosting [11] iteratively learns robust linear features by maximizing the Kullback-Leibler divergence.

Much of the previous work is based on AdaBoost and achieves the asymmetric learning goal by *heuristic* weights manipulations or post-processing techniques. It is not trivial to assess how these heuristics affect the original loss function of AdaBoost. In this work, we construct new boosting algorithms *directly* from asymmetric losses. The optimization process is implemented by column generation. Experiments on toy data and real data show that our algorithms indeed achieve the asymmetric learning goal without any heuristic manipulation, and can outperform previous methods. Therefore, the main contributions of this work are as follows:

(1) We utilize a general and systematic framework (column generation) to construct new asymmetric boosting algorithms, which can be applied to a variety of asymmetric losses. There is no heuristic strategy in our algorithms which may cause suboptimal solutions. In contrast, the global optimal solution is guaranteed for our algorithms.

Unlike Viola-Jones' asymmetric AdaBoost [2], the asymmetric effect of our methods spreads over the entire training process. The coefficients of all weak classifiers are updated at each iteration, which prevents the first weak classifier from absorbing the asymmetry. The number of weak classifiers does not need to be specified before training.

(2) The asymmetric totally-corrective boosting algorithms introduce the asymmetric learning goal into both feature selection and ensemble classifier learning. Both the example weights and the linear classifier coefficients are learned in an asymmetric way.

(3) In practice, L-BFGS-B [12] is used to solve the primal problem, which runs much faster than solving the dual problem and also less memory is needed.

(4) We demonstrate that with the totally corrective optimization, the linear coefficients of some weak classifiers are set to zero by the algorithm such that fewer weak classifiers are needed. We present analysis on the theoretical condition and show how useful the historical information is for the training of successive nodes.

## 2 Asymmetric Losses

In this section, we propose two asymmetric losses, which are motivated by asymmetric AdaBoost [2] and cost-sensitive LogitBoost [13], respectively.

We first introduce an asymmetric cost in the following form:

$$\mathrm{ACost} = \begin{cases} C_1 & \text{if } y = +1 \text{ and } \mathrm{sign}(F(\boldsymbol{x})) = -1, \\ C_2 & \text{if } y = -1 \text{ and } \mathrm{sign}(F(\boldsymbol{x})) = +1, \\ 0 & \text{if } y = \mathrm{sign}(F(\boldsymbol{x})). \end{cases} \tag{1}$$

Here $\boldsymbol{x}$ is the input data, $y$ is the label and $F(\boldsymbol{x})$ is the learned classifier. Viola and Jones [2] directly take the product of ACost and the exponential loss $E_{\mathbf{X},Y}[\exp(-yF(\boldsymbol{x})]$ as the asymmetric loss:

$$E_{\mathbf{X},Y}[(\boldsymbol{I}(y=1)C_1 + \boldsymbol{I}(y=-1)C_2)\exp(-yF(\boldsymbol{x}))], \tag{2}$$

where $\boldsymbol{I}(\cdot)$ is the indicator function. In a similar manner, we can also form an asymmetric loss from the logistic loss $E_{\mathbf{X},Y}[\mathrm{logit}(-yF(\boldsymbol{x}))]$:

$$\mathrm{ALoss}_1 = E_{\mathbf{X},Y}[(\boldsymbol{I}(y=1)C_1 + \boldsymbol{I}(y=-1)C_2)\mathrm{logit}(yF(\boldsymbol{x}))], \tag{3}$$

where $\mathrm{logit}(x) = \log(1+\exp(-x))$ is the logistic loss function.

Masnadi-Shirazi and Vasconcelos [13] proposed cost-sensitive boosting algorithms which optimize different versions of cost-sensitive losses by the means of gradient descent. They proved that the optimal cost-sensitive predictor minimizes the expected loss:

$$-E_{\mathbf{X},Y}[\boldsymbol{I}(y=1)\log(\mathrm{p_c}(\boldsymbol{x})) + \boldsymbol{I}(y=-1)\log(1-\mathrm{p_c}(\boldsymbol{x}))], \tag{4}$$

$$\text{where} \quad \mathrm{p_c}(\boldsymbol{x}) = \frac{e^{\gamma F(x)+\eta}}{e^{\gamma F(x)+\eta} + e^{-\gamma F(x)-\eta}}, \text{with } \gamma = \frac{C_1+C_2}{2}, \eta = \frac{1}{2}\log\frac{C_2}{C_1}.$$

With fixing $\gamma$ to 1, the expected loss can be reformulated to

$$\text{ALoss}_2 = E_{\boldsymbol{X},Y}[\text{logit}(yF(\boldsymbol{x}) + 2y\eta)]. \tag{5}$$

## 3 Asymmetric Totally-Corrective Boosting

In this section, we construct asymmetric totally-corrective boosting algorithms (termed AsymBoost$_{\text{TC}}$ here) from the losses (3) and (5) discussed previously. In contrast to the methods constructing boosting-like algorithms in [13], [14] and [15], we use column generation to design our totally corrective boosting algorithms, inspired by [16] and [17].

Suppose there are $M$ training examples ($M_1$ positives and $M_2$ negatives), and the sequence of examples are arranged according to the labels (positives first). The pool $\mathcal{H}$ contains $N$ available weak classifiers. The matrix $H \in \mathbb{Z}^{M\times N}$ contains binary outputs of weak classifiers in $\mathcal{H}$ for training examples, namely $H_{ij} = h_j(x_i)$. We are aiming to learn a linear combination $F_{\boldsymbol{w}}(\cdot) = \sum_{j=1}^{N} w_i h_j(\cdot)$. $C_1$ and $C_2$ are costs for misclassifying positives and negatives, respectively. We assign the asymmetric factor $k = C_1/C_2$ and restrict $\gamma = (C_1 + C_2)/2$ to 1, thus $C_1$ and $C_2$ are fixed for a given $k$.

The problems of the two AsymBoost$_{\text{TC}}$ algorithms can be expressed as:

$$\min_{\boldsymbol{w}} \sum_{i=1}^{M} l_i \text{logit}(z_i) + \theta \underline{1}^\top \boldsymbol{w} \quad \text{s.t.} \boldsymbol{w} \succcurlyeq \underline{0},\ z_i = y_i H_i \boldsymbol{w}, \tag{6}$$

where $\boldsymbol{l} = [C_1/M_1, \cdots, C_2/M_2, \cdots]^\top$, and

$$\min_{\boldsymbol{w}} \sum_{i=1}^{M} e_i \text{logit}(z_i + 2y_i\eta) + \theta \underline{1}^\top \boldsymbol{w} \quad \text{s.t.} \boldsymbol{w} \succcurlyeq \underline{0},\ z_i = y_i H_i \boldsymbol{w}, \tag{7}$$

where $\boldsymbol{e} = [1/M_1, \cdots, 1/M_2, \cdots]^\top$. In both (6) and (7), $z_i$ stands for the margin of the $i$-th training example. We refer (6) as AsymBoost$_{\text{TC1}}$ and (7) as AsymBoost$_{\text{TC2}}$. Note that here the optimization problems are $\ell_1$-norm regularized. It is possible to use other format of regularization such as the $\ell_2$-norm.

First we introduce the Fenchel conjugate [18] of the logistic loss function logit$(x)$, which is

$$\text{logit}^*(u) = \begin{cases} (-u)\log(-u) + (1+u)\log(1+u), 0 \geq u \geq -1; \\ \infty, \qquad\qquad\qquad\qquad\qquad\qquad \text{otherwise.} \end{cases}$$

We derive the Lagrange dual [18] of AsymBoost$_{\text{TC1}}$. The Lagrangian of (6) is

$$L(\underbrace{\boldsymbol{w}, \boldsymbol{z}}_{\text{primal}}, \underbrace{\boldsymbol{\lambda}, \boldsymbol{u}}_{\text{dual}}) = \sum_{i=1}^{M} l_i \text{logit}(z_i) + \theta \underline{1}^\top \boldsymbol{w} - \boldsymbol{\lambda}^\top \boldsymbol{w} + \sum_{i=1}^{M} u_i (z_i - y_i H_i \boldsymbol{w}). \tag{8}$$

The dual function

$$g(\boldsymbol{\lambda}, \boldsymbol{u}) = \inf_{\boldsymbol{w},\boldsymbol{z}} L(\boldsymbol{w}, \boldsymbol{z}, \boldsymbol{\lambda}, \boldsymbol{u}) \tag{9}$$
$$= -\sum_{i=1}^{M} \underbrace{\sup_{z_i}\Big( - u_i z_i - l_i \mathrm{logit}(z_i)\Big)}_{l_i \mathrm{logit}^*(-u_i/l_i)} + \inf_{\boldsymbol{w}} \underbrace{\Big(\theta \underline{1}^\top - \boldsymbol{\lambda}^\top - \sum_{i=1}^{M} u_i y_i H_i\Big)}_{\text{must be } \underline{0}} \boldsymbol{w}.$$

The dual problem is

$$\max_{\boldsymbol{u}} \quad -\sum_{i=1}^{M} \Big[u_i \log(u_i) + (l_i - u_i)\log(l_i - u_i)\Big]$$
$$\text{s.t.} \sum_{i=1}^{M} u_i y_i H_i \preccurlyeq \theta \underline{1}^\top, \; 0 \preccurlyeq \boldsymbol{u} \preccurlyeq \boldsymbol{l}. \tag{10}$$

Since the problem (6) is convex and the Slater's conditions are satisfied [18], the duality gap between the primal (6) and the dual (10) is zero. Therefore, the solutions of (6) and (10) are the same. Through the KKT condition, the gradient of Lagrangian (8) over primal variable $\boldsymbol{z}$ and dual variable $\boldsymbol{u}$ should vanish at the optimum. Therefore, we can obtain the relationship between the optimal value of $\boldsymbol{z}$ and $\boldsymbol{u}$:

$$u_i^* = \frac{l_i \exp(-z_i^*)}{1 + \exp(-z_i^*)}. \tag{11}$$

Similarly, we can get the dual problem of AsymBoost$_{\mathrm{TC2}}$, which is expressed as:

$$\max_{\boldsymbol{u}} \quad -\sum_{i=1}^{M} \Big[u_i \log(u_i) + (e_i - u_i)\log(e_i - u_i) + 2u_i y_i \eta\Big]$$
$$\text{s.t.} \sum_{i=1}^{M} u_i y_i H_i \preccurlyeq \theta \underline{1}^\top, \; 0 \preccurlyeq \boldsymbol{u} \preccurlyeq \boldsymbol{e}, \tag{12}$$

with

$$u_i^* = \frac{e_i \exp(-z_i^* - 2y_i\eta)}{1 + \exp(-z_i^* - 2y_i\eta)}. \tag{13}$$

In practice, the total number of weak classifiers, $N$, could be extremely large, so we can not solve the primal problems (6) and (7) directly. However equivalently, we can optimize the duals (10) and (12) iteratively using column generation [16]. In each round, we add the most violated constraint by finding a weak classifier satisfying:

$$h'(\cdot) = \mathrm{argmax}_{h(\cdot)} \sum_{i=1}^{M} u_i y_i h(\boldsymbol{x}_i). \tag{14}$$

This step is the same as training a weak classifier in AdaBoost and LPBoost, in which one tries to find a weak classifier with the maximal edge (*i.e.* the minimal weighted error). The *edge* of $h_j$ is defined as $\sum_{i=1}^{M} u_i y_i h_j(x_i)$, which is the inverse of the weighted error. Then we solve the restricted dual problem with one more constraint than the previous round, and update the linear coefficients of weak classifiers ($\boldsymbol{w}$) and the weights of training examples ($\boldsymbol{u}$). Adding one constraint into the dual problem corresponds to adding one variable into the primal problem. Since the primal problem and dual problem are equivalent, we can either solve the restricted dual or the restricted primal in practice. The algorithms of AsymBoost$_{\text{TC1}}$ and AsymBoost$_{\text{TC2}}$ are summarized in Algorithm 1.

---

**Algorithm 1.** The training algorithms of AsymBoost$_{\text{TC1}}$ and AsymBoost$_{\text{TC2}}$.

---

**Input**: A training set with $M$ labeled examples ($M_1$ positives and $M_2$ negatives); termination tolerant $\varepsilon > 0$; regularization parameter $\theta$; asymmetric factor $k$; maximum number of weak classifiers $N_{\max}$.

**Initialization**: $N = 0$; $\boldsymbol{w} = \mathbf{0}$; and $u_i = l_i/2$ or $e_i/(1 + k^{-y_i})$ , $i = 1 \cdots M$.

**for** iteration $= 1 : N_{\max}$ **do**

- Train a weak classifier $h'(\cdot) = \operatorname{argmax}_{h(\cdot)} \sum_{i=1}^{M} u_i y_i h(\boldsymbol{x}_i)$.
- Check for the termination condition:
  **if** iteration $> 1$ and $\sum_{i=1}^{M} u_i y_i h'(\mathbf{x}_i) < \theta + \varepsilon$, **then** break;
- Increment the number of weak classifiers $N = N + 1$.
- Add $h'(\cdot)$ to the restricted master problem;
- Solve the primal problem (6) or (7) (or the dual problem (10) or (12)) and update $u_i$ $(i = 1 \cdots M)$ and $w_j$ $(j = 1 \cdots N)$ .

**Output**: The selected weak classifiers are $h_1, h_2, \ldots, h_N$. The final strong classifier is: $F(\boldsymbol{x}) = \sum_{j=1}^{N} w_j h_j(\boldsymbol{x})$.

---

Note that, in practice, in order to achieve specific false negative rate (FNR) or false positive rate (FPR), an offset $b$ is needed to be added into the final strong classifier: $F(\boldsymbol{x}) = \sum_{j=1}^{n} w_i h_j(\mathbf{x}) - b$, which can be obtained by a simple line search. The new weak classifier $h'(\cdot)$ corresponds to an extra variable to the primal and an extra constraint to the dual. Thus, the minimal value of the primal decreases with growing variables, and the maximal value of the dual problem also decreases with growing constraints. Furthermore, as the optimization problems involved are convex, Algorithm 1 is guaranteed to converge to the global optimum.

Next we show how AsymBoost$_{\text{TC}}$ introduces the asymmetric learning into feature selection and ensemble classifier learning. Decision stumps are the most commonly used type of weak classifiers, and each stump only uses one dimension of the features. So the process of training weak classifiers (decision stumps) is equivalent to feature selection. In our framework, the weak classifier with the maximum edge (*i.e.* the minimal weighted error) is selected. From (11) and (13), the weight of $i$-th example, namely $u_i$, is affected by two factors: the asymmetric factor $k$ and the current margin $z_i$. If we set $k = 1$, the weighting strategy goes back to being symmetric. On the other hand, the coefficients of the linear classifier, namely $\boldsymbol{w}$, are updated by solving the restricted primal problem at each iteration. The asymmetric factor $k$ in the primal is absorbed by all the weak classifiers currently learned. So feature selection and ensemble classifier learning both consider the asymmetric factor $k$.

The number of variables of the primal problem is the number of weak classifiers, while for the dual problem, it is the number of training examples. In the cascade classifiers for face detection, the number of weak classifiers is usually much smaller than the number of training examples, so solving the primal is much cheaper than solving the dual. Since the primal problem has only simple box-bounding constraints, we can employ L-BFGS-B [12] to solve it. L-BFGS-B is a tool based on the quasi-Newton method for bound-constrained optimization.

Instead of maintaining the Hessian matrix, L-BFGS-B only needs the recent several updates of values and gradients for the cost function to approximate the Hessian matrix. Thus, L-BFGS-B requires less memory when running. In column generation, we can use the results from previous iteration as the starting point of current problem, which leads to further reductions in computation time.

The complementary slackness condition [18] suggests that $\lambda_j w_j = 0$. So we can get the conditions of sparseness:

$$\text{If } \lambda = \theta - \textstyle\sum_{i=1}^{M} u_i y_i H_{i,j} > 0, \text{then } w_j = 0. \tag{15}$$

This means that, if the weak classifier $h_j(\cdot)$ is so "weak" that its edge is less than $\theta$ under the current distribution $\boldsymbol{u}$, its contribution to the ensemble classifier is "zero". From another viewpoint, the $\ell_1$-norm regularization term in the primal (6) and (7), leads to a sparse result. The parameter $\theta$ controls the degree of the sparseness. The larger $\theta$ is, the sparser the result would be.

# 4 Experiments

## 4.1 Results on Synthetic Data

To show the behavior of our algorithms, we construct a 2D data set, in which the positive data follow the 2D normal distribution ($N(0, 0.1\mathbf{I})$), and the negative data form a ring with uniformly distributed angles and normally distributed radius ($N(1.0, 0.2)$). Totally 2000 examples are generated (1000 positives and 1000 negatives), 50% of data for training and the other half for test. We compare AdaBoost, AsymBoost$_{\text{TC1}}$ and AsymBoost$_{\text{TC2}}$ on this data set. All the training processes are stopped at 100 decision stumps. For AsymBoost$_{\text{TC1}}$ and AsymBoost$_{\text{TC2}}$, we fix $\theta$ to 0.01, and use a group of $k$'s $\{1.2, 1.4, 1.6, 1.8, 2.0, 2.2, 2.4, 2.6, 2.8, 3.0\}$.

From Figures 1 (1) and (2), we find that the larger $k$ is, the bigger the area for positive output becomes, which means that the asymmetric LogitBoost tends to make a positive decision for the region where positive and negative data are mixed together. Another observation is that AsymBoost$_{\text{TC1}}$ and AsymBoost$_{\text{TC2}}$ have almost the same decision boundaries on this data set with same $k$'s.

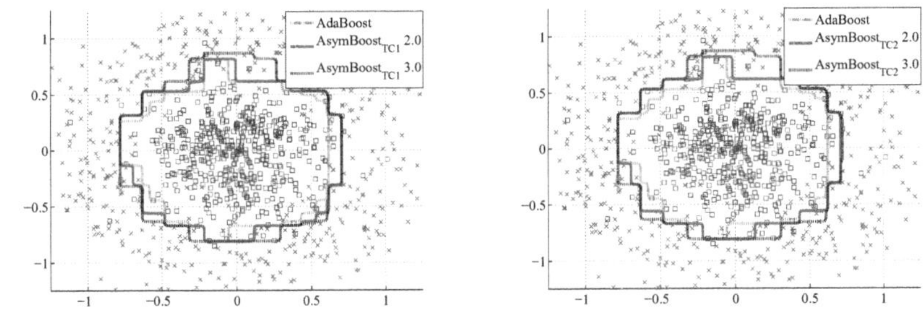


(a) $\text{AsymBoost}_{\text{TC1}}$ vs AdaBoost (b) $\text{AsymBoost}_{\text{TC2}}$ vs AdaBoost

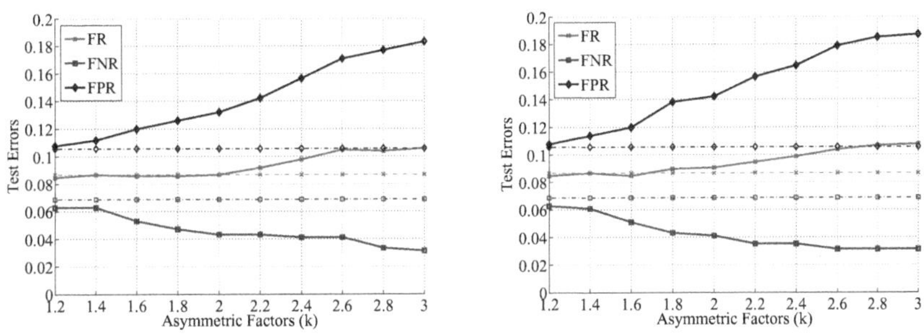


(c) False rates for $\text{AsymBoost}_{\text{TC1}}$ (d) False rates for $\text{AsymBoost}_{\text{TC2}}$

**Fig. 1.** Results on the synthetic data for $\text{AsymBoost}_{\text{TC1}}$ and $\text{AsymBoost}_{\text{TC2}}$, with a group of asymmetric factor $k$s. As the baseline, the results for AdaBoost are also shown in these figures. (1) and (2) demonstrate decision boundaries learned by $\text{AsymBoost}_{\text{TC1}}$ and $\text{AsymBoost}_{\text{TC2}}$, with $k$ is 2.0 or 3.0. The $\times$'s and $\Box$'s stand for training negatives and training positives respectively. (3) and (4) demonstrate false rates (FR), false positive rates (FPR) and false negative rates (FNR) on test set with a group of $k$s $(1.2, 1.4, 1.6, 1.8, 2.0, 2.2, 2.4, 2.6, 2.8$ or $3.0)$, and the corresponding rates for AdaBoost is shown as dashed lines.

Figures 1 (3) and (4) demonstrate trends of false rates with the growth of asymmetric factor ($k$). The results of AdaBoost is considered as the baseline. For all $k$'s, $\text{AsymBoost}_{\text{TC1}}$ and $\text{AsymBoost}_{\text{TC2}}$ achieve lower false negative rates and higher false positive rates than AdaBoost. With the growth of $k$, $\text{AsymBoost}_{\text{TC1}}$ and $\text{AsymBoost}_{\text{TC2}}$ become more aggressive to reduce the false negative rate, with the sacrifice of a higher false positive rate.

### 4.2 Face Detection

We collect 9832 mirrored frontal face images and about 10115 large background images. 5000 face images and 7000 background images are used for training, and 4832 face images and 3115 background images for validation. Five basic types of Haar features are calculated on each $24 \times 24$ image, and totally generate

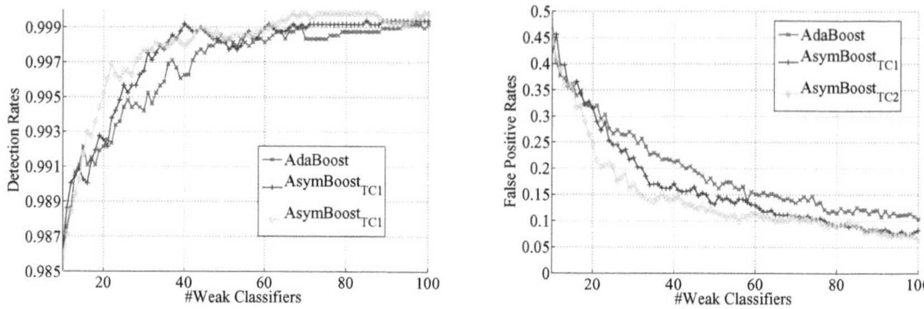


(a) DR with fixed FPR (b) FPR with fixed DR

**Fig. 2.** Testing curves of single-node classifiers for AdaBoost, AsymBoost$_{\mathrm{TC1}}$ and AsymBoost$_{\mathrm{TC2}}$. All the classifiers use the same training and test data sets. (1) shows curves of detection rates (DR) with false positive rates (FPR) fixed to 0.25, (2) shows curves of FPR with DR fixed to 0.995. FPR or DR are evaluated at each weak classifier.

162336 features. Decision stumps on those 162336 features construct the pool of weak classifiers.

**Single-node detectors.** Single-node classifiers with AdaBoost, AsymBoost$_{\mathrm{TC1}}$ and AsymBoost$_{\mathrm{TC2}}$ are trained. The parameters $\theta$ and $k$ are simply set to 0.001 and 7.0. 5000 faces and 5000 non-faces are used for training, while 4832 faces and 5000 non-faces are used for test. The training/validation non-faces are randomly cropped from training/validation background images.

Figure 2 (1) shows curves of detection rate with the false positive rate fixed at 0.25, while curves of false positive rates with 0.995 detection rate are shown in Fig. 2 (2). We set the false positive rate fixed to 0.25 rather than the commonly used 0.5 in order to slow down the increasing speed of detection rates, otherwise detection rates would converge to 1.0 immediately. The increasing/decreasing speed of detection rate/false positive rate is faster than reported in [4] and [5]. The reason is possibly that we use 10000 examples for training and 9832 for testing, which are smaller than the data used in [4] and [5] (18000 training examples and 15000 test examples). We can see that under both situations, our algorithms achieve better performances than AdaBoost in most cases.

The benefits of our algorithms can be expressed in two-fold: (1) Given the same learning goal, our algorithms tend to use smaller number of weak classifiers. For example, from Fig. 2 (2), if we want a classifier with a 0.995 detection rate and a 0.2 false positive rate, AdaBoost needs at least 43 weak classifiers while AsymBoost$_{\mathrm{TC1}}$ needs 32 and AsymBoost$_{\mathrm{TC2}}$ needs only 22. (2) Using the same number of weak classifiers, our algorithms achieve a higher detection rate or a lower false positive rate. For example, from Fig. 2 (2), using 30 weak classifiers, both AsymBoost$_{\mathrm{TC1}}$ and AsymBoost$_{\mathrm{TC2}}$ achieve higher detection rates (0.9965 and 0.9975) than AdaBoost (0.9945).

**Complete detectors.** Secondly, we train complete face detectors with AdaBoost, asymmetric AdaBoost, AsymBoost$_{\text{TC1}}$ and AsymBoost$_{\text{TC2}}$. All detectors are trained using the same training set. We use two types of cascade framework for the detector training: the traditional cascade of Viola and Jones [1] and the multi-exit cascade presented in [7]. The latter utilizes decision information of previous nodes when judging instances in the current node. For fair comparison, all detectors use 24 nodes and 3332 weak classifiers. For each node, 5000 faces + 5000 non-faces are used for training, and 4832 faces + 5000 non-faces are used for validation. All non-faces are cropped from background images. The asymmetric factor $k$ for asymmetric AdaBoost, AsymBoost$_{\text{TC1}}$ and AsymBoost$_{\text{TC2}}$ are selected from $\{1.2, 1.5, 2.0, 3.0, 4.0, 5.0, 6.0\}$. The regularization factor $\theta$ for AsymBoost$_{\text{TC1}}$ and AsymBoost$_{\text{TC2}}$ are chosen from $\{\frac{1}{50}, \frac{1}{60}, \frac{1}{70}, \frac{1}{80}, \frac{1}{90}, \frac{1}{100}, \frac{1}{200}, \frac{1}{400}, \frac{1}{800}, \frac{1}{1000}\}$. It takes about four hours to train a AsymBoost$_{\text{TC}}$ face detector on a machine with 8 Intel Xeon E5520 cores and 32GB memory. Comparing with AdaBoost, only around 0.5 hour extra time is spent on solving the primal problem at each iteration. We can say that, in the context of face detection, the training time of AsymBoost$_{\text{TC}}$ is nearly the same as AdaBoost.

ROC curves on the CMU/MIT data set are shown in Fig. 3. Those images containing ambiguous faces are removed and 120 images are retained. From the figure, we can see that, asymmetric AdaBoost outperforms AdaBoost in both Viola-Jones cascade and multi-exit cascade, which coincide with what reported in [2]. Our algorithms have better performances than all other methods in all points and the improvements are more significant when the false positives are less than 100, which is the most commonly used region in practice.

As mentioned in the previous section, our algorithms produce sparse results to some extent. Some linear coefficients are zero when the corresponding weak classifiers satisfy the condition (15). In the multi-exit cascade, the sparse phenomenon becomes more clear. Since correctly classified negative data are discarded after each node is trained, the training data for each node are different. The "closer" nodes share more common training examples, while the nodes

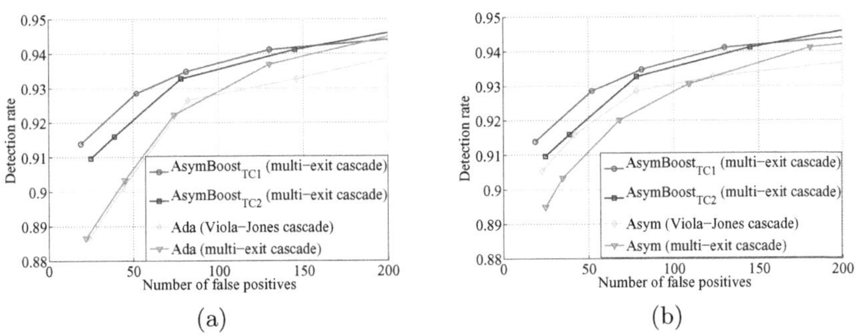


**Fig. 3.** Performances of cascades evaluated by ROC curves on the MIT+CMU data set. AdaBoost is referred to "Ada", and Asymmetric AdaBoost [1] is referred to "Asym". "Viola-Jones cascade" means the traditional cascade used in [2] .

**Table 1.** The ratio of weak classifiers selected at the $i$-th node (column) appearing with non-zero coefficients in the $j$-th node (row). The ratios decrease along with the growth of the node index in each column.

| Node Index | 1 | 2 | 3 | 4 | 5 | 6 | 7 | 8 | 9 | 10 | 11 | 12 | 13 | 14 | 15 |
|---|---|---|---|---|---|---|---|---|---|---|---|---|---|---|---|
| 1 | 1.00 | | | | | | | | | | | | | | |
| 2 | 1.00 | 1.00 | | | | | | | | | | | | | |
| 3 | 1.00 | 1.00 | 1.00 | | | | | | | | | | | | |
| 4 | 0.86 | 1.00 | 0.97 | 1.00 | | | | | | | | | | | |
| 5 | 0.43 | 0.93 | 0.97 | 0.97 | 1.00 | | | | | | | | | | |
| 6 | 0.71 | 0.93 | 0.90 | 1.00 | 0.96 | 1.00 | | | | | | | | | |
| 7 | 0.43 | 0.87 | 0.87 | 0.97 | 0.92 | 0.92 | 1.00 | | | | | | | | |
| 8 | 0.29 | 0.40 | 0.70 | 0.73 | 0.74 | 0.88 | 0.74 | 1.00 | | | | | | | |
| 9 | 0.00 | 0.27 | 0.50 | 0.60 | 0.76 | 0.72 | 0.66 | 0.67 | 1.00 | | | | | | |
| 10 | 0.14 | 0.27 | 0.43 | 0.60 | 0.62 | 0.70 | 0.62 | 0.66 | 0.60 | 1.00 | | | | | |
| 11 | 0.00 | 0.20 | 0.33 | 0.50 | 0.52 | 0.54 | 0.60 | 0.59 | 0.56 | 0.48 | 1.00 | | | | |
| 12 | 0.14 | 0.20 | 0.40 | 0.40 | 0.56 | 0.50 | 0.54 | 0.61 | 0.55 | 0.46 | 0.36 | 1.00 | | | |
| 13 | 0.00 | 0.13 | 0.33 | 0.37 | 0.36 | 0.54 | 0.40 | 0.47 | 0.47 | 0.46 | 0.43 | 0.25 | 1.00 | | |
| 14 | 0.00 | 0.07 | 0.17 | 0.40 | 0.28 | 0.50 | 0.42 | 0.49 | 0.50 | 0.53 | 0.45 | 0.43 | 0.35 | 1.00 | |
| 15 | 0.00 | 0.13 | 0.20 | 0.27 | 0.36 | 0.38 | 0.46 | 0.41 | 0.52 | 0.42 | 0.49 | 0.44 | 0.34 | 0.27 | 1.00 |

**Table 2.** Comparison of the numbers of the effective weak classifiers for the stage-wise boosting (SWB) and the totally-corrective boosting (TCB). We take AdaBoost and $\text{AsymBoost}_{\text{TC1}}$ as representative types of SWB and TCB, both of which are trained in the multi-exit cascade for face detection.

| Node Index | 1 | 2 | 3 | 4 | 5 | 6 | 7 | 8 | 9 | 10 | 11 | 12 | 13 | 14 | 15 | 16 | 17 | 18 |
|---|---|---|---|---|---|---|---|---|---|---|---|---|---|---|---|---|---|---|
| SWB | 7 | 22 | 52 | 82 | 132 | 182 | 232 | 332 | 452 | 592 | 752 | 932 | 1132 | 1332 | 1532 | 1732 | 1932 | 2132 |
| TCB | 7 | 22 | 52 | 80 | 125 | 174 | 213 | 269 | 331 | 441 | 464 | 538 | 570 | 681 | 717 | 744 | 742 | 879 |

"far away" from each other have distinct training data. The greater the distance between two nodes, the more uncorrelated they become. Therefore, the weak classifiers in the early nodes may perform poorly on the last node, thus tending to obtain zero coefficients. We call those weak classifiers with non-zero coefficients "effective" weak classifiers. Table 1 shows the ratios of "effective" weak classifiers contributed by one node to a specific successive node. To save space, only the first 15 nodes are demonstrated. We can see that, the ratio decreases with the growth of the node index, which means that the farther the preceding node is from the current node, the less useful it is for the current node. For example, the first node has almost no contribution after the eighth node. Table 2 shows the number of effective weak classifiers used by our algorithm and the traditional stage-wise boosting. All weak classifiers in stage-wise boosting have non-zero coefficients, while our totally-corrective algorithm uses much less effective weak classifiers.

## 5 Conclusion

We have proposed two asymmetric totally-corrective boosting algorithms for object detection, which are implemented by the column generation technique in convex optimization. Our algorithms introduce asymmetry into both feature selection and ensemble classifier learning in a systematic way.

Both our algorithms achieve better results for face detection than AdaBoost and Viola-Jones' asymmetric AdaBoost. An observation is that we can not see great differences on performances between $\text{AsymBoost}_{\text{TC1}}$ and $\text{AsymBoost}_{\text{TC2}}$ in our experiments. For the face detection task, AdaBoost already achieves a very promising result, so the improvements of our method are not very significant.

One drawback of our algorithms is there are two parameters to be tuned. For different nodes, the optimal parameters should not be the same. In this work, we have used the same parameters for all nodes. Nevertheless, since the probability of negative examples decreases with the node index, the degree of the asymmetry between positive and negative examples also deceases. The optimal $k$ may decline with the node index.

The framework for constructing totally-corrective boosting algorithms is general, so we can consider other asymmetric losses (*e.g.*, asymmetric exponential loss) to form new asymmetric boosting algorithms. In column generation, there is no restriction that only one constraint is added at each iteration. Actually, we can add several violated constraints at each iteration, which means that we can produce multiple weak classifiers in one round. By doing this, we can speed up the learning process. Motivated by the analysis of sparseness, we find that the very early nodes contribute little information for training the later nodes. Based on this, we can exclude some useless nodes when the node index grows, which will simplify the multi-exit structure and shorten the testing time.

**Acknowledgements.** Work was done when P. W. was visiting NICTA Canberra Research Laboratory and Australian National University. NICTA is funded by the Australian Government's Department of Communications, IT, and the Arts and the Australian Research Council through *Backing Australia's Ability* initiative and the ICT Research Center of Excellence programs.

## References

1. Viola, P., Jones, M.J.: Robust real-time face detection. Int. J. Comp. Vis. 57, 137–154 (2004)
2. Viola, P., Jones, M.: Fast and robust classification using asymmetric AdaBoost and a detector cascade. In: Proc. Adv. Neural Inf. Process. Syst., pp. 1311–1318. MIT Press, Cambridge (2002)
3. Fan, W., Stolfo, S., Zhang, J., Chan, P.: Adacost: Misclassification cost-sensitive boosting. In: Proc. Int. Conf. Mach. Learn., pp. 97–105 (1999)
4. Li, S.Z., Zhang, Z.: Float Boost learning and statistical face detection. IEEE Trans. Pattern Anal. Mach. Intell. 26, 1112–1123 (2004)
5. Xiao, R., Zhu, L., Zhang, H.: Boosting chain learning for object detection. In: Proc. IEEE Int. Conf. Comp. Vis., pp. 709–715 (2003)

6. Hou, X., Liu, C., Tan, T.: Learning boosted asymmetric classifiers for object detection. In: Proc. IEEE Conf. Comp. Vis. Patt. Recogn. (2006)
7. Pham, M.T., Hoang, V.D.D., Cham, T.J.: Detection with multi-exit asymmetric boosting. In: Proc. IEEE Conf. Comp. Vis. Patt. Recogn. (2008)
8. Bourdev, L., Brandt, J.: Robust object detection via soft cascade. In: Proc. IEEE Conf. Comp. Vis. Patt. Recogn., San Diego, CA, US, pp. 236–243 (2005)
9. Xiao, R., Zhu, H., Sun, H., Tang, X.: Dynamic cascades for face detection. In: Proc. IEEE Int. Conf. Comp. Vis., Rio de Janeiro, Brazil (2007)
10. Wu, J., Brubaker, S.C., Mullin, M.D., Rehg, J.M.: Fast asymmetric learning for cascade face detection. IEEE Trans. Pattern Anal. Mach. Intell. 30, 369–382 (2008)
11. Liu, C., Shum, H.Y.: Kullback-Leibler boosting. In: Proc. IEEE Conf. Comp. Vis. Patt. Recogn., Madison, Wisconsin, vol. 1, pp. 587–594 (2003)
12. Zhu, C., Byrd, R.H., Nocedal, J.: L-BFGS-B: Algorithm 778: L-BFGS-B, FORTRAN routines for large scale bound constrained optimization. ACM Trans. Mathematical Software 23, 550–560 (1997)
13. Masnadi-Shirazi, H., Vasconcelos, N.: Cost-sensitive boosting. IEEE Trans. Pattern Anal. Mach. Intell. (2010)
14. Friedman, J., Hastie, T., Tibshirani, R.: Additive logistic regression: a statistical view of boosting. Ann. Statist. 28, 337–407 (2000)
15. Rätsch, G., Mika, S., Schölkopf, B., Müller, K.R.: Constructing boosting algorithms from SVMs: An application to one-class classification. IEEE Trans. Pattern Anal. Mach. Intell. 24, 1184–1199 (2002)
16. Demiriz, A., Bennett, K., Shawe-Taylor, J.: Linear programming boosting via column generation. Mach. Learn. 46, 225–254 (2002)
17. Shen, C., Li, H.: On the dual formulation of boosting algorithms. IEEE Trans. Pattern Anal. Mach. Intell. (2010)
18. Boyd, S., Vandenberghe, L.: Convex Optimization. Cambridge University Press, Cambridge (2004)

# The Application of Vision Algorithms to Visual Effects Production

Sebastian Sylwan

Weta Digital Ltd., Wellington, New Zealand

**Abstract.** Visual Effects Production, despite being fundamentally an artistic endeavor, is a very technically oriented business. Like many other businesses in their early ages, it still requires constant innovation and in many cases a great amount of craftsmanship. Given these premises, most VFX companies (especially those at the leading edge) are hungry for novel and better solutions to the problems they encounter everyday. Many published machine vision algorithms are designed to be real-time and fully automatic with low computational complexity. These attributes are essential for applications such as robotic vision, but in most cases overkill or ill-conditioned for Motion Picture Digital Visual Effects facilities, where massive computation resources are commonplace and expert human interaction is available to initialise algorithms and to guide them towards an optimal solution. Conversely, motion pictures have significantly higher accuracy requirements and other unique challenges. Not all machine vision algorithms can readily be adapted to this environment. This talk outlines the requirements of visual effects and indicate several challenges and possible solutions involved in adopting image processing and machine vision algorithms for motion picture visual effects.

## 1 Introduction

Motion Picture Visual Effects (VFX) arguably involve challenges that are complex beyond those present in other computer-based industries and, in many cases, are constantly evolving. VFX production is —as a business— in the early, steep, phase of innovation where a lot of the apparently impossible challenges of last year are this year's run-rate business and become rapidly adopted throughout the industry. This rapid technology adoption is characteristic of a desire to produce always improving results in a very competitive environment.

The quality of the results depends directly in many cases on advancements measurable either in a novel visual result or in increased efficiencies that conduct to more artistic iterations. These are normally the drivers for research implementation. Adoption of research results has in many cases resulted from a reactive approach. When a new challenge is encountered, a solution is sought, often exploring the literature and iterating through solutions, occasionally adapting and improving the techniques. It is relatively rare to find a novel algorithm developed ex-novo inside a VFX facility. Most of the reasons for this limited approach stem from the relatively short timelines of VFX projects. In general, modern Visual

R. Kimmel, R. Klette, and A. Sugimoto (Eds.): ACCV 2010, Part I, LNCS 6492, pp. 189–199, 2011.

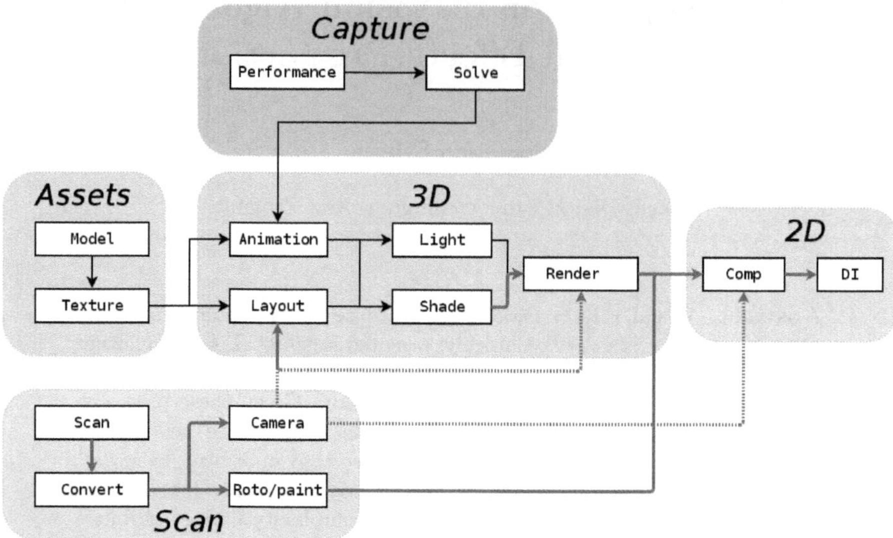


**Fig. 1.** Diagram of a simplified "pipeline" (workflow and data paths) for visual effects movie production, combining live action and CG (Computer Generated) images. The solid red arrows indicate images, dashed red arrows denote camera information, and black arrows indicate 3D model data.

Effects could not exist without most of the methods inherited from computer vision, yet there is in many cases a significant lag between the publication of research results and their adoption in industry.

On the other hand, many techniques that are in academia considered the state-of-the-art (even to the point that they are considered to have "solved the problem") are sometimes insufficient for the needs of visual effects.

**The VFX pipeline.** Visual Effect creation is in part an exercise in large scale data and infrastructure management. The VFX *pipeline* refers to the stages that data takes as it flows through the company, from original source material through to film or digital output. Each stage typically corresponds to a department having people with skills in the tasks and software particular to that stage. 'Pipeline' also describes the programs that are written to manage the data flow and format conversion between different departments and programs. The pipeline code is a significant investment (in the order of one million lines of code in an established facility). It also is a major source of efficiency for the studio, as it automates tasks that are otherwise manual.

Figure 1 shows a typical VFX pipeline. Footage from the set is imported (film is scanned or digital footage is offloaded from tapes or disks) and the images are converted to a working colourspace. The Camera Department tracks these *plates* (individual film-resolution frames) to recover the $4 \times 3$ projection matrices for each camera [7] and, where necessary and possible, recover the structure of objects in the scene, such as the position of buildings. The roto/paint department paints out tracking markers and rigs in each shot, and manually segments (or *rotoscopes*) the plates into elements so that computer graphic objects can later be inserted between them. Meanwhile, models are created, rigged for animation, and textured. Each shot is then built, by laying out the required models in the scene, animating them, setting up the lights and shader parameters, and is finally rendered, using the camera recovered from the plates. Compositors combine the rendered elements with the live action plates. The final shots are then delivered to "DI" (Digital Intermediate) to be colour graded, edited and conformed to the required output formats for the various 3-D projection technologies.

Appropriate stages of the pipeline interface with the *render farm* in order to run computationally intensive tasks such as rendering, as well as data intensive tasks such as data format conversions. The render farm at larger facilities includes a few thousand individual compute nodes. Generally, each node renders a single image frame independently of the others. Therefore there is little need for inter-node communication, so the render farm is optimised for fast access to the file servers rather than for fast inter-node communication.

Different facilities tend to have similar pipelines and use the same commercial packages. This is advantageous, as it makes it easier for new artists to adapt when they join a company. It also helps companies to share assets such as models when they work on collaborative projects.

## 2 The Use of Machine Vision Techniques in VFX

The use of some common Machine Vision algorithms is ubiquitous throughout the Visual Effects pipeline and has been for quite some time [13]. Some fundamental techniques have become widely adopted and commonplace in VFX production. However, even those adoptions are not unchallenged by new acquisition techniques (such as the shift from film to digital cameras) or other changes in the context. Even the most established techniques need —at times— a radical rethinking to account for a different environment or to take advantage of new processing platforms or different data types.

The areas where Computer Vision techniques are commonplace and have had the most impact are *Feature Tracking*, *Segmentation*, *Photogrammetry*, *Computational Photography* and *Motion Capture*, but there are many other areas of application where fundamental techniques have been in use and that would benefit from novel methods. One of these areas, that has recently received a fair amount of attention is Stereoscopic 3D VFX.

While these areas have benefited the most from existing research, in most cases there are still open challenges, and further improvements in the robustness, performance and precision of these algorithms would certainly be useful [13].

The following are some examples of applications of Computer Vision tools in the VFX pipeline. It can readily be noted that some of these applications are fundamental and essential to the process, however, they are seldom sufficient on their own. In most cases, these techniques have been in use for almost the entire history of visual effects. Some of these methods have proven their efficacy in a predominately image based visual effects production and are being modified or adapted to a more predominately 3D CG based environment. It is often the case that established tools do not transfer as well to the new environment.

**De-warping.** The integration of CGI and real world cinematography, requires that any distortion introduced by real world lenses be eliminated, and the images (or plates) be brought to a model that is as close as possible to the pinhole projection used for the creation of 3D CG images. Camera and lens calibration [16] is used to dewarp film images, allowing their seamless overlay. While not necessarily a Computer Vision technique per-se, this is a shared requirement with most vision applications, and needs to be applied as a pre-processing step to any image to be processed. The requirements for rectification in VFX are often more stringent than in many other applications, since the final result of the integration are the final images, seen by the audience, where a minimal discrepancy would be easily and immediately identified and noted as an issue. Often manual late stage corrections are applied to the final images to provide continuity, even at the cost of real-world accuracy. Calibration of camera parameters is commonplace nowadays, and has become routine. Not much is done, however to characterize the full behavior of real world lenses throughout their range, since most vision algorithms assume more simplistic camera models, and static parameters, whereas VFX production follows necessarily any artistic use of motion picture lenses the cinematography mandates, with little -if any- control over the creative choices made during *principal photography*.

**Feature tracking.** Two-dimensional tracking [8] is a built-in feature of commercial graphics packages. It is often the case that a filmed object will be the base for the movement of a CG one, that a specific section of the scene will be replaced with CG or that an effect will be tied to a specific position in the filmed space. The scene, will almost always be reconstructed as a 3D environment, and the real world camera movement reconstructed from the principal photography plates. Three-dimensional matchmoving packages are now in standard use for reconstructing camera motion [3,1]. All these techniques are rooted as is evident in feature tracking. In most cases, the sets are or can be purposely constructed to provide high-gradient features that ease the tracking process. The feature identification is typically sufficient, but does not rely on additional information that could be easily made available to the tools.

**Optical flow.** Optical flow [2] has been in use for a long time in VFX production. This idea is one that brings infinite promise to the field, since an accurate solution would greatly simplify or aid most of the processes Visual Effects artists face everyday. However, while excellent results have been shown, the more prevalent uses of optical flow have probably been video retiming and painterly image manipulations.

**Segmentation.** Segmentation plays a significant role in the separation of layers for compositing. Automatic segmentation has had numerous advances in recent years, but its adoption in feature film VFX has not followed the same path as neighboring applications. Part of the reason for this can be found in the inevitably present imprecisions in the process, that are not easily manipulatable, since the results of these methods are often a matte, rather than a modifiable contour. On the other hand, fully automatic segmentation techniques [14] have seen little adoption, probably both because they lack the necessary user guidance and because they do not scale to the resolution required for film work.

**Photogrammetric reconstruction.** While dense point clouds generated from photogrammetric reconstruction [12] can provide a very useful guide to modelers and is a practical (and sometimes the only possible) way to survey an environment, it doesn't, in most cases, provide the necessary amount of data to serve as a final model. Often the data coming from photogrammetric reconstruction is sparse, noisy or has holes, and is not in a form suitable for human editing. Thus it is most commonly used as a reference for manual modeling rather than as finished geometry.

**Motion capture.** A combination of Vision techniques go into the capture of motion and its application to a skeleton or a shell. This is probably one of the areas where the adoption of Vision techniques has been more prominent, continuous and successful. The future is also promising, since many of the advancements in the field rely on further improvements in vision methods. While much industry practice is far behind the research frontier as it relies on physical markers and active illumination, recent major movies have made use of vision-based body motion capture [6] and markerless facial capture [10].

**Other techniques.** There are numerous other examples of techniques that have an application in VFX production . For example, Ensemble classifiers [4] have been used in conjunction with inpainting techniques for removal of fiduciary markers. Other research topics have not yet seen much adoption in film production. These include shape-from-X, super-resolution, and computational photography among others.

The use of more complex vision based systems like the Lightstage [5] to model reflectance and capture accurate reference images and geometry of actors' skin is becoming widespread. While these are more sophisticated systems rather than individual vision technique, they still suffer from many of the same issues that general tools have. Better optical flow results, or stereo correlation would have an immediate impact on the accuracy of such results and would easily justify a more manual approach.

**Additional challenges of stereoscopic workflows.** In the past few years there has been a resurgence in the production of Stereoscopic 3D movies. These raise a number of different and specific challenges that —given the nature of the source material— can be addressed or helped with Computer Vision techniques. At the same time, many established methods that rely on 2D material can no longer be used in a stereo production environment where parallax and stereo vision are in play. In most cases, camera based techniques will be difficult to adapt, unless they specifically account for stereo parameters.

For example, one of the areas where working with stereoscopic pairs is significantly more challenging is Paint and Roto work, as discrepancies between the two views will be immediately evident as they would break the stereo perception even if they were individually acceptable. If the parallax of the painted image is not correct, the viewer will perceive that the painted area is "shimmering" or "swimming in depth". The only way to correct the problem is a painstaking manual operation. In much the same way, it is important to maintain or restore local color equalization between left and right views. Different electronic configurations or optical paths (beam splitters, lenses, filters, optics, etc.) will create discrepancies that need to be corrected to achieve an effective stereo effect. Another source of issues is the different specular response from eye to eye – a polarisation effect that is also described as "shimmer".

Certain corrections for stereo, for example modifications to the zero parallax plane depth, keystone correction, lens distortion correction or vertical alignment corrections, can also introduce blank areas in images. The blank areas revealed by such operations might need to be filled with extrapolated image data. While this is a problem that can be encountered in monocular VFX, stereo production introduces the additional challenge that the extrapolated data must be spatially and temporally consistent in both views.

In theory stereo correlation could provide depth and disparity, that can be used to transfer image manipulations from one eye to the other. In most cases, however, the accuracy of the correspondence between stereo pairs is not sufficient and the technique would generate artifacts and call attention to the visual effects work rather than making it seamless.

Current commercial packages offer some tools for attempting this (for example, cross-image registration and colour matching tools). These tools are often inadequate for the task, however, since the accuracy of dense disparity solves obtained from automated analysis is hardly ever sufficient.

Better algorithms, possibly relying on user interaction for the refinement of results are required to reach the level of precision, robustness and interactivity required -as an example- for correlation data to be useful for these scenarios.

## 3 Issues in Adapting Research Results

In many cases, the results of Vision algorithms are sufficient to solve most of the challenges posed by industrial and robotic machine vision. However, most if not all of the techniques mentioned above have not been necessarily experiencing the same level of adoption, acceptance and success during the history of VFX since they have not yet provided the necessary level of accuracy. Conversely, however, there is also an apparently conflicting separate issue in the density of the data that most of these algorithms produce.

As an example, while the dense point clouds derived from photogrammetric reconstruction provide a very useful representation of a scene or object, the data is often not usable in production as the density of the point cloud and its error would impose more manual labor on modelers than recreating the model at a level of abstraction that is appropriate for the task. Decimation of generated models, often reduces the precision (for example samples corresponding to corners may be dropped, thus rounding the surface), and while the generated meshes may have the correct level of detail, it is not represented in a structured form, usable in subsequent stages of the pipeline. In many cases it would be more useful to rely on user input to generate a simplified, yet structured mesh that can be further manipulated. Often researchers are biased towards a fully automated solution or assume the user interaction needs to be targeted to a novice user.

In many cases when research is applied proactively to solve problems (whether for a specific production or generic future challenges) there is a cultural gap: Given the reactive nature described above, simpler implementations will be preferred to more complex ones. The time to implement them is an important but not exclusive factor in that decision. More complex solutions tend to have a wider impact over the pipeline, and as such will encounter higher resistance to implementation. Also in many cases, rapid prototyping of tools is more useful than aiming toward a perfectly functioning program, since efficiencies are often gained in the overall process when the interaction with the users is taken into account. Generally however, these conditions need to be matched with a good possibility of being parallelized and a general stability to small changes in user input, to changes in resolution, and temporally. These factors are extremely important for an interactive tool and often ignored in search of an optimal unsupervised solution in academic research. The time an artist spends providing input to an interactive system needs to be optimized since it is the most valuable asset in the process, but the efficiency needs to be considered over the entire process, including the turnaround time. Academic research will endorse and rightly celebrate the achievement of results of a problem statement that is often isolated, is fully automated or assumes simplistic interfaces. In a high-skills professional environment, the efficiency of the interface and the ability to be easily integrated with other tools, often trump the quality of the results in evaluating results of a particular technique.

**Resolution requirements.** Nowadays VFX are generally produced at an approximate resolution of $2000 \times 1000$ or higher which maps under typical cinema viewing conditions a standard visual acuity to roughly one pixel.

Due to hyperacuity [15], however, the resolution requirements for computer vision algorithms is sometimes higher[1]. VFX production somewhat changes the formulation of some of the problems in computer vision. The resolution, precision and interaction requirements may be such that algorithms otherwise considered mature and automatic in the computer vision community may be regarded as labour intensive or unsuitable in film work.

[1] For example, assuming a 2000 pixel wide image projected on a screen that is 7 meters wide and viewed at a distance of 10 meters, and using the standard estimate of one minute of arc per receptor, we estimate 1.2 pixels per receptor.

For example, a common success measure for model-based tracking is to automatically track a target to within several pixels following a one-time training procedure. In visual effects the requirement for highly precise results may require per-shot calibration and training, followed by post-tracking cleanup if the tracking model uses a rank constraint that by definition cannot exactly match the real movement. Conversely, optical flow became widely adopted in some processes once schemes were developed to permit operation on high resolution images (e.g. [2]). Optical flow's intrinsic sub-pixel registration makes it successful even if it is only applicable to a minority of problems (such as temporal interpolation). This is typical of the VFX environment, where sophisticated atomic tools that solve some problems well and can be selected and inserted in the pipeline as needed will be preferred to more complex tools that try to solve a more generic problem formulation but provide overall less reliable results. Another important factor in the adoption of tools is a progressive user-interaction model that allows for a continuous and flexible refinement of the results, rather than a more locked process, even if the individual results of the latter tool were sometimes better.

**Pipeline inertia.** While a sophisticated pipeline is essential for undertaking large projects, it can also become a source of inertia. Rearranging the pipeline is a slow and risky operation due to the amount of code needed and the fact that the whole facility relies on its correct operation. Even adding a new datatype to an existing pipeline is a slow and expensive operation. Algorithms that are easy to adopt, therefore, are ones which do not require major changes to the pipeline. These fall into two categories: those which are used by only one stage of the pipeline, and those that are sufficiently special purpose that they will be used on only a few shots entirely separated from the pipeline.

**Modular production toolsets.** VFX companies often rely on general tools (such as The Foundry's *Nuke* or Autodesk's *Maya*) for the core of their production pipeline and generally develop new solutions as plugins or add-ons to those. The choice of plugins provides VFX artists with a familiar interface and a common framework, where tools can be chained. For developers there is a lower barrier to entry, as many base operations are already available as part of the API, and some processing steps can be assumed to happen outside the individual tool.

Generating a depth map from a stereo camera, for example, might require a denoiser and a color correction step prior to applying the depth-from-stereo plugin. The depth solver may compute an edge mask to help detect discontinuities. Compositing packages are likely to have an array of edge detection techniques, including a paint tool to allow artists to mark edges manually, or remove spurious lines from a computed edge mask. Adapting the solver to accept this edge mask as a separate input, rather than computing it internally, will make the tool more robust, since the most suitable edge detection technique for the input can be selected by the artist to optimise performance.

Published algorithms that implement a complete system may be hindered by inflexible implementations of incidental stages of the process, even if the core algorithm is superior and would be better accepted if they were modular in their construction.

Often, only parts of a published method are extracted and implemented. For example, depth-from-stereo solvers often have an initial step which computes the epipolar geometry and rectifies the image to vertically align the epipolar lines with the image scanlines. Since the epipolar geometry can be derived from the camera projection matrices, which are already available, the step need not be re-implemented as the depth solver would be adapted to work with the existing tools and data rather than redoing parts of the pipeline. This leads to a more versatile system that combines multiple algorithms. Again, in the previous example, a scene may contain a large number of planar surfaces but the depth solve might be unsatisfactory. A number of different planar homography detectors could be employed to improve the quality of the depth solve in each planar region. For example the technique of Lourakis [9] would occasionally be useful where a surface has coplanar circles. Given a suite of planar homography solvers, the artist can simply select the most appropriate tool for each surface and feed the resultant homographies into the depth solver. These tools would be expanded to solve specific circumstances in different projects and as more techniques are published.

**Limited control on source images.** Visual Effects facilities cannot generally guarantee any degree of control over on-set shooting, and in fact they may not become involved in a project until after filming is completed. Often nowadays a VFX supervisor will be involved in most shoots that will be later put through the VFX pipeline, but given the unpredictable nature of the process that may not always be the case. Computer vision tools —by their own nature— will be used in difficult situations where simpler approaches could have been employed but an unforeseeable issue occurred.

The nature of the shoot will often dictate the amount of influence a VFX supervisor can have over filming and what extra data can be captured. When the capture of data can be made in an asynchronous way to *principal photography* it is easier to have a separate crew, trained to the specific capture process, dedicated to the task.

**User guidance.** In most cases, Visual Effects production techniques rely heavily on the input of users, both creatively and for operations that are trivial. This input is generally very artistically skilled. Interfaces should be designed allowing the artists to understand how parameters affect the final output without needing an in-depth understanding of the algorithm. Artists will often prefer tools that provide interactive feedback while manipulating parameters. For example, an "artist-friendly" implementation of a camera solve system based on corner detection in each image followed by matching across images and RANSAC to identify a noise-free set of features might have an interface that presents the artist with a view of only the corners, and the parameters that control the corner finder, so they can see intuitively how the parameters affect the result. They may also view the correspondence vectors and those selected by the RANSAC step, and adjust the corresponding parameters. Each step can be visually portrayed and is easy to understand. Approaches that do not provide such intuitive control are less useful even if they offer superior results in theory.

A common trade-off to user input is computation power. There is normally a fair amount of compute power available to algorithms that provide a sufficiently accurate results automatically. Processes are sometimes separated in parameter exploration phases where interactive visualizations provide an approximation of the results (perhaps on only a few keyframes), leaving the final calculation to run unsupervised on a *render-farm* once an acceptable set of parameters have been found. Compute times of an hour per frame or even more are acceptable in this case.

**Time to adoption.** The requirement for modular tools is also rooted in the contention for resources. Once a project is in production, there are normally a number of challenges to be solved, and development of new solutions is necessarily time and resource constrained.

## 4 Conclusions

The needs of the VFX industry are somewhat different from other fields of application for computer vision algorithms. The environment provides highly skilled artists that can combine or choose different techniques for the task at hand, but may not be aware of the inner workings of the algorithms, and —in most cases— benefit from interactive feedback.

Additionally, relatively large processing resources are normally available and longer processing times are generally acceptable once an acceptable parameter set has been found. However, algorithms for visual effects must produce temporally coherent results at high resolutions. This is quite removed from (for example) machine vision algorithms in other fields that must cover all possible use cases and be real-time, fully automatic, robust, low cost and relatively low-power but that may not need such high precision.

VFX facilities are constantly in search of tools that would provide better efficiency or novel looks. Beyond this motivation, there are normally three major acceptance criteria for adoption of a particular algorithm: it will reduce labour costs sufficiently to offset the development time and resources, it will integrate sufficiently with the existing pipeline, and the normal software development risks should be acceptable. Additionally, given the feedback intensive nature of many of the tools, the cost of training artists should be taken into account. Image processing and computer vision research have played a very important role in the visual effects industry (much of the basic vocabulary in this field is adapted from it). The adoption of research findings could benefit from a better understanding of the environment and peculiar requirements (and available resources) in the production of VFX.

**Acknowledgement.** Many thanks to John Lewis and Peter Hillman for the very valuable feedback and contributions.

## References


1. 3D-Equalizer matchmoving software, http://www.sci-d-vis.com
2. Black, M.J., Anandan, P.: A framework for the robust estimation of optical flow. In: Fourth International Conference on Computer Vision, ICCV, pp. 231–236 (1993)
3. boujou 3D matchmoving software, http://www.2d3.com
4. Breiman, L.: Bagging Predictors. Machine Learning 24(2), 123–140 (1996)
5. Debevec, P., Hawkins, T., Tchou, C., Duiker, H.P., Sarokin, W., Sagar, M.: Acquiring the reflectance field of a human face. In: Proceedings of ACM SIGGRAPH 2000. Computer Graphics Proceedings, Annual Conference Series, pp. 145–156 (2000)
6. Debruge, P.: More than one way to capture motion. Variety (May 30, 2008)
7. Hartley, R.I., Zisserman, A.: Multiple View Geometry in Computer Vision, 2nd edn. Cambridge University Press, Cambridge (2004)
8. Lewis, J.: Fast template matching. In: Vision Interface, pp. 120–123 (1995)
9. Lourakis, M.I.: Plane metric rectification from a single view of multiple coplanar circles. In: Proceedings of ICIP, pp. 509–512 (2009)
10. Contour markerless surface capture, http://www.mova.com
11. Organic Motion markerless motion capture, http://www.organicmotion.com
12. Pollefeys, M., Van Gool, L., Vergauwen, M., Verbiest, F., Cornelis, K., Tops, J., Koch, R.: Visual modeling with a hand-held camera. Int. J. Comput. Vision 59, 207–232 (2004)
13. Roble, D.: Vision in film and special effects. SIGGRAPH Comput. Graph. 33, 58–60 (2000)
14. Shi, J., Malik, J.: Normalized cuts and image segmentation. IEEE Trans. Pattern Analysis and Machine Intelligence 22(8), 888–905 (2000)
15. Westheimer, G.: Visual hyperacuity. Prog. Sensory Physiol. 37, 1–30 (1981)
16. Zhang, Z.: Flexible Camera Calibration by Viewing a Plane from Unknown Orientations. In: Proc. ICCV, pp. 666–673 (1999)

# Automatic Workflow Monitoring in Industrial Environments

Galina Veres[1], Helmut Grabner[2], Lee Middleton[1], and Luc Van Gool[2,3]

[1] University of Southampton, IT Innovation Centre, UK
[2] Computer Vision Laboratory, ETH Zurich, Switzerland
[3] ESAT - PSI / IBBT, K.U. Leuven, Belgium

**Abstract.** Robust automatic workflow monitoring using visual sensors in industrial environments is still an unsolved problem. This is mainly due to the difficulties of recording data in work settings and the environmental conditions (large occlusions, similar background/foreground) which do not allow object detection/tracking algorithms to perform robustly. Hence approaches analysing trajectories are limited in such environments. However, workflow monitoring is especially needed due to quality and safety requirements. In this paper we propose a robust approach for workflow classification in industrial environments. The proposed approach consists of a robust scene descriptor and an efficient time series analysis method. Experimental results on a challenging car manufacturing dataset showed that the proposed scene descriptor is able to detect both human and machinery related motion robustly and the used time series analysis method can classify tasks in a given workflow automatically.

## 1 Introduction

Intelligent visual surveillance is an important area of computer vision research. In this work we focus on real-time workflow monitoring in industrial environments. The aim is to recognise tasks happening in the scene, to monitor the smooth running of the workflow and to recognise any abnormal behaviours. Any deviation from the workflow either by the workers or by the machinery may cause severe deterioration of the quality of the product or may be dangerous.

An example of such an industrial scenario is shown in Fig. 1. By monitoring industrial scenes, one faces several challenges such as recording data in work areas (camera positions and viewing area), industrial working conditions (sparks and vibrations), difficult structured background (upright racks and heavy occlusion of the workers), high similarity of the individual workers (nearly all of them wearing a similar utility uniform), and other moving objects (welding machines and forklifts). Furthermore, the dynamics of workflow can be quite complex and deviations typically occur. Several tasks within a workflow can have different lengths and no clear definition of beginning/ending. Moreover, the tasks can include both human actions and motions of machinery in the observed process; whereas other motions not related to the workflow have to be suppressed (e.g., other people passing or people restocking or replacing racks).

R. Kimmel, R. Klette, and A. Sugimoto (Eds.): ACCV 2010, Part I, LNCS 6492, pp. 200–213, 2011.

**Fig. 1.** Example of an industrial scenario: a workflow consists of several tasks; different length tasks; no clear definition of beginning/ending; coping with irrelevant motion and occlusions

**Related work.** A whole field of operational research is concerned with the design and data mining from workflows. In the computer vision community, this is mainly addressed in applications such as abnormal behaviour recognition or unusual event detection. Many approaches have been suggested over the past years and a complete review is far beyond the scope of this paper. Typically they build a model of normality and the methods can differ in (i) the model used, (ii) the employed learning algorithm for learning the model parameters, and (iii) the features used. Models might be previously trained and kept fixed [11,18,19] or adapt over time [3] to cope with changing conditions. Various machine learning and statistical methods have been used: from clustering [2] and density estimation [8] to Hidden Markov Models for time series analysis commonly used in action recognition [12,14]. Also a broad variety of extracted image features are used such as global scene descriptors [3], optical flow [1,10], 3D motion [14] or object trajectories [8,17].

Depending on the used model and prior knowledge, a lot of training data has to be collected. For example, the purely data driven approach of [3] has to observe every situation and is not able to generalise a person walking on a slightly different path. On the other hand, using more sophisticated "high-level" features (e.g., a person's trajectory) usually requires less data to generalise well. However, these features have to be extracted robustly. Recent advances in object detection and tracking showed that these methods do not perform very well in a cluttered/occluded environment such as those presented in this paper.

**Contribution.** To our knowledge, no state-of-the-art approach is able to cope with the challenges of workfow analysis within industrial environments as reviewed above. We tried several state-of-the-art methods for person detection/

(a) Person detection [4] (b) Person tracking [5]

**Fig. 2.** Examples of state-of-the art methods for detection and tracking

tracking [4,5][1]. However, none of them showed stable and robust results in the industrial environment. Fig. 2 (a) shows typical failures of the detector on our dataset with a recall of 24% and a precision of only 9%. Thus, tracking-by-detection approaches (e.g., [9]) cannot be used to generate trajectories. Also, the person could be hardly tracked as displayed in Fig. 2 (b). The tracker starts very well, however it soon loses the person and drifts away.

The reasons for the failures are due to the nature of the environment such as significant occlusions, clutter similar in structure/shape to a person, the workers coloured similarly to the racks, and the unstable background due to welding flare, machinery operation, and lighting changes. Any of these in isolation would cause problems for person detection and tracking, but all of them together make the problem especially difficult for both detection and tracking and violates the use of approaches which are based on the analysis of trajectories.

Hence, in this paper we propose a novel approach for *real-time workflow monitoring*. It consists of a robust, yet simple, scene descriptor and an efficient time series analysis method. The model is easy to train and update, i.e. it allows adaptation to new workflows and inclusion of prior knowledge. Several activities happening simultaneously can be incorporated in the model while the algorithm is still robust to irrelevant motions.The experimental results on a car manufacturing database show that our approach is suitable to monitor workflows robustly and thus allows for (i) obtaining statistics of the workflow in order to optimise it, as well as (ii) detecting abnormalities (for safety reasons).

The remainder of the paper is organised as follows. In Sec. 2 we present image descriptors and model learning techniques, which are then used by our novel approach for real-time workflow analysis in Sec. 3. Detailed experiments and comparisons are shown in Sec. 4. Sec. 5 concludes the paper.

## 2 Problem Formulation and Modelling

Our goal is to monitor a pre-defined repetitive workflow. We describe a workflow as a sequence of defined tasks that have to be executed in some order (permutations

[1] The code was downloaded from the authors webpages.

are allowed) due to safety or quality reasons. A task is a sequence of observations that corresponds to a physical action like "take object and place it somewhere".

**Problem modeling.** Let $I_t \in \mathbb{R}^{n \times m}$ be gray scale image at time $t$. Given an image sequence $\mathcal{I} = \{I_0, \ldots, I_t\}$ and a set of $L+1$ possible tasks, $\mathcal{L} = \{1, \ldots, L, \#\}$, where $\#$ corresponds to a task not related to the workflow. We want to associate a task $l_t^\star \in \mathcal{L}$ with each image $I_t$ at time $t$, using measurements obtained up to time $t$. This can be seen as temporal $L+1$ class classification problem.

In summary, the goal is to identify for each frame whether the frame belongs to the workflow or not. If the frame belongs to the workflow, then we need to identify which task takes place. Our approach is based on using local motion monitors which serve as scene features for a powerful time series analysis network. In the following we define and review these two components before introducing our approach.

### 2.1 Motion-Grid

Features extracted from the raw pixel values should be discriminative enough to capture relevant changes with respect to the tasks but at the same time be invariant to irrelevant variations. Motivated by the fact that workflow consists of actions of an object (walking of a person) and interactions of objects within the environment (using a welding machine), we use motion as our primary feature cue. Inspired by [1] we introduce local motion monitors.

**Local Motion Monitor.** A local motion monitor $M^{(x,y)}$ observes a position $(x, y)$ and the surrounding $(n \times m)$ pixel neighbourhood $\Omega^{(x,y)}$ of the image (see Fig. 3(a)). The binary output of a motion monitor applied on the image $I_t$ is defined as

$$M^{(x,y)}(I_t) = \begin{cases} 1 & \text{if} \sum_{(i,j)\in\Omega^{(x,y)}} |I_t(i,j) - I_{t-1}(i,j)| > \theta_M \\ -1 & \text{otherwise} \end{cases} \quad (1)$$

In other words, simple frame differencing is used to get changes of the image. If the changes are significant (specified by the parameter $\theta_M$) within the region $\Omega^{(x,y)}$ the local motion monitor $M^{(x,y)}$ "fires", i.e. returns a positive response.

**Motion-Grid.** A motion grid is defined as a set of local motion monitors. The idea of motion grids is to sample an input image sequence by using a fixed overlapping grid. Each grid element corresponds to a local motion monitor as illustrated in Fig. 3(b).

We use the motion grid as a scene descriptor. In other words, the local motion monitors can be seen as features which extract high level information from each image. For one time instance $t$ we concatenate the output of the local motion monitors within the grid into a vector.

### 2.2 Echo State Network

Often Hidden Markov Models [12,14] are used for analysing temporal data. However, HMM require well-defined states formed by robustly extracted features.

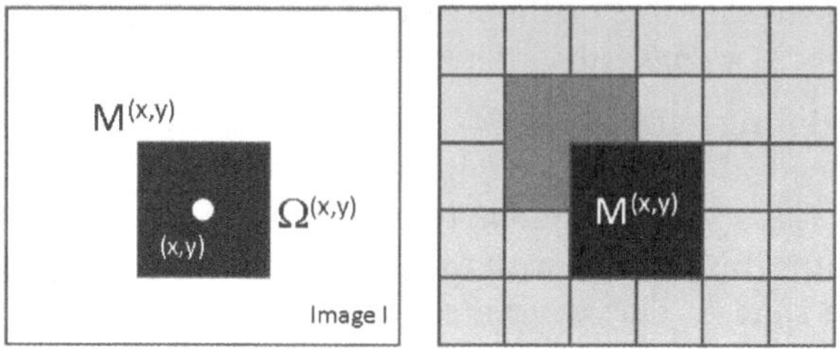


(a) local motion monitor (b) motion-grid

**Fig. 3.** (a) a local motion monitor $M^{(x,y)}$ at position $(x, y)$ analyses the intensity changes in its local neighborhood $\Omega^{(x,y)}$. (b) an overlapping grid of local monitors.

This is not always possible in industrial settings. In this paper we propose to use an alternative approach: Echo State Networks (ESN) [6] which offer several benefits such as (i) fast and simple learning of many outputs simultaneously, (ii) possibilities of both off-line and on-line learning, (iii) ability to learn complex dynamic behaviours, and (iv) directly dealing with high dimensional input data. They have been shown to perform very well in the context of supervised learning, for example emotion recognition [15] and speech recognition [16].

**Principle.** An ESN is a discrete time, continuous state, recurrent neural network proposed by Jaeger [6]. Learning complexity is kept low while good generalisation can be achieved on various dynamic problems. Fig. 4 shows a typical ESN[2]. The hidden layer consists of $N$ randomly connected neurons. If the connectivity is low, this layer provides independent output trajectories. For this reason, the hidden layer is also called a "reservoir". Furthermore, there are neurons which are connected to cycles in the reservoir, so that past states "echo" in the reservoir. The neurons within the hidden layer are also randomly connected to the $k$-dimensional input signal, which drives the network.

The essential condition for the successful application of the ESN is the "echo state" property of their state space. If this condition is met, *read-out* neurons can be trained, which take a linear combination of the input and the hidden layer [13]. In other words, only network output weight adaptation is sufficient to train the network. Consequently an ESN can be seen as a universal dynamical system approximator, which combines the elementary dynamics contained in the reservoir. For a large and rich reservoir of dynamics hundreds of hidden units are required.

Let $\mathbf{u}_t = (u_{1,t}, \ldots, u_{K,t})$ be the input to the network at time $t$. Activations of hidden units are $\mathbf{x}_t = (x_{1,t}, \ldots, x_{N,t})$, and of output units are $\mathbf{y}_t = (y_{1,t}, \ldots, y_{L,t})$. Further, let $\mathbf{W}^{in}_{N\times K}$ be the weights for the input-hidden connection, $\mathbf{W}_{N\times N}$ be the weights for the hidden-hidden connections, $\mathbf{W}^{back}_{N\times L}$ be the weights for the output-hidden connection, and $\mathbf{W}^{out}_{L\times(K+N+L)}$ be the weights for the read-out neurons,

[2] Figure adapted from `http://www.scholarpedia.org/article/Echo_state_network`, 2010/06/05 with the author's permission.

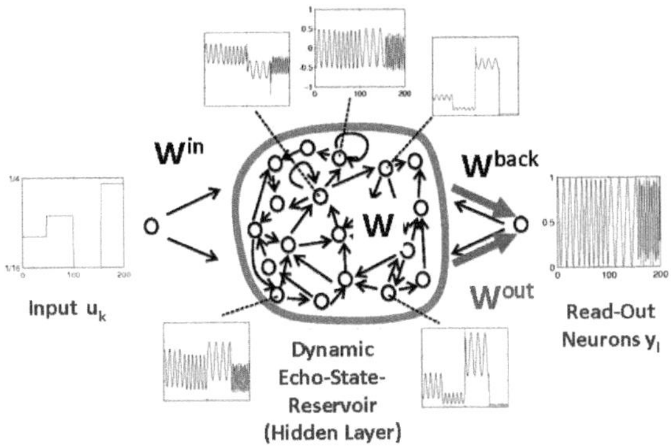


**Fig. 4.** Schematic view of an Echo State Network. Black arrows indicate weights chosen randomly and fixed;red arrows represent the weights to be optimised.

i.e. the connection from all units to the respective read-out neurons. The activation of internal and output units are updated at every time step by:

$$\mathbf{x}_t = f(\mathbf{W}^{in}\mathbf{u}_t + \mathbf{W}\mathbf{x}_{t-1} + \mathbf{W}^{back}\mathbf{y}_{t-1}), \tag{2}$$

where $f = (f_1, \ldots, f_N)$ are the hidden unit's activation functions. The outputs are calculated as:

$$\mathbf{y}_t = f^{out}(\mathbf{W}^{out}[\mathbf{u}_t, \mathbf{x}_t, \mathbf{y}_{t-1}]), \tag{3}$$

where $f^{out} = (f_1^{out}, \ldots, f_L^{out})$ are the output unit's activation functions. The term $[\mathbf{u}_t, \mathbf{x}_t, \mathbf{y}_{t-1}]$ is the concatenation of the input, hidden and previous output activation vectors.

**Training.** Assume a given training sequence $\mathcal{T} = \{(\mathbf{u}_0^{teach}, \mathbf{y}_0^{teach}), \ldots, (\mathbf{u}_T^{teach}, \mathbf{y}_T^{teach})\}$ with known input-output relations. When the network is updated according to Eq. (2), then under certain conditions the network state becomes asymptotically independent of initial conditions. So, the network will asymptotically depend only on the input history, and the network is said to be an echo state network. A condition for the echo state property is to ensure that the maximum eigenvalue (the spectral radius) $|\lambda_{max}|$ of $\mathbf{W}$ is less than 1 [7].

The hidden-hidden weights $\mathbf{W}$, the input-hidden weights $\mathbf{W}^{in}$ and the output-hidden weights $\mathbf{W}^{back}$ are set randomly (typically in a sparse random connectivity pattern) and kept fixed during training. Only the output weights $\mathbf{W}^{out}$ are learned so that the teaching output is approximated by the network. In the ESN approach this task is formulated as minimising the training error:

$$\epsilon_{train}(t) = (f^{out})^{-1}\mathbf{y}_t^{teach} - \mathbf{W}^{out}(\mathbf{u}_t^{teach}, \mathbf{x}_t) \tag{4}$$

in the mean square sense. For example, when considering the read-out as linear combination (i.e., $f^{out}(x) = (f^{out}(x))^{-1} = x$) this can be done quite easily by standard methods of linear regression.

**Testing.** For usage on test sequences, the trained network can be driven by new input sequences and the output is computed as:

$$\hat{\mathbf{y}}_t = \mathbf{W}^{out}\mathbf{x}_t. \tag{5}$$

# 3 Workflow Monitoring

In the ideal world the surveillance system would monitor workflows and detect abnormalities automatically by observing the scene. However, this is a very challenging goal. Therefore we make use of the specifics of workflows in manufacturing environments. In an industrial environment, a lot of time is spent on setting up machinery, designing work cycles and testing the smooth operation of designed workflows. We use this time and the accumulated knowledge to build the mode of normality which is then used during run-time. Hence, our proposed approach consists of two phases: (i) an off-line setup/maintenance phase and (ii) an on-line run-time phase,which are described in the following.

## 3.1 Setup/Maintaince

Given a labelled training set $\{(\mathbf{u}_t, \mathbf{y}_t)\}_{t=1,\dots,T}$ of input/output pairs we train an ESN. One read-out neuron $y_l$ is trained for each individual task $l = 1, \dots, L$, according to the annotation for each frame. Additionally, one read-out neuron $y_{L+1}$ is trained where no specific task definition is given to capture the remaining "class".

**Prior knowledge.** Unfortunately in real-life applications it is not always possible to set up cameras which are focusing only on the relevant areas. As a result, our scene descriptor will detect motion both associated with the workflow and usual behaviour not associated with it. An extra motion in the feature space can result in an increase of intraclass variations for some tasks and potentially in reduction of the correct classification rates. One might use feature selection methods to remove the redundancy and focus on the relevant parts of the images. However, when the training set is relatively small (and the intra-class variance compared to it is quite high) it is hard to establish the robust statistics required.

As an alternative, we suggest to use prior information to identify regions in the frame where the actions related to the workflow can potentially take place. Note, the user is only needed once during the setup of the monitoring process. As soon as no significant changes are happening to the workflow for a given camera view no further interaction is necessary. That is, we mask out local motion monitors, which carry no significant information for the workflow description, i.e. are not within a relevant region $R_{rel}$. The remaining motion monitors form then the motion grid matrix and are used as inputs for the ESN

$$\mathbf{u}_t = [u_t^{(1,1)}, \dots, u_t^{(x,y)}, \dots, u_t^{(n,m)}], \text{where } u_t^{(x,y)} = \begin{cases} M^{(x,y)} & \text{if } (x,y) \in R_{rel} \\ \text{not used} & \text{otherwise} \end{cases} . \tag{6}$$

**Algorithm 1.** Setup/Maintaince (Off-line Training)

**Data**: Given labelled training image sequence $(\mathbf{x}_t^{teach}, \mathbf{y}_t^{teach})$
**Result**: Trained ESN, $(\mathbf{W}^{in}, \mathbf{W}, \mathbf{W}^{back}, \mathbf{W}^{out})$
1 Generate randomly the matrices $(\mathbf{W}^{in}, \mathbf{W}, \mathbf{W}^{back})$;
2 Scale $\mathbf{W}$ such that its maximum eigenvalue $|\lambda_{max}| \leq 1$;
3 Apply the motion grid to obtain the scene descriptor $\mathbf{u}_t^{teach}$ for image $\mathbf{x}_t^{teach}$.;
4 Run this ESN by driving it with the teaching input signal $(\mathbf{u}_t^{teach}, \mathbf{x}_t^{teach})$;
5 Dismiss data from the initial transient. Collect remaining input and network states row-wise into a matrix $\mathbf{M}$. Simultanously, collect the remaining training pre-signals $(f^{out})^{-1}\mathbf{y}_t^{teach}$ into a teacher collection matrix $\mathbf{R}$.;
6 Calculate the output weights $\mathbf{W}^{out} = (\mathbf{M}^{+}\mathbf{R})^{T}$, where $\mathbf{M}^{+}$ is the pseudo inverse of $M$.;

**Algorithm 2.** Run-Time (On-line Testing)

**Data**: Trained ESN, $(\mathbf{W}^{in}, \mathbf{W}, \mathbf{W}^{back}, \mathbf{W}^{out})$, new input image $\mathbf{x}_t$
**Result**: Task $\hat{y}(t)$ for $\mathbf{x}(t)$
1 Apply the motion grid to obtain the scene descriptor $\mathbf{u}_t$ for image $\mathbf{x}_t$.;
2 Update the ESN states with $\mathbf{u}_t$;
3 Calculate the read outs, i.e., $\hat{\mathbf{y}}_t = \mathbf{W}^{out}\mathbf{x}_t$.;
4 Assign concrete task $\hat{y}$ using Eq. (7) and post-processing.

### 3.2 Run-Time

During run-time, the scene descriptors are obtained as in Eq. (6) and processed by the ESN. The predicted output produced by the ESN will not be binary in the general case and more than one output can have non-zero values. Furthermore, since the individual read-out neurons are trained independently and usually from highly unbalanced data, we propose to normalize the response with respect to their mean responses $\bar{\mathbf{y}}$, calculated on the training data. To identify a task for each time instance the significant maximum is taken by

$$\hat{y}_t = \begin{cases} l^{\star} & \text{if } \dfrac{\max\limits_{l=1,\dots,L+1} \mathbf{y}_t(l) - \bar{\mathbf{y}}(l)}{\max\limits_{\substack{l'=1,\dots,L+1 \\ l^{\star}\neq l'}} \mathbf{y}_t(l') - \bar{\mathbf{y}}(l')} > \theta_L \\ L+1 & \text{otherwise} \end{cases}, \text{where } l^{\star} = \arg \max_{l=1,\dots L+1} \mathbf{y}_t(l) - \bar{\mathbf{y}}(l) \tag{7}$$

In other words, the maximum of the $L+1$ outputs is considered to be significant, if the ratio to the second highest value is above some defined threshold $\theta_L$. This threshold influences the precision of our method.

**Post-processing.** Since the tasks do not switch within short time intervals we apply median filtering to remove outliers.

Training and testing are summarised in Alg. 1 and Alg. 2, respectively. Please note, no temporal pre-segmentation has to be done for testing. Moreover, testing is performed on-line and is very efficient, which allows for real-time processing.

# 4 Experimental Results

While our approach is quite general, we focus on an concrete example within a car assembly environment. During each day the same workflow is performed many times. The purpose of the surveillance in this application is to monitor the smooth running of assigned operations.

## 4.1 Car Assembly Example Setup

We recorded approximately 8 hours of video from a single working cell (including gaps between workflows and breaks) due to environment sensitivity. Additonally, restrictions on the possible areas of camera installation were also imposed. A dataset was captured by a PTZ camera at a workcell inside a car manufacturing plant. Omnidirectional cameras cannot be used in this setting due to height restrictions on the camera position. We recorded data at 25 frames per second with relative jitter bounded by 1.6% on frame rate with resolution of 704 by 576 pixels. The camera view for workflow and task recognition and a schematic top-down view of the setting are given in Figure 5.According to the manufacturing requirements each workflow consists of the following 7 tasks, which are usually, however not always, executed sequentially:

*Task1:* A part from Rack 1 (upper) is placed on the welding spot by a worker(s).
*Task2:* A part from Rack 2 is placed on the welding spot by worker(s).
*Task3:* A part from Rack 3 is placed on the welding spot by worker(s).
*Task4:* Two parts from Rack 4 are placed on the welding spot by worker(s).
*Task5:* A part from Rack 1 (lower) is placed on the welding spot by worker(s).
*Task6:* A part from Rack 5 is placed on the welding spot by worker(s).
*Task7:* Worker(s) grab(s) the welding tools and weld the parts together.
*Task8:* Any frame, where no actions from Tasks 1-7 take place.

The tasks are strongly overlapped: all tasks (except Task1) will start/finish at the welding machine. It is difficult to identify, even by eye, which task the frame belongs to at the start/end of the task. Moreover, some tasks can have overlapping paths for a number of frames. In the workflow, the duration of the tasks

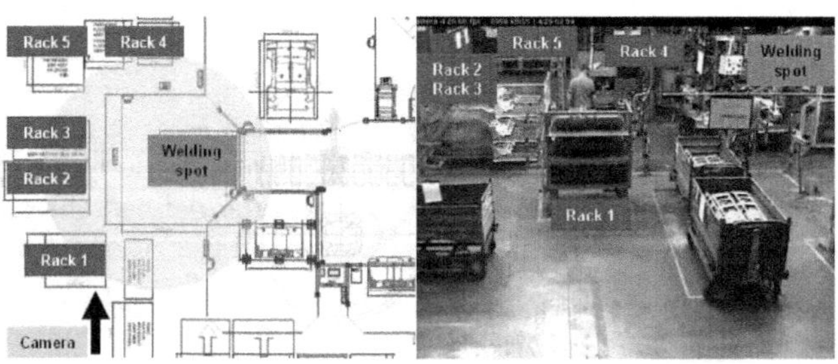


(a) Schematic top-down view (b) Camera view

**Fig. 5.** Schematic and camera view in the car assembly environment is different. Moreover, the duration of the same task changes from the workflow to the workflow. All of these difficulties and the high level of occlusions in the car assembly environment present challenges to workflow monitoring and task recognition (cf. Fig. 1).

### 4.2 Implementation Details

Here we give specific implementation details of our method, which allow reproduction of our experimental results easily. Where the parameters have been specified manually, most of them are not particularly sensitive.

**Motion Grid.** The initial motion grid matrix was calculated for 140 patches overlaid onto the whole image. The size of patches (local motion region) were selected as $\Omega = 100 \times 100$ with an overlap of 0.5. Activation motion threshold is set to $\theta_M = 250$.

**Prior knowledge.** The human operative manually specifies the region where the workflows can potentially take place including welding machine. In fact, 65 local motion monitors in the top half of the image were selected. These monitors form the scene descriptor for each frame. Fig. 6 depicts the descriptors over time for three consecutive tasks and three workflows. Red vertical lines indicate the beginning of the new task according to the annotation.

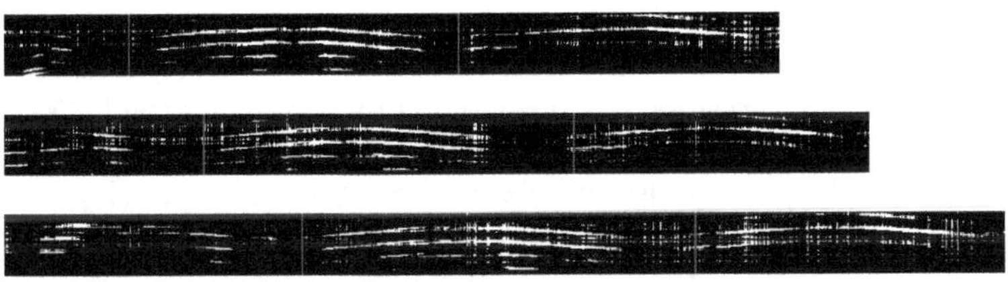

**Fig. 6.** Examples of scene descriptors over time (each column represents a motion grid feature vector) for the first three tasks of three workflows

**Echo-State-Network.** We apply a plain ESN with 12,000 hidden units, 65 inputs and 8 outputs. We used the ESN toolbox written in MATLAB by H. Jaeger and group members[3] and the recommendations given in the on-line tutorial on how to select some ESN's parameters. Furthermore, we investigated dependencies between the ESN parameters and normalised mean-square error. There are many suboptimal sets of parameters which will yield very similar results. Here the spectral radius is $|\lambda| = 0.98$, input scaling and teacher scaling are chosen as 0.1 and 0.3 respectively. Furthermore, additional noise is added to the ESN during the training process to improve the stability.

**Post-processing.** Median filtering with a filter length of 51 is performed.

---

[3] `http://www.reservoir-computing.org/node/129`, 2009/08/05.

### 4.3 Evaluation and Results

On the dataset, 19 workflows and their tasks are manually labelled. Each workflow consists of 3,550 to 7,300 frames, which correspond to 2 to 5 minutes. Ten of them (51,520 frames) were used for training, the remaining nine workflows (48,950 frames) were used for testing. Training takes approximately 50 hours using the MATLAB implementation on a 2.83 GHz computer running Windows Vista. However, testing can be done very efficiently online at 20 fps. For a quantitative evaluation, we use recall-precision measurements. The recall corresponds to the correct classification rate, whereas the precision relates to the trust in a classification. The F-measure is the harmonic mean of these two measurements.

**Comparison.** At first, we extract features from images using the proposed motion-grid approach. As previously stated, no current state-of-the-art method is able to produce robust results on this kind of data. Hence, for a comparison, we implemented a simple data driven approach inspired by the recent work of [3]. Its principle is to store all training data in respective clusters and detect outliers using the principle of meaningful nearest neighbours from these clusters. Since we have labelled training data, we use the following approach. All training images are stored in a large database and during run-time we look for the best matching one. The label of the frame is returned if it is meaningful. Similar to Eq. (7) of our proposed approach this stabilizes results and allows us to specify the precision.

Results are presented in Fig. 7 and in more detail by a confusion matrix in Fig. 8. Whereas the performance of the frame based nearest neighbour voting already achieved good results, our proposed approach, using the ESN as a complex dynamics time series predictor, gains a further 15% during monitoring of tasks. Additionally, it is significantly faster. Confusion matrices (Fig. 8) indicate that the most difficult task for monitoring is Task 1. Though this task is well separated from manufacturing requirements point of view, it is not that easy to distinguish from other tasks using the video recordings, since it shares the same paths as other tasks for some periods of time. Task 7 is the best recognised task from the workflow when the proposed approach is used. It consists of the smaller proportion of human motion (the majority of the time the humans are stationary) and larger proportion of machinery motion and sparks from the welding process. In many cases the tasks are misclassified either as belonging to Task 7 or Task 8. Misclassifications to Task 7 happen since start and end of all tasks are concentrated around the welding machine. Overall, our approach is able to practically achieve above 50% correct classification rate for each of the tasks. Additionally, the raw output of the ESN and the results after post-processing of the our approach with respect to the ground truth is shown in Fig. 9.

Some examples of successful estimated task and typical cases of failures are depicted in Fig. 10. The similarity of the tasks can be demonstrated by the last two images in the fourth row. Task 2 is misclassified as Task 3 (the second image), since the Task 2 is ending in this frame, and our system believes that the Task 3 is starting. Task 1 is misclassified as Task 4 in the third image, since for this frame Task 1 has very similar path to Task 4.

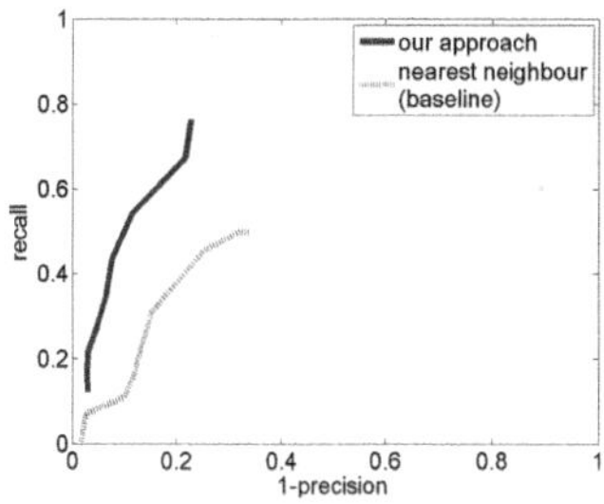


| | Recall | Prec. | F-m. |
|---|---|---|---|
| Motion-grid + NN (baseline) | 54.1 ±14.8% | 68.2 ±10.8% | 60.1 ±13.2% |
| Motion-grid + ESN (our approach) | 78.6 ±11.5% | 77.4 ±10.2% | 78.0 ±10.7% |

**Fig. 7.** Performance for all 9 test workflows

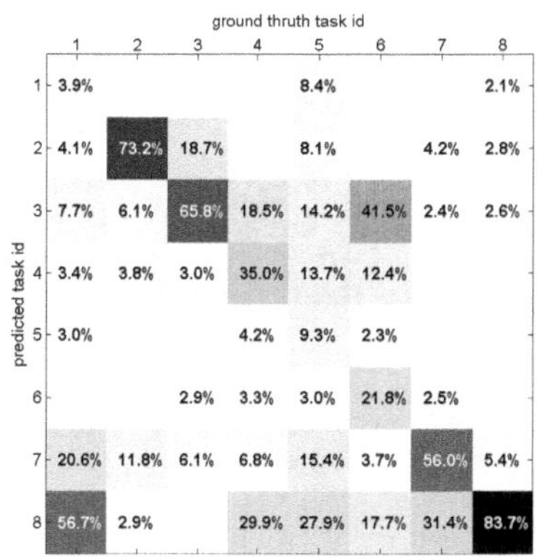


(a) Motion-grid + NN (base-line)

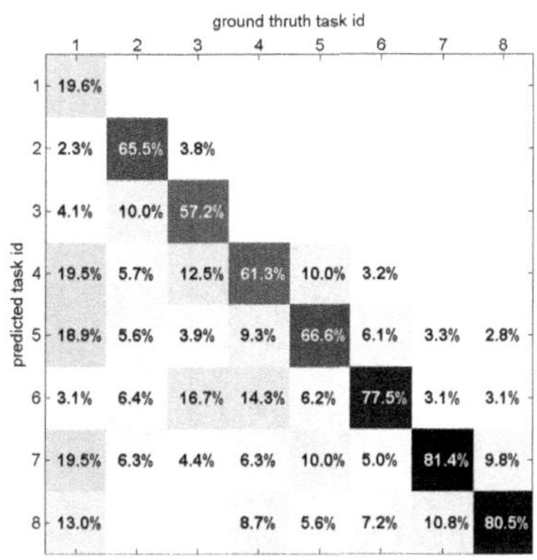


(b) Motion-grid + ESN

**Fig. 8.** Confusion matrices for the whole testing set (at maximum f-Measure)

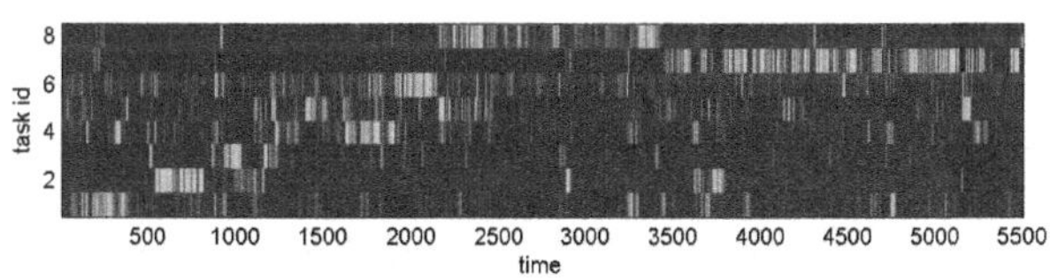


(a) raw output of the ESN

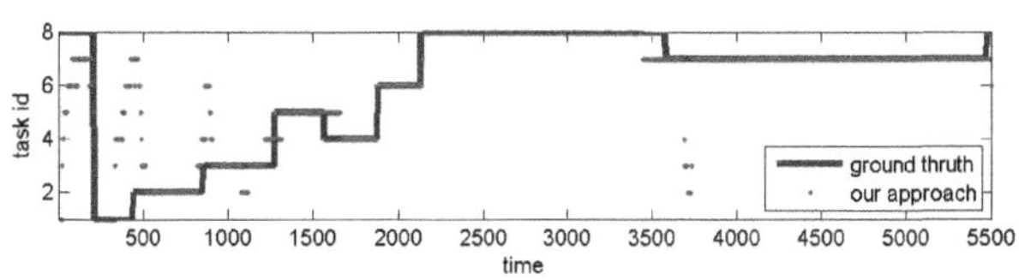


(b) final result

**Fig. 9.** Result of predicted task of our approach for one test workflow sequence

**Fig. 10.** Typical results of successful and failed task recognition over a single working cycle (figure is best viewed in color; full video avalibale on the authors' web-page)

## 5 Conclusion

In industrial environments automatic workflow monitoring faces several challenges: restrictions on where data is recorded and hostile conditions. Object detection/tracking in such settings is a nontrivial matter and not able to produce robust results. Hence, no robust features based on detection/tracking can be extracted. In this paper we proposed an approach based on local motion monitors as a scene descriptor and Echo State Networks as a time series predictor for workflow monitoring. Experimental results show that monitoring the smooth operation of workflow is achievable. With a simple scene detector we are able to robustly detect both human and machine motion relevant to pre-defined workflows. Using the ESN as a classifier of complex dynamics we achieve 78.6% recall rate and 77.4% precision. Further research will investigate how the proposed approach can cope with more complex scenarios when 2 or more tasks occur in parallel.

**Acknowledgement.** This work is supported by the EC Framework 7 Program (FP7/2007-2013) under grant agreement number 216465 (ICT project SCOVIS).

## References


1. Adam, A., Rivlin, E., Shimshoni, I., Reinitz, D.: Robust real-time unusual event detection using multiple fixed-location monitors. PAMI 30, 555–560 (2008)
2. Boiman, O., Irani, M.: Detecting irregularities in images and in video. In: Proc. ICCV (2005)
3. Breitenstein, M., Grabner, H., Gool, L.V.: Hunting nessie: Real time abnormality detection from webcams. In: Proc. ICCV Workshop on Visual Survaillance (2009)
4. Felzenszwalb, P., McAllester, D., Ramanan, D.: A discriminatively trained, multiscale, deformable part model. In: Proc. CVPR (2008)
5. Grabner, H., Bischof, H.: On-line boosting and vision. In: Proc. CVPR, vol. 1, pp. 260–267 (2006)
6. Jaeger, H.: The "echo state" approach to analysing and training recurrent neural networks. Technical Report GMD Report 148, German National Research Institute for Computer Science (2001)
7. Jaeger, H.: Adaptive nonlinear system identification with echo state networks. In: Proc. NIPS, vol. 15, pp. 593–600 (2003)
8. Johnson, N., Hogg, D.: Learning the distribution of object trajectories for event recognition. In: Proc. BMVC (1996)
9. Huang, C., Wu, D., Nevatia, R.: Robust Object Tracking by Hierarchical Association of Detection Responses. In: Forsyth, D., Torr, P., Zisserman, A. (eds.) ECCV 2008, Part II. LNCS, vol. 5303, pp. 788–801. Springer, Heidelberg (2008)
10. Li, J., Gong, S., Xiang, T.: Scene segmentation for behaviour correlation. In: Forsyth, D., Torr, P., Zisserman, A. (eds.) ECCV 2008, Part IV. LNCS, vol. 5305, pp. 383–395. Springer, Heidelberg (2008)
11. Makris, D., Ellis, T.: Learning semantic scene models from observing activity in visual surveillance. Trans. on Systems, Man, and Cybernetics 35, 397–408 (2005)
12. Lv, F., Nevatia, R.: Recognition and segmentation of 3-D human action using HMM and multi-class adaBoost. In: Leonardis, A., Bischof, H., Pinz, A. (eds.) ECCV 2006. LNCS, vol. 3954, pp. 359–372. Springer, Heidelberg (2006)
13. Ozturk, M.C., Xu, D., Principe, J.C.: Analysis and design of echo state networks. Neural Computation 19, 111–138 (2007)
14. Padoy, N., Mateus, D., Weinland, D., Berger, M., Navab, N.: Workflow monitoring based on 3d motion features. In: Proc. ICCV/WS on Video-oriented Object and Event Classification (2009)
15. Scherer, S., Oubbati, M., Schwenker, F., Palm, G.: Real-time emotion recognition from speech using echo state networks. Artificial Intelligence, 205–216 (2008)
16. Skowronski, M., Harris, J.: Automatic speech recognition using a predictive echo state network classifier. Neural Networks 20, 414–423 (2007)
17. Stauffer, C., Grimson, W.: Adaptive background mixture models for real-time tracking. In: Proc. CVPR, vol. 2, pp. 246–252 (1999)
18. Wang, X., Ma, K., Ng, G., Grimson, W.: Trajectory analysis and semantic region modeling using a nonparametric bayesian model. In: Proc. CVPR (2008)
19. Zhong, H., Shi, J., Visontai, M.: Detecting unusual activity in video. In: Proc. CVPR (2004)

# Context-Based Support Vector Machines for Interconnected Image Annotation

Hichem Sahbi[1] and Xi Li[1,2,3]

[1] CNRS Telecom ParisTech, Paris, France
[2] School of Computer Science, The University of Adelaide, Australia
[3] NLPR, CASIA, Beijing, China
hichem.sahbi@telecom-paristech.fr, lixichinanlpr@gmail.com

**Abstract.** We introduce in this paper a novel image annotation approach based on support vector machines (SVMs) and a new class of kernels referred to as context-dependent. The method goes beyond the naive use of the intrinsic low level features (such as color, texture, shape, etc.) and context-free kernels, in order to design a kernel function applicable to interconnected databases such as social networks. The main contribution of our method includes (i) a variational approach which helps designing this function using both intrinsic features and the underlying contextual information resulting from different links and (ii) the proof of convergence of our kernel to a positive definite fixed-point, usable for SVM training and other kernel methods. When plugged in SVMs, our context-dependent kernel consistently improves the performance of image annotation, compared to context-free kernels, on hundreds of thousands of Flickr images.

## 1 Introduction

Recent years have witnessed a rapid increase of image sharing spaces, such as Flickr, due to the spread of digital cameras and mobile devices. An urgent need is how to effectively search these huge amounts of data and how to exploit the structure of these sharing spaces. A possible solution is CBIR (Content-Based Image Retrieval); where images are represented using low-level visual features (color, texture, shape, etc.) and searched by analyzing and comparing those features. However, low-level visual features are usually unable to deliver satisfactory semantics, resulting in a gap between them and the high-level human interpretations. To address this problem, a variety of machine learning techniques were introduced in order to discover the intrinsic correspondence between visual features and semantics of images and allow to predict keywords for images.

### 1.1 Related Work

Conventionally, image annotation is converted into a classification problem. Existing state of the art methods (for instance [1,2]) treat each keyword or concept

R. Kimmel, R. Klette, and A. Sugimoto (Eds.): ACCV 2010, Part I, LNCS 6492, pp. 214–227, 2011.

as an independent class, and then train the corresponding concept-specific classifier to identify images belonging to that class, using a variety of machine learning techniques such as hidden Markov models [2], latent Dirichlet allocation [3], probabilistic latent semantic analysis [4], and support vector machines [5]. The aforementioned annotation methods may also be categorized into two branches; region-based requiring a preliminary step of image segmentation [2,12], and holistic [6,25] operating directly on the whole image space. In both cases, training is achieved in order to learn how to attach keywords with the corresponding visual features.

The above annotation methods heavily rely on their visual features for image annotation. Due to the semantic gap, they are unable to fully explore the semantic information inside images. Another class of annotation methods has emerged that takes advantage of extra information (tags, context, users' feedback, ontologies, etc.) in order to capture the correlations between images and concepts. A representative work is the cross-media relevance model (CMRM) [6,9], which learns joint statistics of visual and concepts and its variants [7,8]. The model uses the keywords shared by similar images to annotate new ones. In [22], the similarity measure between images integrates contextual information for concept propagation. Semi-supervised annotation techniques were also studied and usually rely on graph inference [10,11,12,13]. The original work, in [3,26], is inspired from machine translation and considers images and keywords as two different languages; in that case, image annotation is achieved by translating visual words into keywords.

Other existing annotation methods focus on how to define an effective distance measure for exploring the semantic relationships between concepts in large scale databases. In [19], the Normalized Google similarity Distance (NGD) is proposed by exploring the textual information available on the web. It is a measure of semantic correlations derived from counts returned by Google's search engine for a given set of keywords. Following the idea of [19], the Flickr distance [20] is proposed to precisely characterize the visual relationships between concepts. Each one is represented by a visual language model in order to capture its underlying visual characteristics. Then, a Flickr distance is defined, between two concepts, as the square root of Jensen-Shannon (JS) divergence between the corresponding visual language models. Other techniques consider extra knowledge derived from ontologies (such as the popular WordNet [14,15,16]) in order to enrich annotations [21]. The method in [14] introduces a visual vocabulary in order to improve translation model in the preprocessing stage of visual feature extraction. A directed acyclic graph is used to model the causal strength between concepts, and image annotation is performed by inference on this graph [15]. In [17,18], the semantic ontology information is integrated in the post processing stage in order to further refine initial annotations.

### 1.2 Motivation and Contribution

Among the most successful annotation methods, those based on machine learning and mainly support vector machines; show a particular interest as they are performant and theoretically well grounded [24]. Support vector machines [23], basically require the design of similarity measures, also referred to as *kernels*, which should provide high values when two images share similar structures/appearances and should be invariant, as much as possible, to the linear and non-linear transformations. They also satisfy positive definiteness which ensures, according to Vapnik's SVM theory [24], optimal generalization performance and also the uniqueness of the SVM solution. In practice, kernels should not depend only on intrinsic aspects of images (as images with the same semantic may have different visual and textual features), but also on different sources of knowledge including context.

In this paper, we introduce an image annotation approach based on a new family of kernels which take high values not only when images share the same visual content but also the same context. The context of an image is defined as the set of images, with the same tags, and exhibiting better semantic descriptions, compared to both pure visual and tag based descriptions. The issue of combining context and visual content for image retrieval is not new (see for instance [28,29,32]) but the novel part of this work aims to (i) integrate context, in kernel design useful for classification and annotation, and (ii) plug these kernels in support vector machines in order to take benefit from their well established generalization power [24]. This type of kernels will be referred to as context-dependent (CDK) while those relying only on the intrinsic visual or textual content will be referred to as context-free. Again, our proposed method goes beyond the naive use of low level features and context-free kernels (established as the standard baseline in image retrieval) in order to design a kernel applicable to annotation and suitable to integrate the "contextual" information taken from tagged links in interconnected datasets. In the proposed CDK, two images (even with different visual content and even sharing different tags) will be declared as similar if they share the same visual context[1]. This is usually useful as tags in interconnected data (such social networks) may be noisy and misspelled. Furthermore, the intrinsic visual content of images might not always be relevant especially for categories exhibiting large variation of the underlying visual aspects.

Through this work, an image database is modeled as a graph where nodes are pictures and edges correspond to the shared tagged links. We design our CDK as the fixed point of a constrained energy function mixing (i) a fidelity term which measures visual similarity between images, (ii) a context criterion that captures the similarity between the underlying links and (iii) a regularization term which helps defining a direct analytic solution. In the remainder of this paper we consider $X$ as a random variable standing for all the possible images of the world, here $X$ is drawn from an existing but unknown probability distribution $P$. Terminology and notation will be introduced as we go through different sections of this paper which is organized as follows: Section 2 tackles the issue

[1] For instance, two images I1, I2 (e.g., red Ferrari and black limousine) may be connected to two groups of "visually similar" images (two groups of cars are their contexts). If the two groups of images (groups of cars) are similar then one may conclude that the two images I1, I2 are also similar (see also Fig. 1 in [30]).

of kernel design, followed by some results about the positive definiteness and the convergence of the kernel to a fixed-point. Section 3 shows experimental results and the applicability of CDK in order to handle interconnected datasets including Flickr and NUSWIDE. We will conclude the method in Section 4 while providing promising extensions for a future work.

## 2 Kernel Design

Let us consider $\mathcal{X} = \{x_1, \dots, x_n\}$ as a finite set of images drawn from the same distribution as $X$. Considering $k : \mathcal{X} \times \mathcal{X} \to \mathbb{R}^+$ as a continuous symmetric function which, given two images $(x_i, x_j)$, provides us with a similarity measure; this function will be referred to as kernel. Our goal is to design $k(x_i, x_j)$ by taking into account the properties of $x_i$, $x_j$ and also their links, i.e., the set of images which are connected to $x_i$, $x_j$.

### 2.1 Context and Graph-Links

We model an image database $\mathcal{X}$ using a graph $\mathcal{G} = (\mathcal{V}, \mathcal{E})$ where nodes $\mathcal{V} = \{v_1, \dots, v_n\}$ correspond to pairs $\{(x_i, \psi_f(x_i))\}_i$ and edges $\mathcal{E} = \{e_{i,j,\omega}\}$ are the set of tagged links of $\mathcal{G}$. In the above definition, $\psi_f(x_i)$ corresponds to the features of $x_i$ (color, texture, shape, etc.) while $e_{i,j,\omega} = (v_i, v_j, \omega)$ defines a connection between $v_i$, $v_j$ of type $\omega$. The latter might be any particular tag for instance two images are linked when they share the same semantics, owners, GPS locations, etc. Through this work, two images are connected if they share the same Flickr-tag. *It is worth emphasizing that these tags are different from the concepts (classes) used for training and annotation. Indeed, each image belongs to one or multiple concepts which are different from the tags provided by users (see Section 3).*

Now, introduce the context (or neighborhood) $\mathcal{N}^{\omega}(x_i) = \big\{x_j : (x_i, x_j, \omega) \in \mathcal{E}\big\}$. This definition of context $\{\mathcal{N}^{w}(x)\}_w$ reflects the co-occurrence of different images with particular connection types (again defined using tags).

### 2.2 Context-Dependent Kernel Design

For a finite collection of images, we put some (arbitrary) order on $\mathcal{X}$, we can view a kernel $k$ on $\mathcal{X}$ as a matrix $\mathbf{K}$ in which the "$(x, x')$−element" is the similarity between $x$ and $x'$: $\mathbf{K}_{x,x'} = k(x, x')$. Let $\mathbf{P}_{\omega}$ be the intrinsic adjacency matrices respectively defined as $\mathbf{P}_{\omega,x,x'} = g_{\omega}(x, x')$, where $g$ is a nonnegative decreasing function of any (pseudo) distance involving $(x, x')$, *not necessarily symmetric.* In practice, we consider $g_{\omega}(x, x') = \mathbb{1}_{\{x' \in \mathcal{N}^{\omega}(x)\}}$. Let $\mathbf{D}_{x,x'} = d(x, x')$, ($d(x, x')$ is a dissimilarity metric between $x$ and $x'$). We propose to use the kernel on $\mathcal{X}$ defined by solving

$$\min_{\substack{\mathbf{K} \geq 0 \\ \|\mathbf{K}\|_1 = 1}} \quad \mathrm{Tr}\big(\mathbf{K}\,\mathbf{D}'\big) + \beta\,\mathrm{Tr}\big(\mathbf{K}\,\log\,\mathbf{K}'\big) \; - \; \alpha \sum_{\omega} \mathrm{Tr}\big(\mathbf{K}\,\mathbf{P}_{\omega}\,\mathbf{K}'\,\mathbf{P}'_{\omega}\big)$$

Here the operations log (natural) and $\geq$ are applied individually to every entry of the matrix (for instance, $\log \mathbf{K}$ is the matrix with $(\log \mathbf{K})_{x,x'} = \log k(x,x')$), $\|\cdot\|_1$ is the "entrywise" $L_1$-norm (i.e., the sum of the absolute values of the matrix coefficients) and $\text{Tr}(\cdot)$ denotes matrix trace. The non matrix form of the above objective function may be written $\sum_{x,x'} k(x,x')d(x,x') + \beta \sum_{x,x'} k(x,x') \log(k(x,x')) - \alpha \sum_{\omega,x,x',y,y'} k(x,x')k(y,y')$ (with $y \in \mathcal{N}^{\omega}(x), y' \in \mathcal{N}^{\omega}(x')$). The first term, in the above constrained minimization problem, measures the quality of matching two feature vectors $\psi_f(x)$, $\psi_f(x')$. In the case of visual features, this is considered as the distance, $d(x,x')$, between the visual descriptors (color, texture, shape, etc.) of $x$ and $x'$. A high value of $d(x,x')$ should result into a small value of $k(x,x')$ and vice-versa.

The second term is a regularization criterion which considers that without any a priori knowledge about the visual features, the probability distribution $\{k(x,x')\}$ should be flat so the negative of the entropy is minimized. This term also helps to define a simple solution and solve the constrained minimization problem easily. The third term is a context criterion which considers that a high value of $k(x,x')$ should imply high kernel values in the respective neighborhoods $\mathcal{N}^{\omega}(x)$ and $\mathcal{N}^{\omega}(x')$ of $x$ and $x'$. We formulate the minimization problem by adding an equality constraint and bounds which ensure a normalization of the kernel values and allow to see $\mathbf{K}$ as a joint probability distribution (or P-Kernel [33]).

### 2.3 Solution

The above optimization problem admits a solution $\tilde{\mathbf{K}}$, which is the limit of the context-dependent kernels $\mathbf{K}^{(t)} = G(\mathbf{K}^{(t-1)})/\|G(\mathbf{K}^{(t-1)})\|_1$, with $G(\mathbf{K}) = \exp\{-\frac{\mathbf{D}}{\beta} + \frac{\alpha}{\beta}\sum_{\omega}(\mathbf{P}_{\omega}\mathbf{K}\mathbf{P}'_{\omega} + \mathbf{P}'_{\omega}\mathbf{K}\mathbf{P}_{\omega})\}$, and $\mathbf{K}^{(0)} = \exp(-\mathbf{D}/\beta)/\|\exp(-\mathbf{D}/\beta)\|_1$. By taking small enough $\alpha$, convergence of this kernel to a fixed point is satisfied (see [30]). Note that $\alpha = 0$ corresponds to a kernel which is not context-dependent: the similarities between neighbors are not taken into account to assess the similarity between two images. Besides our choice of $\mathbf{K}^{(0)}$ is exactly the optimum (and fixed point) for $\alpha = 0$. Detailed proof of this solution and its convergence to a fixed point may be found in [30].

### 2.4 Positive Definiteness

A kernel $k : \mathcal{X} \times \mathcal{X} \to \mathbb{R}$ is positive (semi-)definite on $\mathcal{X}$, if and only if the underlying Gram matrix $\mathbf{K}$ is positive (semi-)definite. In other words, it is positive definite if and only if we have $V'\mathbf{K}V > 0$ for any vector $V \in \mathbb{R}^{\mathcal{X}} - \{0\}$. When we just have $V'\mathbf{K}V \geq 0$ for any vector $V \in \mathbb{R}^{\mathcal{X}} - \{0\}$, we just say that it is positive semi-definite. A positive definite kernel guarantees the existence of a Reproducing Kernel Hilbert Space (RKHS) such that $k(x,x') = \langle \phi(x), \phi(x') \rangle$, where $\phi$ is an explicit or implicit mapping function from $\mathcal{X}$ to the RKHS, and $\langle \cdot, \cdot \rangle$ is the dot kernel in the RKHS.

**Proposition 1.** *The context-dependent kernels on $\mathcal{X}$ defined in (2.3) by the matrices $\tilde{\mathbf{K}}$ and $\mathbf{K}^{(t)}$, $t \geq 0$, are positive definite.*

*Proof.* See [30]. □

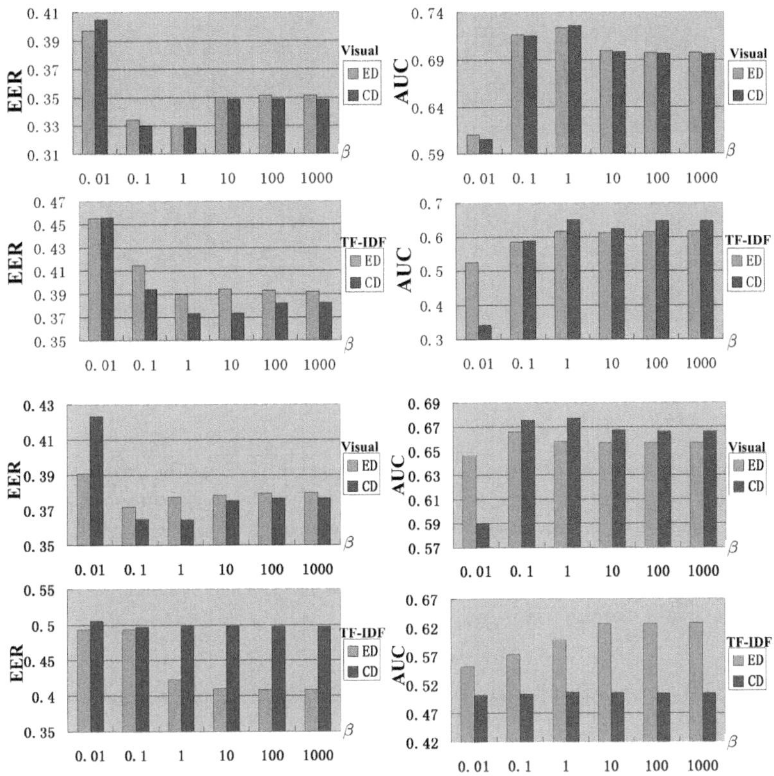


**Fig. 1.** This figure shows the annotation performances of different $\beta$s under different metrics on the MIRFLICKR-25000 (two first rows) and NUSWIDE datasets (other rows)

# 3 Benchmarking

This section evaluates the performance of image annotation tasks and shows the extra advantage of our context-dependent kernel (CDK) with respect to the use of many existing context-free ones such as the gaussian, the polynomial, the chi-square, etc. The point here is also to show the importance of the context in kernel design through different databases and settings.

## 3.1 Databases and Settings

We evaluated CDK on the MIRFLICKR-25000[2] as well as the NUSWIDE[3] datasets. Both sets are challenging; the first one, MIRFLICKR-25000 contains 25,000 images belonging to 24 concepts (for instance "sky, clouds, water, sea,

[2] http://press.liacs.nl/mirflickr/

[3] http://lms.comp.nus.edu.sg/research/NUSWIDE.htm

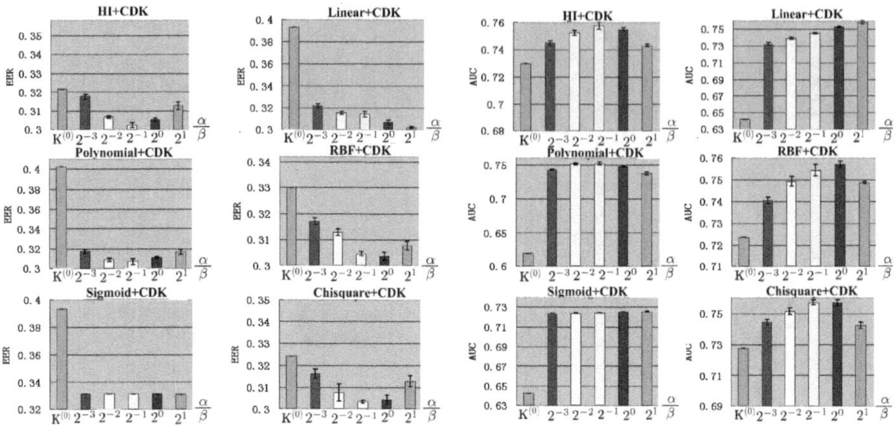


**Fig. 2.** This figure shows the performances of annotation on the MIRFLICKR-25000 dataset (with $\beta = 1$). It includes EER and AUC annotation performances of six context-dependent kernels based on visual features. Compared with the underlying baseline kernels, the six (best) context-dependent kernels achieve a relative gain of respectively 5.91%, 23.09%, 23.69%, 8.03%, 17.34%, and 6.47% for EER. Means and standard deviations are taken over 20 trials.

river,...") while the second one, NUSWIDE database, is larger and contains more than a quarter of a million of images (exactly 269,648) belonging to 81 concepts. Note that both sets were downloaded from Flickr through its public API.

Each image in MIRFLICKR-25000 is processed in order to extract the bag-of-word SIFT representation [27]. Precisely, SIFT features are extracted at three different spatial pyramid levels, and quantized into 200 codewords. Consequently, the visual feature for each image is a 4200-dimensional concatenated histogram of three spatial pyramid levels. Moreover, images in MIRFLICKR-25000 are supplied with tags (which again are different from the concepts used for learning and annotation, see Section 2.1). In total, 1,386 tags are used, each one annotates at least 20 images. As a matter of comparison, textual features are also used. Indeed, each image characterized by its tags, is mapped using the TF-IDF (term frequency-inverse document frequency) resulting into a feature vector of 1,386 dimensions. Images in the NUSWIDE set are also indexed with the bag-of-word SIFT features of 500 dimensions and they are also supplied with 1,000 tags used in order to extract the TF-IDF features.

Let $\Omega$ denote the union of tags over all the images of a given set (either MIRFLICKR-25000 or NUSWIDE). Again, we define the underlying graph $\mathcal{G} = (\mathcal{V}, \mathcal{E})$, here nodes $\mathcal{V}$ are defined similarly as in Section 2.1, whereas edges are defined as $\mathcal{E} = \{e_{i,j,\omega} : \omega \in \Omega, \#\omega \in [T_D, T_U]\}$. Here $\#\omega$ denotes the number of images tagged by $\omega$ and $T_D$, $T_U$ are two fixed thresholds; their setting determines the complexity and topology of the graph and may also affect performance as shown later in this section.

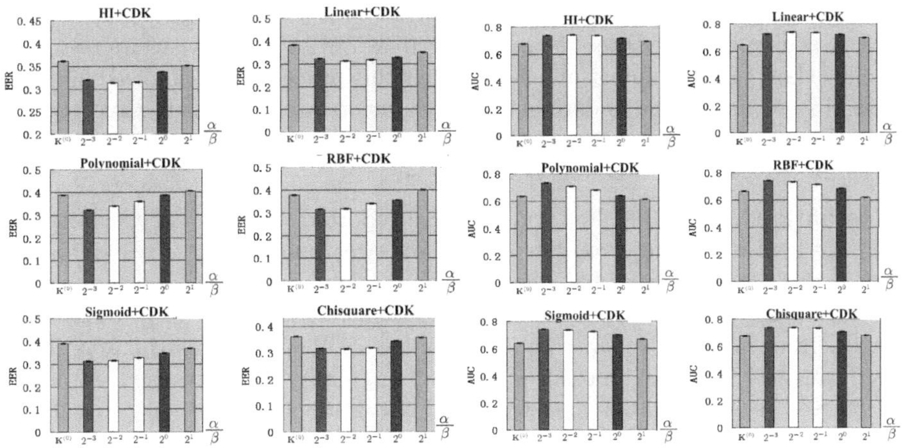


**Fig. 3.** This figure shows the performances of annotation on the NUSWIDE dataset. It includes EER and AUC annotation performances of six context-dependent kernels based on visual features (with $\beta = 1$). Compared with the underlying baseline kernels, the six (best) context-dependent kernels achieve a relative gain of respectively 12.98%, 18.28%, 12.76%, 16.18%, 19.23%, and 12.94% for EER. Means and standard deviations are taken over 20 trials.

### 3.2 Hold-Out Generalization and Comparison

We evaluate $\mathbf{K}^{(t)},\ t \in \mathbb{N}^+$ using six power assist settings

(i) Linear $\mathbf{K}^{(0)}_{x,x'} = \langle \psi_f(x), \psi_f(x') \rangle$,

(ii) Polynomial $\mathbf{K}^{(0)}_{x,x'} = (\langle \psi_f(x), \psi_f(x') \rangle + 1)^2$,

(iii) RBF $\mathbf{K}^{(0)}_{x,x'} = \exp(-\|\psi_f(x) - \psi_f(x')\|_2/\beta)$,

(iv) Histogram intersection $\mathbf{K}^{(0)}_{x,x'} = \sum_i \min(\psi_f(x)_i, \psi_f(x')_i)$,

(v) Chisquare $\mathbf{K}^{(0)}_{x,x'} = 1 - \frac{1}{2} \sum_i \frac{(\psi_f(x)_i - \psi_f(x')_i)^2}{(\psi_f(x)_i + \psi_f(x')_i)}$,

(vi) Sigmoid $\mathbf{K}^{(0)}_{x,x'} = \tanh(\ \langle \psi_f(x), \psi_f(x') \rangle)$.

These context-free kernels are plugged into the underlying CDK kernels $\mathbf{K}^{(t)},\ t \in \mathbb{N}^+$, and the resulting CDKs will be referred to as "Linear+CDK", "Poly+CDK", "RBF+CDK", "HI+CDK", "Chisquare+CDK", and "Sigmoid+CDK" respectively. Our goal is to show the improvement brought when using $\mathbf{K}^{(t)},\ t \in \mathbb{N}^+$, so we tested it against the standard context-free kernels (i.e., $\mathbf{K}^{(t)},\ t = 0$). For this purpose, we trained "one-versus-all" SVM classifiers[4] for each concept in

[4] http://www.csie.ntu.edu.tw/~cjlin/libsvm

the MIRFLICKR-25000 and the NUSWIDE datasets. For each concept, training is achieved using three-random folds ($\sim 75\%$) of the data while testing is achieved on the remaining-fold. Notice that this process is randomized 20 times and the outputs of the underlying SVM classifiers are taken as the average values through these 20 random samplings; this makes classification results less sensitive to sampling and unbalanced classes.

**Evaluation measures.** Performances are reported, on different test sets, using the hold-out equal error rate (EER) and area under ROC (receiver operating characteristics) curve (AUC). The EER is the balanced generalization error which equally weights the positive and the negative errors. It can be easily computed from the ROC curve. The smaller the EER, the better the annotation performance. The AUC measures the ranking quality of a classifier, and it can be viewed as an estimation of the probability that the classifier ranks a randomly selected positive sample higher than a randomly selected negative sample. The larger the AUC, the better the annotation performance. These two measures are evaluated using the standard script provided by the ImageClef evaluation campaigns.

**Context-free kernel setting.** We aim to explore the optimal parameter settings for $\beta$ under an appropriate dissimilarity metric $d(x, x')$. Thus, two popular metrics are used for performance evaluations: (i) Euclidean distance (ED); and (ii) Chisquare distance (CD).

$$\begin{aligned}
&\text{(i)} \quad \text{ED} \quad d(x,x') = \|\psi_f(x) - \psi_f(x')\|_2, \\
&\text{(ii)} \quad \text{CD} \quad d(x,x') = \frac{1}{2}\sum_i \frac{(\psi_f(x)_i - \psi_f(x')_i)^2}{(\psi_f(x)_i + \psi_f(x')_i)}.
\end{aligned}$$

Based on the above two metrics, we tuned the scale parameter $\beta$ to achieve the best EER and AUC annotation performances. Fig. 1 shows the EER and AUC annotation results of different $\beta$s under different metrics using two features on the MIRFLICKR-25000 and the NUSWIDE datasets. Clearly, we found that the best performances are achieved on the MIRFLICKR-25000 dataset when $\beta = 1$ using the CD metric for both visual and TF-IDF features. As for NUSWIDE, it performs best when $\beta = 1$ (resp. $\beta = 1000$) using the CD (resp. ED) metric for visual (resp. TF-IDF) features.

**Influence of the context.** All the reported results show that the influence of the right-hand side of $\mathbf{K}^{(t)}$, $\alpha \neq 0$ increases as $\alpha$ increases (see Fig. 2), nevertheless and as shown in [30], the convergence of $\mathbf{K}^{(t)}$ to a fixed point is guaranteed only if $\alpha$ is bounded. When convergence is not guaranteed, CDK may suffer numerical instabilities resulting into degeneracy of the performance. Therefore it is obvious that $\alpha$ should be set to the highest possible value which also satisfies an upper bound criterion (see [30]). In these diagrams, the weight $\alpha$ is taken from five different values using a logarithmic scale $2^{-3}\beta$, $2^{-2}\beta$, $2^{-1}\beta$, $2^{0}\beta$, and $2^{1}\beta$.

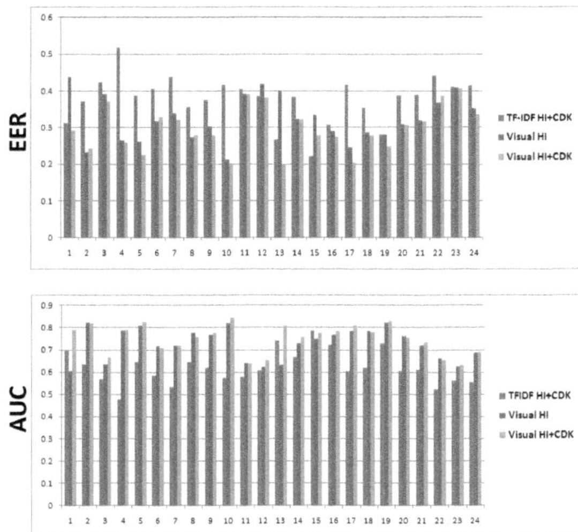


**Fig. 4.** This figure shows the performances of annotation, concept-by-concept, on the MIRFLICKR-25000 set. It includes EER (top) and AUC (bottom) annotation performances of the (best) baseline context-free kernel (**HI**), and the underlying context-dependent kernel (**HI+CDK**). Comparison is also shown with respect to the (best) context-dependent kernel based on TF-IDF (**TF-IDF HI+CDK**). Compared with **TF-IDF HI+CDK** (resp. visual **HI**), the average relative gain of visual **HI+CDK** is 19.83% (resp. 7.20%) for EER and 22.09% (resp. 3.64%) for AUC. The x-axis values correspond to the concept indices (in the same order as the one given in the original database).

**Comparison.** Fig. 2 shows the annotation performance (EER and AUC) of the context-dependent kernel using the six power assist settings defined earlier. Specifically, global annotation performances using bag-of-word visual feature are plotted in Fig. 2 for MIRFLICKR-25000 and Fig. 3 for NUSWIDE, where the x-axis of each sub-figure corresponds to different settings of $\alpha$ and the y-axis shows the underlying error rates ($\mathbf{K}^{(0)}$ corresponds to the baseline context-free kernels). Correspondingly, concept-by-concept annotation performances using bag-of-word visual feature and TF-IDF textual feature are plotted in Fig. 4 for MIRFLICKR-25000 and Fig. 5 for NUSWIDE. According to Fig. 2, performances of context-dependent kernels are mostly better than those of context-free ones, with just two iterations (i.e., $t \geq 2$). From Figs. 4-5, it is seen that the average concept-by-concept EER (resp. AUC) performances for the visual feature are mostly lower (resp. higher) than those for the TF-IDF textual feature. Standard deviations are also shown with respect to different threshold intervals $[T_D, T_U]$ in Fig. 2. This clearly corroborates the statement that improvement is not only due to the nature of the features but also to the integration of the context in kernel design.

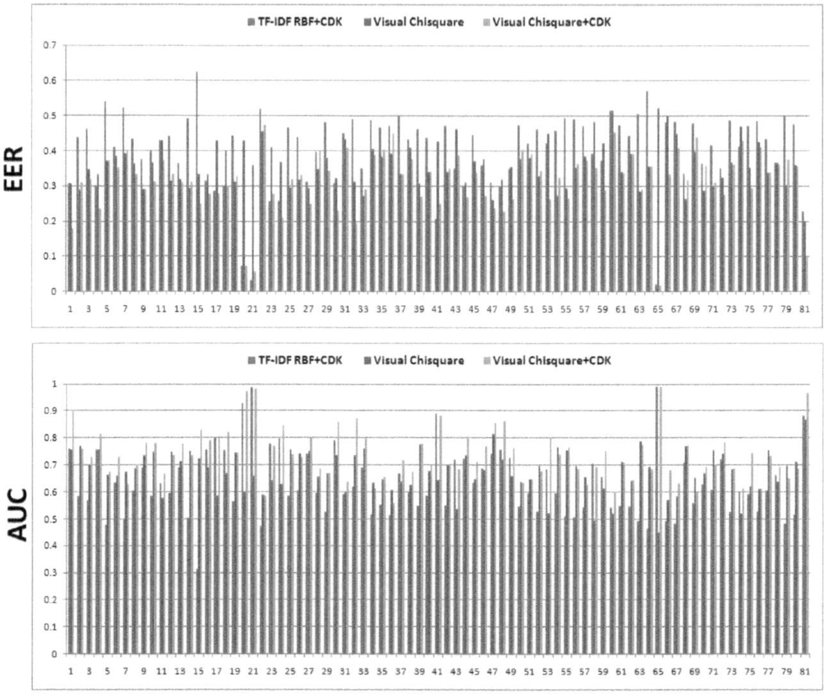


**Fig. 5.** This figure shows the performances of annotation, concept-by-concept, on the NUSWIDE dataset. It includes EER (top) and AUC (bottom) annotation performances of the (best) baseline context-free kernel (**Chisquare**), and the underlying context-dependent kernel (**Chisquare+CDK**). Comparison is also shown with respect to the (best) context-dependent kernel based on TF-IDF (**TF-IDF RBF+CDK**). Compared with **TF-IDF RBF+CDK** (resp. **Chisquare**), the average relative gain of **Chisquare+CDK** is 19.43% (resp. 11.34%) for EER and 19.94% (resp. 10.32%) for AUC. The x-axis values correspond to the concept indices (in the same order as the one given in the original database).

Further illustrations, taken from MIRFLICKR-25000 database, show annotation results of the best context-dependent kernel **HI+CDK** and comparison with respect to the underlying ground truth annotation. The final annotation results of these twelve images are shown in Fig. 6. It is clear that our proposed method achieves reasonable annotation results.

**Runtime.** Note that training time depends on concepts and training set cardinalities. For instance the computation of a $8000^2$ sub-block of the gram matrix $\mathbf{K}^{(t)}$ requires about 7 minutes on a standard 2.6GHZ PC with 2G memory. This time scales linearly w.r.t the number of sub-blocks. Assuming $\mathbf{K}^{(t-1)}$ known for a given pair $x$, $x'$, the worst complexity of evaluating $\mathbf{K}^{(t)}$ is $O(\max(N^2, s))$, where $s$ is the dimension of $\psi_f(x)$ and $N = \max_{x,\omega} \#\{\mathcal{N}^{\omega}(x)\}$. When $N < \sqrt{s}$,

| | | | |
|---|---|---|---|
| Images |  |  |  |
| Our annotation (runtime) | sky sunset tree (~0.181s) | sky structures (~0.191s) | tree sky (~0.157s) |
| Ground truth annotation | clouds sky sunset tree | sky structures | tree sky structures |
| Images |  |  |  |
| Our annotation (runtime) | sky structures (~0.189s) | people (~0.202s) | flower (~0.173s) |
| Ground truth annotation | sky structures | people male | flower |
| Images |  | 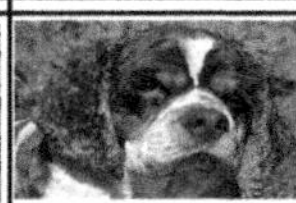 |  |
| Our annotation (runtime) | sky clouds people structures (~0.163s) | dog (~0.151s) | sky clouds people sea river water (~0.156s) |
| Ground truth annotation | sky clouds people structures | dog | sky clouds people sea water |
| Images |  |  |  |
| Our annotation (runtime) | sky car tree people structures (~0.163s) | people structures (~0.133s) | indoor people (~0.15s) |
| Ground truth annotation | sky car tree people male female structures | people male structures | indoor people male female |

**Fig. 6.** This table shows comparisons of ground truth and the best context-dependent kernel (**HI+CDK**) annotations on twelve images from MIRFLICKR-25000

the complexity of evaluating the proposed kernel is strictly equivalent to that of usual kernels such as the linear ($N$ may be forced to a small value by sampling context).

## 4 Conclusion

We introduced in this work a novel approach for kernel design dedicated to interconnected datasets including social networks. The strength of this method resides in the inclusion of context links in kernel design thereby improving annotation performances consistently.

The "Take Home Message" is to show that the information present into a picture can be described *not only* by its intrinsic visual features (suffering the semantic gap) but also by the set of images in its "context". The proposed kernel gathers many fundamental properties (i) second order context criterion which captures links between images (ii) well motivated definition of kernels via an energy function ending with a probabilistic interpretation.

Extensions of this work include the use of ontologies in order to enrich social link types. Other future work will exploit the positive definiteness of CDK in order to use lossless acceleration techniques suitable for even larger scale networks.

**Acknowledgement.** This work is supported by the French National Research Agency (ANR) under the AVEIR project.

## References

1. Carneiro, G., Vasconcelos, N.: Formulating semantic image annotation as a supervised learning problem. In: Proc. of CVPR (2005)
2. Li, J., Wang, J.Z.: Automatic linguistic indexing of pictures by a statistical modeling approach. IEEE Trans. on PAMI 25(9), 1075–1088 (2003)
3. Barnard, K., Duygululu, P., Forsyth, D., Blei, D., Jordan, M.: Matching words and pictures. The Journal of Machine Learning Research (2003)
4. Monay, F., GaticaPerez, D.: PLSA-based Image AutoAnnotation: Constraining the Latent Space. In: Proc. of ACM International Conference on Multimedia (2004)
5. Gao, Y., Fan, J., Xue, X., Jain, R.: Automatic Image Annotation by Incorporating Feature Hierarchy and Boosting to Scale up SVM Classifiers. In: Proc. of ACM MULTIMEDIA (2006)
6. Jeon, J., Lavrenko, V., Manmatha, R.: Automatic image annotation and retrieval using cross-media relevance models. In: Proc. of ACM SIGIR, pp. 119–126 (2003)
7. Lavrenko, V., Manmatha, R., Jeon, J.: A model for learning the semantics of pictures. In: Proc. of NIPS (2004)
8. Feng, S., Manmatha, R., Lavrenko, V.: Multiple Bernoulli relevance models for image and video annotation. In: Proc. of ICCV, pp. 1002–1009 (2004)
9. Liu, J., Wang, B., Li, M., Li, Z., Ma, W., Lu, H., Ma, S.: Dual cross-media relevance model for image annotation. In: Proc. of ACM MULTIMEDIA, pp. 605–614 (2007)
10. Wan, X., Yang, J., Xiao, J.: Manifold-ranking based topic-focused multi-document summarization. In: Proc. of IJCAI, pp. 2903–2908 (2007)
11. Zhou, D., Weston, J., Gretton, A., Bousquet, O., Schölkopf, B.: Ranking on data manifolds. In: Proc. of NIPS (2004)
12. Liu, J., Li, M., Liu, Q., Lu, H., Ma, S.: Image annotation via graph learning. Pattern Recognition 42(2), 218–228 (2009)
13. Liu, J., Li, M., Ma, W., Liu, Q., Lu, H.: An adaptive graph model for automatic image annotation. In: Proc. of ACM International Workshop on Multimedia Information Retrieval, pp. 61–70 (2006)
14. Srikanth, M., Varner, J., Bowden, M., Moldovan, D.: Exploiting ontologies for automatic image annotation. In: Proc. of SIGIR, pp. 552–558 (2005)
15. Wu, Y., Chang, E.Y., Tseng, B.L.: Multimodal metadata fusion using causal strength. In: Proc. of ACM MULTIMEDIA, pp. 872–881 (2005)
16. Miller, G.A.: Wordnet: a lexical database for English. ACM Commun. 38(11), 39–41 (1995)
17. Wang, C., Jing, F., Zhang, L., Zhang, H.J.: Image annotation refinement using random walk with restarts. In: Proc. of ACM MULTIMEDIA, pp. 647–650 (2006)

18. Jin, Y., Khan, L., Wang, L., Awad, M.: Image annotations by combining multiple evidence & wordNet. In: Proc. of ACM MULTIMEDIA, pp. 706–715 (2005)
19. Cilibrasi, R., Vitanyi, P.M.B.: The google similarity distance. IEEE Transactions on Knowledge and Data Engineering (2007)
20. Wu, L., Hua, X., Yu, N., Ma, W., Li, S.: Flickr distance. In: Proc. of ACM MULTIMEDIA (2008)
21. Wang, Y., Gong, S.: Translating Topics to Words for Image Annotation. In: Proc. of ACM CIKM (2007)
22. Lu, Z., Ip, H.H.S., He, Q.: Context-Based Multi-Label Image Annotation. In: Proc. of ACM CIVR (2009)
23. Boser, B., Guyon, I., Vapnik, V.: An training algorithm for optimal margin classifiers. In: Fifth Annual ACM Workshop on Computational Learning Theory, Pittsburgh (1992)
24. Vapnik, V.: Statistical Learning Theory. A Wiley-Interscience Publication, Hoboken (1998)
25. Wang, C., Yan, S., Zhang, L., Zhang, H.: Multi-Label Sparse Coding for Automatic Image Annotation. In: Proc. of CVPR (2009)
26. Duygulu, P., Barnard, K., de Freitas, J.F.G., Forsyth, D.: Object recognition as machine translation: Learning a lexicon for a fixed image vocabulary. In: Heyden, A., Sparr, G., Nielsen, M., Johansen, P. (eds.) ECCV 2002. LNCS, vol. 2353, pp. 97–112. Springer, Heidelberg (2002)
27. Lazebnik, S., Schmid, C., Ponce, J.: Beyond Bags of Features: Spatial Pyramid Matching for Recognizing Natural Scene Categories. In: Proc. of CVPR (2006)
28. Gallagher, A.C., Neustaedter, C.G., Cao, L., Luo, J., Chen, T.: Image Annotation Using Personal Calendars as Context. In: Proc. of ACM Multimedia (2008)
29. Cao, L., Luo, J., Huang, T.S.: Annotating Photo Collection by Label Propagation According to Multiple Similarity Cues. In: Proc. of ACM Multimedia (2008)
30. Sahbi, H., Audibert, J.-Y.: Social network kernels for image ranking and retrieval. In Technical Report, N 2009D009, TELECOM ParisTech (March 2009)
31. Shawe-Taylor, J., Cristianini, N.: Support Vector Machines and other kernel-based learning methods. Cambridge University Press, Cambridge (2000)
32. Yang, Y.H., Wu, P.T., Lee, C.W., Lin, K.H., Hsu, W.H., Chen, H.: ContextSeer: Context Search and Recommendation at Query Time for Shared Consumer Photos. In: Proc. of ACM Multimedia (2008)
33. Haussler, D.: Convolution Kernels on Discrete Structures. In Technical Report UCSC-CRL-99-10, University of California in Santa Cruz, Computer Science Department (July 1999)

# Finding Human Poses in Videos Using Concurrent Matching and Segmentation

Hao Jiang

Boston College, Chestnut Hill, MA 02467, USA

**Abstract.** We propose a novel method to detect human poses in videos by concurrently optimizing body part matching and object segmentation. With a single exemplar image, the proposed method detects the poses of a specific human subject in long video sequences. Matching and segmentation support each other and therefore the simultaneous optimization enables more reliable results. However, efficient concurrent optimization is a great challenge due to its huge search space. We propose an efficient linear method that solves the problem. In this method, the optimal body part matching conforms to local appearances and a human body plan, and the body part configuration is consistent with the object foreground estimated by simultaneous superpixel labeling. Our experiments on a variety of videos show that the proposed method is efficient and more reliable than previous locally constrained approaches.

## 1 Introduction

Detecting human poses in videos has important applications in video editing, movement analysis, action recognition, and human computer interaction, with challenges due to body part articulation, self-occlusion, or background clutter.

In this paper, we detect 2D poses of a specific human subject in monocular videos using a cardboard model built from a single exemplar image. Fig.1 illustrates the problem we tackle: we *concurrently optimize body part matching and*

(a) (b) (c) (d) (e)

**Fig. 1.** To detect the human pose in (a), we first detect body part candidates (small set of samples are illustrated in (b)), partition the image into superpixels in (c), and then we concurrently optimize body part matching in (d) and foreground segmentation by superpixel labeling in (e). The cardboard model in (d) is extracted from a single exemplar image; it contains only the information about object foreground colors and body part rectangle shapes.

R. Kimmel, R. Klette, and A. Sugimoto (Eds.): ACCV 2010, Part I, LNCS 6492, pp. 228–243, 2011.

*object segmentation* for robust pose detection. Concurrently optimizing object matching and segmentation enables more robust results since the two closely related tasks support each other. However, the concurrent optimization is a great challenge due to huge number of feasible configurations. To make things worse, for our application it is difficult to obtain good initializations for both tasks. We therefore have to solve a hard combinatorial problem. In this paper, we propose a highly efficient linear relaxation method to optimize matching and segmentation concurrently for robust human pose detection in videos.

Previous research on simultaneous matching and segmentation [25] focuses on active contours and partial differential equation approaches. These methods require initial contours to be close to real targets; deformable models are used and it is hard to extend them to articulated object pose detection, which our method tackles. PoseCut [18] is the first method for simultaneous human pose estimation and figure-ground separation. It has been successfully applied to 3D human pose tracking in a multiple-camera setting. PoseCut uses a parametric 3D human body model. It explicitly enumerates a small set of poses around an initial guess and uses an efficient dynamic graph-cuts method to compute the optimal foreground estimation for each hypothesis. A gradient descent method is further used to find the optimal pose. PoseCut requires good pose initialization in each video frame. Our proposed method removes this constraint.

Shape prior scheme [19,26] is a popular way to combine pose estimation and segmentation in restricted pose domains such as walking and running. With the object shape prior, pose estimation is able to achieve reliable results; the estimated poses can be further combined with a segmentation algorithm to obtain accurate object foreground estimation. Unfortunately, for unconstrained human poses, the shape prior is weak. The widely used iterative approach [19,26] that alternates between shape matching and segmentation does not work for our problem, since neither matching nor segmentation provides a good initialization for the other. We need methods that optimize object matching and segmentation concurrently instead of iteratively.

Due to the overwhelming computational complexity, concurrent optimization of matching and segmentation for unconstrained human pose detection has not yet been achieved. The contribution of this paper is that we propose a novel linear method that efficiently solves the problem. In this method, body part matching finds the optimal pose that follows local appearances, resembles a human body plan, and the covering region is consistent with the object foreground estimated simultaneously by superpixel labeling. The linear optimization can be relaxed and efficiently solved using a branch and bound method. This linear approach is general and can be easily extended to generic object matching and segmentation.

### 1.1 Related Work

Different methods have been proposed for detecting human poses in videos and images. With multiple view videos [1,23], 3D poses can be detected. Extracting 3D poses in single view videos [2,24] is currently limited to movements in specific domains. In this paper, we focus on finding 2D human poses in single view videos.

Previous methods for 2D human pose estimation use holistic models or body part graph models.

The holistic approach treats a human body as a whole entity. Poses are estimated either by using pose classification [13] if the object can be segmented from the background, or by matching exemplars in databases [3,4,5]. Dynamic models have also been combined with exemplar methods [6] to improve the performance in pose tracking. To estimate unconstrained poses, exemplar methods become increasingly complex since we have to deal with huge pose databases.

Human poses can also be estimated by matching a body part graph model to target images. Body part matching scheme is flexible; it is able to model complex poses with a compact representation. The challenge is how to search for the optimal pose in a huge number of feasible configurations. If the body part relation graph is a tree, polynomial algorithms exist. Felzenszwalb et al. propose an efficient dynamic programming method [9] for pose estimation. Ramanan et al. [7,20] estimate human poses in cluttered images using efficient message passing on trees. Recently, Andriluka et al. [28] extend the tree methods and devise a strong pose detector. Tree structure methods sometimes over-count local evidences because no constraints are directly applied among tree branches.

To solve the over-counting problem, non-tree graph models have also been studied. Pairwise constraints between body parts are introduced to form loopy relation graphs. Searching for poses using non-tree models is NP-hard. A branch and bound method [29] is proposed to solve problems with extra constraints on legs. Approximation methods based on belief propagation [14,16], mathematical programming [11,15] and probability sampling [10,12] have also been proposed. In these methods, overlapping body parts are uniformly penalized, which relieves the over-counting issue, but at the same time, also introduces an undesired penalty to true overlapping body part configurations.

Image segmentation has been used to support body part graph matching in different fashions. Mori [17] uses superpixels to guide pose detection. Ramanan [20] proposes an effective learning method to generate strong body part detectors using soft image segmentation. Johnson et al. [27] use image segmentation to enhance local part detection. Ferrari et al. [21] pre-segment images to obtain rough foregrounds in human upper body pose estimation. In these methods, segmentation is not jointly optimized with the tree structure body part matching. In [8], a pre-segmented object foreground is used to constrain the layout of body parts in the optimization. In the sequential process, the segmentation result greatly affects the performance of pose estimation.

Even though intensively studied, finding human poses in cluttered images and videos is still largely unsolved. In this paper, we focus on detecting a specific subject's poses in single view videos. We propose an efficient linear method that concurrently optimizes body part matching and foreground segmentation. It works for dynamic background videos and unconstrained human poses, and to our knowledge no previous methods are able to achieve the concurrent optimization efficiently in such settings.

# 2 Concurrent Matching and Segmentation

Our task is to detect the poses of a specific human subject in videos. To distinguish the target subject, we extract a cardboard model from a single exemplar image. This procedure is necessary since the target subject may be among a group of people in videos; in this case, a generic pose detector will not work. The cardboard model includes 9 body parts, i.e., a torso and 8 half limbs; each part's appearance is represented by the average RGB color; the foreground color histogram is also stored. We jointly optimize body part matching and foreground estimation for robust pose detection. Formally, we try to find body part matching $\mathcal{X}$ and foreground estimation $\mathcal{Y}$ in a constrained optimization:

$$\min_{\mathcal{X},\mathcal{Y}}\{B(\mathcal{X}) + S(\mathcal{Y}) + \sum_{(u,v)\in I} |h_{u,v} - g_{u,v}|\} \tag{1}$$

$$\text{s.t. } h_{u,v} = 1 \text{ if body parts cover point } (u,v), \text{ otherwise } h_{u,v} = 0.$$
$$g_{u,v} = 1 \text{ if point } (u,v) \text{ is in a foreground superpixel, otherwise } g_{u,v} = 0.$$

Each feasible $\mathcal{X}$ determines a body part configuration in the target image and its cost is $B(\mathcal{X})$. $B(\mathcal{X})$ is small if we detect the true pose. Foreground estimation $\mathcal{Y}$ is obtained by superpixel labeling, and its cost is $S(\mathcal{Y})$, which is small if we label the real object foreground. In Eqn.(1), $(u,v)$ represents points in the target image, $h_{u,v}$ is the body part covering map determined by $\mathcal{X}$, and $g_{u,v}$ is the foreground map determined by $\mathcal{Y}$. The term $\sum_{(u,v)\in I} |h_{u,v} - g_{u,v}|$, where $I$ is the target image point set, penalizes the discrepancy between the body part covering and the foreground estimation. By minimizing the objective function in Eqn.(1), we find the optimal body part matching and foreground segmentation that are also consistent with each other.

## 2.1 Body Part Candidates and Superpixels

Before proceeding to concurrent optimization, we find body part candidates and partition the target image into superpixels. We search for body part candidates using Chamfer matching and color matching. The model shape and colors are extracted from a single exemplar image. Chamfer matching correlates body part bars to the distance transform of the target image edge map. The color differences of each body part with the target image at different locations and orientations are also computed. Chamfer matching costs and color matching costs are linearly combined to form the local body part matching costs. Using non-minimum suppression, we locate body part candidates.

Each half limb candidate is represented by two end points, a rotation angle and a rectangle of specific size. We further group the half limb candidates into full limb candidates and reject apparent wrong pairs: if the distance between the end points of two candidates is greater than a threshold, they cannot be connected together. For the combined limb candidate, its cost is the linear combination of the upper and lower body part matching costs, the distance between the connection joints and the difference between the two sub-limb angles.

We use a graph-cuts method [22] to over-segment images into superpixels. A superpixel contains image pixels that have similar appearance. The superpixels do not consistently partition target objects into body parts. However, points on each limb tend to be in the same superpixel. The overall object coverage, consisting of a bunch of smaller patches, forms a stable foreground region.

The concurrent optimization in Eqn.(1) is a hard combinatorial problem. Due to huge number of feasible configurations for body part assignment and superpixel labeling, exhaustive search is not feasible. Our strategy is to construct a linear formulation and devise an efficient solution.

## 2.2 The Linear Optimization

We express Eqn.(1) as a linear optimization in the following three steps:

**First**, We express the body part matching cost $B(\mathcal{X})$ in Eqn.(1) using linear functions and determine how body part covering map $h_{u,v}$ is related to body part assignments. We introduce indicator variables $x_{n,i}$. If body part $n$ selects candidate $i$, $x_{n,i} = 1$, and otherwise $x_{n,i} = 0$. We also use $(n, i)$ to denote the candidate $i$ of body part $n$. After merging the upper and lower limbs, we have 5 parts. Note that the pose estimation still gives 9-part matching result.

Let the cost of assigning candidate $i$ to body part $n$ be $c_{n,i}$, which can be computed as discussed in §2.1. The overall body part assignment cost is $\sum_{n \in \mathcal{P}} \sum_{i \in \mathcal{A}(n)} (c_{n,i} \cdot x_{n,i})$, where $\mathcal{P}$ is the set of body parts and $\mathcal{A}(n)$ is the candidate set of part $n$. Since each body part selects one and only one candidate, we have $\sum_{i \in \mathcal{A}(n)} x_{n,i} = 1, \forall n \in \mathcal{P}$.

Apart from matching local appearances, body parts also need to follow a body plan: the end points of limbs should be close to the appropriate torso end point. Fig.2(a) illustrates the relation among body parts. The degree that a body part configuration follows a valid body plan can be quantified as:

$$\sum_{n \in \mathcal{P}, n \neq t} || \sum_{i \in \mathcal{A}(n)} \mathbf{p}_{n,i} x_{n,i} - \sum_{k \in \mathcal{A}(t)} \mathbf{t}_{n,k} x_{t,k} || , \qquad (2)$$

where $\mathbf{p}_{n,i}$ is the upper end point of candidate $(n, i)$; $\mathbf{t}_{n,k}$ is the end point of torso candidate $k$ and the end point is adjacent to part $n$; $t$ is the torso. The

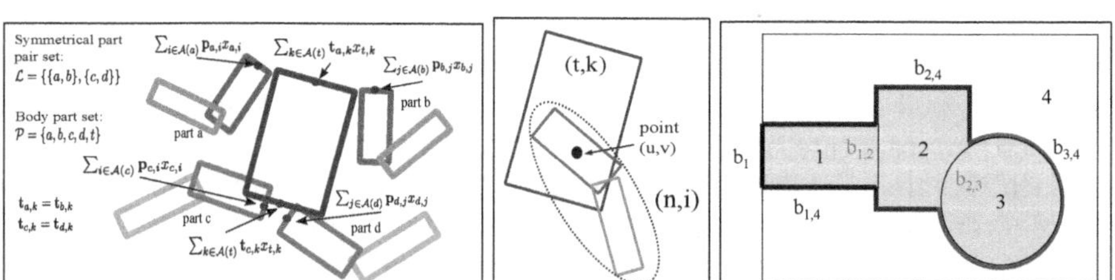


(a) (b) (c)

**Fig. 2.** (a) Notations for body part matching. (b) Part covering. (c) A toy example of superpixel labeling; the gray region is the foreground.

notations are illustrated in Fig.2(a). $||.||$ is the $L_1$ norm. The $L_1$ norm terms can be linearized using auxiliary variables: $\min|\xi|$ is equivalent to $\min(\eta), \text{s.t.} -\eta \leq \xi \leq \eta, \eta \geq 0$. The complete linear form is in Eqn.(3).

Limbs also tend to be symmetrical in spatial locations relative to the torso. If we draw a line segment between the upper arm or the upper leg joints, the center should be close to one suitable end of the torso. The following term is included to quantify the degree of symmetry:

$$\sum_{\{n,m\}\in\mathcal{L}} \left|\left| \sum_{i\in\mathcal{A}(n)} \mathbf{p}_{n,i}x_{n,i} + \sum_{j\in\mathcal{A}(m)} \mathbf{p}_{m,j}x_{m,j} - 2\sum_{k\in\mathcal{A}(t)} \mathbf{t}_{n,k}x_{t,k} \right|\right| ,$$

where $\mathcal{L}$ is the set of symmetrical body part pairs. The notations are illustrated in Fig.2(a). We also use the $L_1$ norm so that this term can be linearized using the auxiliary variable trick. The body part matching cost can then be represented as the linear combination of the local matching cost, the degree that it follows a body plan and the symmetry cost.

For human pose detection, simply optimizing the above body part matching energy is insufficient because it has a strong bias towards single limb detection. To solve the problem, we assemble body parts so that their overall covering is similar to the object foreground, which, as discussed later, is obtained simultaneously by superpixel labeling. To this end, we introduce auxiliary variables $h_{u,v}$ to represent the body part covering map. Here $(u, v)$ is a point in the target image point set $I$. If point $(u, v)$ is covered by the estimated body configuration, we wish $h_{u,v}$ to be 1, and otherwise, 0. $h_{u,v}$ is constrained by the body part assignment variables $x_{n,i}$:

$$\sum_{\forall(n,i) \text{ covers } (u,v)} x_{n,i} \geq h_{u,v}, \; 0 \leq h_{u,v} \leq 1, \; \forall(u,v) \in I.$$

If $(u, v)$ is not covered by any part candidates, $h_{u,v}$ is set to 0. With such constraints, if no body part covers $(u, v)$, $h_{u,v}$ has to be 0; if at least one body part covers $(u, v)$, $h_{u,v}$ can be as big as 1, but it still can be 0. We therefore need to further make sure that $h_{u,v}$ must be 1 if at least one body part covers the pixel:

$$h_{u,v} \geq x_{n,i}, \; \forall(n,i) \text{ covers } (u,v) .$$

As an example, in Fig.2(b), there are two part candidates covering point $(u, v)$. The relation between $h_{u,v}$ and $x$ is: $x_{n,i} + x_{t,k} \geq h_{u,v}$, $h_{u,v} \geq x_{n,i}$, $h_{u,v} \geq x_{t,k}$ and $0 \leq h_{u,v} \leq 1$. It is easy to verify that $h_{u,v}$ is indeed the body part covering map.

**Next**, we represent term $S(\mathcal{Y})$ in Eqn.(1) in linear form and relate it to the foreground map $g_{u,v}$. We introduce binary variable $y_i$ to indicate whether superpixel $i$ is on the foreground or background. If superpixel $i$ is on the foreground, $y_i = 1$, and otherwise $y_i = 0$. To quantify the cost of labeling a superpixel as foreground, we compute the smallest distance from each color in the superpixel to the foreground colors in the template and sum all the color distances to form the superpixel labeling cost. Denoting the cost as the same $c$ as the body part labeling cost but with a single index, the overall cost of the foreground estimation is $\sum_{i \in \mathcal{V}} (c_i \cdot y_i)$, where $\mathcal{V}$ is the set of superpixels in the target image.

Simply minimizing the superpixel assignment cost would result in a small foreground estimation. We need to constrain the size of the foreground segmentation to remove the bias. Assuming that the area of superpixel $i$ is $r_i$, we constrain the object foreground to have an approximate area $s_f$, which is the exemplar foreground area. We therefore need to minimize $|\sum_{i \in \mathcal{V}} (r_i \cdot y_i) - s_f|$. The absolute value of the area difference can be linearized using auxiliary variables.

Besides, we hope that an object foreground contains a group of connected superpixels. Since connected regions tend to have small perimeter, we minimize the overall boundary length of the foreground superpixels to implicitly enforce this constraint. Let $b_{i,j}$ be the length of the common boundary between the neighboring superpixels $i$ and $j$, and $b_i$ be the length of the common boundary between superpixel $i$ and the image bounding box. The perimeter of the foreground region is: $\sum_{\{i,j\} \in \mathcal{N}_s} (b_{i,j} \cdot |y_i - y_j|) + \sum_{i \in \mathcal{D}} (b_i \cdot y_i)$, where $\mathcal{N}_s$ is the set of neighboring superpixel pairs; $\mathcal{D}$ is the set of superpixels adjacent to the image bounding box. Fig.2(c) illustrates a toy example in which $\mathcal{N}_s = \{\{1,2\},\{2,3\},\{1,4\},\{2,4\},\{3,4\}\}$ and $\mathcal{D} = \{1\}$ and the above equation computes the foreground perimeter. The above connectivity term can also be linearized using auxiliary variable tricks. The superpixel labeling energy $S(\mathcal{Y})$ is therefore the linear combination of the three terms: the superpixel color matching term, the size term and the connectivity term.

To facilitate the comparison of foreground estimation with body part covering, we introduce auxiliary variables $g_{u,v}$ to represent the foreground map at the image pixel level. If $(u,v)$ in the target image is covered by a superpixel, $g_{u,v}$ has the same value as the superpixel label: $g_{u,v} = y_i, \forall (u,v) \in R_i, i \in \mathcal{V}$, where $R_i$ is the point set of superpixel $i$.

**Finally**, we are ready to express the complete optimization. we have formulated $B(\mathcal{X})$, $S(\mathcal{Y})$, $h_{u,v}$ and $g_{u,v}$ in Eqn.(1) using linear functions and linear constraints. With the above settings, $\sum_{(u,v) \in I} |h_{u,v} - g_{u,v}|$, where $I$ is the set of points in the target image, equals the difference between the body part covering region and the foreground region estimated in the superpixel labeling. When minimizing the total energy $B(\mathcal{X}) + S(\mathcal{Y}) + \sum_{(u,v) \in I} |h_{u,v} - g_{u,v}|$, we find the optimal body part matching and foreground estimation that are consistent with each other. This consistency criterion is soft, and therefore it allows partial mismatches between the body part rectangles and the foreground superpixels. The concurrent optimization is also a principled way to solve the over-counting issue without introducing an undesired penalty for truly overlapping body parts, since body parts are now encouraged to fit the foreground instead of being simply pushed away from each other.

## 2.3 Relaxation and Branch and Bound Solution

Pose estimation can therefore be formulated as the following linear optimization:

$$\min\{\sum_{(u,v)\in I} z_{u,v} + \alpha_1 \sum_{n\in\mathcal{P}} \sum_{i\in\mathcal{A}(n)} (c_{n,i}\cdot x_{n,i}) + \alpha_2 \sum_{n\in\mathcal{P},n\neq t} \sum_{l=1}^{2} p_n^{(l)} + \tag{3}$$

$$\alpha_3 \sum_{\{n,m\}\in\mathcal{L}} \sum_{l=1}^{2} q_{n,m}^{(l)} + \beta_1 \sum_{i\in\mathcal{V}} (c_i\cdot y_i) + \beta_2 [\sum_{\{i,j\}\in\mathcal{N}_s} (b_{i,j}\cdot y_{i,j}) + \sum_{i\in\mathcal{D}} (b_i\cdot y_i)] + \beta_3 w\}$$

$$\text{s.t.} \sum_{i\in\mathcal{A}(n)} x_{n,i} = 1, \forall n\in\mathcal{P}. \qquad \sum_{\forall(n,i)\text{ covers }(u,v)} x_{n,i} \geq h_{u,v}, 0\leq h_{u,v}\leq 1, \forall(u,v)\in I.$$

$$h_{u,v} \geq x_{n,i}, \forall(n,i) \text{ covers } (u,v), \forall(u,v)\in I.$$

$$g_{u,v} = y_i, \forall(u,v)\in R_i, \forall i\in\mathcal{V}. \quad -z_{u,v} \leq g_{u,v} - h_{u,v} \leq z_{u,v}, \forall(u,v)\in I.$$

$$-p_n^{(l)} \leq \sum_i p_{n,i}^{(l)} x_{n,i} - \sum_k t_{n,k}^{(l)} x_{t,k} \leq p_n^{(l)},\ l=1..2, n\in\mathcal{P},\ n\neq t.$$

$$-q_{n,m}^{(l)} \leq \sum_i p_{n,i}^{(l)} x_{n,i} + \sum_j p_{m,j}^{(l)} x_{m,j} - 2\sum_k t_{n,k}^{(l)} x_{t,k} \leq q_{n,m}^{(l)},$$

$l = 1..2, \{n,m\}\in\mathcal{L}$ which includes two arms and two legs, $t$ is the torso.

$$-y_{i,j} \leq y_i - y_j \leq y_{i,j}, \forall\{i,j\}\in\mathcal{N}_s. \quad -w \leq \sum_{i\in\mathcal{V}} (r_i\cdot y_i) - s_f \leq w.$$

All variables $\geq 0, x, y$ are binaries.

The variables $x_{n,i}, y_i, h_{u,v}$ and $g_{u,v}$ follow the previous definitions. The auxiliary variables $z_{u,v}$, $p_n^{(l)}$, $q_{n,m}^{(l)}$, $y_{i,j}$ and $w$ are included to help turn the $L_1$ norm terms into linear functions. Coefficients $p_{n,i}^{(l)}, l = 1..2$, are the elements of $\mathbf{p}_{n,i}$, and $t_{n,k}^{(l)}, l = 1..2$, are the elements of $\mathbf{t}_{n,k}$; $\mathbf{p}$ and $\mathbf{t}$ are defined in Eqn.(2). In the objective function, the terms with $\alpha$ coefficients correspond to the body part assignment cost $B(\mathcal{X})$ in Eqn.(1); $\beta$ coefficient terms correspond to the superpixel labeling cost $S(\mathcal{Y})$; and the $z$ term is the covering consistency $\sum_{(u,v)\in I} |h_{u,v} - g_{u,v}|$ in Eqn.(1). The $\alpha$ and $\beta$ coefficients are selected manually by trial and error; they are fixed in all the experiments. Typical values are $\alpha_1 = 1$, $\alpha_2 = \alpha_3 = 0.1$ and $\beta_1 = \beta_2 = \beta_3 = 0.01$.

In this formulation, the variables $x$ and $y$ for body parts and superpixels are binary. The map variables $g$ and $h$ are continuous. Directly solving the mixed integer program is not feasible. We relax it for an approximate solution. A relaxation of both $x$ and $y$ into continuous variables yields weak results, in which the superpixel indicator variables $y$ often obtain equal value and body part assignment does not benefit from the decisions on $y$. We therefore only relax $x$ to continuous variables in [0,1] and keep $y$ as binary variables. The relaxed problem can be efficiently solved by a branch and bound method.

An initial random superpixel labeling is used to estimate an upper bound of the optimization. The branch and bound method picks up a superpixel and generates two branches: one labels the superpixel 1, and the other labels it 0. The lower bound of the optimization for each branch is computed using the linear program by relaxing all the other variables in Eqn.(3). If the lower bound is greater than the current upper bound, the branch is cut; otherwise it is expanded by including two branches for another superpixel. The upper bound is updated whenever an integer solution for each $y$ is obtained in a branch. This procedure repeats until every superpixel obtains binary solution in each surviving branch.

Our method quickly converges. In the relaxation solution, very few $x$ variables are nonzero. Keeping only the body part candidates that correspond to these non-zero assignment variables, we solve the full integer program. Since there are few variables, the exhaustive search converges quickly. We can further lower the complexity by reducing the number of $g$ and $h$ variables. We define them on coarser image blocks instead of image pixels. We use 2500 $g$ and $h$ variables respectively in the optimization. With about 100 torso candidates, 10 thousand candidates for each full limb and a few hundred superpixels, the average running time for the concurrent optimization is about 25 seconds on a 2.8GHz machine.

## 3 Experimental Results

We evaluate the proposed method on a variety of video sequences. The test data include recorded videos and the videos from the web of total 4413 frames and 755-frame videos from the HumanEva dataset [30]. The four recorded sequences contain complex poses and strong background clutter. We select the sequences of three different subjects in different actions from the HumanEva dataset. These sequences are from camera one, whose view has the strongest background clutter. For each test sequence, we use the proposed concurrent optimization method to match a cardboard model, estimated from a single exemplar image in the sequence, to the target images to estimate human poses. For fair comparison, we use the same "walking" pose exemplar in each sequence for all the testing methods.

To verify the usefulness of the concurrent optimization approach, we compare it with some of the variations. We first test whether using superpixel labeling alone would yield satisfactory foreground estimation. If this were the case, we could use a sequential optimization instead of the more complex concurrent optimization. Fig.3 shows that the superpixel labeling alone cannot yield reliable foreground segmentation. Without a global shape constraint, it gives lots of false

**Fig. 3.** Foreground estimation comparison. Row 1: sample images from sequence labman-I. Row 2: superpixel partitions of images. Row 3: foreground estimation using superpixel labeling alone. Row 4: foreground estimation using the concurrent optimization.

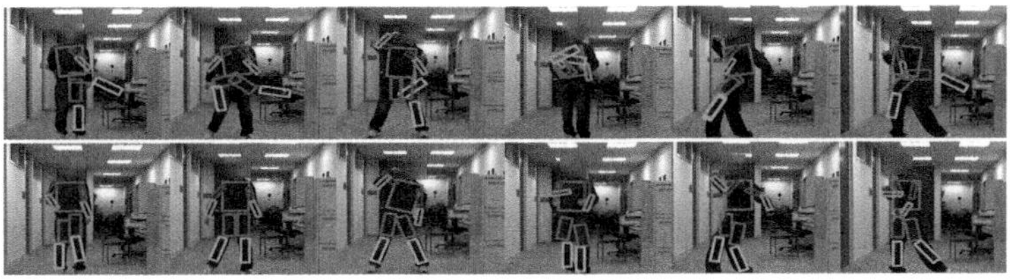

**Fig. 4.** Comparison with the dynamic programming method. The 1st row: the DP sample results for the lab-man-II sequence. The 2nd row: pose estimation using the proposed method.

positives and false negatives. The concurrent optimization is necessary, and it helps to obtain a roughly correct foreground estimation as shown in Fig.3. With a "taller" torso rectangle, the head of the subject is also labeled as foreground in the concurrent optimization.

We proceed to compare the proposed method with a variation that optimizes only the body part matching. If the symmetrical part constraint is also discarded, we have a tree structure graph model. Pose estimation with a tree structure body plan can be exactly solved using dynamic programming (DP). Fig.4 shows sample comparison results for the lab-man-II sequence. Without a global constraint, the dynamic programming method often loses detection of arms and legs, and it is easily distracted by the background clutter. The quantitative comparison is shown in Fig.6 and Fig.7. Fig.6 compares the normalized histograms of per-frame errors. Without ground truth, we use visual inspection to verify the results. The criterion is that a correct body part detection should be closely aligned with the corresponding body part or hallucinate on the occluded one. Since there are 9 body parts, the per-frame error number is from 0 to 9. A good performance is characterized by an error histogram that is high in the low error range and low in the high error range. The proposed method yields much better result than the simple DP approach. As shown in Fig.7, the average per-frame errors of the proposed method are less than half of the errors of DP.

It is indeed useful to use simultaneous segmentation to globally constrain the pose optimization. The question is whether other global constraints would work as well. We set out to test whether a simple max-covering global constraint would be sufficient. We label all the superpixels as 1 to introduce a max-covering constraint: the body parts should cover a region as big as possible. This formulation penalizes the overlapping body parts equally and prefers a stretched pose. The sample comparison results are shown in Fig.5. The max-covering method has a difficult time to decide whether to accept a body part candidate or to reject it as clutter because it does not use the clues from image segmentation. As shown in Fig.5, the errors of max-covering include both false positives and false negatives. Simply adjusting the parameters will reduce one class of errors but increase the other. The proposed method does not have such problems and it achieves much better results on the test image sequences. The quantitative

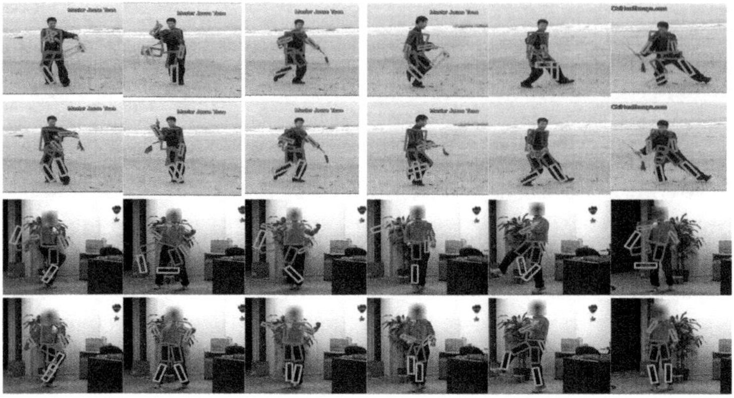

**Fig. 5.** Comparison with max-covering. The odd rows show the results of max-covering on the taichi and lab-man-I sequence. The even rows show how the proposed method improves the results.

comparison in Fig.6 and Fig.7 shows that the proposed method has many fewer errors than max-covering.

The proposed method is indeed better than its variations. But does it have an advantage over other approaches? We first compare the proposed method with the tree inference method [7], a state-of-the-art method for pose detection in videos using a single exemplar. We run the code with [7] on the test videos. The body part detectors are trained from the same walking pose exemplars as those in other testing methods. The sample comparison results are shown in Fig.8. The tree inference method sometimes loses the detections of arms or legs. The proposed method solves the problem by using the global foreground shape constraint. It is also more resistant to clutter. As shown in Fig.8, the proposed method is more reliable in distinguishing two dancers' legs even though they have similar color. The quantitative comparison is shown in Fig.7 and Fig.10.

We further compare the proposed method with a non-tree method [15]. The sample comparison results are shown in Fig.9. The non-tree method uses pairwise prohibition terms to constrain the symmetrical body parts. It uses the same set of body part candidates and costs as the proposed method in the comparison. Compared with the non-tree method, the proposed method works better for

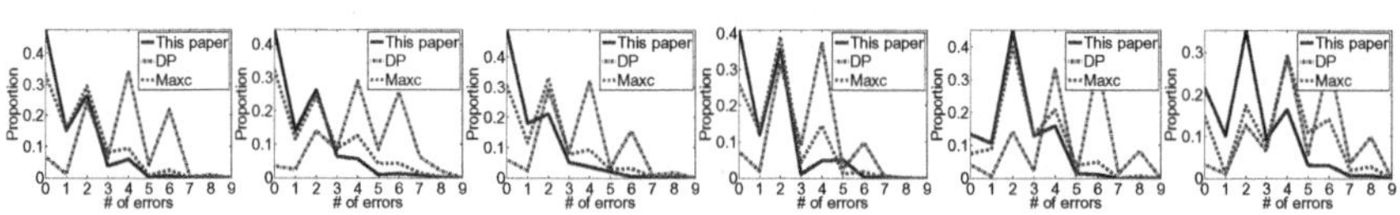


(a) lab-girl (b) lab-man-I (c) lab-man-II (d) lab-man-III (e) taichi (f) dance

**Fig. 6.** The normalized per-frame error histograms of the proposed method (black solid) with the dynamic programming (DP) (red dash-dotted) and max-covering (Maxc) (blue dotted)

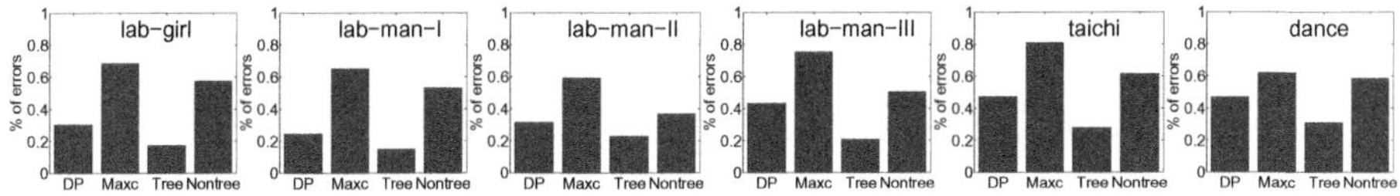


**Fig. 7.** The ratio of the average per-frame errors of the proposed method to other methods on the test sequences

complex poses and is more robust in strong clutter. The quantitative results in Fig.7 and Fig.10 confirm the advantage of the proposed method.

Since we optimize both the body part assignment and the foreground superpixel labeling, the byproduct is a rough object foreground estimation. Rows 1-4 in Fig.11 show some sample results. More pose estimation results randomly sampled from the videos are shown in Fig.11. The proposed method robustly detects poses in the videos. In Fig.11, we also see some part detection errors, especially in the challenging taichi and dance sequences. In our experiments, pose estimation errors are caused mainly by the weak local body part detectors. Using stronger part detectors will further improve the performance.

We also test the proposed method on the ground truth data. Three sequences are selected from the HumanEva [30] dataset. The boxing and walking sequences are down-sampled in time while the jogging sequence includes all the frames. All

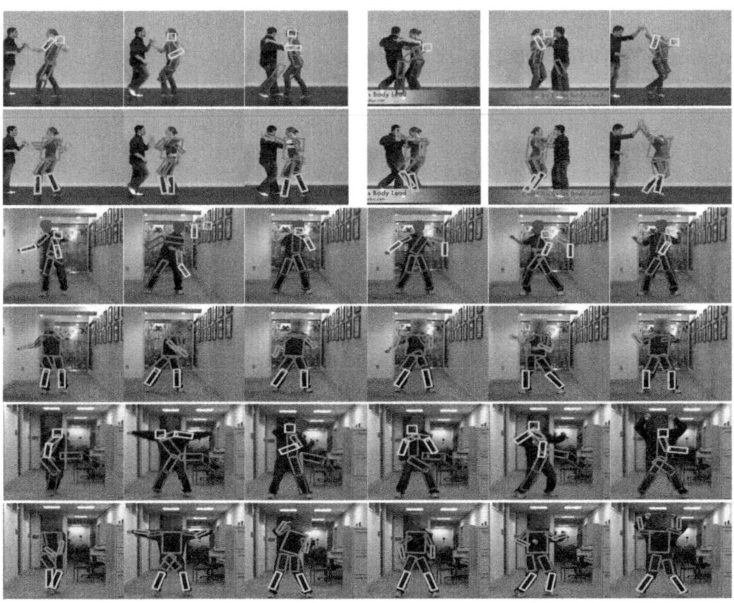

**Fig. 8.** Comparison with the tree inference method [7]. The odd rows show the results of tree method on the dance, lab-man-III and lab-man-II sequences. The even rows show the results of the proposed method.

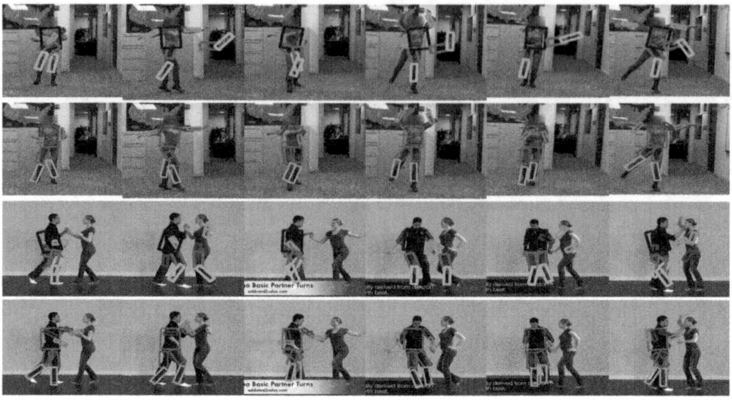

**Fig. 9.** Comparison with a non-tree method [15]. The odd rows show the results of the non-tree method on the lab-girl and dance sequences. The even rows show the results of the proposed method.

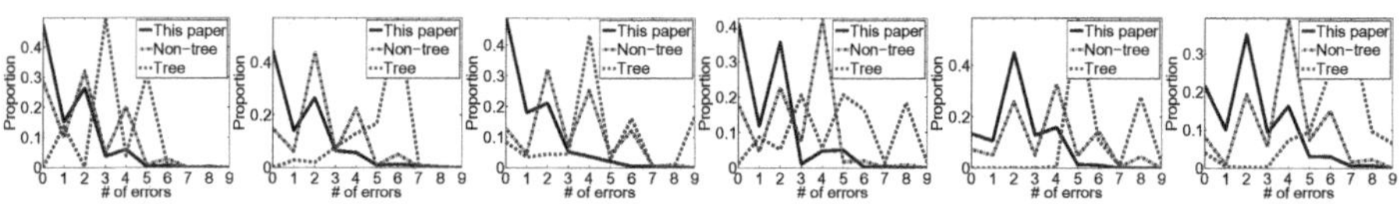


(a) lab-girl (b) lab-man-I (c) lab-man-II (d) lab-man-III (e) taichi (f) dance

**Fig. 10.** The normalized per-frame error histograms of the proposed method (black solid) with the tree inference method [7] (blue dotted) and the non-tree method [15] (red dash-dotted)

the images are pre-scaled so that the target objects have roughly the same size. These sequences have strong background clutter. The comparison of the proposed method, the non-tree method and the tree methods is shown in Fig.12. Besides the tree method in [7] denoted as tree-I, we also compare with a recent tree method in [28], denoted as tree-II, which uses more robust body part detectors. Tree-II is a state-of-the-art generic pose detector. For fair comparison, we modify the code for [28] so that color is also used in local part detection and we tune its parameters to achieve the best performance. We quantify the performance of pose estimation by the overlapping area of body parts and the corresponding ground truth. We compute the overlapping area for arms and legs and do not count the easiest torso. The total overlapping area is normalized by the sum of all the ground truth limb areas to form the pose score. Fig.12 compares the histograms of per-frame pose scores and the average pose scores of different methods. The proposed method has the highest pose scores in all the tests. Visual inspection shows consistent result. Our method greatly improves the pose detection results. The performance improvement is not a surprise. Our model enforces a global shape constraint through simultaneous segmentation

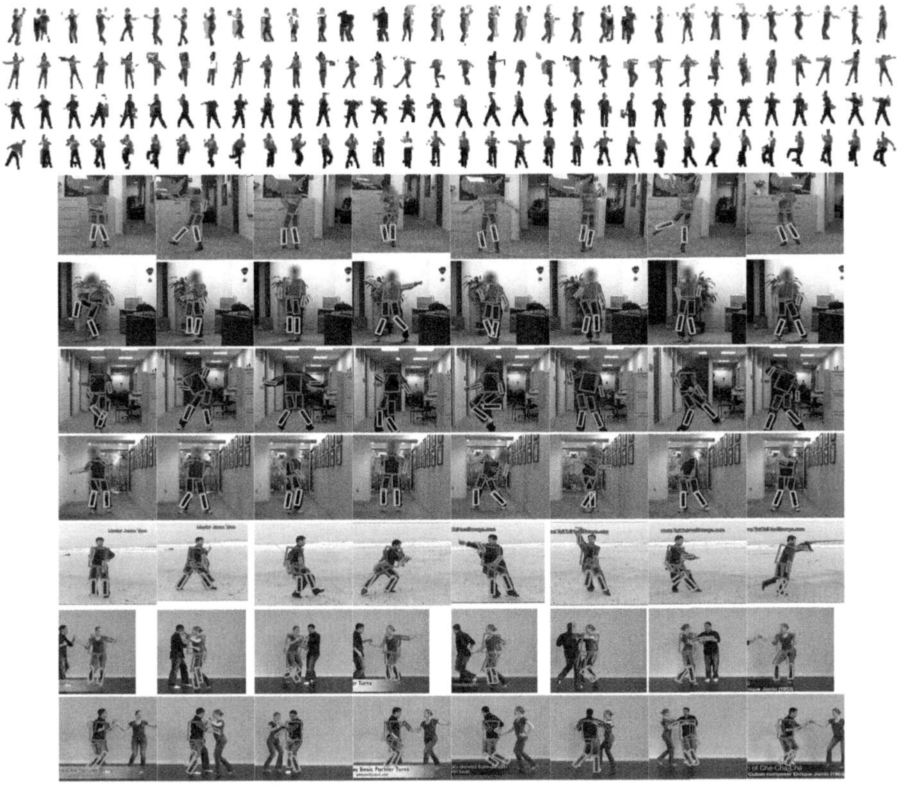

**Fig. 11.** Pose estimation sample results using the proposed method on the test sequences. Row 1-4: object foreground estimation samples. Row 5-11: random samples from lab-girl (548 frames), lab-man-I (779 frames), lab-man-II (1001 frames), lab-man-III (730 frames), taichi (359 frames), dance(woman) (498 frames) and dance(man) (498 frames).

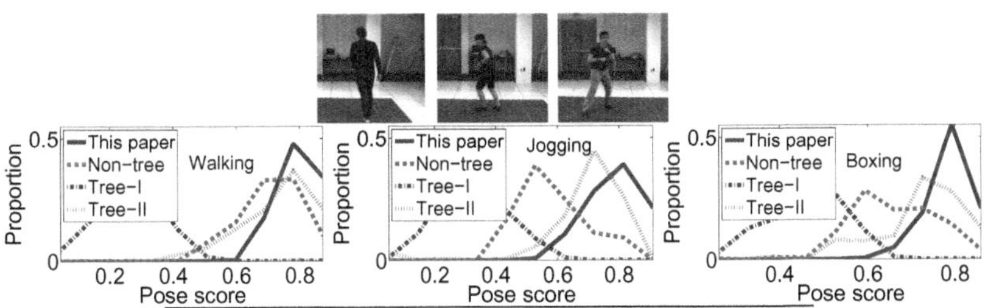


| | This paper | Non-tree[15] | Tree-I[7] | Tree-II[28] |
|---|---|---|---|---|
| Walk | 0.793 | 0.709 | 0.263 | 0.733 |
| Jog | 0.784 | 0.582 | 0.326 | 0.685 |
| Box | 0.788 | 0.660 | 0.460 | 0.725 |

**Fig. 12.** Test on HumanEva walking (222 frames), jogging (362 frames) and boxing (171 frames) sequences. Row 1: sample frames. Row 2: pose score histograms of the proposed method, the non-tree [15], tree-I [7] and tree-II [28] methods. Row 3: average pose scores.

and therefore all the body parts are related through hyper-graph edges. The high order constraint is essential for pose estimation in strong clutter.

## 4 Conclusion

We propose a novel concurrent optimization method to detect human poses in cluttered videos. With a single exemplar image, the proposed method robustly finds human poses in long video sequences. Concurrently optimizing the body part matching and object segmentation is a great challenge due to its huge search space. We efficiently solve the hard combinatorial problem by novel linear relaxation and branch and bound method. Our experiments on a variety of videos show that the proposed method has a clear advantage over locally constrained methods. The linear approach is also general and it can be extended to generic object matching and segmentation.

**Acknowledgement.** The project is supported by NSF grant 1018641.

## References

1. Bregler, C., Malik, J., Pullen, K.: Twist based acquisition and tracking of animal and human kinematics. IJCV 56(3), 179–194 (2004)
2. Sminchisescu, C., Triggs, B.: Estimating articulated human motion with covariance scaled sampling. Inter. J. of Robotics Research 22(6), 371–391 (2003)
3. Mori, G., Malik, J.: Estimating human body configurations using shape context matching. In: Heyden, A., Sparr, G., Nielsen, M., Johansen, P. (eds.) ECCV 2002. LNCS, vol. 2352, pp. 666–680. Springer, Heidelberg (2002)
4. Gavrila, D.M.: A Bayesian, exemplar-based approach to hierarchical shape matching. TPAMI 29(8) (2007)
5. Shakhnarovich, G., Viola, P., Darrell, T.: Fast pose estimation with parameter sensitive hashing. In: ICCV 2003 (2003)
6. Toyama, K., Blake, A.: Probabilistic tracking with exemplars in a metric space. IJCV 48(1), 9–19 (2002)
7. Ramanan, D., Forsyth, D.A., Zisserman, A.: Strike a pose: tracking people by finding stylized poses. In: CVPR 2005 (2005)
8. Jiang, H.: Human pose estimation using consistent max-covering. In: ICCV 2009 (2009)
9. Felzenszwalb, P.F., Huttenlocher, D.P.: Pictorial structures for object recognition. IJCV 61(1) (January 2005)
10. Ioffe, S., Forsyth, D.A.: Probabilistic methods for finding people. IJCV 43(1), 45–68 (2001)
11. Ren, X.F., Berg, A.C., Malik, J.: Recovering human body configurations using pairwise constraints between parts. In: ICCV 2005, vol. 1, pp. 824–831 (2005)
12. Lee, M.W., Cohen, I.: Human upper body pose estimation in static images. In: Pajdla, T., Matas, J(G.) (eds.) ECCV 2004. LNCS, vol. 3022, pp. 126–138. Springer, Heidelberg (2004)
13. Rosales, R., Sclaroff, S.: Inferring body pose without tracking body parts. In: CVPR 2000 (2000)

14. Sigal, L., Black, M.J.: Measure locally, reason globally: occlusion sensitive articulated pose estimation. In: CVPR 2006 (2006)
15. Jiang, H., Martin, D.R.: Global pose estimation using non-tree models. In: CVPR 2008 (2008)
16. Wang, Y., Mori, G.: Multiple tree models for occlusion and spatial constraints in human pose estimation. In: Forsyth, D., Torr, P., Zisserman, A. (eds.) ECCV 2008, Part III. LNCS, vol. 5304, pp. 710–724. Springer, Heidelberg (2008)
17. Mori, G.: Guiding model search using segmentation. In: ICCV 2005 (2005)
18. Kohli, P., Rihan, J., Bray, M., Torr, P.H.S.: Simultaneous segmentation and pose estimation of humans using dynamic graph Cuts. IJCV 79(3), 285–298 (2008)
19. Pawan Kumar, M., Torr, P.H.S., Zisserman, A.: OBJCUT. In: CVPR 2005 (2005)
20. Ramanan, D.: Learning to parse images of articulated objects. In: NIPS 2006 (2006)
21. Ferrari, V., Manuel, M., Zisserman, A.: Pose search: retrieving people using their pose. In: CVPR 2008 (2008)
22. Felzenszwalb, P.F., Huttenlocher, D.P.: Efficient graph-based image segmentation. IJCV 59(2) (2004)
23. Gupta, A., Mittal, A., Davis, L.S.: Constraint integration for efficient multiview pose estimation with self-occlusions. IEEE TPAMI 30(3), 493–506 (2008)
24. Urtasun, R., Fleet, D., Fua, P.: Temporal motion models for monocular and multiview 3D human body tracking. CVIU 104(2), 157–177 (2006)
25. Yezzi, A., Zollei, L., Kapur, T.: A variational framework for joint segmentation and registration. In: IEEE Workshop on Mathematical Methods in Biomedical Image Analysis 2001 (2001)
26. Chen, C., Fan, G.: Hybrid body representation for integrated pose recognition, localization and segmentation. In: CVPR 2008 (2008)
27. Johnson, S., Everingham, M.: Combining discriminative appearance and segmentation cues for articulated human pose estimation. In: IEEE International Workshop on Machine Learning for Vision-based Motion Analysis (2009)
28. Andriluka, M., Roth, S., Schiele, B.: Pictorial structures revisited: people detection and articulated pose estimation. In: CVPR 2009 (2009)
29. Tian, T.P., Sclaroff, S.: Fast globally optimal 2D human detection with loopy graph models. In: CVPR 2010 (2010)
30. HumanEva Dataset, http://vision.cs.brown.edu/humaneva

# Modeling Sense Disambiguation of Human Pose: Recognizing Action at a Distance by Key Poses

Snehasis Mukherjee, Sujoy Kumar Biswas, and Dipti Prasad Mukherjee

Electronics and Communication Sciences Unit, Indian Statistical Institute,
203 B. T. Road, Kolkata, India
{snehasismukho,skbhere}@gmail.com, dipti@isical.ac.in

**Abstract.** We propose a methodology for recognizing actions at a distance by watching the human poses and deriving descriptors that capture the motion patterns of the poses. Human poses often carry a strong visual sense (intended meaning) which describes the related action unambiguously. But identifying the intended meaning of poses is a challenging task because of their variability and such variations in poses lead to visual sense ambiguity. From a large vocabulary of poses (visual words) we prune out ambiguous poses and extract key poses (or key words) using centrality measure of graph connectivity [1]. Under this framework, finding the key poses for a given sense (i.e., action type) amounts to constructing a graph with poses as vertices and then identifying the most "important" vertices in the graph (following centrality theory). The results on four standard activity recognition datasets show the efficacy of our approach when compared to the present state of the art.

## 1 Introduction

Action recognition at a distance typically happens when the performer is far away from the camera and appears very small (approximately 30 to 40 pixels tall) in the action videos. Modeling the body parts of the performer is difficult since the limbs are not distinctly visible. The only reliable cue in such case is the pose information and we bank on the motion pattern of the poses to derive our descriptors. The contribution in this paper is twofold. First, we propose a novel pose descriptor which not only captures the pose information but also takes the motion pattern of the poses into account. Secondly, our contribution lies in developing a framework for modeling the intended meaning associated with the human poses in different contexts. The activity recognition methodology proposed here is based on the premise that human actions are composed of repetitive motion patterns and a sparse set of key poses (and their related movements) often suffice to characterize an action quite well. The proposed methodology follows the bag-of-word approach and "word" here refers to the pose descriptor of the human figure corresponding to a single video frame. Consequently a "document" corresponds to the entire video of a particular action. The poses can often be very ambiguous and they may exhibit confusing interpretations about the nature of associated actions. The inherent variability present in the

R. Kimmel, R. Klette, and A. Sugimoto (Eds.): ACCV 2010, Part I, LNCS 6492, pp. 244–255, 2011.

poses may make a single pose a likely candidate for multiple action categories or sometimes none at all. Variation in poses is the primary source of visual sense ambiguity and in such cases it becomes difficult to infer which pose signifies what kind of visual senses (i.e., human actions). For example, top row in Fig. 1 shows some ambiguous poses and by looking at them one cannot tell for certain the corresponding actions. Whereas, the bottom row illustrates the key poses which unambiguously specify the related actions. Action recognition in videos by bag-of-word based methods either seek right kind of features for video words [2,3,4,5] or model the abstraction behind the video words [6,7,8,9]. There are initiatives which study pose specific features [9,10,11] but modeling visual senses associated with poses in videos is largely an unexplored research area. Our work not only derives novel pose descriptors but, more importantly, seeks to model visual senses exhibited by the human poses. For each visual sense (i.e., action type) we rank the poses in order of "importance" using centrality measure of graph connectivity [12]. Google uses centrality measures to rank webpages [13] and recently this ranking technique has spelled success in feature ranking for object recognition [14] and video-action recognition [8] tasks. Also, the ambiguity of visual poses bears a direct similarity with the word sense disambiguation [1] in Natural Language Processing (NLP), where the senses associated with a text word vary from context to context and the objective comprises identifying its true meaning expressed in a given context. This paper is all about to get rid of such sense ambiguities associated with different human poses and find out a sparse set of key poses that can efficiently distinguish human activities from each other.

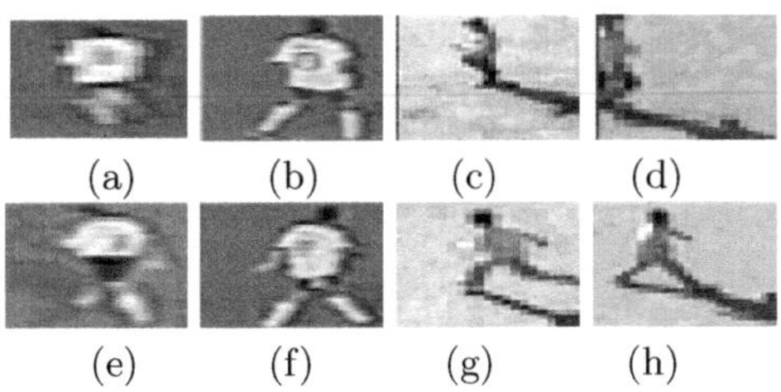


**Fig. 1.** Top row shows some ambiguous poses (labeled by our algorithm) from Soccer (a-b) and Tower (c-d) datasets and the confusion is in deciding whether they represent running or walking. The bottom row shows retrieved key poses (by our algorithm) for walking (e - f) from Soccer dataset, running (g) and walking (h) from Tower dataset.

Section 2 describes the proposed methodology and Section 3 presents results followed by conclusions in Section 4.

## 2 Proposed Methodology

In our approach, each frame of an action video provides a multidimensional vector, which we refer as pose descriptor. The descriptors are derived by combining motion cue from the optical flow field [15] and pose cue from the oriented gradient field of a particular video frame. The pose descriptors, upon clustering (Section 2.2) result into a moderately compact representation $S$. The key poses are extracted (Section 2.3) from this initial large codebook $S$ in a supervised manner and this sparse set of key poses are used for classification of an unknown target video. We elaborate our descriptor extraction process below.

### 2.1 Pose Descriptor: Combining Motion and Pose

Our pose descriptors derive the benefit of motion information from optical flow field and pose information from the orientation field. The optical flow field $\overrightarrow{\boldsymbol{F}}$ is weighted with the strength of the oriented gradient field $\overrightarrow{\boldsymbol{B}}$ to produce a resultant flow field that we call $\overrightarrow{\boldsymbol{V}}$, i.e.,

$$\overrightarrow{\mathbf{V}} = |\overrightarrow{\mathbf{B}}| .* \overrightarrow{\mathbf{F}}, \tag{1}$$

where the symbol '.*' represents the point wise multiplication of the two matrices.

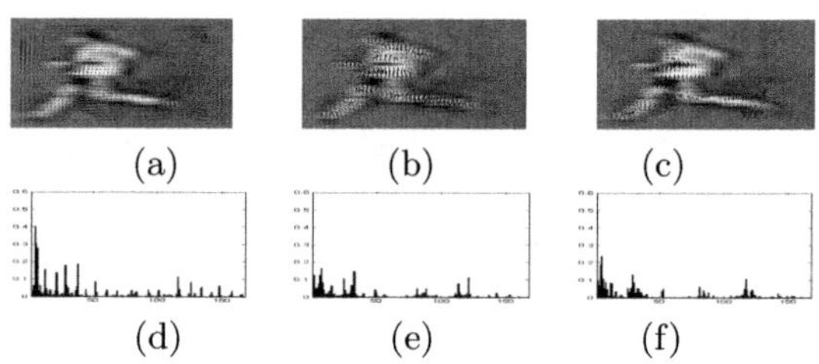


**Fig. 2.** (a) The optical flow field, (b) gradient field, (c) weighted optical flow and (d), (e), (f) show the respective pose descriptor (histograms obtained from (1))

The effect of this weighted optical flow field is best understood if one treats the oriented gradient field as a band pass filter (Fig. 2). The optical flow vectors, appearing in large magnitude on the background (Fig. 2(a)) quite away from the human figure (originated due to signal noise or unstable motion), get "blurred" and eventually filtered out on modulation with the gradient field. This is because the gradient field takes high value where the edge is prominent, preferably along the edge boundary of the foreground figure, but it is very low in magnitude on the uniform background space (Fig. 2(b)). Since gradient strength along the human silhouette is quite high, the optical flow vectors there get a boost upon modulation with gradient field strength. So we filter in the motion information along the silhouette of the human figure and suppress the flow vectors elsewhere in the frame (Fig. 2(c)). So our descriptor is basically a motion-pose descriptor preserving the motion pattern of the human pose. Following section gives the final details for building up the descriptor.

Suppose we are given a collection of $M$ frames in a video sequence of some action $A$ and we represent the sequences by $I_1, I_2, \ldots, I_M$. We define $I$ to be an $m \times n$ grey image defined as a function such that for any pixel $(x, y)$, where $(x, y) \in Z \times Z$, the image $I(x, y) \in \theta, \theta \subset Z$. Corresponding to $I_{t-1}$ and $I_t$ $(t = 2, 3, \ldots, M)$ we compute the optical flow field $\overrightarrow{\boldsymbol{F}}$. Also, we derive the gradient field $\overrightarrow{\boldsymbol{B}}$ corresponding to frame $I_t$ and following (1) we obtain $\overrightarrow{\boldsymbol{V}}$. We consider a three layer image pyramid (Fig. 3(a)) where in the topmost layer we distribute the field vectors of $\overrightarrow{\boldsymbol{V}}$ in an $L$-bin histogram. Here each bin denotes a particular octant in the angular radian space. We take the value of $L$ as 8, because orientation field is quantized enough when resolved in eight directions, i.e., in every 45 degrees. The next layer in the image pyramid splits the image into 4 equal (or almost equal) blocks and each block produces one 8-bin histogram leading to 32-dimensional histogram vector. Similarly, the bottommost layer has 16 blocks and hence 128-dimensional histogram vector. All the histogram vectors are $L1$-normalized separately for each layer and concatenated together resulting in a 168-dimensional pose descriptor. Once we have the pose descriptors we seek to quantize them into a visual codebook of poses. Next section outlines the details of the visual codebook formation process.

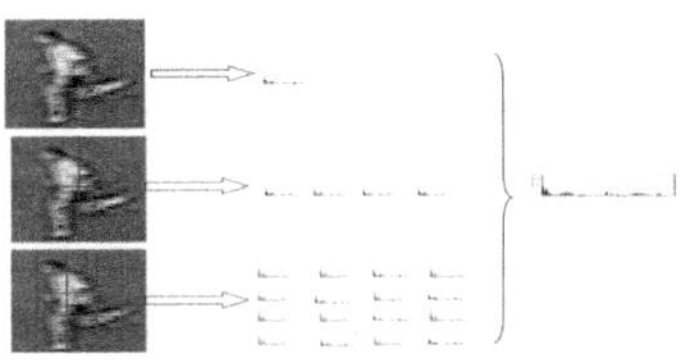

(a)

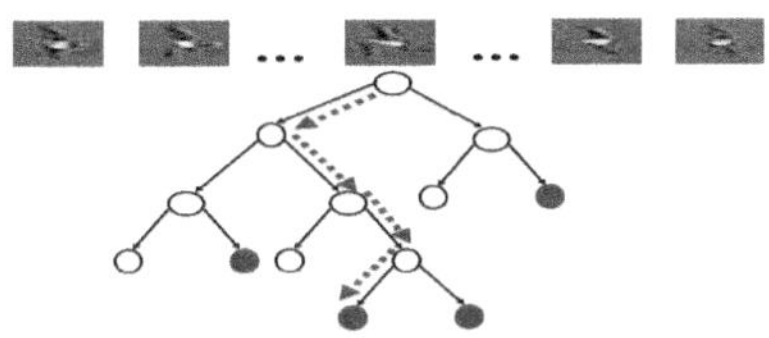

(b)

**Fig. 3.** (a) Formation of 168-dimensional pose descriptor in three layers. (b) Mapping of a pose descriptor to a pose word in the kd-tree; leaf nodes in the tree denote poses and red leaf nodes denote key poses.

## 2.2 Unsupervised Learning of Visual Codebook of Poses

Human activity follows a sequence of pose patterns and such pose sequence occurs in a repetitive fashion throughout the entire action span. The shortest sequence that gets repeated in an entire action video is defined as action cycle $C$ and within an action cycle $C$ let us suppose $T$ poses use to occur where $T$ is the cycle length of the shortest action sequence. For a given action video $A = \{I_1, I_2, \ldots, I_M\}$, we can write a sequence $A = \{C_1, C_2, \ldots, C_H\}$, where $H << M$ and $C_i$ $(i = 1, 2, \ldots, H)$ denotes the action cycle. In action $A$, each frame $I_i$ $(i = 1, 2, \ldots, M)$ produces an action descriptor $q_i$ (1). Clearly $A = \{C_1, C_2, \ldots, C_H\}$ seeks a true partitioning of the pose descriptors $\{q_i\}_{i=1}^{M}$ for all types of action $A$. Ideally $C_1 = C_2 = \ldots = C_H$ as all the action cycles should contain same set of pose patterns. But in practice $C_i$ and $C_j$ $(i \neq j; C_i, C_j \in A)$ are not same because of pose variation, related noise, uncertainty related to cycle length $T$. However, for a given action video $A$, $C_i$ and $C_j$ together contain redundant information as many of the poses that occur in $C_i$ also occur in $C_j$. This follows from the cyclical nature of the human action. Clustering technique removes much of such redundancies present in the pose descriptors $\{q_i\}_{i=1}^{M}$. Please note, our intention here is not to seek true partitioning of pose descriptors but to obtain a large vocabulary $S$ where at least some of the redundancies of the pose descriptors are eliminated. An optimum (local) lower bound on the codebook size of $S$ can be estimated by Akaike Information Criterion (AIC), Bayesian Information Criterion (BIC) [16] or one can directly employ X-means algorithm [17] which is a divisive clustering technique where splitting decision depends on the local BIC score (i.e., does BIC increase or decrease upon splitting the parent cluster into child ones). K-means based clustering techniques [17] rely on Euclidean distance metric which is isotropic in nature and suffer from curse of dimensionality when the dimension of the feature space increases. To alleviate the curse of dimensionality we adopted a kd-tree based clustering algorithm [18] where the leaf nodes of the kd-tree would indicate our pose clusters. One can also choose (depending on computational expense) multiple samples from each leaf node to construct the large pose vocabulary $S = \{p_1, p_2, \ldots, p_R; p \in \Re^d\}$, where $d$ denotes the dimensionality of the pose descriptors.

### 2.3 What Does a Pose Tell Us? Modeling Sense Disambiguation of Human Pose

The pose codebook $S$ contains poses which are often ambiguous. We first outline the motivation behind the choice of our model and then explain how we model the visual senses and prune out the ambiguous poses.

**Motivation behind Sense Disambiguation of Pose:** Fig. 1 shows examples of some poses and the following relations illustrate what sense or action type the poses in Fig. 1 depict. Poses in relation (i) confuse between two senses (i.e., waking and running) but poses in relation (ii) and (iii) strongly express the associated sense. (i) $\{Fig.1(a)\}or\{Fig.1(c)\} \rightarrow \{running, walking\}$; (ii) $\{Fig.1(e)\} \rightarrow \{walking\}$; (iii) $\{Fig.1(g)\} \rightarrow \{running\}$.

We identify the key poses separately for each action type. For each action $A$ we construct a pose graph which captures the very basic essence - how poses in $S$ occur over the entire action span. Each action cycle $C$ in $A$ is defined as a pose sequence, i.e., $C = \{p_1, p_2, \ldots, p_T\}$, $C \in A$ and $p_i \in S$ $(i = 1, 2, \ldots, T)$. When we see a pose $p'$ that is strongly suggestive of some particular action $A$, we can expect that most of the action cycles $C$ in $A$ contain that pose; for example, Fig. 1(g) illustrates one such key pose which is difficult to miss in any action cycle of a running video. Please note such key pose $p'$ not necessarily has maximum number of occurrences in action video $A$. Rather such pose must occur at uniform interval, neither only in the beginning nor only at the end and nor intermittently in between. A pose $p'$ which has a strong sense associated (indicating that it represents action $A$) must have the highest cardinality of the following set $\lambda_{p'|A}$ (set $\lambda$ of $p'$ given action $A$) given by

$$\lambda_{p'|A} = \{C|C \in A \text{ and } p' \in C\}. \tag{2}$$

Poses which occur all of a sudden or are not prominent in each action cycle can be considered as deviations from the regular poses. Such irregular pose patterns are ambiguous and they need to be pruned out from $S$. They happen primarily because of tremendous variability of human poses and secondarily, though to a lesser extent, because of associated noise (for example, shadows in tower dataset, transmission noise in soccer dataset).

**Construction of Pose Graph G:** The pose graph for each action type $A$ contains poses from $S$ as vertices and an edge between two pose vertices explains the joint behavior of two poses - how well they "describe" the action together. By "describe" we of course mean how regularly the pose $u$ (or $v$) occurs in an action video $A$. It is quite reasonable that a pose $u$ that occurs regularly in $A$ has a high cardinality $\lambda_{u|A}$. This is because more regularly it occurs, the more action cycles it will belong to. Next we define the pose graph.

***Definition 1.*** *A pose graph for a given action type $A$ is an undirected edge-labeled graph $G = (S, E)$ where each vertex in $G$ corresponds to a pose belonging to the initial codebook $S$; $E$ is the set of edges and $\omega : E \rightarrow (0, 1]$ is the edge weight function. There is an undirected edge between the poses $u$ and $v$ ($u \neq v$ and $u, v \in S$), with edge weight $\omega(u, v)$, iff $0 < \omega(u, v) \leq 1$. It is assumed that $\omega$ is symmetric i.e., $\omega(u, v) = \omega(v, u)$, for all $u$, $v$ and $\omega(u, u) = 0$ for all $u$.*

$$\omega(u, v) = \begin{cases} \frac{1}{\eta(u,v)} & \text{when } \eta(u, v) \neq 0 \\ \infty & \text{otherwise,} \end{cases} \tag{3}$$

where $\eta(u, v) = |\lambda_{u|A} \bigcap \lambda_{v|A}|$.

In practical setting $\infty$ can be replaced by a large constant when $u$ and $v$ do not have an edge in between them. Construction of $G$ requires computation of $\eta(u, v)$ which again depends on $\lambda_{u|A}$ (2). For computing this set one has to map each video frame (actually the pose descriptor $p$ derived from this frame) to the most appropriate element in $S$ (either by nearest neighbor computation or a suitable Kd-tree traversal algorithm illustrated in Fig. 3(b)).

**Key Pose Selection by Pose Ranking:** Given the pose graph $G$ for a particular action the task next is to evaluate which one of the poses is most "important" in characterizing an action. An equivalent problem exists in social network analysis [1,12] (viz. search in matrimonial websites or business/social network websites) where importance of a node (may be a webpage or more specifically a person) in a network is identified using centrality measure of graph connectivity. Such ranking technique is used in web search (Google's PageRank algorithm [13]) and it has been recently used with success in feature ranking for object recognition [14] and feature mining task [8]. Inspired by their success we seek to rank poses and choose the $N$-best key poses for a particular kind of action. Grouping all such key poses together we build our discriminatory codebook $\xi$.

There are various centrality measures of graph connectivity to accomplish the ranking task in a pose graph [1]. Basically the choice of connectivity measure influences the selection process for the highly ranked poses. Given a graph connectivity measure "$e$" and the set of vertices $S$ and $u, v \in S$, we induce a ranking $rank_e$ of the vertices $u$ and $v$ such that $rank_e(u) \leq rank_e(v)$ iff $e(u) \geq e(v)$. Then for each action type $A$, we select the best ranking pose $v$ according to $e(v)$. To make $e$ explicit we adopt eccentricity as a measure of graph connectivity [12]. We choose eccentricity because it is simple and fast to compute. Like other centrality measures it relies on the following notion of centrality - a node is central if it is maximally connected to all other nodes. Centrality measure determines the degree of relevance of a single vertex $u$ in a graph $G$ and hence can be viewed as a measure of influence over the network.

***Definition 2.*** *Given a pose graph $G = (S, E)$, where the vertex set is represented by initial codebook $S$ and $E$ stands for edges, the distance $d(u, v)$ between two pose words $u$ and $v$ (where $u, v \in S$) is the sum of the edge weights on a shortest path from $u$ to $v$ in $G$. Eccentricity $e(u)$ of a vertex $u \in S$ is the maximum distance from $u$ to any other vertex $v \in S$, i.e., $e(u) = max\{d(u, v)|v \in S\}$.*

Ranking of poses is strictly based on eccentricity score and the implementation is straightforward. The Floyd-Warshall algorithm [19] computes all-pair-shortest path to evaluate the eccentricity $e(u)$ of each pose $u \in S$. For each action, we choose its $N$-best key poses by selecting poses with $N$-lowest eccentricity in a pose graph. Once we identify the key poses for a particular kind of action we repeat the same process for all kinds of actions. The key poses $p_1, p_2, \ldots, p_k$ extracted from all the action types are grouped together and serve as key pose codebook $\xi$

$$\xi = \{p_1, p_2, \ldots, p_k\} \;\; \forall \;\; p_i \in S, \;\; where \;\; i = 1, 2, \ldots, k. \tag{4}$$

### 2.4 Action Descriptor for Target Video Classification

Once we have a video of frames $I_1, I_2, \ldots, I_M$ and the key pose codebook $\xi$, the traditional bag-of-words implementation requires extracting pose descriptor $q_r$ for each frame $I_r$ and then mapping it to some key pose $p_i$, $\forall i = 1, 2, \ldots, k$, in a hard way and thereby building the codebook histogram $AD$ ($AD$ stands for action descriptor), where each bin $i$ of $AD$ keeps the occurrence count of key pose $p_i$, $i = 1, 2, \ldots, k$. The histogram $AD$ will be used for classification task by a suitable classifier. In [20], Gemert *et. al.* showed soft allocation in codebook increases the classification accuracy. We studied the four codebook models (as described in [20]) and documented their performance in the results section.

Fig. 5 shows the effect of the number of key poses on the overall accuracy for different types of codebook model - plausibility, uncertainty, kernel codebook and traditional codebook. Plausibility performs the best among all and we get the best overall accuracy when the average number of key poses selected for each action is, three. Number of key poses in Soccer is slightly lower than other dataset because in soccer, pose ambiguity is high and only a handful of key poses exist in each action class. We use support vector machines (libSVM [24]) for classification of histogram features $AD$ following "one versus all" framework for multi-class classification.

## 3 Experiments, Results and Discussions

**Video Dataset:** The choice of dataset is made keeping in mind the focus of our paper - recognizing action at a distance. Soccer [4], tower [21], hockey [22] datasets contain human performer far away from the camera and 40 pixels tall approximately. Only exception is KTH [23] dataset where we evaluated our proposed methodology on medium size ($\tilde{1}00$ pixels tall) human figure. All the action videos are clearly labeled and we used the labels as provided with the dataset. The soccer dataset contains several video sequences of digitized World Cup football game from an NTSC video tape [4]. We arranged this entire dataset into a total of 34 different video sequences of 8 different actions - run left angular, run left, walk left, walk in/out, run in/out, walk right, run right and run right angular. The Texas Austin (tower) dataset for human action recognition consists of 108 video sequences of nine different actions performed by six different people, each person showing every action twice. The nine actions are pointing, standing, digging, walking, carrying, running, wave 1(one hand), wave 2 (both hands), jumping. The bounding rectangles of the human performer (as well as foreground filter-masks which we did not require) are supplied with the dataset. The hockey dataset consists of 70 video tracks of hockey players with 8 different actions, e.g., skate down, skate left, skate leftdown, skate leftup, skate right, skate rightdown, skate rightup and skate up. The KTH dataset of human motion contains six different types of human actions (boxing, handclapping, hand waving, jogging, running, walking) performed by 25 different persons for 4 times each in the following environments - outdoor, outdoor with scale variation, outdoor with different cloths and indoor.

**Table 1.** Classification accuracy of proposed approach compared to state-of-the-art

| Activity | Overall accuracy (%) | | | |
|---|---|---|---|---|
| | Soccer data | Tower data | Hockey data | KTH data |
| S-LDA [6] | 77.81 | 93.52* | 87.50 | 91.20 |
| S-CTM [6] | 78.64 | 94.44* | 76.04 | 90.33 |
| Effros [4] | 49.23 | – | – | – |
| Niebles [7] | – | – | – | 81.50 |
| Performance over $\xi$ | 79.41 | 95.37 | 88.50 | 91.33 |
| Performance over $S$ | 58.83 | 72.22 | 68.75 | 72.22 |

* based on our implementation of [6]

### 3.1 Experimental Setup

We use support vector machines [24] with radial basis function for classification task following a "Leave-one-out" scheme. A 10-fold cross validation is performed on the training set to tune the parameters of the radial basis function. Holding one sequence out we build our codebooks and action descriptors from the training set and then try to classify the action descriptor of the held out video sequence. This is repeated for all datasets. Please note that used the datasets as available and no preprocessing step is involved. Our approach is efficient both in terms of consumed time and accuracy in detecting human actions. Though building the key pose vocabulary separately for each action takes longest time, but this is done once and we reap benefit later while classifying video with a small set of just 5 or 6 key poses per actions. The average time consumed by key pose dictionary construction amounts to little less than one minute. We assumed the cycle length $T$ constant at 10 because most of the video actions complete a full cycle within 10 frames. The classification task takes a few seconds on a machine with processor speed 2.37 GHz. 512 MB RAM.

### 3.2 Results and Discussions

We report the overall accuracy of our proposed method and compare them with the state-of-the-art (Table 1). One important achievement of the proposed method is that $\xi$ is built with 5/6 key poses per action. Our codebook is much smaller than the codebook used by [6]. The confusion matrices in Fig. 4 illustrate the class wise recognition rate for each action and it is apparent that our model confuses when two action types have a number of similar poses. In soccer dataset, variation in poses is quite high and it is often difficult (even for humans) to distinguish between two very similar actions (like *run-left* and *run-left-angular*). Consequently some mistakes are made by the proposed approach because of the ambiguous nature of poses. In the Tower dataset, the video quality is relatively better than soccer but shadows are very prominent in the images. We did not remove the shadows; instead we allowed them to provide extra cues about the action type. In hockey video, the major confusion is between left up and left down (or right up and right down). The poses are quite similar in these two actions. In KTH most of the confusions occurred in classifying running and jogging because of their almost similar patterns of poses.

The graphs in Fig. 5 reveal an important fact - with more poses per action recognition accuracy drops down. This is expected as liberal choice of poses from $S$ adds ambiguity to the key pose vocabulary $\xi$ and confuses the recognition task. We build $\xi$ with (on an average) 3, 4, 3, 5 key poses selected from $S$ for soccer, tower, hockey and KTH dataset. Our design decision to use $\xi$ as a refined codebook (built from $S$) is substantiated by the recognition accuracies with $S$ as our codebook (Table 1). The performance - when $S$ is used in place of $\xi$ for deriving action descriptor $AD$ followed by classification - is much worse suggesting that pruning out ambiguous poses does pay off.

| | run left angular | run left | walk left | walk in/out | run in/out | walk right | run right | run right angular |
|---|---|---|---|---|---|---|---|---|
| run left angular | 60 | 20 | 20 | 0 | 0 | 0 | 0 | 0 |
| run left | 20 | 80 | 0 | 0 | 0 | 0 | 0 | 0 |
| walk left | 0 | 0 | 100 | 0 | 0 | 0 | 0 | 0 |
| walk in/out | 0 | 0 | 0 | 80 | 20 | 0 | 0 | 0 |
| run in/out | 0 | 0 | 0 | 20 | 80 | 0 | 0 | 0 |
| walk right | 0 | 0 | 0 | 0 | 0 | 100 | 0 | 0 |
| run right | 0 | 0 | 0 | 0 | 0 | 0 | 67 | 33 |
| run right angular | 0 | 0 | 0 | 0 | 0 | 0 | 33 | 67 |

(a)

| | pointing | standing | digging | walking | carrying | running | wave1 | wave2 | jogging |
|---|---|---|---|---|---|---|---|---|---|
| pointing | 100 | 0 | 0 | 0 | 0 | 0 | 0 | 0 | 0 |
| standing | 0 | 83 | 0 | 0 | 0 | 0 | 0 | 0 | 17 |
| digging | 0 | 0 | 92 | 0 | 0 | 08 | 0 | 0 | 0 |
| walking | 0 | 0 | 0 | 100 | 0 | 0 | 0 | 0 | 0 |
| carrying | 0 | 0 | 0 | 0 | 100 | 0 | 0 | 0 | 0 |
| running | 0 | 0 | 0 | 0 | 0 | 100 | 0 | 0 | 0 |
| wave1 | 0 | 0 | 0 | 0 | 0 | 0 | 100 | 0 | 0 |
| wave2 | 08 | 0 | 0 | 0 | 0 | 0 | 0 | 92 | 0 |
| jogging | 0 | 08 | 0 | 0 | 0 | 0 | 0 | 0 | 92 |

(b)

| | skate down | skate left | skate left down | skate left up | skate right | skate right down | skate right up | skate up |
|---|---|---|---|---|---|---|---|---|
| skate down | 100 | 0 | 0 | 0 | 0 | 0 | 0 | 0 |
| skate left | 0 | 100 | 0 | 0 | 0 | 0 | 0 | 0 |
| skate left down | 0 | 0 | 83 | 17 | 0 | 0 | 0 | 0 |
| skate left up | 0 | 0 | 17 | 83 | 0 | 0 | 0 | 0 |
| skate right | 0 | 0 | 0 | 0 | 100 | 0 | 0 | 0 |
| skate right down | 17 | 0 | 0 | 0 | 0 | 50 | 33 | 0 |
| skate right up | 0 | 0 | 0 | 0 | 0 | 17 | 83 | 0 |
| skate up | 0 | 0 | 0 | 0 | 0 | 0 | 0 | 100 |

(c)

| | boxing | hand clapping | hand waving | jogging | running | walking |
|---|---|---|---|---|---|---|
| boxing | 96 | 0 | 0 | 3 | 1 | 0 |
| hand clapping | 0 | 97 | 3 | 0 | 0 | 0 |
| hand waving | 0 | 0 | 100 | 0 | 0 | 0 |
| jogging | 1 | 0 | 0 | 83 | 13 | 3 |
| running | 4 | 0 | 0 | 18 | 72 | 6 |
| walking | 0 | 0 | 0 | 0 | 0 | 100 |

(d)

**Fig. 4.** Accuracy plot with average number of key poses per action for (a) Soccer dataset, (b) Tower dataset, (c) Hockey dataset and (d) KTH dataset

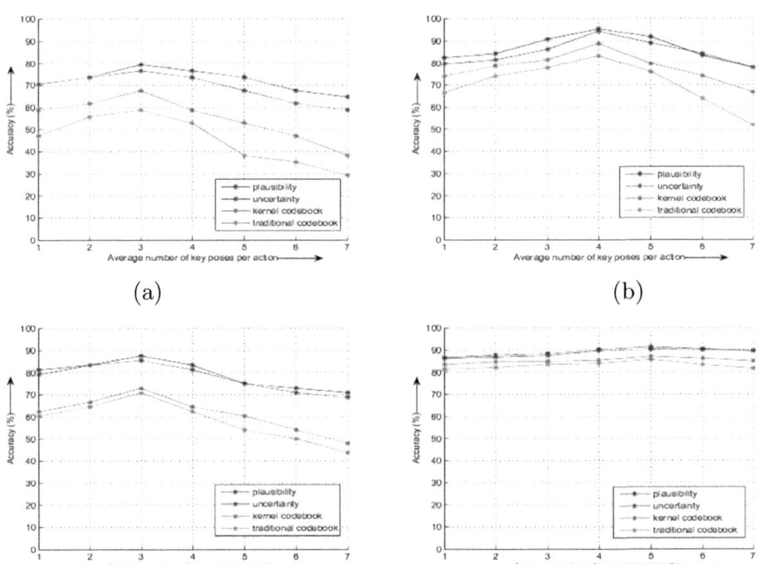


(a) (b) (c) (d)

**Fig. 5.** Accuracy plot with average number of key poses per action for (a) Soccer dataset, (b) Tower dataset, (c) Hockey dataset and (d) KTH dataset

Tables 2 and 3 show some key poses retrieved by our algorithm along with their centrality measures. Each of the key poses clearly indicates the intended sense and the related human action becomes immediately apparent.

**Table 2.** Key poses of Soccer dataset (4 actions) with corresponding eccentricity value

| Activity | rla | rl | wl | rio |
|---|---|---|---|---|
| Key poses (top three key poses from each activity) with corresponding eccentricity value | 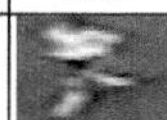 |  |  | 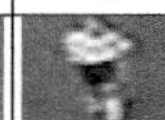 |
| | 1.89 | 2.0 | 1.73 | 1.42 |
| |  | 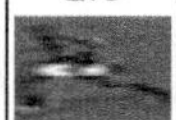 |  |  |
| | 2.10 | 2.5 | 1.98 | 1.49 |
| |  |  |  |  |
| | 2.33 | 2.5 | 2.15 | 1.67 |

**Table 3.** Key poses of Tower dataset (4 actions) with corresponding eccentricity value

| Activity | C | D | W2 | J |
|---|---|---|---|---|
| Key poses (top three key poses from each activity) with corresponding eccentricity value |  |  |  |  |
| | 1.27 | 1.47 | 1.5 | 1.3 |
| |  |  |  |  |
| | 1.29 | 1.59 | 1.52 | 1.45 |
| |  |  |  |  |
| | 1.29 | 1.72 | 1.6 | 1.49 |

## 4 Conclusions and Future Scope

This paper studies the problem of key pose retrieval for definite action patterns by modeling the visual senses expressed by different human poses. From a large vocabulary of poses the proposed methodology prunes out ambiguous poses and builds a small but highly discriminatory codebook of key poses. In selecting the key poses we made a ranking of poses for a given action type using centrality theory and choose $N$-best poses among them. The reported accuracy with our small codebook size is slightly superior to the state of the art. It is demonstrated that identifying key poses can provide vital clue about the kind of human activity.

## References


1. Navigli, R., Lapata, M.: An experimental study of graph connectivity for unsupervised word sense disambiguation. IEEE Trans. on PAMI 32(4), 678–692 (2010)
2. Dollar, P., Rabaud, V., Cotrell, G., Belongie, S.: Behavior Recognition via Sparse Spatio-Temporal Features. In: IEEE Int. Workshop on VS-PETS, pp. 65–72 (2005)
3. Laptev, I., Lindeberg, T.: Space-time Interest Points. In: 9th ICCV, vol. 1, pp. 432–439 (2003)
4. Efros, A.A., Berg, A.C., Mori, G., Malik, J.: Recognizing Action at a Distance. In: 9th ICCV, vol. 2, pp. 726–733 (2003)
5. Ikizler, N., Duygulu, P.: Histogram of Oriented Rectangles: A New Pose Descriptor for Human Action Recognition. Image and Vision Computing 27, 1515–1526 (2009)
6. Wang, Y., Mori, G.: Human Action Recognition by Semi-Latent Topic Models. IEEE Trans. on PAMI 31(10), 1762–1774 (2009)
7. Niebles, J.C., Wang, H., Li, F.-F.: Unsupervised Learning of Human Action Categories Using Spatial-Temporal Words. IJCV 79(3), 299–318 (2008)
8. Liu, J., Luo, J., Shah, M.: Recognizing Realistic Actions from Videos "in the Wild". In: CVPR (2009)
9. Niebles, J., Le, F.F.: A hierarchical model of shape and appearance for human action classification. In: CVPR (2007)
10. Bissacco, A., Yang, M.H., Soatto, S.: Detecting humans with their pose. In: NIPS (2007)
11. Fengjun, L., Nevatia, R.: Single View Human Action Recognition using Key Pose Matching and Viterbi Path Seraching. In: CVPR (2007)
12. Wasserman, S., Faust, K.: Social Network Analysis: Methods and Applications. Cambridge University Press, Cambridge (1994)
13. Brin, S., Page, L.: The anatomy of a large-scale hyper-textual web search engine. Computer Networks and ISDN Systems 30(1-7), 107–117 (1998)
14. Kim, G., Faloutsos, C., Hebert, M.: Unsupervised modeling of object categories using link analysis technique. In: CVPR (2008)
15. Lucas, B.D., Kanade, T.: An Iterative Image Registration Technique with an Application to Stereo Vision. In: 7th IJCAI, pp. 674–679 (1981)
16. Bishop, C.M.: Pattern Recognition and Machine Learning. Springer, Heidelberg (2006)
17. Pelleg, D., Moore, A.W.: X-means: Extending K-means with efficient Estimation of the Number of Clusters. In: ICML (2000)
18. Narayan, B.L., Murthy, C.A., Pal, S.K.: Maxdiff kd-trees for Data Condensation. PRL 27(3), 187–200 (2005)
19. Cormen, T.H., Leiserson, C.E., Rivest, R.L., Stein, C.: Introduction to Algorithms. MIT Press, Cambridge (2003)
20. Gemert, J.C.V., Veenman, C.J., Smeulders, A.W.M., Geusebroek, J.M.: Visual Word Ambiguity. IEEE Trans. on PAMI 32(7), 1271–1283 (2010)
21. Chen, C.C., Ryoo, M.S., Aggarwal, J.K.: UT-Tower Dataset: Aerial View Activity Classification Challenge (2010), `http://cvrc.ece.utexas.edu/SDHA2010/Aerial_View_Activity.html`
22. Lu, W.L., Okuma, K., Little, J.J.: Tracking and Recognizing Actions of Multiple Hockey Players Using the Boosted Particle Filter. Image and Vision Computing 27(1-2), 189–205 (2009)
23. Schuldt, C., Laptev, I., Caputo, B.: Recognizing Human Actions: A Local SVM Approach. In: 17th ICPR, pp. 32–36 (2004)
24. `http://www.csie.ntu.edu.tw/~cjlin/libsvm/` (June 2010)

# Social Interactive Human Video Synthesis

Dumebi Okwechime, Eng-Jon Ong, Andrew Gilbert, and Richard Bowden

CVSSP, University of Surrey, Guildford, Surrey, GU17XH, UK
{d.okwechime,e.ong,a.gilbert,r.bowden}@surrey.ac.uk

**Abstract.** In this paper, we propose a computational model for social interaction between three people in a conversation, and demonstrate results using human video motion synthesis. We utilised semi-supervised computer vision techniques to label social signals between the people, like *laughing*, *head nod* and *gaze direction*. Data mining is used to deduce frequently occurring patterns of social signals between a speaker and a listener in both *interested* and *not interested* social scenarios, and the mined *confidence* values are used as conditional probabilities to animate social responses. The human video motion synthesis is done using an appearance model to learn a multivariate probability distribution, combined with a transition matrix to derive the likelihood of motion given a pose configuration. Our system uses social labels to more accurately define motion transitions and build a texture motion graph. Traditional motion synthesis algorithms are best suited to large human movements like walking and running, where motion variations are large and prominent. Our method focuses on generating more subtle human movement like head nods. The user can then control who speaks and the interest level of the individual listeners resulting in social interactive conversational agents.

## 1 Introduction

Human motion synthesis has extensive applications in the movie and gaming industry. The ability to control the movement of a character in a video scene can provide an attractive alternative to post-production filming, providing movie editors with the means to edit an actor's performance without having to re-record the scene, which can be very expensive and time consuming. Although CGI (Computer Generated Imagery) is the vastly popular medium for computer game characters, photo-realistic human animation has proven to heighten realism in the gaming experience especially in combat platforms. Though there has been extensive research in the field for motion capture and human video texture synthesis, little work has been done in developing video based socially interactive avatars, capable of responding appropriately to non-verbal communication. This issue is addressed in this paper.

Our aim is to develop a human video *Motion Model*, specifically tailored for synthesising social interactive behaviour. This is done by combining a Probability Density Function (PDF) with a Markov Transition Matrix to derive the likelihood of pose and the probability of transitions respectfully. To increase reliability of the motion model, we introduce *Texture Motion Graph*, akin to Motion

R. Kimmel, R. Klette, and A. Sugimoto (Eds.): ACCV 2010, Part I, LNCS 6492, pp. 256–270, 2011.

Graphs by Kovar et al [1]. We extend the approach to allow multiple identical subgraphs to increase connectivity.

We propose a novel approach to derive social dynamics using data mining [2] to efficiently identify social trends between a group of three people in a conversation. The *confidence* values extracted from the mining are used as weighted conditional probabilities to generate appropriate non-verbal responses, resulting in a fully automated social interactive system. The user can select who speaks and subsequently who listens, and control the listeners level of interest in the conversation, effectively changing their social dynamics. Although the control of social dynamics is demonstrated on video synthesis, the elements of the computational model of interaction (i.e. *gaze*, *nod*, *laugh* etc) can be readily extended to other synthesis approaches such as *motion captured skeleton* animations.

The paper is divided into the following sections. Section 2 briefly details a background in the field of motion texture synthesis. Section 3 presents an overview of the entire system. Section 4 and 5 describes the approach of generating the motion model and deriving trends in the social dynamics respectfully. Section 6 presents the social interaction motion control, and the remainder of the paper describes the results and conclusion.

## 2 Background

Synthesis has extensive applications in graphics and computer vision, and can be categorised into three groups: textures synthesis of discrete images, temporal texture synthesis in videos, and motion synthesis in motion captured data. Early approaches to texture synthesis were based on parametric [3] and non-parametric [4] methods, which create novel textures from example inputs. Kwatra et al [5] generate perceptually similar patterns from a small training data set, using a graph cut technique based on Markov Random Fields (MRF). Approaches to static texture synthesis paved the way for temporal texture synthesis methods, often used in the movie and gaming industries for animating photo-realistic characters and editing video scenery. An example is presented by Bhat et al [6] who used texture particles to capture dynamics and texture variation travelling along user defined flow lines. This was used to edit dynamic textures in video scenery.

A number of researchers have used statistical models to learn generalised motion characteristics for synthesis of novel motion [7][8][9]. Unfortunately all these systems use a generalisation of the motion rather than the original data, and cannot guarantee that the synthesised motion looks natural.

Motion synthesis using example-based methods, i.e. retaining the original motion data to use in synthesis, provides an attractive alternative as there is no loss of detail from the original data [10][11][12][13]. Representing motion transitions using a *motion graph* [14][15][16][17], originally introduced by Kovar et al [1], provides additional user-control on positioning, using both pieces of original data and automatically generated transitions to perform an optimal graph walk that satisfies user-defined constraints. Treuille et al [13] developed a system that synthesizes kinematic controllers which blend subsequences of precaptured motion clips to achieve a desired animation in real-time.

In some cases, techniques used for motion synthesis of motion captured data, are similar to the techniques used for temporal texture synthesis of videos. By substituting pixel intensities (or other texture features) with marker co-ordinates, and applying motion constraints suited to the desired output, a similar framework can be extended to both domains. Schödl et al [18] introduced *Video Textures* which computes the distances between frames to derive appropriate transition points to generate a continuous stream of video images from a small amount of training video. Similarly, Flagg et al [19] presented *Human Video Textures*, where, given a video of a martial artist performing various actions, they produce a photo-realistic avatar which can be controlled, akin to a combat game character. In these cases, human texture synthesis is performed on periodic data, or constrained to guarantee the actor returns to a neutral pose.

Research in social interaction can be grouped into two main categories: emotion based on cognitive psychology [20], and linguistics based on dialogue understanding [21]. Though emotion understanding using high level deduction of social behaviour, like tone of voice and facial expressions, is of vital importance in how people socially interact, emotion recognition in a natural conversation, especially amongst adults, is a very complex problem and would require extensive data and research in deducing social trends. Also, structured dialogue can not be easily interpreted to observe generalised unconscious and non-verbal social behaviour.

## 3 Overview

Our proposed system consists of two stages: **Human Video Texture Synthesis**, and a **Social Interaction Model**.

*Video Texture Synthesis* allows a user to reproduce motion in a novel way by specifying which type of motion inherent in the original sequence to perform. Given a data set of a full body video sequence, following dimensional reduction via PCA, an unsupervised segmentation derives cut point clusters, where each cluster represents groups of similar frames that can seamlessly blend together. A texture motion graph is built to guarantee connectivity. Finally, a dynamic model is learnt, combining kernel density estimation with a markov transition matrix to derive the likelihood of transitioning from one cut point to another to generate novel motion sequences.

Developing a *Social Interaction Model* starts with a social behaviour experiment, whereby scenarios when the listener is *interested* and *not interested* in the topic of conversation can be determined. Using various action recognition techniques, social signals of the video are labelled such as *head nods*, *laughing* etc. Data mining is used to derive trends between the listener and the speakers given these social signals, producing conditional probabilities of a listener's social behaviour given a speaker's social signal. These conditional probabilities are used as weights to drive the motion model, given the user control over the social dynamics.

# 4 Human Video Texture Synthesis

## 4.1 Video Data

The data set consists of approximately 30 minutes of video and audio recording of the full-body frontal view (516×340, 25 frames per second, 48kHz) and the close-up frontal face view (720×576, 25 frames per second, 48kHz) of 3 individuals having a conversation with each other. Each person remained in a stationary position relative to the camera as shown in Figure 1, although they were not constrained to do so, so there exists considerable ambient motion in their pose. Only the full-body video sequence is used for synthesis. The close-up fontal face video was only recorded to assist with the semi-supervised social signal labelling, which will be discussed in more detail later.

Each full-body video consists of approximately 43000 frames. To reduce computation complexity, the videos were reduced to grayscale and resized to a quarter of their original size. Given a video sequence $\mathbf{X}$, each frame is represented as a vector $\mathbf{x}_i$ where $\mathbf{X} = \{\mathbf{x}_1, ..., \mathbf{x}_{N_T}\}$ and $N_T$ is the number of frames and $\mathbf{x}_i = (x_{i1}, y_{i1}, ..., x_{ix}, y_{iy}) \in \Re^{xy}$ for an x × y image.

To further reduce the complexity, Principal Component Analysis (PCA) is used for dimensionality reduction. The dimension of the feature space $|x_i|$ is reduced by projecting into the eigenspace $d$, where $d$ is the chosen lower dimension $d \leq |x_i|$ such that $\sum_{i=1}^{d} \frac{\lambda_i}{\Sigma \forall \lambda} \geq .98$ or 98% of the energy is retained. $\mathbf{Y}$ is defined as a set of all points in the dimensionally reduced data where $\mathbf{Y} = \{\mathbf{y}_1, ..., \mathbf{y}_{N_T}\}$ and $\mathbf{y}_i \in \Re^d$.

## 4.2 Identifying Cut Points

Similar to most work on motion synthesis, the motion data needs to be analysed to compute some measure of similarity between frames and derive points of intersection within the data. These points will allow transitions, either by 'motion blending' or 'switching', between different subsequences, producing motion paths not inherent in the original data.

**Fig. 1.** Image showing full-body view of recorded video data of three people having a conversation

The common approach is to compute the L2 distance over a window of frames in time and use a user defined threshold to balance between the quality of the transitions and the number of candidate transition points [1][19][18][14][15]. This approach works well, however, for very large data sets, it can be tedious to compute the distance between every frame. In our case, we would have to compute 3 separate $43000 \times 43000$ similarity matrices (one for each person), which would be time consuming and computationally expensive. Balci et al [16] proposed an iterative clustering procedure based on k-means to define clusters of poses suitable for transitions. However, k-means produces cluster centres not embedded in the data which can result in noise and outliers.

Instead, we adopt a k-medoid cluster algorithm to define $N_C$ k-medoid points, where $N_C < N_T$. Each k-medoid point is defined as the local median in regions of high density, and can be used to define regions where appropriate transitions are possible. By only computing the L2 distance at these points, we reduce the amount of computation required to define candidate transitions, focusing attention on regions where transitions are most likely.

We define each k-medoid point as $\boldsymbol{\delta}_k \in \mathbf{Y}$. To preserve dynamics and account for temporal shape similarity, we compute a linearly weighted average of similarity over a fixed window of 0.25 seconds, centred on the k-medoids. Using a user defined threshold $\theta$, the nearest points to each k-medoid points are identified to form clusters of cut points. The set containing the cut points of the $n^{th}$ cluster is defined as $\mathbf{Y}_n^c = \{\mathbf{y}_{n,1}^c, ..., \mathbf{y}_{n,Q_n}^c\}$, where the number of cut points of the $n^{th}$ cluster is denoted as $Q_n$. $N_C$ is empirically determined based on the number of candidate cut points versus the quality of transitions and the amount of computation. In this work $N_C = 165$, greatly reducing the number L2 distance calculations.

Formally, each cut point belongs to a single cluster where plausible transitions can only be made between group members. This approach works well in data sets with high connectivity, however, less so for human video data where connectivity is limited. To overcome this problem, we extended the approach to allow cut points to belong to more than one cluster, providing the cut point clusters with more opportunities to perform novel movement.

For simplicity we define the transitions from the $n^{th}$ cluster contents in the form $\{(\mathbf{y}_{n,1}^c, \mathbf{z}_{n,1}^c, \iota_{n,1}, C_{n,1}^z), ..., (\mathbf{y}_{n,Q_n}^c, \mathbf{z}_{n,Q_n}^c, \iota_{n,Q_n}, C_{n,Q_n}^z)\}$, where $\mathbf{y}_n^c$ is a cut point in the $n^{th}$ cluster acting as the start transition point, $\mathbf{z}_n^c$ is the end transition point denoting the end of the subsequence between $\mathbf{y}_n^c$ and $\mathbf{z}_n^c$ (where $\mathbf{z}_n^c \in \mathbf{Y}^c$ and $\mathbf{z}_n^c \neq \mathbf{y}_n^c$), $\iota_n$ is the frame number of $\mathbf{y}_n^c$ in the original data, and $C_n^z$ is the index of the cluster $\mathbf{z}_n^c$ belongs to.

### 4.3 Texture Motion Graph

So far, the cut points can be used to transition between different subsequences available in the data, however, there is no consideration to whether the transitions can perform all the available motion types or whether it leads to a dead-end. As a result, we pre-compute a *Texture Motion Graph* to guarantee global connectivity to different types of movements in the video data set.

Motion graph, proposed by Kovar et al [1], essentially connects various subsequences together to a form a directed graph, whereby the edges are the generated cut points. By assembling the graph, we can identify and eliminate cut points with low connectivity, improving reliability in the sequence selection process. Various forms of motion graph have been proposed in recent years [16][15][14][22], built for animating motion captured data. *Texture Motion Graph* is specifically tailored for video textured data and extended to overcome ambiguities in transitioning between video frames.

Generating smooth blends between human video textures is a very challenging topic, since as human beings, we can easily recognise unnatural human movement or textures. Our similarity measure performs well with distinguishing different body poses, however does not account for facial gestures like laughing, talking and subtle changes in gaze direction. To overcome this, we use the social signal labels to assist the similarity measure. We prune the clusters of cut points to have the same gaze, talking, and laughing labels, as those of its k-medoid cluster centre, discarding cut points which do not match. This reduces the occurrences of rapid and unnatural changes in facial expressions out of context to the social interaction.

We define $n$ strongly connected subgraphs for each unique set of social signals. This structure efficiently populates the graph with various links to social behaviour, making them easily accessible from any subgraph pose configuration. Social signals, such as head nods and head shakes, can occur in short quick bursts, lasting only a few seconds. By making available varying occurrences of a set of labels, we increase the opportunity of transitioning to a social behaviour quickly, and easily, making the graph more responsive. The Tarjan algorithm is used to derive the strongly connected subgraphs, and in our experiments we found $n = 3$ sufficient to populate our graph with varying social behaviour.

### 4.4 Dynamic Model

Conventional motion graph synthesis traverses the graph, connecting motion segments based on user specified constraints such as position, orientation and timing [1][15]. Little interest is given to how common or likely the connecting nodes are given the data set. In smaller data sets, the user has limited choices of nodes to traverse, hence the quality of the chosen node of traversal is of little importance. However, in a densely populated data set, better quality transitions can be produced by computing the likelihood of a pose or frame as an additional parameterised weight. Hence, a dynamic model is learnt to derive the likelihood of pose in eigenspace based on the data set of video frames.

A statistical model of the constraints and dynamics present within the data can be created using a Probability Density Function (PDF). An appearance PDF is created using kernel estimation where each kernel $p(\mathbf{y}_i)$ is a Gaussian centred on a data example $p(\mathbf{y}_i) = G(\mathbf{y}_i, \Sigma)$. The likelihood of a posture or pose in eigenspace is modelled as a mixture of Gaussians using a multivariate normal distribution.

$$P(\mathbf{y}) = \frac{1}{N_T}\sum_{i=1}^{N_T} p(\mathbf{y}_i) \tag{1}$$

where the covariance of the Gaussian is:

$$\Sigma = \alpha \begin{pmatrix} \sqrt{\lambda_1} & \cdots & 0 \\ \vdots & \ddots & \vdots \\ 0 & \cdots & \sqrt{\lambda_d} \end{pmatrix} \tag{2}$$

The width of the Gaussian in the $i^{th}$ dimension is set to $\alpha\sqrt{\lambda_i}$. For all experiments $\alpha = 0.25$.

To reduce estimation time without sacrificing accuracy, a kd-tree is used to localise queries to neighbouring kernels, assuming the kernel estimation outside a local region contributes nominally to the local density estimation. Equation 1 is simplified to:

$$P'(\mathbf{y}) = \frac{1}{|\mathbf{Y}'|}\sum_{\forall \mathbf{y}_i \in \mathbf{Y}'} p(\mathbf{y}_i) \tag{3}$$

where $\mathbf{Y}' \subseteq \mathbf{Y}$, and $\mathbf{Y}'$ is a set containing the nearest neighbour kernels to $\mathbf{y}_i$ found efficiently with the kd-tree.

By learning a PDF, the data is represented in a generalised form which is analogous to a generative model. Using this form on its own, it is possible to generate novel motion frames, using pre-computed motion derivatives for a global approximation, combined with a gradient decent for optimisation. However, such a model runs the risk of smoothing out subtle motion details, and is not suitable for video. Instead, we combine the PDF with a Markov Transition Matrix to determine the likelihood of transitioning between cut points. This allows motion generation based on the original data, retaining important motion information.

Since transitions from one state to the next are dependent only on the current state, we define the conditional probability of moving from one cluster to another as $P(C_t|C_{t-1}) = p_{C_{t-1},C_t}$ where $C_t$ is defined as the index for a cluster at time $t$.

### 4.5 Motion Synthesis

To generate novel motion sequences, the procedure is:

1. Given the current position in eigenspace $\mathbf{y}_t^c$, find all adjacent cut point neighbours in $\mathbf{Y}_t^c$ as defined in Section 4.2, to represent start transition points.
2. Find all associated end transition points $\mathbf{z}_{t,m}^c | m = \{1, ..., Q_t\}$. This gives a set of $Q_t$ possible transitions from the starting point $\mathbf{y}_t^c$ in eigenspace.
3. Denote the cut point group index that $\mathbf{y}_t^c$ belongs to as $C_t$.
4. Calculate the likelihood of each transition as:

$$\phi_m = P(C_{C_t,m}^z | C_t) P'(\mathbf{z}_{C_t,m}^c) \tag{4}$$

   where $\boldsymbol{\Phi} = \{\phi_1, .., \phi_{Q_t}\}$.

5. Normalise the likelihoods such that $\sum_{i=1}^{Q_t} \phi_i = 1$.
6. Since a maximum likelihood approach will result in repetitive animations, we randomly select a new start transition point $\mathbf{y}_{t,k}^c$ from $\boldsymbol{\Phi}$ based upon its likelihood as:

$$\arg\min_k \left( \sum_{j=1}^{k} \phi_j \geq r \right) \tag{5}$$

where $k$ is the index of the newly chosen end transition point, $k \in m$, and r is a random number between 0 and 1, $r \in [(0, 1)]$.
7. All frames associated to the transition sequence between $\mathbf{y}_{t,k}^c$ and $\mathbf{z}_{t,k}^c$ are rendered.
8. The process then repeats from step (1) where $\mathbf{y}_{t+1}^c = \mathbf{z}_{t,k}^c$.

# 5 Social Interaction Model

## 5.1 Social Behaviour Experiment

Our data set[1] consists of recordings of 3 people in a conversation. We refer to the three individuals as person A, B, and C. Prior to capture, each person was given a questionnaire and asked to score from $1 - 3$ their general interest on a given set of book genres, film genres, and music genres. They were also given specific questions like: favourite sports, language(s) they spoke fluently, favourite music concerts, favourite theatre show etc. Their questionnaires were analysed to define 4 generic scenarios:

1. All interested in topic
2. Two people interested in topic, one person is not
3. One person interested in topic, two people are not
4. None are interested in topic

These 4 generic scenarios where derived from 8 topics of conversation as detailed in Table 1. The sixth column of Table 1 shows the limited duration of each topic, chosen to suite the scenario. A projector displayed the topic of conversation for discussion, and a quiet bell would ring to make the subjects aware of the change in topic. The subjects were unaware of the nature of the experiment, and were simple asked to discuss the topic displayed on the screen.

The aim of this experiment was to observe the social dynamics between the three people in scenarios when *interested* or *not interested* in the topics. To achieve this, we need to quantise their social behaviour in some form in order to obtain a clear distinction in social behaviour.

## 5.2 Semi-supervised Social Signal Labelling

Pentland [23] proposed measuring non-linguistic social signals using four main observations: *activity level*, *engagement*, *emphasis* and *mirroring*. Using this as

[1] The data set along with annotation can be made available upon request. Please email {d.okwechime@surrey.ac.uk}.

**Table 1.** Table showing 8 different social scenarios dictated by the topic of conversation. The three people are referred to as person **A**, **B**, **C**. The numbers indicate their interest in the topics where 3 is a high interest and 1 is a low interest.

| Scenario | A | B | C | Topic | Duration |
|---|---|---|---|---|---|
| 1 | 3 | 3 | 3 | Classical Music | 5 minutes |
| 2 | 3 | 3 | 1 | Adventure Novels | 5 minutes |
| 3 | 3 | 1 | 3 | Philosophy Novels | 5 minutes |
| 4 | 1 | 3 | 3 | Rock Music | 5 minutes |
| 5 | 3 | 1 | 1 | Sailing (Spoken in French) | 2.5 minutes |
| 6 | 1 | 3 | 1 | Triathlon/Les Miserables (Spoken in Afrikaans) | 2.5 minutes |
| 7 | 1 | 1 | 3 | Radio Head Concert | 2.5 minutes |
| 8 | 1 | 1 | 1 | Horror Novels | 1.5 minute |

our base, we chose to observe 7 social signals in the conversation: *Voiced*, *Talking*, *Laughing*, *Head Shake*, *Head Nod*, *Activity Measure*, and *Gaze Direction*.

We use a variety of techniques to derive each label.

1. **Voiced[V]:** Audio stream represented using 12 MFCCs (Mel-Frequency Cepstral Coefficients) and a single energy feature of the standard HTK setup [24]. For each person, a few voiced segments were labelled and a Mahalanobis distance measure was used to derive a correlation between the voiced and non-voiced regions.
2. **Talking[T]:** With the voiced segments labelled, it was a simple process of labelling the voiced segments which were talking. This was done by hand.
3. **Laughing[L]:** The Viola-Jones face detector [25] was used to segment the face region in each frame. The lip region was localised by cropping the lower-centre region of the face. An AdaBoost classifier was then trained for laughing and used to label the remaining data.
4. **Head Shake[S]:** The Viola-Jones face detector was used to determine the movement of the face. Fast Fourier transform (FFT) was used to define high frequency movement along the x-axis
5. **Nod[N]:** Similar to head shakes, FFT was used to define high frequency movement along the y-axis.
6. **Activity Measure[A]:** The torso region of the full body video was segmented using colour and the mean-scaled standard deviation of velocity was measured. The leg and head regions are ignored because, there was no leg movement (subjects are stationary), and since we are more interested in gesture activity, changes in head posture/gaze would bias the activity measure.
7. **Gaze Direction[G]:** The eye pupils and the corners of the eyes were tracked using a Linear Predictor tracker [26]. The corners of the eyes were normalised to 0 and 1, and the position of the eye pupil within this region was used to determine if the person was gazing left [GL], right [GR] or centre [GC].

This produces $N_T$ sets of social signal labels of 27 dimensions, where $1-9$ is for person A, $10-18$ for person B and $19-27$ for person C. We define 2 complete sets of social signal vectors for *interested* and *not interested* scenarios as $\mathcal{F}_{(Int)}$, and $\mathcal{F}_{(NoInt)}$ such that $\mathcal{F} = \{\mathbf{f}_i\}_{i=1}^{N_T}$ where $\mathbf{f}_i$ is a 27 dimensional binary vector.

## 5.3 Data Mining for Social Trends

This framework is driven by the speaker. At any given time, there is only one speaker and two listeners. We are interested in the combination of social signals a listener performs given a speaker's social behaviour when the listener is *interested* and *not interested* in the conversation. Manually observing all combinations of listener and speaker behaviours in such a large data set would be virtually impossible. A solution would be to make some common sense prior assumptions of expected trends (i.e. an *interested* listener would gaze more at the speaker than when they are *not interested*) and focus primarily on these assumptions. However, there is no way of proving or disproving such assumptions, and, there is a large list to chose from.

We propose a novel approach to deriving social dynamics and trends between the subjects based on data mining [2]. Data mining allows for large data sets to be mined to identify the reoccurring patterns within the data in an efficient manner. In this framework, Apriori Association rule [2][27] mining is used. Formally developed for supermarkets to analyse millions of customer's shopping trends, we aim to find *association rules* within the numerous combinations of social trends between the subjects in an *interested* and *not interested* scenario given the speaker's social behaviour.

An association rule is a relationship of the form $\{R_i^A\} \Rightarrow R_i^C$ where $R_i^A$ is a set of social signals of the speaker, and $R_i^C$ a sets of social signals of the listener. $R_i^A = \{r_{i,1}^A, ..., r_{i,|R_i^A|}^A\}$ is the antecedent where $r_i^A$ denotes a speaker's social signal, and $R_i^C = \{r_{i,1}^C, ..., r_{i,|R_i^C|}^C\}$ the consequence where $r_i^C$ is a listener's social signal. An example would be, if $R_1^A = \{[T], [N]\}$, and $R_1^C = \{[N]\}$ as defined in Section 5.2, then, $\{R_1^A\} \Rightarrow R_1^C$ would imply 'when person A is talking and nods, person B is very likely to also nod'. The belief of each rule is measured by a *support* and *confidence* value. The *support* measures the statistical significance of a rule, it is the probability that a transaction contains itemset $R_i^A$.

$$sup(\{R_i^A\} \Rightarrow R_i^C) = sup(\{R_i^A\} \cup R_i^C) \tag{6}$$

The *confidence* is the number of occurrences in which the rule is correct, relative to the number of cases in which it is applicable.

$$conf = \frac{sup(\{R_i^A\} \cup R_i^C)}{sup(R_i^A)} * 100 \tag{7}$$

Apriori Association mining is applied to the social signal labels for both *interested* listener and *not interested* listener scenarios, to derive frequently occurring association rules. We define the set of all rules extracted using data mining as:

$$R = \{(R_i^A \Rightarrow R_i^C, conf_i)\}_{i=1}^{|R|} \tag{8}$$

where the total number of rules is $|R|$.

Traditionally, data mining looks for a combination of symbols that occur simultaneously, however, a listener's social behaviour is always a response to the speaker's social signals, hence, co-articuation is not possible. To account for this, *temporal bagging* within a set temporal window is used to enforce a temporal coherence between features. Given a speaker's social signal, we observe the listener's social behaviour $s = 10$ frames in the future (approx $\frac{1}{2}$ a second).

## 6 Social Interactive Motion Control Using Apriori Mining

Since we are only interested in deriving animations of the listener, we compute the conditional probability of the listener's social response given the speaker's social signals, as weighted variables to control the motion model.

Given the chosen speaker's 9 dimensional binary vector $\mathbf{f}_t$ (as explained in Section 5.2) at time $t$, where $\mathbf{f}_t \subset \mathbf{f}_\iota$ ($\iota$ is the frame index of the current query cut point as detailed in Section 4.2), we derive the power set $2^{\mathbf{f}_t}$ for all combinations of the speaker's active social signals. We find a suitable matching set of rules $R^t \subset R$ such that $\forall(R_j^{A,t} \Rightarrow R_j^{C,t}, conf_j) \in R^t$, where there exists $\mathbf{f} \in 2^t$ and $\mathbf{f} \subset R_j^A$. The weighted combination of the results are obtained as follows:

$$W = \sum_{j=1}^{|R^t|} conf_j I(R_j^{C,t}, \mathbf{f}_\iota) \tag{9}$$

where

$$I(R_j^C, \mathbf{f}) = \begin{cases} 1 \text{ if } \mathbf{f} \subset R_j^C \\ 0 \text{ otherwise} \end{cases} \tag{10}$$

In the motion synthesis process, Equations 4 is altered as follows, for all:

$$\phi_m = P(C_{C_t,m}^z | C_t).P'(\mathbf{z}_{C_t,m}^c).W \tag{11}$$

## 7 Animation/Results

To validate our experiment, we observe the trends in the mined *confidence* values, which details the likelihood of a listener's social response given a speaker's social signal. 1350 rules were extracted from the mining in the *interested* scenario, and 1400 in the *not interested* scenario, which resulted in 1034 matching rules in both scenarios. Dividing the *confidence* values in the *interested* scenario by those in the *not interested* scenario for matching rules, we obtain results of greater than 1 when the rules occur more frequently in the *interested* scenario, and less than 1 when they occur more in the *not interested* scenario. The results are shown in Table 2. We are unable to show all combinations of association rules so we show the highest trends greater than 10.

The association rule labels are as detailed in Section 5.2. To add clarity to the gaze labels, instead of [GL], [GR] and [GC], we use [GA], [GB], [GC], [GN],

**Table 2.** Table showing the highest trends (*approx* > 10) for a set of rules for confidence values from the *interested* scenarios cv(int) divided by the confidence values for the *not interested* scenario cv(not), for the individual people. Association rule labels are as detailed in Section 5.2.

| No. | Association Rules | $\frac{\text{cv(int)}}{\text{cv(not)}}$ | No. | Association Rules | $\frac{\text{cv(int)}}{\text{cv(not)}}$ |
|---|---|---|---|---|---|
| 1 | $\{C = [S] \Rightarrow A = [\mathbf{T}]\}$ | 11 | 11 | $\{B = [N] \Rightarrow C = [\mathbf{T}]\}$ | 13 |
| 2 | $\{B = [GN] \Rightarrow A = [\mathbf{T}]\}$ | 48 | 12 | $\{B = [GN] \Rightarrow C = [\mathbf{T}]\}$ | 11.6 |
| 3 | $\{C = [GA] + [S] \Rightarrow A = [\mathbf{V}]\}$ | 74 | 13 | $\{B = [A] \Rightarrow C = [\mathbf{T}]\}$ | 11 |
| 4 | $\{C = [GN] + [N] \Rightarrow A = [\mathbf{N}]\}$ | 40.2 | 14 | $\{B = [GC] \Rightarrow A = [\mathbf{T}]\}$ | 18 |
| 5 | $\{C = [A] + [N] \Rightarrow A = [\mathbf{N}]\}$ | 12 | 15 | $\{A = [L] + [S] \Rightarrow B = [\mathbf{T}]\}$ | 11.9 |
| 6 | $\{A = [GN] + [A] \Rightarrow B = [\mathbf{A}]\}$ | 38.5 | 16 | $\{B = [GC] + [N] \Rightarrow C = [\mathbf{T}]\}$ | 13.3 |
| 7 | $\{A = [L] + [GB] \Rightarrow C = [\mathbf{T}]\}$ | 33 | 17 | $\{B = [A] + [N] \Rightarrow C = [\mathbf{L}]\}$ | 13.4 |
| 8 | $\{B = [GC] + [A] \Rightarrow C = [\mathbf{T}]\}$ | 25.3 | 18 | $\{C = [GN] + [S] \Rightarrow A = [\mathbf{N}]\}$ | 11 |
| 9 | $\{B = [GC] + [A] \Rightarrow C = [\mathbf{N}]\}$ | 46 | 19 | $\{A = [GN] + [N] \Rightarrow C = [\mathbf{T}]\}$ | 11 |
| 10 | $\{C = [GN] \Rightarrow A = [\mathbf{S}]\}$ | 10 | 20 | $\{C = [L] + [S] \Rightarrow A = [\mathbf{N}]\}$ | 12 |

representing *gazing at person A*, *gazing at person B*, *gazing at person C*, and *gazing at no one*, respectively. This allows us to know who the speaker gazes at.

Rows 2 and 3 in Table 2 present the most prominent trends. Row 2 suggests that 'when person B is talking and gazing at no one, person A talks'. In this context, person A talking suggests turn-taking, showing more interest in participating in the conversation in an *interested* scenario as opposed to a *not interested* scenario. Row 3 suggest that 'when person C is talking, gazing at person A and shaking their head, person A is voiced'. Voiced regions imply an exchange of short single words like 'uh-huh' or 'yea', used by a listener to express acknowledgement and understand to the speaker. Other high trends like in rows 4, 5, and 6 also suggests mirroring, where the listener mimics the speaker's social signal such as *nods*, and *active* body movement. Looking at these general contrasts, we see that *talking* is a highly common response from an *interested* listeners, so we can safely assume that turn-taking is an important measure of social interest.

These quantitative results prove there is a clear distinction between an *interested* and *not interested* listener in a social context, and our social experiment provides an appropriate means of modelling levels of social interest.

Using the motion model, the user is given control over the various combinations of social behaviour of the human video avatars, however, not all combinations of social behaviour controls are possible. This is the case for all avatars with regards to performing *head shakes*. This is mostly due to the limited availability of the particular social behaviour in the data set, resulting in very limited connectivity in the texture motion graph. Regardless, most of the popular combinations of social behaviour like *laughing* and *head nods* are connected and responsive.

Not only can the user control the social behaviour of the video avatars but, using the data mined *confidence* values, autonomous interaction is possible. As shown in Figure 2, the user has interactive control over who speaks, and the level of interests of the listeners. With these set parameters, the avatars interact appropriate, traversing the texture motion graph to attach video subsequences

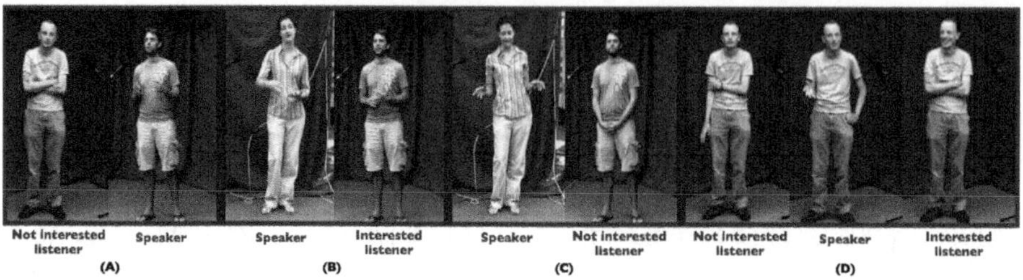


**Fig. 2.** Image showing human video texture synthesis of conversational avatars. (A), (B) and (C) show different generated videos given the scenarios of an *interest* and *not interested* listener in the conversation. (D) demonstrates the diversity of the approach, allowing identical avatars to socially interact together.

together, guided by the data mined conditional weights. Figure 2 (D), further demonstrate the approach by producing autonomous interaction generated by identical avatars.

## 8 Conclusion

Our social dynamics model is able to derive trends between a speaker and listener in a conversation. We successfully parameterise these trends using data mining to derive the conditional probability of a listener's behaviour given a speaker's social signal. Human video motion modelling using a texture motion graph produces plausible transitions between cut points, allowing interactive control over a video avatar's social behaviour. Future work will include video blending strategies for photorealistic data to fix glitches occurring on the seams between video segments.

Utilising the social dynamics model to drive the animation, the user can alter the interest level of participants in the conversation, effectively changing their social responses. This approach can be extended to other social scenarios such as *conflicts*, and other synthesis approaches such as *motion captured* animations, and future work will extend the observed social signals within the system.

**Acknowledgements.** This work was supported by the EPSRC project LILiR (EP/E027946/1) and the European Community's Seventh Framework Programme (FP7/2007-2013) under grant agreement number 231135 - DictaSign.

## References

1. Kovar, L., Gleicher, M., Pighin, F.: Motion graphs. In: Proc. of ACM SIGGRAPH, July 2002, vol. 21(3), pp. 473–482 (2002)
2. Agrawal, A., Imielinski, T., Swami, A.: Mining association rules between sets of items in large databases. In: Proc. of the 1993 ACM SIGMOD Int. Conf. on Management of Data SIGMOD 1993 (1993)

3. Szummer, M., Picard, R.: Temporal texture modeling. In: Proc. of IEEE Int. Conf. on Image Processing, pp. 823–826 (1996)
4. Efros, A., Leung, T.: Texture synthesis by non-paramteric sampling. In: Int. Conf. on Computer Vision, pp. 1033–1038 (1999)
5. Kwatra, V., Schodl, A., Essa, I., Turk, G., Bobick, A.: Graphcut textures. In: ACM Trans. on Graphics, SIGGRAPH 2003, vol. 22(3), pp. 277–286 (2003)
6. Bhat, K., Seitz, S., Hodgins, J., Khosla, P.: Flow-based video synthesis and editing. In: ACM Trans. on Graphics, SIGGRAPH 2004 (2004)
7. Troje, N.F.: Decomposing biological motion: A framework for analysis and synthesis of human gait patterns. J. Vis. 2, 371–387 (2002)
8. Pullen, K., Bregler, C.: Synthesis of cyclic motions with texture (2002)
9. Okwechime, D., Bowden, R.: A generative model for motion synthesis and blending using probability density estimation. In: Fifth Conference on Articulated Motion and Deformable Objects, Mallorca, Spain, (July 9-11, 2008)
10. Tanco, L.M., Hilton, A.: Realistic synthesis of novel human movements from a database of motion captured examples. In: Proc. of the IEE Workshop on Human Motion (HUMO 2000) (2000)
11. Arikan, O., Forsyth, D., O'Brien, J.: Motion synthesis from annotation. In: ACM Transaction on Graphics, SIGGRAPH 2003, July 2003, vol. 22(3), pp. 402–408 (2003)
12. Okwechime, D., Ong, E.J., Bowden, R.: Real-time motion control using pose space probability density estimation. In: IEE Int. Workshop on Human-Computer Interaction (2009)
13. Treuille, A., Lee, Y., Popovic, Z.: Near-optimal character animation with continuous control. In: Proceedings of SIGGRAPH 2007, vol. 26(3) (2007)
14. Rachel, H., Gleicher, M.: Parametric motion graph. In: 24th Int. Symposium on Interactive 3D Graphics and Games, pp. 129–136 (2007)
15. Shin, H., Oh, H.: Fat graphs: Constructing an interactive character with continuous controls. In: Proc. of the 2006 ACM SIGGRAPH/Eurographics Symposium on Computer Animation, vol. 298 (2006)
16. Balci, K., Akarun, L.: Generating motion graphs from clusters of individual poses. In: 24th Int. Symposium on Computer and Information Sciences, pp. 436–441 (2009)
17. Lee, J., Chai, J., Reitsma, P., Hodgins, J., Pollard, N.: Interactive control of avatars animated with human motion data. ACM Trans. on Graphics 21, 491–500 (2002)
18. Schödl, A., Szeliski, R., Salesin, D., Essa, I.: Video textures. In: Proc. of the 27th Annual Conf. on Computer Graphics and Interactive Techniques, SIGGRAPH 2000, pp. 489–498. ACM Press/Addison-Wesley Publishing Co., New York (2000)
19. Flagg, M., Nakazawa, A., Zhang, Q., Kang, S., Ryu, Y., Essa, I., Rehg, J.: Human video textures. In: Proc. of the 2009 Symposium on Interactive 3D Graphics and Games, pp. 199–206. ACM, New York (2009)
20. Ekman, P., Friesen, W.: Facial action coding system. Consulting Psychologists Press, Palo Alto (1977)
21. Argyle, M.: Bodily communication. Methuen (1987)
22. Beaudoin, P., Coros, S., van de Panne, M., Poulin, P.: Motion-motif graphs. In: Proc. of the 2008 ACM SIGGRAPH/Eurographics Symposium on Computer Animation, pp. 117–126 (2008)
23. Pentland, A.: A computational model of social signaling. In: 18th Int. Conf. on Pattern Recognition, ICPR (2006)

24. Mertins, A., Rademacher, J.: Frequency-warping invariant features for automatic speech recognition. In: Proceedings of 2006 IEEE International Conference on Acoustics, Speech and Signal Processing, ICASSP 2006, vol. 5 (2006)
25. Viola, P., Jones, M.: Rapid Object Detection using a Boosted Cascade of Simple Features. In: Proc. IEEE CVPR 2001 (2002)
26. Ong, E.J., Lan, Y., Theobald, B.J., Harvey, R., Bowden, R.: Robust facial feature tracking using selected multi-resolution linear predictors. In: Int. Conf. Computer Vision. ICCV 2009 (2009)
27. Agrawal, R., Srikant, R.: Fast algorithms for mining association rules in large databases. In: Proc. of 20th Int. Conf. on Very Large Data Bases, VLDB 1994, pp. 487–499 (1994)

# Efficient Visual Object Tracking with Online Nearest Neighbor Classifier

Steve Gu, Ying Zheng, and Carlo Tomasi

Department of Computer Science, Duke University

**Abstract.** A tracking-by-detection framework is proposed that combines nearest-neighbor classification of bags of features, efficient subwindow search, and a novel feature selection and pruning method to achieve stability and plasticity in tracking targets of changing appearance. Experiments show that near-frame-rate performance is achieved (*sans* feature detection), and that the state of the art is improved in terms of handling occlusions, clutter, changes of scale, and of appearance. A theoretical analysis shows why nearest neighbor works better than more sophisticated classifiers in the context of tracking.

## 1 Introduction

Visual object tracking is crucial to visual understanding in general, and to many computer vision applications ranging from surveillance and robotics to gesture and motion recognition. The state of this art has advanced significantly in the past 30 years [1,2,3,4,5,6,7,8]. Recently, advances in apparently unrelated areas have given tracking a fresh impulse: Specifically, progress in the definition of features invariant to various imaging transformations [9,10], online learning [11,12], and object detection [13,14,15,16] have spawned the approach of *tracking by detection* [17,18,19,20,21], in which a target object identified by the user in the first frame is described by a set of features. A separate set of features describes the background, and a binary classifier separates target from background in successive frames. To handle appearance changes, the classifier is updated incrementally over time. Motion constraints restrict the space of boxes to be searched for the target.

In a recent example of this approach, Babenko *et al.* [20] adapt Multiple Instance Learning (MIL) [11,12] by building an evolving boosting classifier that tracks bags of image patches, and report excellent tracking results on challenging video sequences. The main advantages of tracking by detection come from the flexibility and resilience of its underlying representation of appearance. Several parametric learning techniques such as Support Vector Machines (SVM, [22]), boosting [20], generative models [23], and fragments [24] have been used successfully in tracking by detection. More recently, Santner *et al.* propose a sophisticated tracking system called PROST [21] that achieves top performance with a smart combination of three trackers: template matching based on normalized cross correlation, mean shift optical flow [25], and online random forests [26] to predict the target location.

R. Kimmel, R. Klette, and A. Sugimoto (Eds.): ACCV 2010, Part I, LNCS 6492, pp. 271–282, 2011.

However, since computation occurs at frame rate, efficiency in both appearance learning and target/background classification is a paramount consideration for practical systems. In addition, image boxes that might contain the target must be enumerated quickly. Finally, and perhaps most fundamentally, the so-called *stability-plasticity dilemma* [21] must be addressed: a *stable* description of target appearance, based only on the first frame, can handle occlusions well, but fails to track an object whose appearance changes over time. A more *plastic* description can be obtained by updating features from observations in subsequent frames, but at the cost of potential confusion between target and foreground when incorrectly classified features contaminate the training sets.

To address these issues, we propose to use Nearest Neighbor (NN) as the underlying, non-parametric classifier; Efficient Subwindow Search (ESS, [15]) to hypothesize target locations; and a novel feature updating and pruning method to achieve a proper balance between plasticity and stability. Despite the simplicity of the NN classifier, Boiman *et al.* [27] demonstrate its state-of-art performance for object categorization. We analyze NN geometrically to suggest why NN captures appearance change better than competing methods. In addition, NN requires no training other than data collection, and is efficient when the size of the data sample is small, as is the case in tracking by detection. Further efficiency can be achieved by the use of KD trees. The use of ESS leads to improved handling of scale changes over existing state-of-the-art trackers [20,21], which cannot search variable-size windows. For features, we use SIFT [9], although other, more recent methods such as SURF [10], Self-Similarity [28] or Critical Nets [29] could be used as well. Together with our feature update and pruning method, this combination leads to a unified tracking-by-detection framework that handles appearance changes, occlusion, background clutter, and scale changes in a principled and effective way, as our experiments demonstrate.

## 2 Tracking with an Online Nearest-Neighbor Classifier

For ease of exposition, we describe our appearance and motion model separately first. We then show how to integrate them seamlessly via online NN classification into a simple and efficient algorithm. Finally, we discuss why our tracking framework can handle significant background clutter, scale changes, occlusion and appearance change.

### 2.1 Appearance Model

Let $V(I) = \{(\mathbf{x}_1, \mathbf{v}_1), \cdots, (\mathbf{x}_n, \mathbf{v}_n)\}$ be the set of (SIFT) key points of an image $I$ where $\mathbf{x}_i \in \mathbb{R}^2$ is the 2D coordinate and $\mathbf{v}_i \in \mathbb{R}^d$ is the $d$-dimensional descriptor vector of the $i^{th}$ feature. We use $\Theta(W; I)$ to represent the set of key point descriptors of $I$ within the window $W$: $\Theta(W; I) \triangleq \big\{\mathbf{v} \in \mathbb{R}^d \mid (\mathbf{x}, \mathbf{v}) \in V(I), \mathbf{x} \in W\big\}$. Given an image sequence $(I_0, W_0), (I_1, W_1), \cdots, (I_k, W_k)$ where $W_i$ is the tracked window in image $I_i$, we describe how to compute the features belonging to the object and the background respectively. Let $\mathcal{B} \subset \mathbb{R}^d$ be the static background model and let $\mathcal{O}_k \subset \mathbb{R}^d$ be the dynamic object model updated up to the frame index $k$. Initially we set $\mathcal{O}_0 = \Theta(W_0; I_0)$ and put the rest of the key point descriptors into the background model $\mathcal{B}$. Given $\mathcal{O}_{k-1}$ and $W_k$, we compute:

$$\mathcal{O}_k \leftarrow \mathcal{O}_{k-1} \bigcup \mathcal{F}_\lambda \left[\Theta(W_k; I_k), \mathcal{O}_{k-1}, \mathcal{B}\right] \tag{1}$$

where $\mathcal{F}_\lambda : 2^{\mathbb{R}^d} \times 2^{\mathbb{R}^d} \times 2^{\mathbb{R}^d} \to 2^{\mathbb{R}^d}$ is the *filter* operator on the feature set:

$$\mathcal{F}_\lambda \left[A, B, C\right] \triangleq \{\mathbf{v} \in A \mid \|\mathbf{v} - NN_B(\mathbf{v})\| < \lambda \|\mathbf{v} - NN_C(\mathbf{v})\|\} \tag{2}$$

where $NN_B(\mathbf{v}) \triangleq \arg\min_{\mathbf{u}\in B} \|\mathbf{u} - \mathbf{v}\|$ is the nearest neighbor of $\mathbf{v}$ in the set $B$.

The idea behind the design of $\mathcal{F}_\lambda$ is simple: for each feature $a \in \Theta(W_k; I_k)$, we apply the ratio test in (2) to see if $a$ is close enough to the target set $\mathcal{O}_{k-1}$ when compared to its distance to the background set $\mathcal{B}$, and we only preserve those features in $\Theta(W_k; I_k)$ that pass the ratio test. Here $\lambda$ is the selection criterion and is fixed to $2/3$, analogously to the familiar matching criterion used in SIFT[9]. We thus have an model updating scheme that keeps adjusting the appearance change while avoiding confusion between object and background features.

### 2.2 Motion Model

Given $\mathcal{O}_{k-1}, \mathcal{B}, W_{k-1}$ and $I_k$, the motion model aims to locate an optimal window $W_k$ that encloses the current object. To this end, we need to evaluate a window based on its appearance. We propose to use the following score function:

$$\mu(W; \mathcal{O}_{k-1}, \mathcal{B}, W_{k-1}, I_k) = \underbrace{\sum_{\mathbf{v}\in\Theta(W;I_k)} \text{sign}\left(\mathcal{F}_\lambda\left[\{\mathbf{v}\}, \mathcal{O}_{k-1}, \mathcal{B}\right]\right)}_{\text{appearance similarity}} - \underbrace{\kappa(W, W_{k-1})}_{\text{motion penalty}} \tag{3}$$

where $\text{sign}(A) = 1$ if $A \neq \emptyset$ and $-1$ otherwise. In the implementation, we can also promote matched features by assigning scores greater than 1. The penalty function $\kappa(W, W_{k-1})$ measures the difference between the window $W$ and the window $W_{k-1}$ in terms of position drift and shape change. We set:

$$\kappa(W_1, W_2) = \gamma \left( \underbrace{\|O_1 - O_2\|}_{\text{position drift}} + \underbrace{|h_1 - h_2|}_{\text{height}} + \underbrace{|w_1 - w_2|}_{\text{width}} + s(W_1, W_2) \right) \tag{4}$$

where $\|O_1 - O_2\|$ is the distance between the centroids $O_1$ and $O_2$ of the windows $W_1$ and $W_2$, and $(w_1, h_1), (w_2, h_2)$ are the width and height of $W_1$ and $W_2$. The term $s(W_1, W_2) \triangleq \max\left\{|\frac{h1}{w1} - \frac{h2}{w_2}|, |\frac{w_1}{h_1} - \frac{w_2}{h_2}|\right\}$ then penalizes changes in the aspect ratio although other more sophisticated penalties are appropriate here as well. Finally, $\gamma$ measures the relative importance of the penalty function in the score $\mu(W; \mathcal{O}_{k-1}, \mathcal{B}, W_{k-1}, I_k)$. The optimal window $W_k$ in $I_k$ is then

$$W_k = \arg\max_{W} \mu(W; \mathcal{O}_{k-1}, \mathcal{B}, W_{k-1}, I_k) \tag{5}$$

where $W$ can be an arbitrary window within the domain of the image $I_k$. The presence of the penalty function $\kappa(W, W_{k-1})$ ensures that the current tracked window cannot drift or change its shape arbitrarily relative to the previous tracked window. Notice that the motion model does not take advantage of any motion continuity or locality. In practice, as is typical to many trackers, we could restrict the search region to a local image patch instead of the entire image domain. Although we believe that the motion model can be improved by introducing advanced filtering techniques such as a Kalman filter or a particle filter, we do not include those techniques in order to show that our tracker already yields excellent tracking results even without them.

### 2.3 Algorithm and Implementation

The algorithm for tracking is very simple:

**Input:** the object model $\mathcal{O}_{k-1}$, previous window $W_{k-1}$ and the image $I_k$.
**Output:** the updated object model $\mathcal{O}_k$ and the current tracked window $W_k$.
**Step.1:** $W_k \leftarrow \arg\max_W \mu(W; \mathcal{O}_{k-1}, \mathcal{B}, W_{k-1}, I_k)$
**Step.2:** $\mathcal{F}_\lambda \leftarrow \{\mathbf{v} \in \Theta(W_k; I_k) \mid \|\mathbf{v} - NN_{\mathcal{O}_{k-1}}(\mathbf{v})\| < \lambda \|\mathbf{v} - NN_{\mathcal{B}}(\mathbf{v})\|\}$
**Step.3:** $\mathcal{O}_k \leftarrow \mathcal{O}_{k-1} \bigcup \mathcal{F}_\lambda$

The nearest neighbor search can be computed quite efficiently with KD trees if approximation is allowed. We use published software [30] to compute both the SIFT descriptors and the nearest neighbor query, with two modifications for efficiency: First, at any time we keep only features from the most recent $\tau$ frames, so that the total number of features in $\mathcal{O}_k$ ranges only from hundreds to thousands, enabling real time performance. Second, we do not explicitly modify the KD tree data structure during the update. Instead, we always maintain $1 + \tau$ KD trees, each corresponding to a frame, and whenever a frame is added or deleted, we add or remove the entire tree associated to that frame. The KD tree associated to the first frame is never deleted. The background model is assumed to be static in our implementation. However, we could easily update the background model as well, similarly to what we do with the object model.

**Fig. 1.** The method of efficient subwindow search ensures that our tracker can handle significant scale change efficiently. The video sequence is downloaded from YouTube.

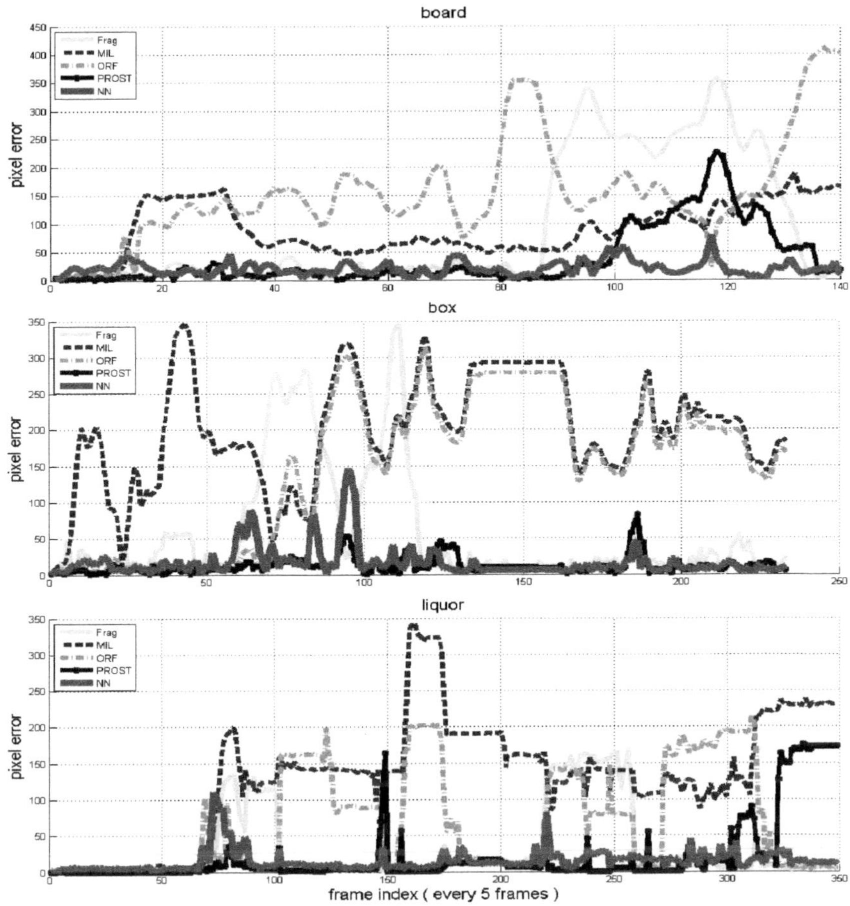


**Fig. 2.** Mean distance error for the selected sequences frame by frame

Exhaustive subwindow search is needed in Step.1. Although the worst running time is $O(n^2)$ where $n$ is the number of pixels in the search area, the actual performance of the optimal window search is often close to sublinear time thanks to branch and bound [15]. We adapt this Efficient Subwindow Search (ESS) method by applying a penalty on position drift and shape change. Specifically, let $\mathcal{R}$ be the range space containing all possible windows and $W_{k-1}$ the tracked window in the previous frame. The quality function $f^{\text{new}}(\mathcal{R})$ is modified from the quality function $f^{\text{old}}(\mathcal{R})$ by Lampert et al.[15] so that

$$f^{\text{new}}(\mathcal{R}) = f^{\text{old}}(\mathcal{R}) - \min_{W \in \mathcal{R}} \kappa(W, W_{k-1}) \tag{6}$$

It is not difficult to prove that this quality function satisfies the two constraints for branch and bound: (1) $f^{\text{new}}(\mathcal{R}) \geq \mu(W; \mathcal{O}_{k-1}, \mathcal{B}, W_{k-1}, I_k)$ for each window $W \in \mathcal{R}$; and (2)$f^{\text{new}}(\mathcal{R})$ converges to the true score when there is a unique window in $\mathcal{R}$. The quality function $f^{\text{new}}(\mathcal{R})$ can be evaluated in constant time by pre-computing the *integral image* in linear time.

The running time of our algorithm, excluding the time for computing SIFT descriptors, is close to frame rate in our MATLAB implementation on a single-core laptop. Computing the key point descriptors is relatively slow ( several frames per second ) compared to tracking time. However, we compute the SIFT descriptors for the entire image for the ease of experiment, and this is not necessary. One speed up is that we only compute features in a local image area. Another speed up comes from other choices of key point descriptors. Finally, we could port the program to C/C++ and run it on a more advanced computer.

# 3 Analysis

We explain why our tracking framework can handle background clutter, occlusion, scale and appearance change in a principled way.

*Background Clutter.* Our tracking framework avoids the confusion between target and background features by applying the filter operator $\mathcal{F}_\lambda$ on the feature set within the tracked window. Therefore, only those features that resemble the features in the current object model are updated.

*Scale changes.* The tracking model by equation (5) ensures that we search windows with all possible locations and shapes, and this covers even significant scale changes. The other contributing factor comes from the use of SIFT descriptors that have been demonstrated to handle partial scale changes well. Notice that our tracking framework does not depend on a specific key point descriptor, and should benefit from any improvement in this area. In Figure 1 we show that our tracker can handle significant scale change.

*Occlusion.* We follow the bag-of-features approach which naturally handles occlusion. The score of a window is high if there are enough matched features inside the window. If only few matched features are present, there is a strong indication that the object is occluded. The current window $W_k$ stays stable relative to the previous window $W_{k-1}$ when occluded because of the penalty enforced by $\kappa(W_k, W_{k-1})$. We exhibit different cases of occlusion in Figure 4.

*Appearance Changes.* The most challenging part of tracking-by-detection systems is perhaps how to adapt to the appearance change incrementally. We show in the left column of Figure 6 that our tracker can adjust to appearance change fairly well. Interestingly we can directly estimate the "shape" of the *object feature space*, the set $\mathcal{F}_\lambda(\mathbb{R}^d, O, B)$ where $O$ and $B$ are the set of object and background features. As a result, we can predict the set of features that is allowed in the object model $O$. We recall from classical geometry that for $\lambda < 1$, the inequality $\|\mathbf{x} - \mathbf{a}\| < \lambda\|\mathbf{x} - \mathbf{b}\|$ defines a $d$-dimensional hyper ball: $\mathfrak{B}_\lambda(\mathbf{a}, \mathbf{b}) \triangleq \left\{\mathbf{x} \in \mathbb{R}^d \mid \|\mathbf{x} - \frac{\mathbf{a}-\lambda^2\mathbf{b}}{1-\lambda^2}\| < \frac{\lambda}{1-\lambda^2}\|\mathbf{a} - \mathbf{b}\|\right\}$ with centroid located

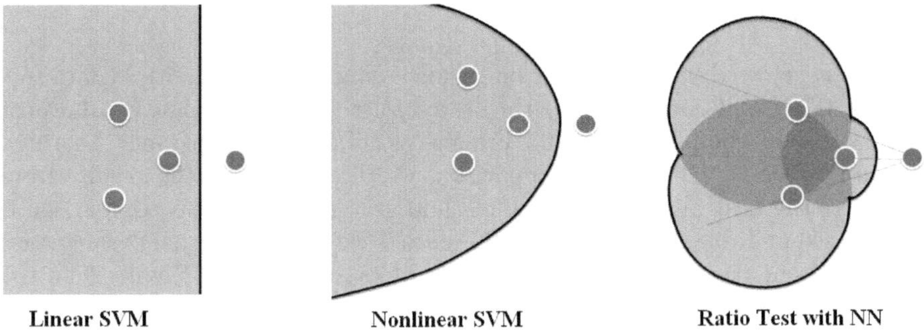


**Fig. 3.** The two dimensional object feature space ( the pink shaded area) generated by linear (left) and nonlinear (middle) support vector machines and the nearest neighbor classifier (right) where the blue and red dots are the features belonging to the object and the background. The ratio test with the NN classifier produces an object feature space that is bounded by the union of a set of two dimensional disks.

at $\frac{\mathbf{a}-\lambda^2\mathbf{b}}{1-\lambda^2}$ and radius equal to $\frac{\lambda}{1-\lambda^2}\|\mathbf{a}-\mathbf{b}\|$. We therefore can quantify the volume of the object feature space $\mathcal{F}_\lambda(\mathbb{R}^d, O, B)$ using the intersection of Voronoi regions. Let $\mathrm{Vor}(\mathbf{v}; S) \triangleq \{\mathbf{x} \in \mathbb{R}^d \mid NN_S(\mathbf{x}) = \mathbf{v}\}$ be the Voronoi region of $\mathbf{v}$ spanned by $S$ and let $\mathcal{I}(S_1, S_2) = \{(\mathbf{a}, \mathbf{b}) \in S_1 \times S_2 \mid \mathrm{Vor}(\mathbf{a}; S_1) \bigcap \mathrm{Vor}(\mathbf{b}; S_2) \neq \emptyset\}$ be the intersection of the Voronoi regions of the set $S_1$ and $S_2$ respectively. We have:

**Theorem 1 (Ball Cover).** *The object feature space is bounded by the union of a set of d-dimensional hyper balls:* $\mathcal{F}_\lambda(\mathbb{R}^d, O, B) \subseteq \bigcup_{(\mathbf{a},\mathbf{b})\in\mathcal{I}(O,B)} \mathfrak{B}_\lambda(\mathbf{a}, \mathbf{b})$

*Proof.* Since the set of nearest neighbors to the feature $\mathbf{v} \in O$ is expressed by $\mathbf{v}$'s Voronoi region spanned by $O$, we can rewrite the object feature space:

$$\mathcal{F}_\lambda(\mathbb{R}^d, O, B) = \bigcup_{\mathbf{a}\in O} \bigcup_{\mathbf{b}\in B} \left[\mathrm{Vor}(\mathbf{a}; O) \bigcap \mathrm{Vor}(\mathbf{v}; B) \bigcap \mathfrak{B}_\lambda(\mathbf{a}, \mathbf{b})\right] \tag{7}$$

$$= \bigcup_{(\mathbf{a},\mathbf{b})\in\mathcal{I}(O,B)} \left[\left(\mathrm{Vor}(\mathbf{a}; O) \bigcap \mathrm{Vor}(\mathbf{b}; B)\right) \bigcap \mathfrak{B}_\lambda(\mathbf{a}, \mathbf{b})\right] \tag{8}$$

$$\subseteq \bigcup_{(\mathbf{a},\mathbf{b})\in\mathcal{I}(O,B)} \mathfrak{B}_\lambda(\mathbf{a}, \mathbf{b}) \,. \tag{9}$$

The ball cover theorem has many appealing properties: First, it shows that the object feature space is well bounded by our selection criterion. This is in contrast to the typical use of classifiers such as linear or non-linear SVM where the object feature space is unbounded. We illustrate this difference in the 2D cartoon in Figure 3. Second, for $\mathbf{a}, \mathbf{b} \in O$ with $\|\mathbf{a} - NN_B(\mathbf{a})\| > \|\mathbf{b} - NN_B(\mathbf{b})\|$, $\mathbf{a}$ contributes more than $\mathbf{b}$ to the object feature space. In other words, the more discriminative (between target and background) the feature is, the larger the feature space it generates (*i.e.*, the larger the radius of the hyper ball).

## 4 Experiments

We test our new NN-based tracker on 8 video sequences from multiple data sets [5,20,21,31,32] and compare it to the state-of-art tracking methods: AdaBoost [18], Online Random Forests [26], Fragments [5], Multiple Instance Learning [20] and PROST [21]. In the comparison, we directly quote the results from [21] where the best results from each method were reported. This comparison is summarized in Table 1. In Figure 4 and Figure 6, we show the actual performance of the different trackers. The selected error plots are shown in Figure 2. In the experiment, we fix the parameters once for all (i.e. $\lambda = \frac{2}{3}, \gamma = 0.1$) and use the default SIFT parameters in [30]. The evaluation criterion is the same as what is used in [20,21] except that we only compute the mean distance error $e$:

$$e = \frac{1}{n} \sum_{i=1}^{n} \| O_i - O_i^g \| \tag{10}$$

where $n$ is the number of frames and $\|O_i - O_i^g\|$ is the Euclidean distance between the tracked window centroid $O_i$ and the ground truth window centroid $O_i^g$.

Table 1 shows that despite the simplicity of the proposed method, our tracker yields excellent tracking results often close to or even better than the state-of-art methods, which rely on more sophisticated online learning methods. More specifically, our tracker achieves the best result in the sequences Girl, Faceocc2, Board and Liquor, and the second best result in the sequences David and Box. These results verify that our tracking framework can handle significant occlusion, background clutter and appearance change. Our tracker can successfully and closely follow the object in 7 out of the 8 video sequences and wins in the highest number of trials. PROST achieves excellent results as well, followed by the MIL and Frag tracker.

### 4.1 Failure Modes

Our tracker is not without its limitations. For instance, it failed to follow the target closely in the Lemming sequence ( Figure 5 ). There are three major

**Table 1.** Mean distance error to the ground truth. Bold: best. Underlined: second best

| Sequences | # Frames | AdaBoost | ORF | FragTrack | MILTrack | PROST | NN |
|---|---|---|---|---|---|---|---|
| Girl [31] | 502 | 43.3 | – | 26.5 | 31.6 | <u>19.0</u> | **18.0** |
| David [32] | 462 | 51.0 | – | 46.0 | <u>15.6</u> | **15.3** | <u>15.6</u> |
| Faceocc1 [5] | 886 | 49.0 | – | **6.5** | 18.4 | <u>7.0</u> | 10.0 |
| Faceocc2 [20] | 812 | 19.6 | – | 45.1 | <u>14.3</u> | 17.2 | **12.9** |
| Board [21] | 698 | – | 154.5 | 90.1 | 51.2 | <u>37.0</u> | **20.0** |
| Box [21] | 1161 | – | 145.4 | 57.4 | 104.6 | **12.1** | <u>16.9</u> |
| Lemming [21] | 1336 | – | 166.3 | 82.8 | **14.9** | <u>25.4</u> | 79.1 |
| Liquor [21] | 1741 | – | 67.3 | 30.7 | 165.1 | <u>21.6</u> | **15.0** |

**Fig. 4.** From top to bottom: Girl, Faceocc1, Faceocc2 and David. Faceocc1 and Faceocc2 have significant occlusion. David experiences appearance change. Girl has both occlusion and appearance change. For ease of visualization, we show only the comparison between the NN-based tracker ( solid red ) and the MIL tracker ( dashed blue ). NN follows the object more closely and handles occlusion better than MIL.

**Fig. 5.** Our tracker failed in the Lemming sequence for the reason that the object contains significant motion blur which is not well captured by the SIFT descriptors

limiting factors: First, experiments show that the SIFT descriptor cannot handle motion blur well, because no features are found in regions of uniform texture. For the same reason, very few features can be reliably found on the body of the lemming even when the object is static. Therefore, the current NN tracker would benefit from descriptors that capture uniform regions and motion blur better. Second, our tracker does not utilize any advanced motion models. Consequently, if the feature matching step fails completely, i.e. generates uniform matching

**Fig. 6.** Each column (left: board; middle:liquor; right:box) shows the performance of NN, MIL, Frag, PROST and ORF on selected frames. NN typically outperforms all other methods in cases with significant occlusion, scale and appearance change.

scores at all the pixels, our tracker stays still due to the presence of a motion penalty. How to use advanced filtering technique to cope with matching failure under heavy background clutter is left for future work. Third, the current tracker cannot localize objects very precisely when the object's shape deforms. This is because a rectangle that is axis-parallel to the image boundaries is used to bound the target. How to localize regions of varying shape and orientations in an efficient manner is an interesting challenge.

## 5 Conclusions and Future Work

The combination of nearest-neighbor classification of bags of features and efficient subwindow search yields a simple and efficient tracking-by-detection algorithm that handles occlusions, clutter, and significant changes of scale and appearance. Performance quality in terms of stability and plasticity is competitive and often better than the previous state of the art. Our theoretical analysis suggests some of the reasons why nearest neighbor works better than more sophisticated classifiers in this context. Immediate future work entails improvements of implementation, as suggested earlier, for true real-time performance, and the use of more recent feature detection schemes. Longer term questions concern the incorporation of more detailed motion models, backup search strategies for lost objects, tracking complex objects by parts and the exploitation of *a priori* appearance models for certain categories of targets (e.g., people, cars).

**Acknowledgement.** This material is based upon work supported by the NSF under Grant IIS-1017017 and by the ARO under Grant W911NF-10-1-0387.

## References

1. Lucas, B., Kanade, T.: An iterative image registration technique with an application to stereo vision. In: IJCAI, pp. 674–679 (1981)
2. Shi, J., Tomasi, C.: Good features to track. In: IEEE CVPR, pp. 593–600 (1994)
3. Isard, M., Blake, A.: A smoothing filter for CONDENSATION. In: Burkhardt, H.-J., Neumann, B. (eds.) ECCV 1998. LNCS, vol. 1406, pp. 767–781. Springer, Heidelberg (1998)
4. Comaniciu, D., Ramesh, V., Meer, P.: Real-time tracking of non-rigid objects using mean shift. In: IEEE CVPR, vol. 2, pp. 142–149 (2000)
5. Adam, A., Rivlin, E., Shimshoni, I.: Robust fragments-based tracking using the integral histogram. In: IEEE CVPR, pp. 798–805 (2006)
6. Avidan, S.: Ensemble tracking. IEEE PAMI 29, 261–271 (2007)
7. Li, Y., Ai, H., Yamashita, T., Lao, S., Kawade, M.: Tracking in low frame rate video: A cascade particle filter with discriminative observers of different life spans. IEEE PAMI 30, 1728–1740 (2008)
8. Özuysal, M., Calonder, M., Lepetit, V., Fua, P.: Fast keypoint recognition using random ferns. IEEE Trans. Pattern Anal. Mach. Intell. 32, 448–461 (2010)
9. Lowe, D.: Object recognition from local scale-invariant features. In: ICCV, pp. 1150–1157 (1999)

10. Bay, H., Tuytelaars, T., Van Gool, L.: SURF: Speeded up robust features. In: Leonardis, A., Bischof, H., Pinz, A. (eds.) ECCV 2006. LNCS, vol. 3951, pp. 404–417. Springer, Heidelberg (2006)
11. Viola, P., Platt, J., Zhang, C.: Multiple instance boosting for object detection. In: NIPS (2005)
12. Dietterich, T., Lathrop, R., Lozano-Pérez, T.: Solving the multiple instance problem with axis-parallel rectangles. Artif. Intell. 89, 31–71 (1997)
13. Dalal, N., Triggs, B.: Histograms of oriented gradients for human detection. In: IEEE CVPR, pp. 886–893 (2005)
14. Felzenszwalb, P., McAllester, D., Ramanan, D.: A discriminatively trained, multiscale, deformable part model. In: IEEE CVPR (2008)
15. Lampert, C., Blaschko, M., Hofmann, T.: Efficient subwindow search: A branch and bound framework for object localization. IEEE PAMI 31, 2129–2142 (2009)
16. Everingham, M., Gool, L., Williams, C., Winn, J., Zisserman, A.: The PASCAL Visual Object Classes Challenge (VOC 2009) Results (2009), `http://www.pascal-network.org/challenges/VOC/voc2009/workshop/index.html`
17. Tomasi, C., Petrov, S., Sastry, A.: 3d tracking = classification + interpolation. In: ICCV, pp. 1441–1448 (2003)
18. Grabner, H., Bischof, H.: On-line boosting and vision. In: IEEE CVPR, pp. 260–267 (2006)
19. Grabner, H., Leistner, C., Bischof, H.: Semi-supervised on-line boosting for robust tracking. In: Forsyth, D., Torr, P., Zisserman, A. (eds.) ECCV 2008, Part I. LNCS, vol. 5302, pp. 234–247. Springer, Heidelberg (2008)
20. Babenko, B., Yang, M., Belongie, S.: Visual tracking with online multiple instance learning. In: IEEE CVPR, pp. 983–990 (2009)
21. Santner, J., Leistner, C., Saffari, A., Pock, T., Bischof, H.: PROST Parallel Robust Online Simple Tracking. In: IEEE CVPR (2010)
22. Tian, M., Zhang, W., Liu, F.: On-line ensemble SVM for robust object tracking. In: Yagi, Y., Kang, S.B., Kweon, I.S., Zha, H. (eds.) ACCV 2007, Part I. LNCS, vol. 4843, pp. 355–364. Springer, Heidelberg (2007)
23. Zhao, X., Liu, Y.: Generative estimation of 3D human pose using shape contexts matching. In: Yagi, Y., Kang, S.B., Kweon, I.S., Zha, H. (eds.) ACCV 2007, Part I. LNCS, vol. 4843, pp. 419–429. Springer, Heidelberg (2007)
24. Prakash, C., Paluri, B., Nalin Pradeep, S., Shah, H.: Fragments based parametric tracking. In: Yagi, Y., Kang, S.B., Kweon, I.S., Zha, H. (eds.) ACCV 2007, Part I. LNCS, vol. 4843, pp. 522–531. Springer, Heidelberg (2007)
25. Werlberger, M., Trobin, W., Pock, T., Wedel, A., Cremers, D., Bischof, H.: Anisotropic huber-l1 optical flow. In: BMVC (2009)
26. Breiman, L.: Random forests. Mach. Learning 45, 5–32 (2001)
27. Boiman, O., Shechtman, E., Irani, M.: In defense of nearest-neighbor based image classification. In: IEEE CVPR (2008)
28. Shechtman, E., Irani, M.: Matching local self-similarities across images and videos. In: IEEE CVPR (2007)
29. Gu, S., Zheng, Y., Tomasi, C.: Critical nets and beta-stable features for image matching. In: ECCV, pp. 663–676 (2010)
30. Vedaldi, A., Fulkerson, B.: VLFeat: An open and portable library of computer vision algorithms (2008), `http://www.vlfeat.org/`
31. Birchfield, S.: Elliptical head tracking using intensity gradients and color histograms. In: IEEE CVPR, pp. 232–237 (1998)
32. Ross, D., Lim, J., Lin, R., Yang, M.: Incremental learning for robust visual tracking. IJCV 77, 125–141 (2008)

# Robust Tracking with Discriminative Ranking Lists

Ming Tang, Xi Peng, and Duowen Chen

National Lab of Pattern Recognition, Institute of Automation,
Chinese Academy of Sciences, Beijing 100190, China

**Abstract.** In this paper, we propose to model the target object by using discriminative ranking lists of object patches under different scales. The ranking list of each object patch is its $K$ nearest neighbors. The object patches of the same scale with ranking lists of high purities (means with high probabilities to be on the target object) constitute the object model under that scale. A pair of object models of two different scales collaborate to determine which patches may be from the target object in the next frame. The superior ability to alleviate the model drift problem over several state-of-the-art tracking approaches is demonstrated quantitatively through extensive experiments.

## 1 Introduction

Visual object tracking plays a key role in a variety of applications of computer vision. Although it has been investigated intensively during recent decades [1,2,3], the accuracy and stability of tracking algorithms are still unsatisfactory, especially when the object and background are partially similar in appearance, or the object undergoes changes in both appearance and shape.

Extensively reviewing the related works on visual tracking is beyond the scope of this paper, here we only focus on one of the main visual tracking categories, i.e. tracking-by-classification, which has been pervasive since Avidan's well-known work [4]. In order to effectively classify object and background, two main kinds of approaches were developed, that is, *feature-based approach* and *learning-based approach.*

In the feature-based approaches, criteria are designed to select a number of most discriminative features to separate object out of background. Collins *et al.* [5] designed a two-class variance ratio measure to select the most discriminative features to locate the object in the next frame. Yin and Collins [6] developed the work in [5] by dividing local background into several sub-regions and employing the approach in [5] for every pair of object and sub-regions. Mahadevan and Vasconcelos [7] proposed to use a novel saliency criterion based on mutual information to select the most discriminative features. The object was located by using the same criterion and selected features in the next frame.

On the contrary, the learning-based approaches usually learn a strong binary classifier based on object and its background, and use the learned classifier to further guide tracking in the later frames. Avidan [8] proposed "ensemble tracking"

R. Kimmel, R. Klette, and A. Sugimoto (Eds.): ACCV 2010, Part I, LNCS 6492, pp. 283–295, 2011.

which was based on the online construction of a binary classifier. The weak classifiers were integrated via AdaBoost. A similar work was proposed by Grabner and Bischof [9] which used the boosted feature selection algorithm to construct strong classifiers. Lu and Hager [10] represented the object and its local background with two sets of randomly sampled image patches. A binary classifier was learned and used to classify patches sampled in the next frame. Tang *et al.* [11] proposed a co-training tracker which tracked the object cooperatively using independent features.

In general, the target object is modeled implicitly or explicitly in different approaches. In the feature-based approaches, there is generally no explicit object model [5,6,7], while the learning-based approaches usually learn strong classifiers as the object model. A single classifier [4,12] is learned to detect the object in the next frame, or several strong classifiers [11,13] are learned and cooperate with each other to reduce the tracking error. A prominent advantage of tracking-by-classification is that they all take both foreground and background into account; however, they are only from the classification point of view. Therefore, such an object model is inadequate in descriptive ability, bias is inevitably introduced into it. And this is one of the main reasons to cause *model drift* problem [5] in tracking by classification approaches.

In this paper, aiming at the aforementioned problem, we developed an effective tracking algorithm, Discriminative Ranking Lists Tracker (DRLTracker), to substantially alleviate model drift.

The algorithm flow is briefed in Fig. 1. The key idea is to construct two object models of two different scales by using k-NN classification. Such pair of object models combine both discriminative and descriptive information of the target object, and advantages of patches of different scales. Generally speaking, patches of large scale are more discriminative than those of small scale, while the accuracy of location for large scale patches could be lower than small ones due to the sparser distribution of large ones. And it will be difficult to discriminate pixels near to the sides of bounding boxes only with large scale patches. Therefore, combining results from both scales can locate the object more discriminatively and accurately. Section 5 illustrates more accurate and stable performance while using double-scale patches instead of single-scale ones.

To demonstrate the notable improvement against *model drift* problem, we quantitatively compared our method with other three state-of-the-art algorithms [5,8,10] in terms of two criteria [14]. The reasons we chose these two algorithms for comparison are that these algorithms adopt a philosophy similar to ours in representing objects thus so that different performances caused by different object representations can be avoided as much as possible, and they also take similar steps. Extensive comparison showed that our method achieved noticeable improvement over them in terms of accuracy and stability.

The rest of the paper is organized as follows. Sections 2, 3, and 4 describe the three main steps of our algorithm: modeling a pair of object models, locating the object, and updating the object models, respectively. Experiments are presented in Section 5, and conclusion in Section 6.

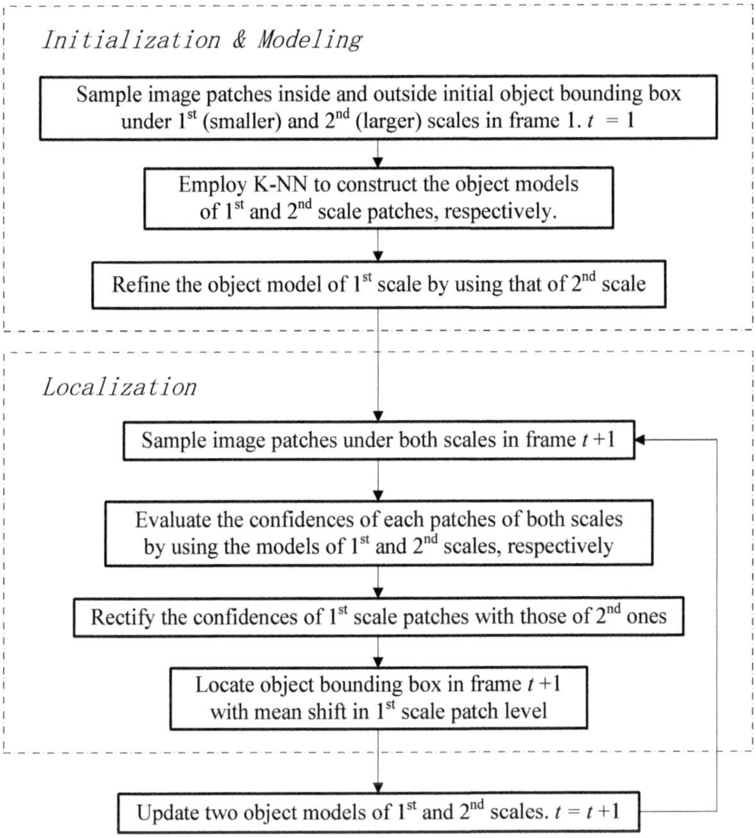


**Fig. 1.** The flow chart of our DRLtracker

## 2 Modeling a Pair of Object Models

Roughly speaking, the object model under a scale is composed of object patches of the same scale as each other. And these patches are very probably located on the target object. Totally, a pair of object models under two different scales are constructed.

Suppose a rectangular target object bounding box $X_t$ is obtained manually or automatically in frame t. Initially, $t = 1$. Image patches of the $1^{st}$ and $2^{nd}$ scales, $s_1$ and $s_2$, are randomly sampled inside and outside $X_t$, respectively to form two sets of object patches $P_t^{o,s_1} = \{p_i^{s_1}\}_{1 \leq i \leq N_1}$ and $P_t^{o,s_2} = \{p_i^{s_2}\}_{1 \leq i \leq N_2}$, respectively, and two sets of local background patches $P_t^{b,s_1}$ and $P_t^{b,s_2}$, respectively, where $s_1 < s_2$. Here, sampling outside $X_t$ means sampling patches in the annular rectangular bounding box of local background of the target object. Illustrations are given with cyan bounding boxes in the bottom row of Fig. 6.

For $\forall p_i^{s_k} \in P_t^{o,s_k}$, K-NN with Euclidean distance metrics is employed to select K nearest neighbors to $p_i^{s_k}$ from $P_t^{o,s_k} \cup P_t^{b,s_k} \setminus p_i^{s_k}$ to generate the *ranking list for modeling under scale* $s_k$, $R_i^{s_k}$, of $p_i^{s_k}$. $R_i^{s_k} = \{p_{i1}^{s_k}, p_{i2}^{s_k}, \ldots, p_{iK}^{s_k}\}$.Where $k = 1, 2$.

It is reasonable to assume that most parts belong to the target object within the object bounding box. Nevertheless, it is common that not all parts do so. Therefore, it is desirable that patches that are not on the target object should be discarded from the object model. According to K-NN classification, it is probable that $p_i^{s_k}$ belongs to the target object if most of the elements of $R_i^{s_k}$ belong to $P_t^{o,s_k}$. On the other hand, it is almost definite that $p_i^{s_k}$ comes from the background if enough elements of $R_i^{s_k}$ are in $P_t^{b,s_k}$. Therefore, the number of elements of $R_i^{s_k}$ coming from $P_t^{o,s_k}$ indicates whether $p_i^{s_k}$ belongs to the target object or not. Consequently, the *purity*, $\alpha_i^{s_k}$, is defined as follows to depict the likelihood that $p_i^{s_k}$ falls onto the target object.

$$\alpha_i^{s_k} = \frac{1}{K} \sum_{j=1}^{K} I(p_{ij}^{s_k} \in P_t^{o,s_k}), \tag{1}$$

where indicator function $I(x) = 1$ if $x$ is true, and 0 otherwise. According to the above discussion, it is clear that the larger $\alpha_i^{s_k}$, the higher the likelihood that $p_i^{s_k}$ falls onto the target object, and vice versa. Purity $\alpha_i^{s_k}$ is a reasonable measure for $p_i^{s_k}$ to belong to the target object or not.

The object model $\Omega_t^{s_2}$ is composed of all $p_i^{s_2}$'s with $\alpha_i^{s_2} > \alpha_\tau^{s_k}$. Formally, $\Omega_t^{s_2} = \{(p_i^{s_2}, \alpha_i^{s_2}) \,|p_i^{s_2} \in P_t^{o,s_2}, \alpha_i^{s_2} > \alpha_\tau^{s_2}\}$. In all our experiments,

$$\alpha_\tau^{s_2} = \frac{1}{|P_t^{o,s_2}|} \sum_{\forall p_i^{s_2} \in P_t^{o,s_2}} \alpha_i^{s_2},$$

where $|Y|$ is the cardinality of set $Y$.

In order to using the advantages of both scales while avoid their disadvantages, the large scale patches, i.e., $\forall p_i^{s_2} \in P_t^{o,s_2}$, are used to filter out some small scale patches, $p_j^{s_1}$'s. We observed that if $p_j^{s_1}$ is not located in any $p_i^{s_2}$, where most parts of $p_i^{s_2}$ are on the target object, $p_j^{s_1}$ is probably not located on the target object. In such case, $\alpha_j^{s_1}$ should be reset to 0. Because there may be many large patches which cover the identical small one, the patch with the largest purity is used to modify the purity of $p_j^{s_1}$. Formally,

$$\alpha_j^{s_1} = \begin{cases} \alpha_j^{s_1}, & \text{if } \left( \max\limits_{p_i^{s_2}:\, p_j^{s_1} \in p_i^{s_2}} \alpha_i^{s_2} \right) > \alpha_\tau^{s_2}; \\ 0, & \text{otherwise.} \end{cases} \tag{2}$$

Then the object model $\Omega_t^{s_1}$ is constructed by using the same method as constructing $\Omega_t^{s_2}$ with modified purities $\alpha_j^{s_1}$'s.

Another factor seemingly relative to the probability that $p_i^{s_k}$ falls onto the target object is to take the distances from $p_i^{s_k}$ to $p_{ij}^{s_k}$'s of $L(p_{ij}^{s_k}) = 1$ into account.

However, our experiments and quantitative comparison show that the effect of this factor on the probability is almost trivial. Therefore, it is ignored in our algorithm.

## 3 Locating the Object

In our algorithm, the key to locate the target object is to assign a reliable confidence to each patch of the first scale, $s_1$, of the next frame according to the object models. Its confidence reflects the probability that the patch belongs to the target object. If a patch of new frame is close enough to many patches of object model, it will be on the target object with high confidence. Otherwise, it comes very probably from background, and its confidence should be set to 0. The formal description is as follows.

Within the next frame $t+1$, $M$ patches are randomly sampled in a uniform distribution under scales $s_k$ from the image region around $X_t$ to form candidate sets $P_{t+1}^{s_k} = \{p_j^{s_k}\}_{1 \leq j \leq M}$, where $k = 1, 2$. For $\forall\, (p_i^{s_k}, \alpha_i^{s_k}) \in \Omega_t^{s_k}$, K-NN is employed to select $K$ nearest patches to $p_i^{s_k}$ from $P_{t+1}^{s_k}$ to generate the *ranking list for locating under scale* $s_k$, $\widetilde{R}_i^{s_k}$, of $p_i^{s_k}$. $\widetilde{R}_i^{s_k} = \{p_{i1}^{s_k}, p_{i2}^{s_k}, \ldots, p_{iK}^{s_k}\}$.Where $K$ is a parameter and identical to that in Section 2.

According to the assumption that the identical objects appearing in two adjacent frames are similar to each other, the following conclusion can be drawn: There exists probability that $p_j^{s_k} \in P_{t+1}^{s_k}$ falls onto the target object, if $p_j^{s_k} \in \widetilde{R}_i^{s_k}$. Furthermore, the more such $\widetilde{R}_i^{s_k}$'s, the more possible that $p_j^{s_k}$ falls onto the target object of frame $t+1$. Therefore, the confidence for $p_j^{s_k}$ to belong to the target object is defined as

$$c_j^{s_k} = \sum_{i=1}^{|\Omega_t^{s_k}|} \alpha_i^{s_k} \cdot I(p_j^{s_k} \in \widetilde{R}_i^{s_k}), \tag{3}$$

Then, two confidence maps, $c^{s_1}$ and $c^{s_2}$, are naturally constructed by assigning the confidence of each patch in $P_{t+1}^{s_1}$ and $P_{t+1}^{s_2}$ to its central pixels. Fig. 2 (a) to (d) and (e) to (h) show examples of confidence maps with yellow marks under scale $s_2$ and $s_1$, respectively.

On one hand, each $p_i^{s_k} \in \Omega_t^{s_k}$ gives likelihood $\alpha_i^{s_k}$ to candidate patches in $P_{t+1}^{s_k}$ based on its own $\widetilde{R}_i^{s_k}$. On the other hand, each $p_j^{s_k} \in P_{t+1}^{s_k}$ cumulates likelihood values from all patches of $\Omega_t^{s_k}$, and forms its confidence of being from the target object. In our experiments, only a small number of patches in $P_{t+1}^{s_k}$ would receive significant confidence from $\Omega_t^{s_k}$, and for most of other patches, their confidences were close or equal to zero.

In principle, the object can be located in the new frame with confidence maps $c^{s_1}$ or $c^{s_2}$. $c^{s_1}$, however, includes more false positive patches than $c^{s_2}$ does, resulting in lower accuracy in tracking. $c^{s_2}$, on the other hand, is much sparser than $c^{s_1}$ is, resulting in less stability. It is observed that a large scale patch with non-zero confidence can often indicate whether small scale patches are positive or not, based on whether it covers those small scale patches or not. Therefore,

$c^{s_1}$ and $c^{s_2}$ should be fused to refine $c^{s_1}$. Because there may be many large patches which cover the identical small one, the patch with the largest $c_i^{s_2}(x_i, y_i)$, centered at $(x_i, y_i)$, will transfer its confidence to the small one. Formally,

$$c_j^{s_1}(x_j, y_j) = \begin{cases} c_j^{s_1}(x_j, y_j) + \max\limits_{p_i^{s_2}:\, p_j^{s_1} \in p_i^{s_2}} c_i^{s_2}(x_i, y_i), & \text{if } \max\limits_{p_i^{s_2}:\, p_j^{s_1} \in p_i^{s_2}} c_i^{s_2}(x_i, y_i) > 0; \\ 0, & \text{otherwise.} \end{cases} \quad (4)$$

Similar to [8,10], the object bounding box $X_{t+1}$ in frame $t+1$ is located through mean-shift [15] with refined $c^{s_1}$.

## 4 Updating Object Models

To update the object models, we again sample patches of two scales from frame $t+1$ inside and outside bounding box $X_{t+1}$ to form two pairs of patches sets, $P_{t+1}^{o,s_1}$, $P_{t+1}^{b,s_1}$, and $P_{t+1}^{o,s_2}$, $P_{t+1}^{b,s_2}$. Then, the same approach as in Sec. 2 is used to construct two temporal object models, $\tilde{\Omega}_{t+1}^{s_1}$ and $\tilde{\Omega}_{t+1}^{s_2}$.

Updating the object models from $\Omega_t^{s_k}$ to $\Omega_{t+1}^{s_k}$ consists of four steps, where $k = 1, 2$. The first is to remove several patches of $\tilde{\Omega}_{t+1}^{s_1}$ if they are not covered by any patches of $\tilde{\Omega}_{t+1}^{s_2}$. These patches are false positive. The second is to remove patches from $\tilde{\Omega}_{t+1}^{s_k}$ who have been well expressed by $\Omega_t^{s_k}$. That is, evaluating the confidences $c_j^{s_k}$ of $\forall p_j^{s_k} \in \{p_j^{s_k} | \{p_j^{s_k}, \alpha_j^{s_k}\} \in \tilde{\Omega}_{t+1}^{s_k}\}$ using Eq. (3). If $c_j^{s_k} > \tau_c$, where $\tau_c$ is a threshold, $\tilde{\Omega}_{t+1}^{s_k} = \tilde{\Omega}_{t+1}^{s_k} \setminus \{p_j^{s_k}, \alpha_j^{s_k}\}$. The third is to remove several patches of $\Omega_t^{s_k}$ if they can not express the variation of target object well. That is, for $\forall \{p_i^{s_k}, \alpha_i^{s_k}\} \in \Omega_t^{s_k}$, K-NN is employed to select K nearest neighbors to $p_i^{s_k}$ from $\forall p_j \in P_{t+1}^{o,s_k} \cup P_{t+1}^{b,s_k}$ to generate the *ranking list for updating under scale* $s_k$, $\bar{R}_i^{s_k}$, of $p_i^{s_k}$, and then the purity $\bar{\alpha}_i^{s_k}$ is calculated by using $\bar{R}_i^{s_k}$ and Eq. (1) with $P_t^{o,s_k} = \{p_i^{s_k} | \{p_i^{s_k}, \alpha_i^{s_k}\} \in \tilde{\Omega}_{t+1}^{s_k}\}$. If $\bar{\alpha}_i^{s_k} < \alpha_T^{s_k}$, $\Omega_t^{s_k} = \Omega_t^{s_k} \setminus \{p_i^{s_k}, \alpha_i^{s_k}\}$. If $|\Omega_t^{s_k} \cup \tilde{\Omega}_{t+1}^{s_k}| < \lambda |\Omega_1^{s_k}|$, where $\lambda \in [0, 1]$ is a parameter, stop removing elements from $\Omega_t^{s_k}$. If $|\Omega_t^{s_k}| < \tau_{\text{inf}}$, there exists significant and abrupt variation, e.g., serious occlusion, in bounding box $X_{t+1}$. In this case, set $\Omega_{t+1}^{s_k}$ to original $\Omega_t^{s_k}$ and quit updating process. And finally, the patches remained in $\Omega_t^{s_k}$ and $\tilde{\Omega}_{t+1}^{s_k}$ constitute the current object models $\Omega_{t+1}^{s_k}$. That is, $\Omega_{t+1}^{s_k} = \Omega_t^{s_k} \cup \tilde{\Omega}_{t+1}^{s_k}$. If $|\Omega_{t+1}^{s_k}| > |\Omega_1^{s_k}|$, randomly remove elements from $\tilde{\Omega}_{t+1}^{s_k}$ until $|\Omega_{t+1}^{s_k}| = |\Omega_1^{s_k}|$.

## 5 Experiments

We implemented the proposed approach in C++ and tested it on many challenging sequences. The ground truth of object bounding box was drawn manually in the original sequences. Two types of features, color histogram (8 bins for R, G, and B, respectively) and histogram of gradient orientation (8 bins), were used as features in our experiments. These two types of features were normalized respectively before concatenating. The patch size was made proportional to the target object, usually 0.25 to 0.5 times the dimension of target object. $K = 15$. The

**Table 1.** The accuracy and stability comparison among single and multiple scale schemes for the sequence in Fig.2. The large and small scale patches are 0.5 and 0.33 times of both sides of bounding box, respectively. The multi-scale scheme performs superiorly to the single scale ones.

| | Single Scale patches | | Multi-Scale patches |
|---|---|---|---|
| | Large scale | Small scale | Both scales |
| Accuracy | 29.53 | 21.10 | **15.29** |
| Stability | 18.77 | 28.31 | **8.94** |

sampling rate (the ratio of the number of sampled patches to that of all spatially eligible patches) is normally $1-16\%$. Generally, the sampling rate within object bounding box (e.g., 16%) is larger than that within its local background (e.g., 2%) in order to have approximately equal numbers of patches in both.

In our experiments, we employed two criteria, accuracy and stability, proposed in [14], to quantitatively evaluate the performance of different trackers. Given the ground truth $(G_x^t, G_y^t)$ of the target center, and the center $(O_x^t, O_y^t)$ of the located bounding box in frame $t$ by a tracker, the **Accuracy** $(A)$ and **Stability** $(S)$ over a sequence of $T+1$ frames are defined, respectively as

$$A = \max_{0 \leq t \leq T} \sqrt{(O_x^t - G_x^t)^2 + (O_y^t - G_y^t)^2},$$

$$S = \frac{1}{2T} \sum_{\substack{i=x,y \\ 1 \leq t \leq T}} [(O_i^t - G_i^t) - (O_i^{t-1} - G_i^{t-1})]^2.$$

Intuitively, accuracy evaluates the maximal offset of the trajectory away from the ground truth, and stability evaluates the consistency of the trajectory variation with the ground truth. We believe that the model drift problem is one of the main reasons to cause the location drift and trajectory offset. Therefore, the smaller the location drift and trajectory offset, the smaller the model drift.

### 5.1 Single-Scale Patch vs. Multi-scale Patch

In this subsection, the positive effectiveness of multi-scale patch sampling on tracking performance is illustrated. Several frames of a challenging sequence are shown in Fig.2. While only using single scale patches in our algorithm, the accuracy and stability are inferior to using two-scale patches. See Table 1 and Fig.2. The reason that multi-scale sampling scheme is superior to single-scale one is that the multi-scale scheme relies on a refined confidence map of small scale patches to locate the object. On the one hand, it efficiently eliminates the "false positive" small scale patches under the help of large ones; on the other hand, it covers target object more densely and widely than confidence maps produced by the object model of single large scale patches.

### 5.2 Ability to Remove Background Outliers

To illustrate the robustness of our algorithm against different initial bounding boxes, two different dimensions of boxes are set to show that background outliers

(a) (b) (c) (d) (e) (f) (g) (h) (i) (j) (k) (l)

**Fig. 2.** Single-scale sampling vs. multi-scale sampling. Nonzero confidences (yellow marks) and localizations (red bounding boxes) on frames 6, 11, 25 and 53 of CMU parking lot video are shown. (a) to (d), (e) to (h), and (i) to (l) are produced with large-, small-, and multi- scale samplings, respectively. It is seen that the bounding boxes in (i) to (l) fit the target object better than those in (a) to (d) and (e) to (h).

(a) (b) (c) (d) (e) (f)

**Fig. 3.** Two kinds of initialization on Seq.2, normal (a) and enlarged (d). In normal initialization, the object bounding box fits the dimension of target object well, while in the enlarged one, the target object is deliberately located in a corner of an enlarged box, almost 2 times the dimension of target object. (a) and (d) show the centers of sampled patches. (b) and (e) show the centers of patches in object model with normal and enlarged initializations, respectively. (c) and (f) are the confidence maps of next frame generated with normal and enlarged initializations, respectively.

can be effectively removed from both of them in modeling and localization steps. The sampled small scale patches in two initializations, which cover the bounding boxes very densely, are shown in Fig. 3 (a) and (d), respectively. It is seen that the centers of patches of object model, marked with yellow signs in Fig. 3(b) and (e), are almost all located on the target object in both initializations. This means that background outliers are effectively eliminated while modeling in both cases. It is also noticed that the centers of patches with nonzero confidences, marked with yellow signs in Fig. 3(c) and (f), are almost all located on the target object in the next frame. This means that there is almost no negative influence for the background outliers on locating target object in the next frame. Consequently, the localization of object is accurate and stable in the next frame.

### 5.3 Necessity of Purity

In this subsection, the necessity of purity $\alpha_i^{s_k}$ is illustrated. Our algorithm is simplified by treating each object patch identically. That is, $\alpha_i^{s_k}$'s are set to be a

(a) (b) (c)

**Fig. 4.** Comparison of confidence maps. (a) is the original segment. The confidence maps in (b) and (c) resulted from our original and simplified algorithms, respectively. Note that much more patches with nonzero confidences are on the target object, and much less patches with nonzero confidences are on the background in (b) than in (c).

constant. By the comparison of confidence maps (see Fig. 4), it is seen that the purity can improve the confidence maps substantially. Therefore, more accurate and stable tracking will be achieved with purity than without it.

### 5.4 Comparison with State-of-the-Art Approaches

Three other state-of-the-art tracking algorithms, Collins *et al.* algorithm [5], Avidan's ensemble tracking [8][1], and Lu and Hager's algorithm [10][2] , were compared with ours. All of them represent object and background with local and separate parts (specifically, pixels or patches), and use mean-shift in confidence maps to locate the object in the next frame without motion prediction and background motion estimation.

**Fig. 5.** Tracking results for frames 1, 38, 63, 89, 109 and 139 on Seq.1 with the deliberately enlarged initial bounding. While our algorithm tracks the object steadily, the ensemble tracker [8] fails at around frame 60.

In this paper, quantitative comparisons on Seqs.1 to 4 among the above three algorithms and ours are presented. Several representative frames of the sequences are shown in Figs. 5 to 8, respectively[3]. In Seq.1 (Fig. 5), the target object (two human bodies) undergoes out of plain rotation, and the portable camera shakes heavily. In Seq.2 (Fig. 6), the walking person was tracked in a low figure-ground contrast and low resolution surveillance video [8]. The person's leg has similar color to the ground, and his upper part is similar to the passing car. The target

[1] The program was downloaded from the programmers united develop net (www.pudn.com).

[2] The authors are grateful to Dr. Lu for providing their code [10].

[3] Whole sequences and more experimental results are in the supplemental materials.

**Fig. 6.** Comparison of DRLTracker with Lu and Hager's [10] on Seq.2. The first row is our results, the second row is from [10]. Frames 22, 42, 48, 49, 103, 109 are shown. The figure is really confused with ground and passing car. For such fitted initial bounding box used in this sequence, DRLTracker and [10] performed almost the same in accuracy and stability (see Table 2). And if the initial bounding box becomes a little bit larger, Lu and Hager's algorithm will almost always fail to track the walking person.

**Fig. 7.** Tracking results with DRLTracking on frames 1, 14, 16, 17, 18, 19, 32, 40, 50 and 76 (from left to right and top to down) of Seq.3. The object undergoes continual scale and appearance change and serious occlusion.

object in Seq.3 (Fig. 7) undergoes continual scale and appearance change and serious occlusion. Very distractive clutter appears in Seq.4 (Fig. 8) when the man closer to cars was tracked.

The performance comparisons were reported in Table 2.

It can be seen from Table 2 that the accuracy and stability of our algorithm are almost consistently superior to those of the other three ones.

For Seqs.5 and 6 (Fig. 9), the backgrounds are extremely cluttered, and the object undergoes drastic deforming and out of plain rotation. Due to the weak ability to eliminating outliers in bounding box, all the algorithms [5,8,10] fails to track the whole sequences, while ours locates the target objects accurately and stably.

We also compared DRLTracker with other recently published approaches [9,12][4] on several public videos. Fig. 10 shows comparisons on several representative frames of the truck video (Seq.9). The target truck is seriously disturbed by other background trucks (about from frames 140 to 257) and undergoes out of plain rotation (about from frames 391 to 500). BoostingTracker [9] can not track the target truck stably and lost it two times. SemiBoostingTracker [12] lost the target from frame 50 on. On the contrary, DRLTracker can always catch the target truck accurately and stably.

[4] The codes were downloaded from http://www.vision.ee.ethz.ch/boostingTrackers/

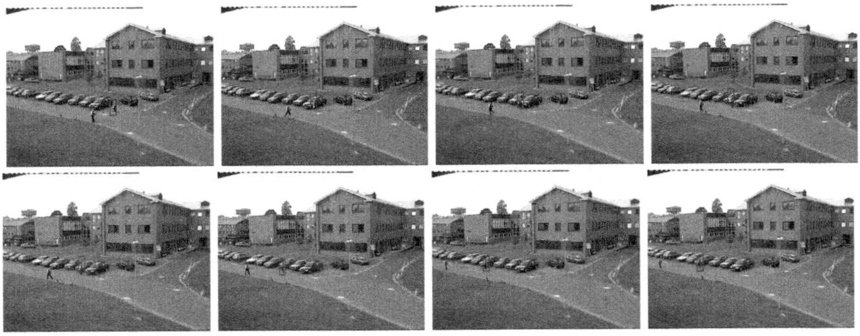

**Fig. 8.** Tracking results with DRLTracking on Seq.4, from a PETS 2001 sequence. The background cars are very distractive for tracking the man closer to the cars. From left to right and top to down, frames 4, 49, 80, 93, 104, 155, 219, 226 are shown.

**Table 2.** The accuracy and stability comparison among three state-of-the-art algorithms [5,8,10] and ours for Seqs, 1 to 4 (Figs. 5 to 8). The top part is for accuracy, and the bottom part for stability.

| | Seq.1 (40 frs) | Seq.2 (119 frs) | Seq.3 (78 frs) | Seq.4 (300 frs) |
|---|---|---|---|---|
| Collins' | fail from $14^{th}$ fr. | fail from $1^{st}$ fr. | fail from $1^{st}$ fr. | fail from $1^{st}$ fr. |
| Avidan's | fail from $5^{th}$ fr. | 18.35 | fail from $25^{th}$ fr. | 26.47 |
| Lu's | 40.36 | **16.86** | 34.31 | fail from $48^{th}$ fr. |
| Ours | **15.03** | 17.11 | **24.33** | **15.29** |
| Collins' | fail from $14^{th}$ fr. | fail from $1^{st}$ fr. | fail from $1^{st}$ fr. | fail from $1^{st}$ fr. |
| Avidan's | fail from $5^{th}$ fr. | 3.84 | fail from $25^{th}$ fr. | 6.67 |
| Lu's | 30.40 | **2.46** | 11.36 | fail from $48^{th}$ fr. |
| Ours | **9.13** | 3.53 | **9.49** | **5.87** |

**Fig. 9.** Tracking results for difficult Seqs.5 and 6. Our algorithm tracked objects accurately and stably, while other three ones failed.

**Fig. 10.** Tracking comparisons between DRLTracker and BoostingTracker [9] on the truck video (Seq.9, 500 frames). While DRLTracker tracked objects accurately and stably, BoostingTracker [9] lost the target truck from frames 161 to 180, catches it again from frames 181 to 428, and lost it again from frame 429 on. The numbers of presented frames are 1, 160, 162, 181, 429, 498, respectively, from left to right. The top and bottom rows are tracking results of BoostingTracker and DRLTracker, respectively.

## 6 Conclusion

Aiming at the model drift problem in track by classification approaches, we presented a novel approach to alleviate the problem. Our DRLTracker utilizes ranking lists and double-scale features to model the object and locate it. The superior ability to alleviate model drift over several state-of-the-art tracking approaches has been demonstrated quantitatively through extensive experiments.

**Acknowledgement.** This work is supported by National Nature Science Foundation of China. Grant No. 60835004, 60572057, and 60873185.

## References

1. Babenko, B., Yang, M., Belongie, S.: Visual Tracking with Online Multiple Instance Learning. In: CVPR (2009)
2. Santner, J., Leistner, C., Saffari, A., Pock, T., Bischof, H.: Prost: Parallel Robust Online Simple Tracking. In: CVPR (2010)
3. Kwon, J., Lee, K.: Visual Tracking Decomposition. In: CVPR (2010)
4. Avidan, S.: Support Vector Tracking. In: CVPR (2001)
5. Collins, R., Liu, Y., Leordeanu, M.: Online Selection of Discriminative Tracking Features. IEEE Trans. on PAMI 27, 1631–1643 (2005)
6. Yin, Z., Collins, R.: Spatial Divide and Conquer with Motion Cues for Tracking through Clutter. In: CVPR (2006)
7. Mahadevan, V., Vasconcelos, N.: Saliency-based Discriminant Tracking. In: CVPR (2009)
8. Avidan, S.: Ensemble Tracking. IEEE Trans. on PAMI 29, 261–271 (2007)
9. Grabner, H., Bischof, H.: On-line Boosting and Vision. In: CVPR (2006)
10. Lu, L., Hager, D.: A Nonparametric Treatment for Location/Segmentation Based Visual Tracking. In: CVPR (2007)
11. Tang, F., Brennan, S., Zhao, Q., Tao, H.: Co-tracking Using Semi-Supervised Support Vector Machines. In: ICCV (2007)

12. Grabner, H., Leistner, C., Bischof, H.: Semi-supervised on-line boosting for robust tracking. In: Forsyth, D., Torr, P., Zisserman, A. (eds.) ECCV 2008, Part I. LNCS, vol. 5302, pp. 234–247. Springer, Heidelberg (2008)
13. Liu, R., Cheng, J., Lu, H.: A Robust Boosting Tracker with Minimum Error Bound in a Co-training Framework. In: ICCV (2009)
14. Zhang, J., Chen, D., Tang, M.: Combining discriminative and descriptive models for tracking. In: Zha, H., Taniguchi, R.-i., Maybank, S. (eds.) ACCV 2009. LNCS, vol. 5994, pp. 113–122. Springer, Heidelberg (2010)
15. Fukunaga, K., Hostetler, I.: The Estimation of the Gradient of a Density Function with Appllications in Pattern Recognition. IEEE Tran. on Information Theory (1975)

# Analytical Dynamic Programming Tracker

Seiichi Uchida, Ikko Fujimura, Hiroki Kawano, and Yaokai Feng

Kyushu University, Fukuoka, Japan

**Abstract.** Visual tracking is formulated as an optimization problem of the position of a target object on video frames. This paper proposes a new tracking method based on dynamic programming (DP). Conventional DP-based tracking methods have utilized DP as an efficient breadth-first search algorithm. Thus, their computational complexity becomes prohibitive if the search breadth becomes large according to the increase of the number of parameters to be optimized. In contrast, the proposed method can avoid this problem by utilizing DP as an analytical solver rather than the conventional breadth-first search algorithm. In addition to experimental evaluations, it will be revealed that the proposed method has a close relation to the well-known KLT tracker.

## 1 Introduction

Generally, visual tracking [1] is formulated as an optimization problem of the target object position (and other posture parameters) at each video frame. Thus, tracking accuracy as well as computational complexity depends on optimization strategy.

Dynamic programming (DP) [2] is a well-known optimization strategy. It has been utilized in visual tracking for over 20 years and is still applied in many recent tracking methods. Since DP guarantees the global optimality of its solution, it can provide the most reliable (i.e., accurate) tracking result for a given optimization problem. This ability also indicates robustness against various distortions, such as occlusion.

Surprisingly, all the conventional DP-based tracking methods have utilized DP in the same manner. That is, DP is always utilized as an efficient *breadth-first search* algorithm; many candidate trajectories representing partial tracking results are branched and unified at each frame.

To the author's best knowledge, all of the conventional methods are commonly built on this breadth-first search algorithm, although they have their own originality at cost definition, computation reduction, etc.

Main contribution of this paper is to propose a novel DP-based tracking method, called *analytical DP tracker*. In the proposed method, we will utilize DP as an *analytical* solver of the visual tracking problem, not as a breadth-first search algorithm. It is very rare that DP is used as an analytic solver in past researches on computer vision and pattern recognition. (A smoothing method by Angel [3] and a contour matching method by Serra and Berthod [4] are the only exceptions.) This forgotten aspect of DP, however, is very beneficial from the viewpoint of computational efficiency.

R. Kimmel, R. Klette, and A. Sugimoto (Eds.): ACCV 2010, Part I, LNCS 6492, pp. 296–309, 2011.

The key idea of developing the analytical DP tracker is a quadratic representation of the cost at each frame. Here, the cost means some cost for locating the target at a certain position on the frame. By the quadratic representation, the objective function of the tracking problem becomes differentiable and it is possible to apply analytical DP to find the optimal tracking result efficiently. The quadratic cost may be found naturally in some kinds of objective functions (as discussed in 4.3), or may be derived intentionally through some approximations. Although quadratic approximation seems to be so rough, there are several techniques (such as bi-directional strategy described in Section 7.2) to have sufficient results under the approximation.

The merits of the analytical DP tracker are summarized as follows:

- Given a quadratic cost at every frame, it can provide the optimal tracking result only by $O(T)$ computations, where $T$ is the number of frames.
- It inherits good properties of the conventional DP-based trackers; that is, it can guarantee the global optimality of the tracking result, it is robust to occlusion, and it is capable of introducing user's interaction.
- It has a strong relation to a reputable tracker called KLT tracker [5,6].
- It can be extended for dealing with other deformations such as rotation and scaling with slight increase of computational complexity. Although the conventional DP-based trackers also can be extended, it causes drastic increase of computational complexity.

## 2 Related Work

Visual tracking methods can be classified into two types: online tracking and offline tracking. The majority belong to the former because it can realize real-time tracking. KLT tracker [5,6] and particle filter are typical online tracking methods. Online tracking methods determine the target position using the current and past frames. The inevitable drawback of online tracking is that if it misses the target at a certain frame due to any distortion (e.g., occlusion), it is very difficult to recover correct tracking in succeeding frames.

Although offline tracking methods are minority, they have great robustness against distortions. This is because it optimizes its tracking result after all video frames are given; in other words, the target position of each frame is determined by using not only past frames but also future frames. Thus, offline tracking will be better than online tracking if real-time processing is not necessary. For example, motion and behavior analysis (such as ball tracking for understanding and annotating soccer games) will be a good application for offline tracking. Motion capture for modeling human motion will also be another application because it requires accuracy rather than real-time process.

DP is one of the most important tools for offline tracking. It is well-known that DP is the most popular tool for nonlinear matching (called DP matching [7] or dynamic time warping), curve detection [8], and Snakes [9,10,11]. DP has also been employed in offline tracking methods [12,13,14,15,16,17,18] for over 20 years. Although there are many modern optimization strategies nowadays, such as graph cut and belief propagation (message passing), DP is still utilized in recent offline tracking methods. This may be because of its good properties, such as global optimality of its solution, numerical stability, and versatility in dealing with various cost functions and constraints.

One drawback of DP is its computational complexity. Since DP has been utilized as a breadth-first search algorithm in computer vision and pattern recognition, its computational complexity becomes large if search breadth becomes large. When we track a target object moving on an $M \times N$ image without rotation and scaling, search breadth becomes $O(MN)$. If we allow $R$-level rotation and $S$-level scaling, search breadth becomes $O(MNRS)$. In these cases, $O(M^2N^2T)$ and $O(M^2N^2R^2S^2T)$ computations are necessary in total, respectively.

In order to reduce large computations of the conventional DP-based tracking methods, various techniques have been employed. A classic technique is beam-search which "prunes" less hopeful candidate trajectories at each frame. More sophisticated techniques can be found in recent methods. In [17], k-d trees are utilized to select possible target positions efficiently at each frame. In [18], AdaBoost is utilized to "recognize" possible target positions. These recent methods are still similar to classic beam-search from the viewpoint that all these methods limit possible candidate trajectories by using some local criteria at each frame.

In order to avoid the struggle against huge search breadth, another aspect of DP is newly exploited in the proposed method as a radical remedy. As already noted in Section 1, DP can be used as a very efficient analytical solver for the tracking problem if the cost at each frame is defined as a quadratic function. The detail of the analytical DP tracker will be discussed in Section 4 after reviewing the conventional DP-based tracking method in Section 3.

# 3 Conventional DP Tracker

## 3.1 General Formulation of Tracking Problem

Generally, visual tracking is formulated as the minimization problem of the following objective function:

$$F(\boldsymbol{w}_1, \ldots, \boldsymbol{w}_T) = \lambda \sum_{i=1}^{T} d_i(\boldsymbol{w}_i) + \sum_{i=1}^{T-1} \|\boldsymbol{w}_{i+1} - \boldsymbol{w}_i\|^2, \tag{1}$$

where $\boldsymbol{w}_i = (x_i, y_i)^T$ is the object position in the $i$-th frame image with $M \times N$ pixels. The first term $d_i(\boldsymbol{w}_i)$ evaluates some cost for locating the target at $\boldsymbol{w}_i$ in the $i$-th frame. It is predetermined for each $\boldsymbol{w}_i$ as shown in Fig. 1 (a). The second term is used for smoothing target motion. The constant $\lambda$ is a weight. The position $\overline{w}_i = w_i$ which minimizes $F$ gives the tracking result.

## 3.2 Solution by Conventional DP

DP has been employed to obtain the globally optimal solution of the above tracking problem. In order to derive the conventional DP algorithm, consider the following function:

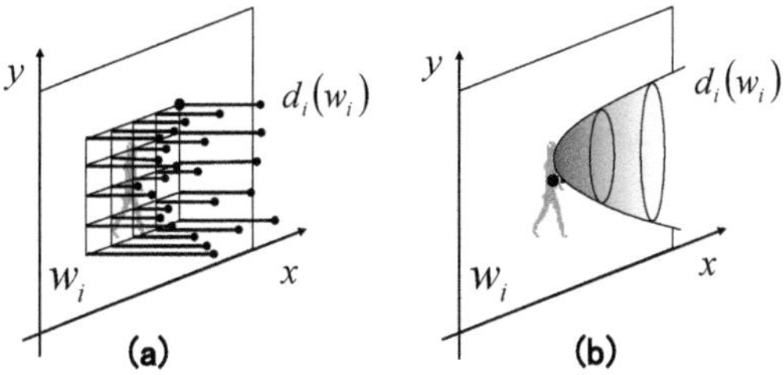


**Fig. 1.** Cost for conventional method (a) and the proposed method (b)

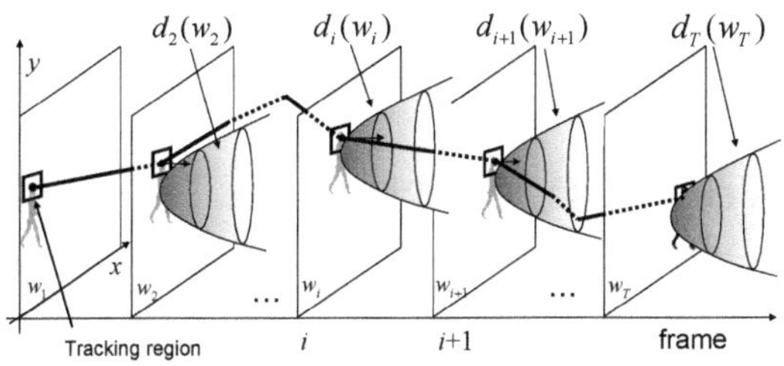


**Fig. 2.** Quadratic function $d_i(w_i)$ and optimization process of $\boldsymbol{w}_i$

$$f_i(\boldsymbol{w}_{i+1}) = \min_{\boldsymbol{w}_1,\ldots,\boldsymbol{w}_i} \sum_{k=1}^{i} \left[\lambda d_k(\boldsymbol{w}_k) + \|\boldsymbol{w}_{k+1} - \boldsymbol{w}_k\|^2\right]. \tag{2}$$

Using $f_i$, the global minimum $\min F$ can be represented as

$$\min F = \min_{\boldsymbol{w}_T} \left[f_{T-1}(\boldsymbol{w}_T) + \lambda d_T(\boldsymbol{w}_T)\right]. \tag{3}$$

From the principle of optimality [2], Eq. (2) can be rewritten as a recursion equation,

$$f_i(\boldsymbol{w}_{i+1}) = \min_{\boldsymbol{w}_i} \left[f_{i-1}(\boldsymbol{w}_i) + \lambda d_i(\boldsymbol{w}_i) + \|\boldsymbol{w}_{i+1} - \boldsymbol{w}_i\|^2\right]. \tag{4}$$

At $i = 1$, if the initial target position is given as $\overline{w}_1$, this equation becomes

$$f_1(\boldsymbol{w}_2) = \lambda d_1(\overline{w}_1) + \|\boldsymbol{w}_2 - \overline{w}_1\|^2 \tag{5}$$

The conventional DP algorithm is based on the calculation of $f_i(\boldsymbol{w}_{i+1})$ from $i = 1$ to $(T-1)$ according to (5) and (4) at every *discretized* position $\boldsymbol{w}_{i+1} \in [1, 2, \ldots, M] \times [1, 2, \ldots, N]$. Then, using $f_{T-1}(\boldsymbol{w}_T)$ and (3), $\min F$ is obtained. Note that the number of possible positions, $MN$, is the search breadth of the conventional DP algorithm. This is because every $\boldsymbol{w}_{i+1}$ has its own candidate trajectory whose accumulated cost is $f_i(\boldsymbol{w}_{i+1})$.

The globally optimal tracking result $\overline{w}_1, \ldots, \overline{w}_T$ is obtained through a back-tracking process. Specifically, the optimal target position $\overline{w}_T$ is first determined as $\boldsymbol{w}_T$ which gives $\min F$. Then, from $i = T-1$ to 1, $\overline{w}_i$ is determined as $\boldsymbol{w}_i$ which minimizes (4) with $\overline{w}_{i+1} = \boldsymbol{w}_{i+1}$.

### 3.3 Computational Complexity of Conventional DP

Let $\mathcal{W}_i$ denote the number of all possible $\boldsymbol{w}_i$, that is, the search breadth at the $i$-th frame. In this paper, we define $\boldsymbol{w}_i = (x_i, y_i)^T$ and thus $\mathcal{W}_i = MN$. Eq. (4) requires $O(\mathcal{W}_{i-1})$ computations for its minimum search. Since the above algorithm calculates (4) for $O(\mathcal{W}_i T)$ times, the total computational complexity becomes $O(\mathcal{W}_{i-1}\mathcal{W}_i T) = O(M^2N^2T)$.

If we allow rotation and scaling, $\boldsymbol{w}_i$ becomes a 4-dimensional vector of $x_i$, $y_i$, rotation angle, and scaling factor. Clearly, by this extension, the conventional method suffers from huge computations due to the drastic increase of $\mathcal{W}_i$. Thus, limiting $\mathcal{W}_i$ is a straightforward remedy. For example, setting an $L \times L$ window on each frame will reduce the computations ($M^2N^2 \to L^4$). Unfortunately, this remedy still suffers from a dilemma that accuracy may be sacrificed.

## 4 The Analytical DP Tracker

### 4.1 Quadratic Representation of Cost

The key idea to derive the analytical solution of the tracking problem of Section 3.1 is a quadratic representation of the cost $d_i(\boldsymbol{w}_i)$, i.e.,

$$d_i(\boldsymbol{w}_i) = \boldsymbol{w}_i^T \boldsymbol{P}_i \boldsymbol{w}_i + \boldsymbol{q}_i^T \boldsymbol{w}_i + r_i, \tag{6}$$

where $\boldsymbol{P}_i$ is a $2\times 2$ matrix, $\boldsymbol{q}_i$ is a two-dimensional vector, and $r_i$ is a scalar. (Unless otherwise mentioned, we assume only $x$-$y$ translation (i.e., $\boldsymbol{w}_i = (x_i, y_i)^T$). We assume these elements are predetermined in some way, whose example will be shown in Section 5. As noted before, this quadratic cost will be found naturally in some kinds of objective function. It will also be derived intentionally through some approximations. Section 4.3 will detail this point.

Figure 1 (b) illustrates a quadratic cost. Figure 2 illustrates a sequence of quadratic costs $d_1(\boldsymbol{w}_1), \ldots, d_i(\boldsymbol{w}_i), \ldots, d_T(\boldsymbol{w}_T)$ in addition to the sequence $\boldsymbol{w}_1, \ldots, \boldsymbol{w}_i, \ldots, \boldsymbol{w}_T$ to be optimized. It should be noted that $\boldsymbol{w}_i$ can be considered as a continuous variable under the quadratic representation of (6). Again, although quadratic representation seems to be rough, it can be practically compensated by using iterative updating (which is also employed in KLT), or bi-directional strategy, or multiple quadratic functions, etc.

### 4.2 Solution by Analytical DP

Now, the objective function (1) becomes differentiable with respect to $\boldsymbol{w}_i = (x_i, y_i)^T$. For the optimal tracking result, we may have a naive idea to solve the system of equations, $\{\partial F/\partial x_i = 0, \partial F/\partial y_i = 0, \forall i\}$. Unfortunately, its coefficient matrix becomes large (a $2T \times 2T$ matrix) and does not have a special structure (like a tridiagonal matrix) which enables fast solution.

Fortunately, DP provides a very efficient solution for the optimization problem. This new solution is different from the breadth-first search of Section 3.2; it is *analytical* solution with just $O(T)$ computations.

An important fact to derive the analytical solution by DP is that $f_i(\boldsymbol{w}_{i+1})$ becomes a quadratic function of $\boldsymbol{w}_i$, as well as $d_i(\boldsymbol{w}_i)$. This fact can be proven inductively by using the quadratic property of $d_i(\boldsymbol{w}_i)$ and the smoothness term. (Its proof is omitted here.) From this fact, $f_i(\boldsymbol{w}_{i+1})$ can also be represented as a quadratic function,

$$f_i(\boldsymbol{w}_{i+1}) = \boldsymbol{w}_{i+1}^T \boldsymbol{A}_i \boldsymbol{w}_{i+1} + \boldsymbol{b}_i^T \boldsymbol{w}_{i+1} + c_i, \tag{7}$$

where $\boldsymbol{A}_i$ is a $2 \times 2$ symmetric matrix, $\boldsymbol{b}_i$ is a two-dimensional vector, and $c_i$ is a scalar. These elements are variables and to be determined during the optimization process. (In contrast, $\boldsymbol{P}_i$, $\boldsymbol{q}_i$, and $r_i$ in (6) are predetermined constants.)

Substituting (7) into (4), we have

$$f_i(\boldsymbol{w}_{i+1}) = \min_{\boldsymbol{w}_i} \Big[ \boldsymbol{w}_i^T \boldsymbol{A}_{i-1} \boldsymbol{w}_i + \boldsymbol{b}_{i-1}^T \boldsymbol{w}_i + c_{i-1} + \lambda(\boldsymbol{w}_i^T \boldsymbol{P}_i \boldsymbol{w}_i + \boldsymbol{q}_i^T \boldsymbol{w}_i + r_i) \\ + \|\boldsymbol{w}_{i+1} - \boldsymbol{w}_i\|^2 \Big]. \tag{8}$$

The right-hand side of the above equation is a quadratic function of $\boldsymbol{w}_i$ and therefore $\overline{w}_i$ which attains its minimum is derived analytically as

$$\overline{w}_i = \left[\boldsymbol{A}_{i-1} + \lambda \boldsymbol{P}_i + \boldsymbol{I}\right]^{-1} \left(\boldsymbol{w}_{i+1} - (\boldsymbol{b}_{i-1} + \lambda \boldsymbol{q}_i)/2\right). \tag{9}$$

This shows the (backtrack) procedure to determine the optimal target position $\overline{w}_i$ for a given $\boldsymbol{w}_{i+1}$.

Substituting $\overline{w}_i$ of (9) into (8) and comparing with (7), we have

$$\left.\begin{array}{ll} \boldsymbol{A}_i & = \boldsymbol{I} - \left[\boldsymbol{A}_{i-1} + \lambda \boldsymbol{P}_i + \boldsymbol{I}\right]^{-1} \\ \boldsymbol{b}_i & = (b_{i-1} + \lambda \boldsymbol{q}_i)^T \left[\boldsymbol{I} - \boldsymbol{A}_i\right] \\ c_i & = -\frac{1}{4} \boldsymbol{b}_i \left(\boldsymbol{b}_{i-1} + \lambda \boldsymbol{q}_i\right) + c_{i-1} + \lambda r_i \end{array}\right\}. \tag{10}$$

This shows the recursive procedure to determine $\boldsymbol{A}_i, \boldsymbol{b}_i, c_i$ from $\boldsymbol{A}_{i-1}, \boldsymbol{b}_{i-1}, c_{i-1}$. Substituting (7) and (6) into (5), the initial condition of this recursive procedure, i.e., $\boldsymbol{A}_1, \boldsymbol{b}_1, c_1$, is given as follows:

$$\left.\begin{array}{ll} \boldsymbol{A}_1 & = \boldsymbol{I} \\ \boldsymbol{b}_1 & = -2\overline{w}_1 \\ c_1 & = \lambda d_1(\overline{w}_1) + \overline{w}_1^T \overline{w}_1 \end{array}\right\}. \tag{11}$$

Figure 3 summarizes the above procedure for the solution of the tracking problem based on analytical DP. After (11), $\{\boldsymbol{A}_i, \boldsymbol{b}_i, c_i | i = 1, \ldots, T-1\}$ is determined by using (10) recursively. Then, $\overline{w}_T$ is determined by substituting $\boldsymbol{A}_{T-1}, \boldsymbol{b}_{T-1}, c_{T-1}$ into the following equation, which is derived by the minimum condition of (3):

$$\overline{w}_T = -\frac{1}{2}[\boldsymbol{A}_{T-1} + \lambda \boldsymbol{P}_T]^{-1}(\boldsymbol{b}_{T-1} + \lambda \boldsymbol{q}_T) \tag{12}$$

The minimum value $\min F$ is given by using $\overline{w}_T$ and (3). Finally, the backtrack procedure (9) will provide the tracking result as $\overline{w}_{T-1}, \ldots, \overline{w}_2$.

**Input:**
Coefficients of cost functions $\{\boldsymbol{P}_i, \boldsymbol{q}_i, r_i \mid i = 1, \ldots, T\}$,
weight $\lambda \in \Re^+$, and initial position: $\overline{w}_1 \in \Re^2$
**Output:**
Optimal location: $\overline{w}_2, \ldots, \overline{w}_i, \ldots, \overline{w}_T$, and minimum distance: $\min F$

**[Step 1: Initial condition]**
Obtain $\boldsymbol{A}_1, \boldsymbol{b}_1, c_1$ by (11).
**[Step 2: DP recursion]**
For $i = 2$ to $T - 1$, obtain $\boldsymbol{A}_i, \boldsymbol{b}_i, c_i$ from $\boldsymbol{A}_{i-1}, \boldsymbol{b}_{i-1}, c_{i-1}$ by (10).
**[Step 3: Termination]**
Obtain $\overline{w}_T$ and $\min F$ by (12) and (3), respectively.
**[Step 4: Backtrack]**
For $i = T - 1$ downto 2, obtain $\overline{w}_i$ from $\overline{w}_{i+1}$ by (9).

**Fig. 3.** Analytical DP for visual tracking

### 4.3 Global Optimality of Solution

The tracking result given by the analytical DP tracker is the globally optimal solution which minimizes (1) with the quadratic cost (6). Thus, if the optimization problem is originally defined with the quadratic cost, the analytical DP tracker will provide the most accurate solution.

As an example, let us consider an objective function $F = \prod_i P_1(\boldsymbol{w}_i)P_2(\boldsymbol{w}_{i+1} - \boldsymbol{w}_i)$ where $P_1(\boldsymbol{w}_i)$ is a likelihood and $P_2(\boldsymbol{w}_{i+1} - \boldsymbol{w}_i)$ is a smoothness prior and both of $P_1$ and $P_2$ are two-dimensional normal distributions. In this case, the logarithm of $F$ results in the optimization problem of (1) with the quadratic cost (6). Thus, the solution by the proposed method will be the truly globally optimal solution of its original problem.

In contrast, for an optimization problem which is NOT originally defined with the quadratic cost, the analytical DP tracker will provide an approximation solution of the original problem. In this case, we must derive the quadratic cost by approximating the original cost. The tracking result $\overline{w}_1, \ldots, \overline{w}_T$ is different from the globally optimal solution of the original problem. In Section 5, a method to derive the quadratic cost (Fig. 1(b)) from a discrete cost (Fig. 1(a)) is discussed.

### 4.4 Computational Complexity

As shown from Fig. 3, the computational complexity of the proposed method is $O(T)$ and thus independent of the image size $M \times N$. This fact reveals the superiority of the proposed method over the conventional DP-based tracking method. This superiority will become more significant if we deal with rotation and scaling. By this extension, $\boldsymbol{w}_i$ becomes a 4-dimensional vector. As noted in Section 3.3, the computational complexity of the conventional method will increase drastically by this extension. In contrast, the computational complexity of the proposed method will increase very slightly; the increase mainly comes from the fact that $2 \times 2$ matrices become $4 \times 4$ matrices. This shows a good property of the analytical DP tracker.

The above computational complexity of the analytical DP tracker does not include the computations for the quadratic representation of $d_i(\boldsymbol{w}_i)$, that is, the computations for predetermining $\boldsymbol{P}_i$, $\boldsymbol{q}_i$, and $r_i$. Thus, to retain the computational superiority, it is important to minimize the computations for the quadratic representation. In Section 5, one feasible method of the quadratic representation will be given.

## 5 Implementation Issues

### 5.1 Quadratic Representation by Approximation

If the cost is not originally quadratic, tracking performance of the proposed method depends on the accuracy of quadratic approximation of $d_i(\boldsymbol{w}_i)$. For example, if the original cost is given by

$$\delta_i(\boldsymbol{w}_i) = \sum_{\epsilon} \left[I_i(\boldsymbol{w}_i + \epsilon) - I_1(\boldsymbol{w}_1 + \epsilon)\right]^2 , \tag{13}$$

we must determine the elements $\boldsymbol{P}_i, \boldsymbol{q}_i, r_i$ so that $\delta_i(\boldsymbol{w}_i) \sim d_i(\boldsymbol{w}_i)$.

In this paper, we derive a quadratic cost $d_i(\boldsymbol{w}_i)$ by Taylor series expansion, referring KLT tracker [5,6]. Specifically, $\delta_i(\boldsymbol{w}_i)$ is expanded to be a second-order polynomial function around a given approximation center $\widetilde{w}_i$.

Now, the remaining problem is how to determine $\widetilde{w}_i$ before the approximation. As noted above, the original cost $\delta_i(\boldsymbol{w}_i)$ is approximated around $\widetilde{w}_i$ by Taylor series expansion. Thus, if $\boldsymbol{w}_i$ becomes more distant from $\widetilde{w}_i$, the approximation error between $\delta_i(\boldsymbol{w}_i)$ and $d_i(\boldsymbol{w}_i)$ becomes more significant. Consequently, in order to determine $\overline{w}_i$ at the "true" target position by the proposed method, we must set $\widetilde{w}_i$ close to the true target position for better quadratic approximation; this is obviously a chicken-and-egg problem. The following Section 5.2 will be devoted to relax this problem.

### 5.2 Iterative Updating

The difficulty on deriving a good quadratic cost $d_i(\boldsymbol{w}_i)$ can be relaxed by iterative updating of the tracking result. The process of the iterative tracking is described as follows:

1. Find a very rough tracking result $\overline{w}_1^0, \ldots, \overline{w}_i^0, \ldots, \overline{w}_T^0$ by using, for example, the conventional DP on down-sampled frames (i.e., images whose size $(M \times N)$ is small enough).
2. Set $k = 1$.
3. $\widetilde{w}_i^k \leftarrow \overline{w}_i^{k-1}$, for all $i$.
4. Obtain quadratic cost function $d_i(\boldsymbol{w}_i)$ by Taylor series expansion around $\widetilde{w}_i^k$, for all $i$.
5. Find updated tracking result $\overline{w}_1^k, \ldots, \overline{w}_i^k, \ldots, \overline{w}_T^k$ by using the analytical DP tracker.

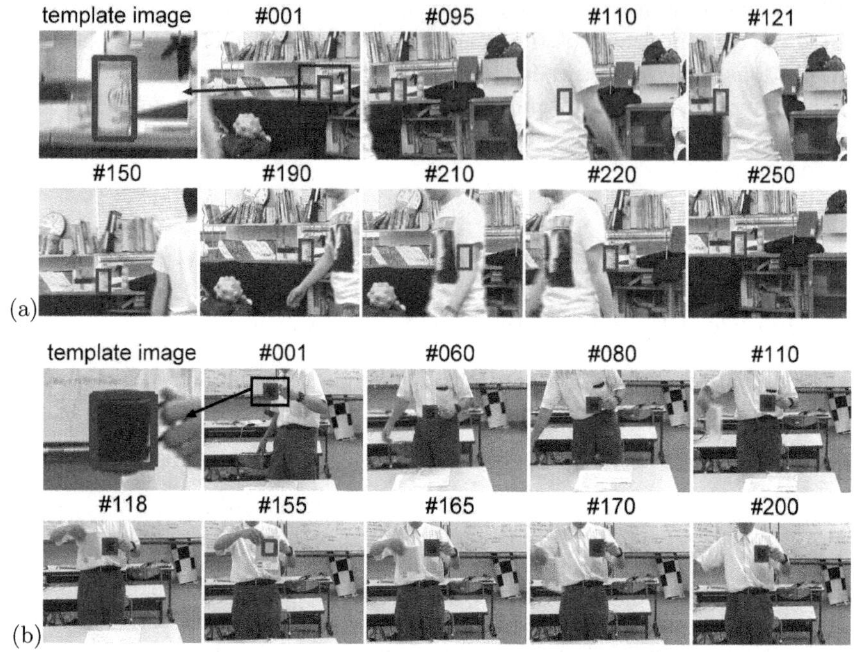


**Fig. 4.** Results by analytical DP tracker

6. Terminate if $k$ reaches a pre-specified maximum iteration number. Otherwise, $k \leftarrow k + 1$ and goto 3.

Even though the initial tracking result $\overline{w}_1^0, \ldots, \overline{w}_i^0, \ldots, \overline{w}_T^0$ is not accurate, it will be updated iteratively and converged around the true target position. Note that the initial rough tracking result can also be given by linear interpolation of the points specified manually at several frames.

# 6 Experimental Results

## 6.1 Tracking Results

Figure 4(a) shows a video sequence captured by a hand-held camera. The target object (an orange color plastic object) was occluded completely by a man during 100$\sim$ 120-th frames and 200 $\sim$ 220-th frames. Figure 4(b) shows another video sequence. The target object (a cup) was occluded completely by a book during 120$\sim$ 160-th frames. In both video sequences, the target object underwent neither scaling nor rotation.

In advance to the optimization, the target object template $I_1$ was manually specified at the first frame (#001). Using this $I_1$, $\delta_i(\boldsymbol{w}_i)$ was evaluated at each frame and then $d_i(\boldsymbol{w}_i)$ was derived according to the procedure of Section 5. The iterative updating process of Section 5.2 was also employed. Unless otherwise mentioned, the pre-specified maximum iteration number was fixed at 20.

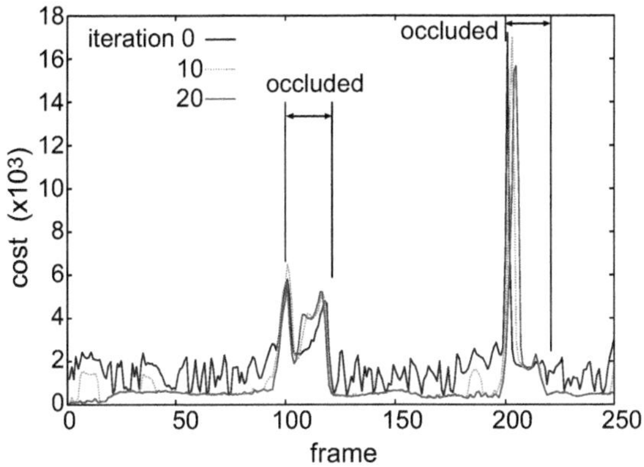


**Fig. 5.** Change of cost $d_i(\overline{w}_i)$ during iterative updating

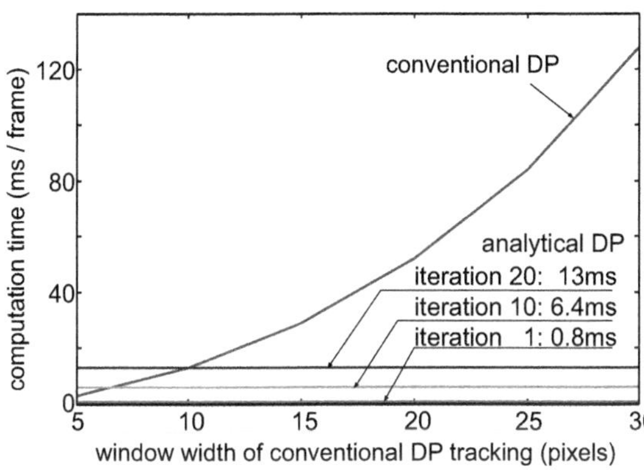


**Fig. 6.** Computation time

In Fig. 4, the red rectangular shows the target position optimized by the analytical DP tracker. Both results reveal that the proposed method could track the target objects accurately. Even if the target object was occluded completely, the tracking result was still stable. (See the 210-th frame in (a) and the 155th frame in (b).) This shows that the offline tracking with DP is robust against occlusion as expected. Note that the mean-shift tracker faild to track the object due to the occlusion. (See the supplementary video.)

Figure 5 shows the change of $d_i(\overline{w}_i)$ during the iterative updating process for the video sequence of Fig. 4(a). The initial tracking result ("iteration 0"), which was obtained by the conventional DP on the down-sampled frames, shows that $d_i(\overline{w}_i)$ was fluctuating and rather large. In contrast, the tracking results after the iterative updating shows that the cost $d_i(\overline{w}_i)$ decreases monotonically according to the iteration and finally converged into a very small value (except for the frames with occlusion). This result shows the accuracy of the tracking result quantitatively.

### 6.2 Computation Time

Figure 6 shows the computation time per a frame of the analytical DP tracking and the conventional DP tracking. They were measured on a PC (Pentium D).

Since the conventional DP tracking required more than 10 hours without any special treatment for reducing computations, a two-step optimization procedure and a window were introduced for reducing its computations. The first step was the same as that of the iterative updating of the proposed method; that is, a very rough tracking result was obtained on down-sampled frame images as the initial tracking result. At the second step, an $L \times L$ window was set around the initial tracking result. The horizontal axis of Fig. 6 was $L$ for the conventional method.

From several comparative experiments, it was known that $L$ should be 15 pixels at least to have the tracking result as accurate as that by the analytical DP tracking. Thus fact confirms the superiority of the analytical DP tracking over the conventional method; at $L = 15$, the computation time of the proposed method was about half of the conventional method. Note again that if rotation and scaling are allowed (that is, the dimension of control variables becomes higher), this superiority will become far more significant.

## 7 Discussion

### 7.1 Relation to KLT Tracker

The quadratic representation of tracking cost can be found in KLT tracker [5]. KLT tracker is a well-known online tracking method and the target position $\boldsymbol{w}_i$ is optimized using a quadratic cost function $d_i(\boldsymbol{w}_i)$ derived by Taylor series expansion of $\delta_i(\boldsymbol{w}_i)$ around $\widetilde{w}_i = \boldsymbol{w}_{i-1}$. (This assumes that the displacement between the $(i-1)$th and $i$th frames is very small.) In order to improve the tracking accuracy, $\overline{w}_i$ giving the minimum of $d_i(\boldsymbol{w}_i)$ is then considered as the updated approximation center $\widetilde{w}_i$ in KLT. This procedure is repeated until convergence and $\overline{w}_i$ is determined as the converged $\boldsymbol{w}_i$. Then the optimization of $\boldsymbol{w}_{i+1}$ starts.

This iterative updating process of KLT tracker is very similar to the iterative updating process of the analytical DP tracker introduced in Section 5.2. Especially, they have the same purpose of compensating poor approximation ability

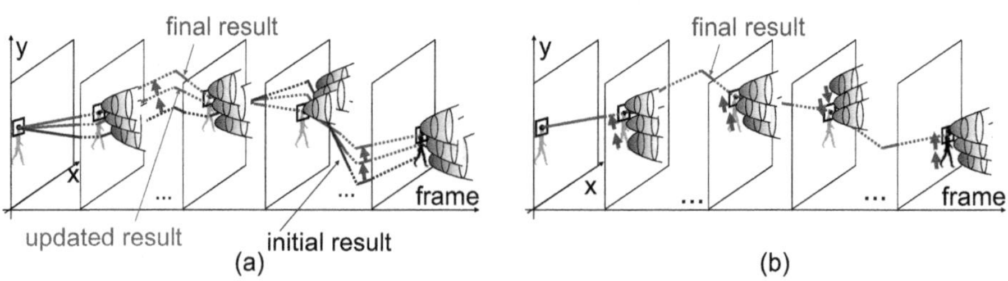


**Fig. 7.** Iterative optimization of (a) analytical DP tracker and (b) KLT tracker

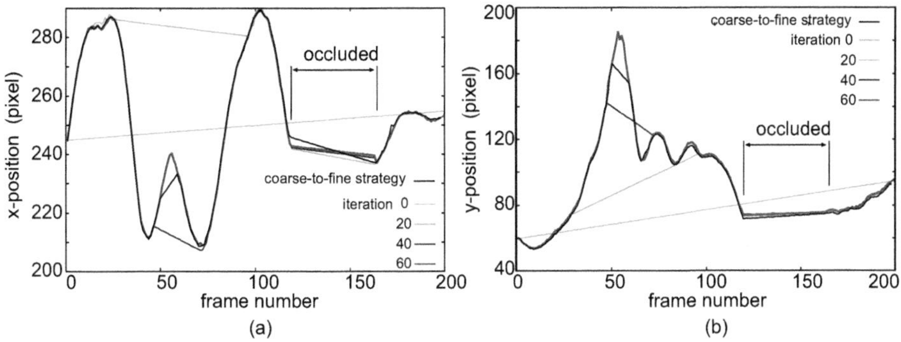


**Fig. 8.** Tracking result by bi-directional strategy. (a) The trajectory of $x_1, \ldots, x_T$. (b) The trajectory of $y_1, \ldots, y_T$.

of Taylor series expansion. They, however, are different in their optimization strategies. Figure 7 show their iterative updating processes. As noted above, the iteration of KLT tracker is performed within a frame. In other words, a "local" iterative optimization is done. In contrast, the iteration of the analytical DP tracker is performed for all the tracking positions $\boldsymbol{w}_1, \ldots, \boldsymbol{w}_i, \ldots, \boldsymbol{w}_T$. Thus a "global" iterative optimization done. Consequently, we can say that the analytical DP tracker with the iterative updating process is a global optimization version of KLT tracker[1].

### 7.2 Bi-directional Strategy

Bi-directional tracking in [15] is a tracking method where not only the initial position but also the final position are specified. Even for offline tracking, the specification of the final position is very beneficial to obtain better tracking accuracy.

This idea can be utilized in the iterative updating for improving computational efficiency. First, a linear trajectory connecting the two specified positions is considered as the initial tracking result. Then, it was used as the approximation center to derive quadratic cost function $d_i(\boldsymbol{w}_i)$ at all $i$. A difference from the iterative updating process in Section 5.2 is that the approximation center $\widetilde{w}_i^k$ is not set at $\overline{w}_i^{k-1}$ but selected from $\overline{w}_{i-1}^{k-1}$ or $\overline{w}_{i+1}^{k-1}$ according to their scores. This selection mechanism is introduced for propagating the reliable tracking results around the both ends ($i = 1$ and $T$) gradually into the middle part ($i \sim T/2$) according to the iteration.

Figure 8 shows the change of the tracking result given by the bi-directional strategy. The propagation of accurate tracking results around both ends can be observed. After 60 iterations, we could have the tracking result which almost

[1] Precisely speaking, there is another difference between analytical DP tracker and KLT tracker. That is, the former evaluates the smoothness of its tracking result whereas the latter does not.

equals the result by the iterative updating of Section 5.2("Coarse-to-fine strategy" in Fig. 8). The bi-directional strategy does not require the initial tracking result by the conventional method on down-sampled frames and therefore it is more computationally efficient.

### 7.3 Multiple Quadratic Functions

In this paper, we assume the cost $d_i(\boldsymbol{w}_i)$ is represented by a quadratic function, i.e., a unimodal function. Unless the cost is originally quadratic, this representation is not perfect. In fact, the cost by (13) will be a multimodal function.

One possible remedy to improve the representation accuracy is multiple quadratic functions. This is analogous to Gaussian mixture. Specifically, $K$ quadratic functions, $d_{i,k}(\boldsymbol{w}_i), k = 1, \ldots, K$, are prepared at the $i$-th frame for approximating the multimodal function. When optimizing $\boldsymbol{w}_i$ by the analytical DP tracker, one quadratic function is selected from the $K$ functions at each frame. It is interesting to note that the selection itself can be done optimally by the conventional DP.

## 8 Conclusion

In this paper, a novel DP-based tracking method, called analytical DP tracker, has been proposed. Quadratic representation of the tracking cost at each frame allows an analytical solution by DP. This solution is very effective; in fact, given the quadratic cost at each frame, we have globally optimal tracking result with only $O(T)$ computations, where $T$ is the number of frames. In addition to this computational feasibility, it has been revealed experimentally that the analytical DP tracker could provide accurate tracking results against complete occlusions by fully utilizing its offline optimization property. As the practical problem, we have also discussed several techniques to derive the quadratic tracking cost by approximating some original tracking cost. Finally, it was shown that the analytical DP tracker can be considered as an offline version of the well-known KLT tracker.

Future work will focus on the extension for dealing with rotation and scaling. Contour models such as Schoenemann and Cremers [19] can also be employed as an extended target model. As noted before, those extensions will enhance the superiority of the proposed method in computational efficiency over the conventional DP-based tracking methods. In addition, we must investigate more accurate and sophisticated cost representation, such as multiple quadratic functions.

## References

1. Yilmaz, A., Javed, O., Shah, M.: Object tracking: A survey. ACM Comput. Surveys 38, 1–45 (2006)
2. Bellman, R., Dreyfus, S.: Applied Dynamic Programming. Princeton University Press, Princeton (1962)

3. Angel, E.: Dynamic programming for noncausal problems. IEEE Trans. Automatic Control AC-26, 1041–1047 (1981)
4. Serra, B., Berthod, M.: Subpixel contour matching using continuous dynamic programming. In: CVPR, pp. 202–207 (1994)
5. Tomasi, C., Kanade, T.: Detection and tracking of point features. Tech. Report, CMU-CS-91-132, Carnegie Mellon Univ. (1991)
6. Shi, J., Tomasi, C.: Good features to track. In: CVPR, pp. 593–600 (1994)
7. Sakoe, H., Chiba, S.: A dynamic programming algorithm optimization for spoken word recognition. IEEE Trans. ASSP ASSP-26, 43–49 (1978)
8. Montanari, U.: On the optimal detection of curves in noisy pictures. ACM Comm. 14, 335–345 (1971)
9. Amini, A.A., Weymouth, T.E., Jain, R.C.: Using dynamic programming for solving variational problems in vision. TPAMI 12, 855–867 (1990)
10. Geiger, D., Gupta, A., Vlontzos, J.: Dynamic programming for detecting, tracking and matching deformable contours. TPAMI 17, 294–302 (1995)
11. Akgul, Y.S., Kambhamettu, C.: A coarse-to-fine deformable contour optimization framework. TPAMI 25, 174–186 (2003)
12. Barniv, Y.: Dynamic programming solution for detecting dim moving target. IEEE Trans. Aerospace and Electronic Systems 21, 144–156 (1985)
13. Arnold, J., Shaw, S., Pasternack, H.: Efficient target tracking using dynamic programming. IEEE Trans. Aerospace and Electronic Systems 29, 44–56 (1993)
14. Han, M., Xu, W., Tao, H., Gong, Y.: An algorithm for multiple object trajectory tracking. In: CVPR, pp. 864–871 (2004)
15. Sun, J., Zhang, W., Tang, X., Shum, H.: Bi-directional tracking using trajectory segment analysis. In: ICCV, pp. 717–724 (2005)
16. Dreuw, P., Deselaers, T., Rybach, D., Keysers, D., Ney, H.: Tracking using dynamic programming for appearance-based sign language recognition. In: FGR, pp. 293–298 (2006)
17. Buchanan, A., Fitzgibbon, A.: Interactive feature tracking using k-d trees and dynamic programming. In: CVPR, pp. 626–633 (2006)
18. Wei, Y., Sun, J., Tang, X., Shum, H.-Y.: Interactive offline tracking for color objects. In: ICCV, pp. 1–8 (2007)
19. Schoenemann, T., Cremers, D.: A combinatorial solution for model-based image segmentation and real-time tracking. TPAMI 32, 1153–1164 (2010)

# A Multi-Scale Learning Framework for Visual Categorization

Shao-Chuan Wang and Yu-Chiang Frank Wang

Research Center for Information Technology Innovation
Academia Sinica, Taipei, Taiwan

**Abstract.** Spatial pyramid matching has recently become a promising technique for image classification. Despite its success and popularity, no prior work has tackled the problem of learning the optimal spatial pyramid representation for the given image data and the associated object category. We propose a Multiple Scale Learning (MSL) framework to learn the best weights for each scale in the pyramid. Our MSL algorithm would produce class-specific spatial pyramid image representations and thus provide improved recognition performance. We approach the MSL problem as solving a multiple kernel learning (MKL) task, which defines the optimal combination of base kernels constructed at different pyramid levels. A wide range of experiments on Oxford flower and Caltech-101 datasets are conducted, including the use of state-of-the-art feature encoding and pooling strategies. Finally, excellent empirical results reported on both datasets validate the feasibility of our proposed method.

## 1 Introduction

Among existing methods for image classification, the bag-of-features model [1,2] has become a very popular technique and has demonstrated its success in recent years. It quantizes image descriptors into distinct visual words, and uses a compact histogram representation to record the numbers of occurrences of each visual word in an image. One of the major problems of this is the determination of visual words, since the widely-used strategy is to cluster local image descriptors into a set of disjoint groups, and thus the representative of each group is considered as a visual word of the given image data [1,2] (the collection of such visual words is called a *dictionary* (or a *codebook*)). However, the major concern of this technique is that it discards the spatial order of local descriptors. Lazebnik et al. [3] proposed a spatial pyramid matching (SPM) technique to address this concern by utilizing a spatial pyramid image representation, in which an image is iteratively divided into grid cells in a top-down way (i.e. from coarse to fine scales). Instead of constructing a codebook by vector quantization, Yang et al. [4] further extended the spatial pyramid representation and proposed a ScSPM framework with sparse coding of image descriptors and max pooling techniques. Since only linear kernels are required in their work, Yang's ScSPM is able to address large-scale classification problems with reasonable computation time.

R. Kimmel, R. Klette, and A. Sugimoto (Eds.): ACCV 2010, Part I, LNCS 6492, pp. 310–322, 2011.

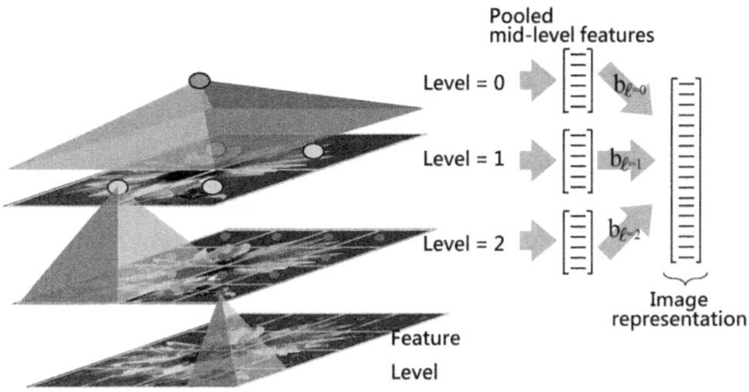


**Fig. 1.** An illustration of spatial pyramid image representation. Red dots are the encoded coefficient vectors of local image descriptors. Gray pyramids represent the *pooling* operations. Blue, yellow, and green dots are the pooled vectors at levels 0, 1 and 2, respectively. Each dot describes the patch statistics within the associated grid region. These pooled vectors are typically concatenated with predetermined weights (i.e. $b_\ell$ are fixed) as a single vector, which is the final spatial pyramid image representation. Given the image data and the associated object category, our multi-scale learning (MSL) framework aims at identifying the optimal weights for improved classification.

To the best of our knowledge, no prior work has addressed the determination of the *best* spatial pyramid representation of the given image data and the associated object category. Existing methods using SPM only focus on the designs of feature encoding methods, pooling strategies and the corresponding classifiers, and all prior work uses predetermined weights to concatenate mid-level representation in each scale (c.f. Fig. 1). It is not surprising that, for visual categorization, some object images are more discriminative at coarse levels, while others contain more descriptive information at finer scales. Therefore, we advocate the *learning* of the best spatial pyramid representation by approaching this problem as solving a multiple kernel learning (MKL) task, and we refer to our proposed method as a Multiple Scale Learning (MSL) framework. More specifically, given the image data and the associated object category, our MSL determines the optimal combination of base kernels constructed at different pyramid levels, and we will show that this task can be posed as a convex optimization problem and guarantees the global optimum. We will also show that the learned weights for each image scale provide descriptive and semantic interpretation of image data, and our proposed spatial pyramid representation significantly improves the recognition performance of image classification.

## 2 Related Work

Our work is built upon the recent development of spatial pyramid representation [3, 5] and kernel learning techniques [6, 7] for image classification. As shown in Figure 1, Lazebnik et al. [3] suggested to partition an image into $2^\ell \times 2^\ell$ grids in different scales $\ell = 0, 1, 2$, etc. The histogram of visual words (or equivalently the vectors pooled by the sum operation) within each grid is calculated. All histograms from different grids and levels are concatenated with a predetermined factor (e.g. 1 or $1/2^{2\ell}$). The final concatenated vector is thus considered as the spatial pyramid representation of the given image. We note that if the coarsest level $\ell = 0$ is used, SPM is simply the standard bag-of-features model.

To the best of our knowledge, existing approaches using SPM for image classification simply integrate visual word histograms generated at different levels of the pyramid in an *ad hoc* way, which might not be practical. We thus propose to construct the *optimal* spatial pyramid representation for each class by *learning* the weighting factors at each level in the pyramid. Our goal is not only to achieve better recognition performance, but provides an effective visualization of sematic and scale information for each object class. While it is possible to use cross validation to determine the optimal weights for each visual word histogram at different levels in the pyramid, it will significantly increases the computation complexity, especially if there is a large number of free parameters to be determined in the entire system. We note that existing work has utilized different learning or optimization strategies to address this type of problem, and the performance can be improved without sacrificing the computation load. More specifically, researchers in machine learning communities have proposed boosting techniques to select the optimal kernel or feature combination for recognition, regression, etc. problems [8, 9, 10, 11, 12]. Other methods like metric/similarity learning [13, 14], distance function learning [15, 16, 17, 18], and descriptor learning [19] also apply the latest optimization strategies to adaptively learn the parameters from the data. Recently, one of the successful examples in image classification and kernel learning is the fusion of heterogeneous features proposed by Gehler and Nowozin [10], and also by Bosch et al. [12]. Gehler and Nowozin proposed to combine heterogeneous features via multiple kernel learning as well as linear programming boosting methods (LPBoost), while Bosch fused shape and appearance features via MKL combined with a regions of interest preprocessing. Both reported attractive results on Caltech datasets.

Inspired by the above work, we propose to use a MKL framework to identify discriminating image scales, and thus weight the image representations accordingly. We will show in Sect. 4 that the performance of our proposed framework outperforms state-of-the-art methods using bag-of-features or SPM models using predetermined weighting schemes. It is worth repeating that, once the optimal weights for each scale are determined, one can easily extract significant scale information for each image object class. This provides an effective semantic interpretation for the given image data.

## 3 Multi-Scale Learning for Image Classification

Previously, Gehler et al. [10] associated image features with kernel functions, and transformed the feature selection/combination problem into a task of kernel selection. Similarly, Subrahmanya and Shin [20] performed a feature selection procedure by constructing base kernels using different group of features. Our proposed

MSL framework incorporates multi-scale spatial and appearance information to learn the optimal spatial pyramid representation for image classification.

In our MSL, we define a multi-scale kernel matrix, which is positive semi-definite and satisfies

$$K_{ij} \equiv K(\boldsymbol{x}_i, \boldsymbol{x}_j) = \sum_{\ell=0}^{L} b_\ell k^\ell(\boldsymbol{v}_i^\ell, \boldsymbol{v}_j^\ell), \tag{1}$$

where $\boldsymbol{x}_i$ is the image representation of $i$-th image, $k^\ell$ is the kernel function constructed at level $\ell$ in the spatial pyramid, $b_\ell$ is the associated weight, and $\boldsymbol{v}^\ell \in \mathbf{R}^{(2^{2\ell})\mathcal{K}}$ is the vector produced by concatenating all $2^{2\ell}$ pooled vectors at level $\ell$. We note that if the base kernel is linear (as we did in this paper), $b_\ell$ will be super-linearly proportional to the number of grids in level $\ell$, since the kernel output is the inner product between the two pooled vectors from each level.

The determination of the optimal weights in the above equation is known as the multiple kernel learning problem. Several algorithms have been proposed to solve the MKL problem and its variants. The reviews of MKL from an optimization viewpoint can be seen in [6, 7, 21, 22, 23], and we particularly employ the algorithm proposed by Sonnenburg et al. [23] due to its efficiency and simplicity of implementation.

In order to learn the optimal kernels over image scales to represent an image, we convert the original MKL problem into the following optimization problem (in its primal form),

$$\begin{aligned} (P) \min_{\boldsymbol{w}_\ell, w_0, \boldsymbol{\xi}, \boldsymbol{b}} \quad & \tfrac{1}{2}(\textstyle\sum_{\ell=0}^{L} b_\ell \langle \boldsymbol{w}_\ell, \boldsymbol{w}_\ell \rangle)^2 + C \sum_{i=1}^{N} \xi_i \\ \text{subject to} \quad & y_i(\textstyle\sum_{\ell=0}^{L} b_\ell \langle \boldsymbol{w}_\ell, \Phi(\boldsymbol{v}_i^\ell)\rangle + w_0) \geq 1 - \xi_i \\ & \textstyle\sum_{\ell=0}^{L} b_\ell = 1, \boldsymbol{b} \succeq 0, \boldsymbol{\xi} \succeq 0, \end{aligned} \tag{2}$$

where $\langle \cdot, \cdot \rangle$ represents the inner product in the $L_2$ Hilbert space, $b = (b_0, b_1, \ldots, b_L)^T$, and $\boldsymbol{\xi} = (\xi_1, \xi_2, \ldots, \xi_N)^T$. However, similar to the standard SVM optimization problem, the above optimization problem is not as explicit as its dual problem, which is shown as follows,

$$\begin{aligned} (D) \min_{\mathbf{a}, \gamma} \quad & \gamma - \textstyle\sum_i^N a_i \\ \text{subject to} \quad & 0 \preceq \boldsymbol{a} \preceq C, \textstyle\sum_i^N a_i y_i = 0 \\ & \tfrac{1}{2} \textstyle\sum_{ij}^N a_i a_j y_i y_j k_{ij}^\ell \leq \gamma, \forall \ell = 0, 1, ..., L, \end{aligned} \tag{3}$$

where $k_{ij}^\ell = k^\ell(\boldsymbol{v}_i^\ell, \boldsymbol{v}_j^\ell) = \langle \Phi(\boldsymbol{v}_i^\ell), \Phi(\boldsymbol{v}_j^\ell) \rangle$, and $\boldsymbol{a} = (a_1, a_2, \ldots, a_N)^T$. If the kernel is linear (as ours in this paper), $\langle \Phi(\boldsymbol{v}_i^\ell), \Phi(\boldsymbol{v}_j^\ell) \rangle$ is simply $\langle \boldsymbol{v}_i^\ell, \boldsymbol{v}_j^\ell \rangle$. Note that we have one quadratic constraint for each kernel $k^\ell$, i.e., we have $L+1$ constraints in total. Sonnenburg et al. [23] have shown that the above problem can be reformulated as a semi-infinite linear program (SILP),

$$\begin{aligned} \max_{\boldsymbol{b},\theta} \quad & \theta \\ \text{subject to} \quad & \textstyle\sum_{\ell=0}^{L} b_\ell = 1, \boldsymbol{b} \succeq 0, \sum_{\ell=0}^{L} b_\ell S_\ell(\boldsymbol{a}) \geq \theta \\ & \forall \boldsymbol{a} \in \mathbf{R}^N \text{ with } 0 \preceq \boldsymbol{a} \preceq C \text{ and } \textstyle\sum_i y_i a_i = 0, \end{aligned} \tag{4}$$

where $S_\ell(\boldsymbol{a}) \equiv \frac{1}{2}\sum_{ij}^{N} a_i a_j y_i y_j k_{ij}^\ell - \sum_i^N a_i$.

Note that the above SILP is actually a *linear* programming problem due to the fact that $\theta$ and $\boldsymbol{b}$ are linearly constrained with *infinite* constraints, i.e. there will be a constraint for each $\boldsymbol{a} \in \mathbf{R}^N$ satisfying $0 \preceq \boldsymbol{a} \preceq C$ and $\sum_i y_i a_i = 0$. To solve this problem, a wrapper algorithm [23] is proposed to alternatively optimize $\boldsymbol{a}$ and $\boldsymbol{b}$ in each iteration. When $\boldsymbol{b}$ is fixed, SILP turns into a single kernel SVM problem, which can be efficiently solved by many SVM solvers such as LibSVM [24]. On the other hand, when $\boldsymbol{a}$ is fixed, we need to solve a linear programming problem with finite constraints, which can be also efficiently solved by many linear programming solvers. As a result, this wrapper algorithm enjoys the benefit of easy and efficient implementation.

**Algorithm 1.** Multi-scale learning for class-specific spatial pyramid representation

```
{1}. Building the kernels in all scales:
for ℓ = 0 to L do
  for all i, j do
    k_ij^ℓ ← ⟨v_i^ℓ, v_j^ℓ⟩ {for linear kernel}
  end for
end for
k^ℓ ← k^ℓ/Tr(k^ℓ) {trace normalization}
{2}. Learning b_ℓ by solving a semi-infinite linear program [23]:
(a, b_ℓ) ← SILP(y, k)
```

Note that all kernel matrices have been normalized to unit trace in order to balance the contributions of base kernels. Algorithm 1 shows our proposed algorithm for learning class-specific spatial pyramid representations. Note that $\boldsymbol{a}, \boldsymbol{b_\ell}$ in Algorithm 1 are the MKL parameters; $\boldsymbol{a}$ represent the Lagrange multipliers for SVM, and $\boldsymbol{b_\ell}$ describe the optimal weights of each base kernel, indicating the preferable spatial pyramid image representation for each object category. In our implementation of the SILP solver, we integrate LibSVM [24] and the MATLAB function `linprog` for solving the single kernel SVM and the linear programming problems, respectively.

After solving the above optimization problem for the given image data, we obtain the estimated optimal weighting factors $b_\ell$ and equivalently acquire the significance of concatenated pooled vectors from different scales. This weighted and concatenated feature vector will be the final form for our spatial pyramid image representation.

**Fig. 2.** Example images from the Oxford flower dataset [25]

**Fig. 3.** Example images from the Caltech 101 dataset [26]

# 4 Experiments

## 4.1 Datasets

We conduct experiments on Oxford flower [25] and Caltech-101 [26] datasets in this paper. The Oxford flower dataset is a small-scale dataset containing 17 different types of flowers (80 images each). Fig. 2 shows some example images from this dataset. We randomly pick 40 images per category for training, and the remaining 40 for testing. To evaluate the feasibility and scalability of our proposed method, we further consider the Caltech-101 dataset. This dataset consists of 101 different object classes with variant numbers of images per object category (see Fig. 3 for example images). To compare our results to those reported in prior work, we use the same experimental setups such as the selection of training and test sets (15 to 30 training images per object category, and up to 50 images per category for testing), and the choice of the evaluation metric (i.e. the mean average precision (MAP)).

In our experiments on both datasets, SIFT descriptors are extracted from 16 $\times$ 16 pixel patches of an image, and the spacing between adjacent patches is 6 pixels (horizontally and vertically). We further resize the longer side of the image to 300 pixels if its width or height exceeds 300 pixels. Prior work on the Caltech-101 dataset also did similar operations [2, 4].

### 4.2 Dictionary Learning and Local Descriptor Encoding

We choose two dictionary learning scenarios for comparisons: vector quantization (VQ) and sparse coding (SC). We select $\mathcal{K} = 225$ and 900 as the sizes of the dictionary. To perform sparse coding, we use the SPAMS software package developed by Mairal et al. [27], and the parameter $\lambda$, which controls the sparsity of the encoded coefficient vectors $\boldsymbol{\alpha}$, is 0.2. We note that only training images are involved during the phase of dictionary learning.

### 4.3 Training

In our experiment, we adopt the one-vs-rest scheme to design multi-class MSL classifiers. Each classifier recognizes one class against all others, and thus learns the optimal weights $b_\ell$ of different image scales for the corresponding object category. Fig. 5 shows a visualization example of the learned $b_\ell$, as well as the predetermined ones used in prior work. We consider only linear kernels for a major advantage that the computation complexity for training and testing will be significantly reduced compared to the cases using nonlinear kernels. Therefore, our proposed method is scalable to large-scale classification problems. The only free parameter to be determined is the regularization term $C$, and we apply a five-fold cross validation to search for its optimal value.

### 4.4 Results of the Oxford Flower Dataset

To compare our proposed MSL method with existing methods for image classification, we consider two different bag-of-features models as the baselines: the standard one without pyramid representation (i.e. level $L = 0$), and the SPM which concatenates pooled vectors from each scale with constant weights. Sum

**Table 1.** Mean average precision (MAP) comparison table for Oxford flower dataset. L: the maximal level in the spatial pyramid.

| Encoding method | | L | Pooling method | MSL | MAP $\mathcal{K}$=225 | MAP $\mathcal{K}$=900 |
|---|---|---|---|---|---|---|
| VQ | (a) | 0 | Sum Pooling | No | 36.76% | 40.00% |
| | (b) | 2 | Pyramid Sum Pooling | No | 48.09% | 49.26% |
| | (c) | 2 | Pyramid Sum Pooling | Yes | 53.68% | 55.74% |
| | (d) | 0 | Max Pooling | No | 19.12% | 40.59% |
| | (e) | 2 | Pyramid Max Pooling | No | 55.29% | 55.59% |
| | (f) | 2 | Pyramid Max Pooling | Yes | 53.82% | 57.35% |
| SC | (g) | 0 | Sum Pooling | No | 42.94% | 47.50% |
| | (h) | 2 | Pyramid Sum Pooling | No | 50.74% | 55.00% |
| | (i) | 2 | Pyramid Sum Pooling | Yes | 55.00% | 58.68% |
| | (j) | 0 | Max Pooling | No | 40.74% | 53.38% |
| | (k) | 2 | Pyramid Max Pooling | No | 60.30% | 62.79% |
| | (m) | 2 | Pyramid Max Pooling | Yes | 60.15% | 65.29% |

and max pooling operations are used in each baseline method for the completeness of the comparison. The number of levels in the spatial pyramid is chosen as 3 (i.e. $\ell = 0$, 1 and 2) for the experiments on this dataset. The complete results and comparisons on the Oxford flower dataset are shown in Table 1.

As can been seen in Table 1, the MAP for all cases increases when the size of the dictionary (i.e. the number of visual words $\mathcal{K}$) grows. When using VQ to learn the dictionary, the cases using the max pooling strategy obtained better MAP values than those using the sum pooling one, except for the case of the standard bag-of-features model (MAP = 36.76% in Table 1(a) vs. 19.12% in (d)). We note that when the sum pooling method is applied to construct feature vectors for classification (i.e. Table 1(a) to (c)), the use of SPM improves the recognition performance, while our approach outperforms the one using predetermined weights (53.68% vs. 48.09% when $\mathcal{K} = 225$, and 55.74% vs. 49.26% when $\mathcal{K} = 900$). The max pooling strategy is also observed the same advantage of applying SPM for classification (see Table 1(d) to (f)), while both SPM methods obtained comparable MAP values.

When the dictionary is learned by SC (Table 1(g) to (m)), we observe significant improvements in MAP for all cases. Our method with max pooling strategy resulted in the *highest* MAP = 65.29% when a larger size of dictionary $\mathcal{K} = 900$ was used. Comparing to the standard SPM with constant weights, we obtained comparable MAP when $\mathcal{K} = 225$. We expect to see a significant improvement in MAP when a larger-scale classification problem is of concern (our test results on Caltech-101 support this).

### 4.5 Results of the Caltech-101 Dataset

Our test results on the Caltech-101 dataset and the comparison with different dictionary learning and feature pooling strategies are shown in Fig. 4. We note that the number of levels in the spatial pyramid is 4 (i.e. $\ell = 0, 1, 2, 3$). We now summarize our findings as follows:

a. The use of sparse coding for dictionary learning outperforms that learned by vector quantization. More specifically, the dictionary learned by sparse coding together with different feature pooling techniques consistently improves the recognition performance than those learned by vector quantization. This confirms the observation in [4].

b. Spatial pyramid representation significantly improves recognition accuracy. From Fig. 4, we see that the use of spatial pyramid representation with either sparse coding or vector quantization technique outperforms the pooled features from a single image scale. This is consistent with the findings in [3] and [4].

c. A larger size of the dictionary improves the performance when a single level of image representation is used. However, it produces negligible improvements when spatial pyramid representation is considered.

d. Our muti-scale learning (MSL) improves MAP with different feature encoding (sparse coding and vector quantization) and pooling strategies. In particular, our proposed framework together with sparse coding and the pyramid max pooling (PMP) strategy achieved the best MAP among all methods.

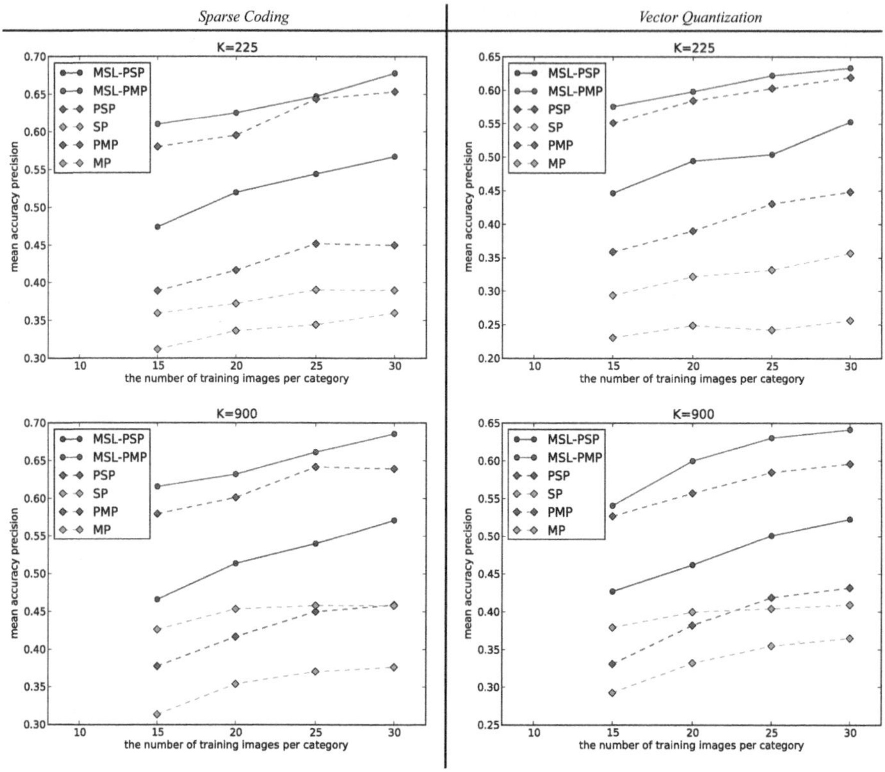


**Fig. 4.** Mean average precision comparison table for the Caltech-101 dataset. $\mathcal{K}$: the size of dictionary. MSL: Our Multiple Scale Learning (L=3). PSP: Pyramid Sum Pooling (L=3). SP: Sum Pooling (L=0). PMP: Pyramid Max Pooling (L=3). MP: Max Pooling (L=0). Best viewed in color.

We note that the method `PMP` in the left column of Fig. 4 is our implementation of ScSPM, which is proposed by Yang et al. [4]. We did not reproduce exactly the same results as reported in [4], probably due to the SIFT descriptor extraction, feature normalization process, etc. engineering details. Therefore, we choose to use the same implementation details for all methods on both datasets for comparisons.

To show the competitiveness of our methods, we also compare our approach with prior work on the Caltech 101 dataset. Our method outperforms many previously proposed methods, and we believe that this is because our approach is able to extract more salient properties of class-specific visual patterns across different image scales from different object categories. It is worth repeating that, different from many previous methods, our MSL framework only requires linear kernels and thus provides excellent scalability to large-scale image classification problems.

Fig. 5 illustrates the values of $b_\ell$ for all 101 object classes using different methods. The top row is the case of the standard bag-of-features model. Since

**Table 2.** Comparison with prior work on the Caltech-101 dataset. The number of training images per class for all methods is 15.

| Method | MAP |
|---|---|
| Raina et al. [28] | 46.6% |
| Berg et al. [29] | 48% |
| Mutch and Lowe [30] | 51% |
| Lazebnik et al. [3] | 56.40% |
| Zhang et al. [31] | 59.10% |
| Frome and Singer [16] | 60.30 % |
| Lin et al. [32] | 61.25 % |
| Our method | **61.43** % |

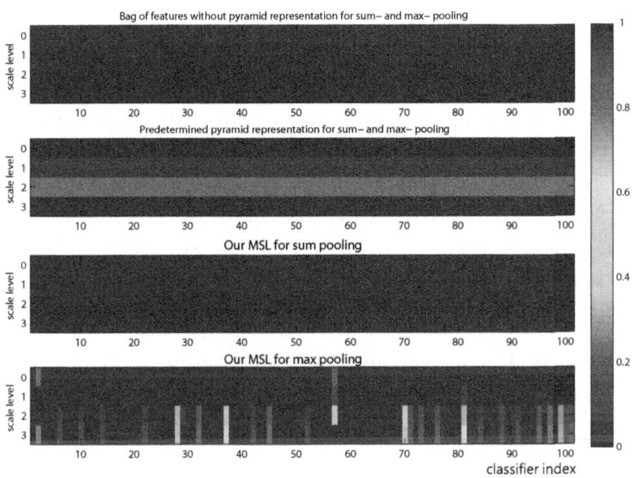


**Fig. 5.** Visualization of $b_\ell$ for different methods on the Caltech-101 dataset. The encoding method considered is vector quantization with $\mathcal{K} = 900$. Best viewed in color.

no pyramid information is used, we simply have $b_0 = 1$, and $b_\ell = 0$ otherwise. As for the SPM framework adopted by Yang et al. [4], in which the pooled vectors from each grid at each level are simply concatenated as the final image representation. It can been seen that finer scales in images are generally assigned larger (and fixed) weights due to the increasing number of grids in those pyramid levels. Finally, the last two rows in Fig. 5 present the $b_\ell$ learned by our method using pyramid sum and max pooling strategies, respectively. Together with the recognition performance reported, this visualization of our $b_\ell$ confirms that we are able to learn the optimal spatial pyramid representation given the image data, and our method can capture class-dependent salient properties of visual patterns in different image scales.

We would like to point out that we are aware of recent work which proposed to combine multiple types of descriptors or features for classification, and thus very promising results were reported [10,11,12,33]. Our MSL framework can be easily combined with these ideas, since multiple feature descriptors can be integrated into our proposed framework and can still be solved by MKL techniques. In such cases, we expect a significantly greater improvement on the recognition accuracy over state-of-the-art classification methods.

## 5 Conclusion

We presented a novel MSL framework that automatically learns the optimal spatial pyramid image representation for visual categorization, which is done by solving a MKL problem which determines the optimal combination of base kernels constructed by features pooled from different image scales. Our proposed method is able to capture class-specific salient properties of visual patterns in different image scales, and thus improves the recognition performance. Among different dictionary learning and pooling strategies, our proposed framework based on sparse coding and pyramid max pooling strategies outperforms prior methods on Oxford flower and Caltech 101-datasets. In addition, through the visualization of the weights learned for each image scale and for each object category, our MSL framework produces a class-specific spatial pyramid image representation, which cannot be achieved by the standard SPM. Finally, since only linear kernels are required in our proposed learning framework, our method is computationally feasible for large-scale image classification problems.

**Acknowledgement.** We are grateful for the anonymous reviewers for their helpful comments. This work is supported in part by the National Science Council of Taiwan under NSC98-2218-E-001-004 and NSC99-2631-H-001-018.

## References

1. Csurka, G., Bray, C., Dance, C., Fan, L.: Visual categorization with bags of keypoints. In: ECCV Workshop on Statistical Learning in Computer Vision, pp. 1–22 (2004)
2. Jurie, F., Triggs, B.: Creating efficient codebooks for visual recognition. In: ICCV 2005: Proceedings of the Tenth IEEE International Conference on Computer Vision (ICCV 2005), Washington, DC, USA, vol. 1, pp. 604–610. IEEE Computer Society, Los Alamitos (2005)
3. Lazebnik, S., Schmid, C., Ponce, J.: Beyond bags of features: Spatial pyramid matching for recognizing natural scene categories. In: CVPR 2006: Proceedings of the 2006 IEEE Computer Society Conference on Computer Vision and Pattern Recognition, Washington, DC, USA, pp. 2169–2178. IEEE Computer Society, Los Alamitos (2006)
4. Yang, J., Yu, K., Gong, Y., Huang, T.S.: Linear spatial pyramid matching using sparse coding for image classification. In: CVPR 2009: Proceedings of the 2009 IEEE Computer Society Conference on Computer Vision and Pattern Recognition, pp. 1794–1801. IEEE Computer Society, Los Alamitos (2009)

5. Grauman, K., Darrell, T.: The pyramid match kernel: discriminative classification with sets of image features. In: ICCV 2005: Proceedings of the Tenth IEEE International Conference on Computer Vision (ICCV 2005), vol. 2, pp. 1458–1465. IEEE Computer Society, Los Alamitos (2005)
6. Lanckriet, G.R.G., Cristianini, N., Bartlett, P., Ghaoui, L.E., Jordan, M.I.: Learning the kernel matrix with semidefinite programming. J. Mach. Learn. Res. 5, 27–72 (2004)
7. Bach, F.R., Lanckriet, G.R.G., Jordan, M.I.: Multiple kernel learning, conic duality, and the SMO algorithm. In: ICML 2004: Proceedings of the Twenty-First International Conference on Machine Learning, p. 6. ACM, New York (2004)
8. Crammer, K., Keshet, J., Singer, Y.: Kernel design using boosting. In: Advances in Neural Information Processing Systems 15, pp. 537–544. MIT Press, Cambridge (2003)
9. Hertz, T., Hillel, A.B., Weinshall, D.: Learning a kernel function for classification with small training samples. In: ICML 2006: Proceedings of the 23rd International Conference on Machine Learning, pp. 401–408. ACM, New York (2006)
10. Gehler, P.V., Nowozin, S.: On feature combination for multiclass object classification. In: ICCV 2009: Proceedings of the Tenth IEEE International Conference on Computer Vision (ICCV 2009). IEEE Computer Society, Los Alamitos (2009)
11. Varma, M., Ray, D.: Learning the discriminative power-invariance trade-off. In: Proceedings of the IEEE International Conference on Computer Vision. Rio de Janeiro, Brazil (2007)
12. Bosch, A., Zisserman, A., Munoz, X.: Image classification using ROIs and multiple kernel learning. In: IJCV 2008 (2008) (submitted)
13. Chopra, S., Hadsell, R., LeCun, Y.: Learning a similarity metric discriminatively, with application to face verification. In: CVPR 2005: Proceedings of the 2005 IEEE Computer Society Conference on Computer Vision and Pattern Recognition. IEEE Computer Society, Los Alamitos (2005)
14. Babenko, B., Branson, S., Belongie, S.: Similarity metrics for categorization: from monolithic to category specific. In: ICCV 2009: Proceedings of the Tenth IEEE International Conference on Computer Vision (ICCV 2009), Kyoto, Japan. IEEE Computer Society, Los Alamitos (2009)
15. Hertz, T., Bar-Hillel, A., Weinshall, D.: Learning distance functions for image retrieval. In: CVPR 2004: Proceedings of the 2004 IEEE Computer Society Conference on Computer Vision and Pattern Recognition, vol. 2 (2004)
16. Frome, A., Singer, Y., Malik, J.: Image retrieval and classification using local distance functions. In: Schölkopf, B., Platt, J., Hoffman, T. (eds.) Advances in Neural Information Processing Systems 19, pp. 417–424. MIT Press, Cambridge (2007)
17. Weinberger, K.Q., Saul, L.K.: Distance metric learning for large margin nearest neighbor classification. J. Mach. Learn. Res. 10, 207–244 (2009)
18. Yang, L., Jin, R., Sukthankar, R., Liu, Y.: An efficient algorithm for local distance metric learning. In: AAAI 2006: Proceedings of the 21st National Conference on Artificial Intelligence, pp. 543–548. AAAI Press, Menlo Park (2006)
19. Winder, S., Brown, M.: Learning local image descriptors. In: CVPR 2007: Proceedings of the 2007 IEEE Computer Society Conference on Computer Vision and Pattern Recognition, pp. 1–8. IEEE Computer Society, Los Alamitos (2007)
20. Subrahmanya, N., Shin, Y.C.: Sparse multiple kernel learning for signal processing applications. IEEE Transactions on Pattern Analysis and Machine Intelligence 99 (2009)

21. Bach, F.R., Thibaux, R., Jordan, M.I.: Computing regularization paths for learning multiple kernels. In: Saul, L.K., Weiss, Y., Bottou, L. (eds.) Advances in Neural Information Processing Systems 17, pp. 73–80. MIT Press, Cambridge (2005)
22. Rakotomamonjy, A., Bach, F., Canu, S., Grandvalet, Y.: More efficiency in multiple kernel learning. In: ICML 2007: Proceedings of the 24th International Conference on Machine Learning, pp. 775–782. ACM, New York (2007)
23. Sonnenburg, S., Rätsch, G., Schäfer, C., Schölkopf, B.: Large scale multiple kernel learning. J. Mach. Learn. Res. 7, 1531–1565 (2006)
24. Chang, C.C., Lin, C.J.: LIBSVM: a library for support vector machines (2001), `http://www.csie.ntu.edu.tw/~cjlin/libsvm`
25. Nilsback, M.E., Zisserman, A.: A visual vocabulary for flower classification. In: Proceedings of the IEEE Conference on Computer Vision and Pattern Recognition, vol. 2, pp. 1447–1454 (2006)
26. Fei-Fei, L., Fergus, R., Perona, P.: Learning generative visual models from few training examples: An incremental bayesian approach tested on 101 object categories. Comput. Vis. Image Underst. 106, 59–70 (2007)
27. Mairal, J., Bach, F., Ponce, J., Sapiro, G.: Online dictionary learning for sparse coding. In: ICML 2009: Proceedings of the 26th Annual International Conference on Machine Learning, pp. 689–696. ACM, New York (2009)
28. Raina, R., Battle, A., Lee, H., Packer, B., Ng, A.Y.: Self-taught learning: transfer learning from unlabeled data. In: ICML 2007: Proceedings of the 24th International Conference on Machine Learning, pp. 759–766. ACM, New York (2007)
29. Berg, A.C., Berg, T.L., Malik, J.: Shape matching and object recognition using low distortion correspondence. In: CVPR 2005: Proceedings of the 2005 IEEE Computer Society Conference on Computer Vision and Pattern Recognition, vol. 1, pp. 26–33. IEEE Computer Society, Los Alamitos (2005)
30. Mutch, J., Lowe, D.G.: Multiclass object recognition with sparse, localized features. In: CVPR 2006: Proceedings of the 2006 IEEE Computer Society Conference on Computer Vision and Pattern Recognition, Washington, DC, USA, pp. 11–18. IEEE Computer Society, Los Alamitos (2006)
31. Zhang, H., Berg, A.C., Maire, M., Malik, J.: SVM-KNN: Discriminative nearest neighbor classification for visual category recognition. In: CVPR 2006: Proceedings of the 2006 IEEE Computer Society Conference on Computer Vision and Pattern Recognition, pp. 2126–2136. IEEE Computer Society, Los Alamitos (2006)
32. Lin, Y.Y., Liu, T.L., Fuh, C.S.: Local ensemble kernel learning for object category recognition. In: CVPR 2007: Proceedings of the 2007 IEEE Computer Society Conference on Computer Vision and Pattern Recognition, pp. 1–8. IEEE Computer Society, Los Alamitos (2007)
33. Cao, L., Luo, J., Liang, F., Huang, T.S.: Heterogeneous feature machines for visual recognition. In: ICCV 2009: Proceedings of the Tenth IEEE International Conference on Computer Vision (ICCV 2009). IEEE Computer Society, Los Alamitos (2009)

# Fast Spectral Reflectance Recovery Using DLP Projector

Shuai Han[1], Imari Sato[2], Takahiro Okabe[1], and Yoichi Sato[1]

[1] Institute of Industrial Science, The University of Tokyo, Japan
[2] National Institute of Informatics, Japan
{hanshuai,takahiro,ysato}@iis.u-tokyo.ac.jp, imarik@nii.ac.jp

**Abstract.** Spectral reflectance is an intrinsic characteristic of objects which is useful for solving a variety of computer vision problems. In this work, we present a novel system for spectral reflectance recovery with a high temporal resolution by exploiting the unique color-forming mechanism of DLP projectors. DLP projectors use color wheels to produce desired light. Since the color wheels consist of several color segments and rotate fast, a DLP projector can be used as a light source with spectrally distinct illuminations. And, the appearance of a scene under the projector's irradiation can be captured by a high-speed camera. Our system is built on easily available devices and capable of taking spectral measurements at 100Hz. Based on the measurements, spectral reflectance of the scene is recovered using a linear model approximation. We carefully evaluate the accuracy of our system and demonstrate its effectiveness by spectral relighting of dynamic scenes.

## 1 Introduction

The amount of light reflected on an object's surface varies for different wavelengths. The ratio of the spectral intensity of reflected light to incident light is known as the spectral reflectance. It is an intrinsic characteristic of objects that is independent of illuminations and imaging sensors. Therefore, spectral reflectance offers direct descriptions about objects that are useful to computer vision tasks, such as color constancy, object discrimination, relighting etc.

Several methods have been proposed for spectral reflectance recovery. Maloney used an RGB camera to recover the spectral reflectance under ambient illumination [1]. This method is limited by a low recovery accuracy due to its RGB 3-channel measurements. To get measurements that contain more than 3 channels, some works attach filters to a light source to modulate the illumination [2] or sequentially place a set of band-pass filters in front of a monochromatic camera to produce a multi-channel camera [3]. Since switching among filters is time-consuming, these methods are unsuitable for dynamic scenes. To increase temporal resolution, specially designed clusters of different types of LEDs were created [4]. The LED clusters work synchronously with an RGB camera for conducting spectral measurements at 30 fps. Since such self-made light sources, as

R. Kimmel, R. Klette, and A. Sugimoto (Eds.): ACCV 2010, Part I, LNCS 6492, pp. 323–335, 2011.

well as the controller for synchronization, are not easily available, a level of effort is required to build a similar system.

What we seek is a practical system for fast spectral reflectance recovery built on easily available devices. In this work, we exploit the unique color-forming mechanism of Digital Light Processing (DLP) projectors and apply it for spectral measurements. DLP projectors use color wheels to produce the desired light. The color wheels are composed of several color segments, and the light that gets through these segments has specific spectral distributions. In other words, DLP projectors provide several spectrally distinct illuminations. When the color wheels rotate quickly, the light emitted from the DLP projectors rapidly switches among these illuminations. Making use of this switch, we built an imaging system that takes spectral measurements with a high temporal resolution.

In the system, a DLP projector is used as a light source, and a high-speed camera is used to capture the scenes' appearance under the projector's irradiation. A standard diffuse white board is placed in the scene to recover the illumination spectra of the captured frames. In order to reduce the number of required measurements for an accurate spectral reflectance recovery, we represent the spectral reflectance as a linear combination of a limited number of spectral bases, which was done in previous studies [5, 6]. Using this linear model, the spectral reflectance of the scene points can be reconstructed by using every five consecutive captured frames.

The contributions of this work are summarized below.

- **Dense temporal spectral measurement:** Our system is capable of taking spectral measurements at 100 Hz. This enables measurement for the fast-moving objects, and the recovered results are degraded little by motion blur.
- **Easily built imaging system:** Considering that high-speed cameras are becoming readily available in end-user markets and no synchronization between the projector and the camera is required, our system can be easily replicated by others. Furthermore, using the DLP projectors as light sources, the irradiation uniformity within the entire projection plane can be guaranteed, so the calibrations are simple and the working volume is large.

This paper is organized as follows. Section 2 gives a brief review of the related works. Section 3 presents our imaging system and its use for spectral reflectance recovery. Section 4 verifies its accuracy. Section 5 shows the relighting results of a static scene and a moving object. We conclude this work in Section 6.

## 2 Related Work

Spectral reflectance can be recovered under passive illumination. Maloney and Wandell used color constancy and an RGB camera for spectral reflectance recovery [1], but the accuracy of their method was low due to the RGB 3-channel measurement. For accurate results, Tominaga put a set of band-pass filters in front of a monochromatic camera, so that more than 3 channels can be measured [3]. However, this method trades off temporal resolution for the spectral resolution, and thus, is unsuitable for dynamic scenes.

Other existing methods for spectral reflectance recovery rely on active illumination. DiCalro and Wandell recovered the spectral reflectance as an intermediate result [7], but the accuracy was limited by the expression of the spectral reflectance as a combination of three spectral bases. To recover spectral reflectance with high accuracy, D'Zmura proposed a method using distinct illuminations [8], but the author only showed the results using synthetic data, and how well the proposed method works for real scenes was left unknown. Cui et al. proposed an algorithm for selecting an optimized set of wide-band filters and built a multi-illumination system [2]. They attached the selected filters to a light source, and used it as an additional light source for spectral reflectance recovery under ambient illumination. This method works well for static scenes. However, switching among different illuminations is time-consuming, so the system is not applicable for moving objects.

To measure the dynamic scenes, Park et al. built an imaging system based on multiplexed illumination [4]. They focused on the combinations of different LEDs and built LED clusters to capture 30 fps multi-spectral videos. However, their system requires specially built LED clusters and synchronization between the LED clusters and a camera. Accordingly, their system is not easily available. Moreover, using these self-made LED clusters, irradiation uniformity can be guaranteed only in a small area, so the working volume is quite limited.

Our work is also related to DLP-based active vision. Nayar et al. implemented a programmable imaging system using a modified DLP projector-camera pair [9]. Users can control the radiometric and geometric characteristics of the captured images by using this system. Narasimhan et al. exploited the temporal dithering of DLP projectors for a wide range of applications [10]. Zhang and Huang used the fast illumination variation of a DLP projector for real-time 3D shape measurements [11]. These three works only utilized the fast alternation between the "on" and "off" statuses of the digital micromirror device in a DLP projector; the spectral information was disregarded. In contrast, we use the spectral information from the emitted light for the spectral reflectance recovery. Our work is the first to recover spectral reflectance using a DLP projector.

# 3 Spectral Reflectance Recovery

## 3.1 Three Steps for Spectral Reflectance Recovery

There are three factors related to image brightness: the incident light, the scene, and the camera. Suppose the camera has a linear intensity response, this relationship can be expressed as

$$I_{m,n} = \int s(\lambda)c_m(\lambda)l_n(\lambda)d\lambda, \quad (1)$$

where $\lambda$ is the wavelength, $I_{m,n}$ is the intensity of a scene point in a captured frame, $s(\lambda)$ is the spectral reflectance of that point, $c_m(\lambda)$ is the spectral response function of the camera at the $m$th color channel, and $l_n(\lambda)$ is the spectrum of the $n$th illumination.

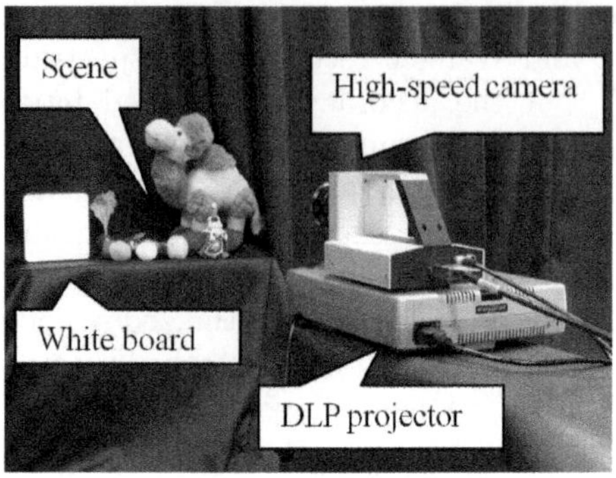


**Fig. 1.** Prototype System. Composed of a DLP projector (PLUS™U2-1130), a high-speed camera (PointGrey™ Lightning) and a white board (labsphere™ SRT-99).

The goal of this work is to recover spectral reflectance $s(\lambda)$ in a visible range (400−−700[nm]). From Eq. 1, we can see that a large set of spectrally distinct measurements are required if we want to recover $s(\lambda)$ with high spectral resolution. To reduce the number of required measurements without sacrificing spectral resolution, we approximate the spectral reflectance as a combination of a limited number spectral basis functions. This approximation procedure was also used in former works [4, 7].

Several linear models [5, 6] and a nonlinear model [12] have been built by using principal component analysis [13] or other tools (see Ref. [14] for a review about surface reflectance approximation). With regard to how many bases are required for accurate reconstruction, different works have different conclusions [5, 6, 15, 16, 17]. We adopt an 8-dimension linear model for spectral reflectance derived from Ref. [6] on account of its high reconstruction accuracy. On the basis of this linear model, the spectral reflectance is represented as

$$s(\lambda) = \sum_{j=1}^{8} \alpha_j b_j(\lambda), \tag{2}$$

where $b_j(\lambda)(j = 1, 2, .., 8)$ is the $j$th spectral basis from Ref. [6] (spectral resolution:10nm), $\alpha_j$ is the corresponding coefficient. Substituting Eq.2 for Eq.1, we obtain

$$I_{m,n} = \sum_{j=1}^{8} \alpha_j \int b_j(\lambda) c_m(\lambda) l_n(\lambda) d\lambda \tag{3}$$

In this work, we first estimate $\alpha_j$ from observed $I_{m,n}$. Then, spectral reflectance $s(\lambda)$ is reconstructed by substituting $\alpha_j$ into Eq. 2.

As shown in Fig. 1, our imaging system is composed of a one-chip DLP projector, a high-speed RGB camera with a linear intensity response and a standard diffuse white board. Using this system, we do spectral reflectance recovery in the follow three steps.

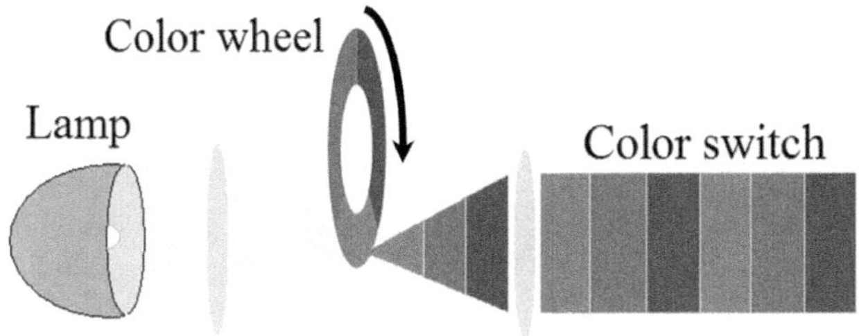


**Fig. 2.** Color switch caused by rotation of color wheel

**1. Image acquisition:** Scene's appearance under the projector's irradiation $I_{m,n}$, is acquired by using the high-speed camera. Every five consecutive frames are used as one measurement for the spectral reflectance recovery. (Section 3.2)
**2. Illumination recovery:** Illumination spectra $l_n(\lambda)$, changes from frame to frame. We use the diffuse white board as a calibration target to recover the illumination of captured frames. (Section 3.3)
**3. Spectral reflectance reconstruction:** Based on the 8-dimensional linear model, spectral reflectance $s(\lambda)$, can be reconstructed from the acquired images and recovered illuminations. (Section 3.4)

We explain each of these steps in detail in the following parts.

### 3.2 Image Acquisition by Color Switch

Different from other kinds of projectors, DLP projectors use color wheels to produce the desired light. The color wheel consists of several color segments, and these segments only allow light in a specific wavelength range to get through. When the color wheel quickly rotates, the light emitted from DLP projectors changes rapidly. In our work, this temporal variation in light is referred to as "color switch". A diagrammatic sketch is shown in Fig. 2. In our system, a DLP projector equipped with a 3-segment color wheel has been used (PLUS$^{TM}$U2-1130). since the color wheel rotates at 120 rps (round per second), color switch occurs at 360 Hz ($3 \times 120$).

The human eyes, and common video cameras work at low rates (24−−30 [Hz]), and thus they cannot detect the color switch. In this work, a 500 fps camera (PointGrey$^{TM}$ Lightning) is adopted to take images of scenes under the projector's irradiation. The camera outputs 24bit ($8 \times 3$) color images at a SXGA resolution ($1280 \times 1024$), and its linear intensity response can be verified by adjusting the shutter speed. In addition, the spectral response function of the camera $c_m(\lambda)$ ($m = 1, 2, 3$), was measured by using a monochromator and a spectrometer. The monochromator is used to generate a sequence of narrow-band lights. The spectral radiance of these lights is measured by the spectrometer. We expose the camera's sensor to the narrow-band lights and capture images. The relationship between the RGB values in the captured images and the spectral radiance of the corresponding lights, i.e., spectral response function, is shown in Fig. 3. During one rotation of the color wheel, the high-speed camera can

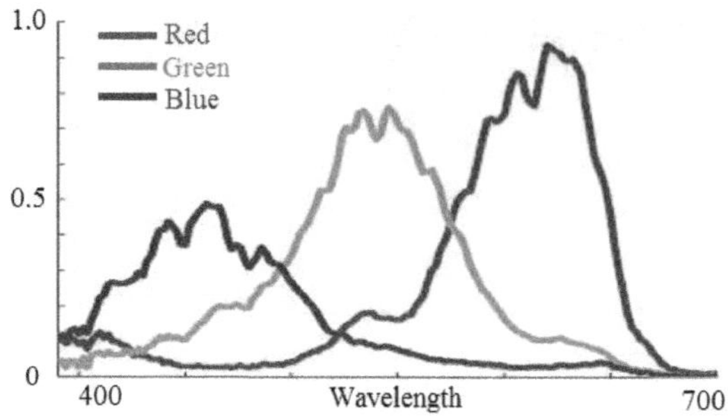


**Fig. 3.** Camera's spectral response function for RGB 3 channels

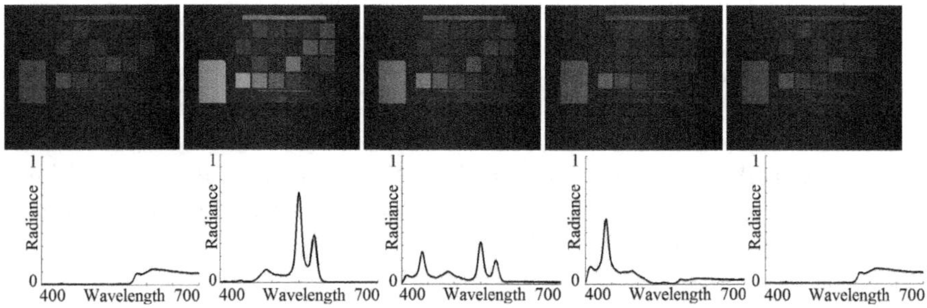


**Fig. 4.** One measurement of Macbeth ColorChecker and corresponding illumination spectra. Top row: 5 frames captured sequentially by the 500 fps camera in 1/100s. Bottom row: recovered illuminations of corresponding frames.

capture 4.17 frames. So, we use five consecutive frames as one measurement for the spectral reflectance recovery. Fig. 4 shows one measurement about Macbeth ColorChecker. We can see that the scene's appearance clearly changes under the color switch of the DLP projector. It should be noted that the color switch occurs at 360Hz, but the camera operates at 500 fps, so the projector and the camera work asynchronously.

### 3.3 Illumination Recovery

Our system does not require synchronization between the projector and the camera. Due to the asynchronism, the illumination changes from frame to frame. In this section, we describe how to recover the illumination spectrum $l_n(\lambda)$ of every frame using a standard diffuse white board (labsphere$^{TM}$ SRT-99) placed within the scene as a calibration target.

As mentioned above, light that gets through different segments on color wheels has distinct spectral distributions. If we use these spectral distributions as the illumination bases, light emitted from the DLP projectors can be expressed by a linear combination of these bases. In our system, since the three segments of the color wheel correspond to the RGB color filters, we can acquire these three distinct illuminations by inputting the projector (255, 0, 0), (0, 255, 0), and

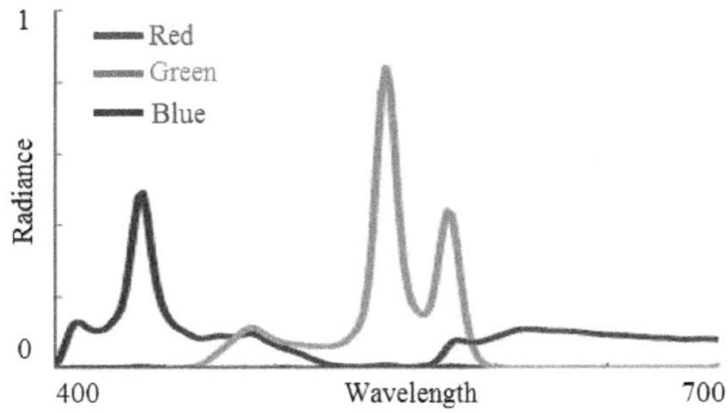


**Fig. 5.** Spectra of three distinct illuminations of the DLP projector

(0, 0, 255) respectively. Their spectra, which are measured by a spectrometer, are shown in Fig. 5. For each frame, its illumination spectrum ,$l_n(\lambda)$, can be represented as

$$l_n(\lambda) = \sum_{k=1}^{3} \beta_{n,k} p_k(\lambda), \text{ subject to } \beta_{n,k} > 0, \tag{4}$$

where $p_k(\lambda)$ is the spectrum of the $kth$ illumination basis of the DLP projector, $\beta_{n,k}$ is the corresponding coefficient.

By using Eqs.1 and 4, the brightness of a surface point on the white board is

$$I^w_{m,n} = \sum_{k=1}^{3} \beta_{n,k} \int p_k(\lambda) s^w(\lambda) c_m(\lambda) d\lambda, \tag{5}$$

where $I^w_{m,n}$ is the intensity of that point, and $s^w(\lambda)$ means its spectral reflectance. Use $P_{k,m}$ to represent the intensity of the point at the $mth$ channel under the $kth$ illumination basis

$$P_{k,m} = \int p_k(\lambda) s^w(\lambda) c_m(\lambda) d\lambda \ (k = 1, 2, 3), \tag{6}$$

Eq. 5 can be rewritten as

$$I^w_{m,n} = \sum_{k=1}^{3} \beta_{n,k} P_{k,m}, \tag{7}$$

$P_{k,m}(k = 1, 2, 3)$ can be measured by using the high-speed camera to capture images of the white board under three distinct illuminations of the projector. We only need to measure them once in advance.

From Eq. 7, we see that the intensity of a surface point on the white board under illumination $l_n(\lambda)$ is a linear combination of its intensities under three illumination bases

$$\boldsymbol{I}^w_n = \left[I^w_{1,n} \ I^w_{2,n} \ I^w_{3,n}\right]^T = \boldsymbol{P}^w \boldsymbol{\beta}_n, \tag{8}$$

where $\boldsymbol{I}_n^w$ represents the RGB value of a surface point on the white board under the $n$th illumination, $\boldsymbol{P}^w$ is a $3 \times 3$ matrix consists of $P_{k,m}(k = 1, 2, 3,\ m = 1, 2, 3)$, $\boldsymbol{\beta}_n$ is the corresponding $3 \times 1$ coefficient vector.

In principle, $\boldsymbol{\beta}_n$ can be easily calculated by $\boldsymbol{\beta}_n = (\boldsymbol{P}^w)^{-1}\boldsymbol{I}_n^w$. However, due to the noise , $\beta_{n,k}$ $(k = 1, 2, 3)$ may sometimes be negative. This conflicts with the non-negative constraint of Eq. 4. Thus, we solve $\boldsymbol{\beta_n}$ as a non-negative least squares problem:

$$\boldsymbol{\beta}_n = \underset{\boldsymbol{\beta}_n}{argmin} \mid \boldsymbol{I}_n^w - \boldsymbol{P}^w \boldsymbol{\beta}_n \mid^2, \quad \text{subject to } \beta_{n,k} \geq 0 \ (k = 1, 2, 3) \tag{9}$$

Using calculated $\boldsymbol{\beta_n}$, illumination spectrum $l_n(\lambda)$ can be reconstructed by using Eq.4.

### 3.4 Spectral Reflectance Reconstruction Using Constrained Model

Since $l_n(\lambda)$ is recovered in Section 3.3, the integral in Eq. 3 can be represented as known coefficients: $f_{j,m,n} = \int b_j(\lambda)c_m(\lambda)l_n(\lambda)d\lambda$. One measurement that contains five consecutive frames can be written in matrix form as

$$\boldsymbol{I} = \boldsymbol{F}\boldsymbol{\alpha}, \tag{10}$$

where $\boldsymbol{I}$ is a $15 \times 1$ vector (15 measurements: RGB 3 channels $\times$ 5 frames), $\boldsymbol{F}$ is a $15 \times 8$ matrix (15 measurements $\times$ 8 spectral bases), and $\boldsymbol{\alpha}$ is an $8 \times 1$ coefficient vector.

If $\boldsymbol{\alpha}$ is estimated from $\boldsymbol{I}$, spectral reflectance $s(\lambda)$ can be reconstructed by Eq. 2. In this way, the problem of spectral reflectance recovery can be solved by the 8 coefficients estimation. The DLP projector in our system has three spectrally distinct illuminations, and the high-speed camera provides a 3-channel measurement under each illumination. In total, we can obtain $3 \times 3$, i.e., 9 effective channels. Thus, the problem of estimating 8 coefficients is over-determined. However, using the least squares solution in Eq. 10, the reconstructed spectral reflectance does not always satisfy the non-negative constraint and the solutions tend to be unstable. Therefore, we adopted the constrained minimization method proposed in Ref. [4]. We use the first derivative of the spectral reflectance respective to $\lambda$ as the constraint:

$$\boldsymbol{\alpha} = \underset{\boldsymbol{\alpha}}{argmin} \left[ \mid \boldsymbol{I} - \boldsymbol{F}\boldsymbol{\alpha} \mid^2 + \gamma \left| \frac{\partial s(\lambda)}{\partial \lambda} \right|^2 \right], \text{subject to } \mathrm{b}_{\boldsymbol{m}}\boldsymbol{\alpha} \geq 0 \text{ for all } \lambda, \tag{11}$$

where $\gamma$ is a weight for the constraint term. $\boldsymbol{b_m}$ is a $31 \times 8$ matrix whose columns are the 8 spectral bases.

## 4 Accuracy Evaluation

In this section, we evaluate the accuracy of our system by using Macbeth ColorChecker. In the system, every five consecutive frames captured by the

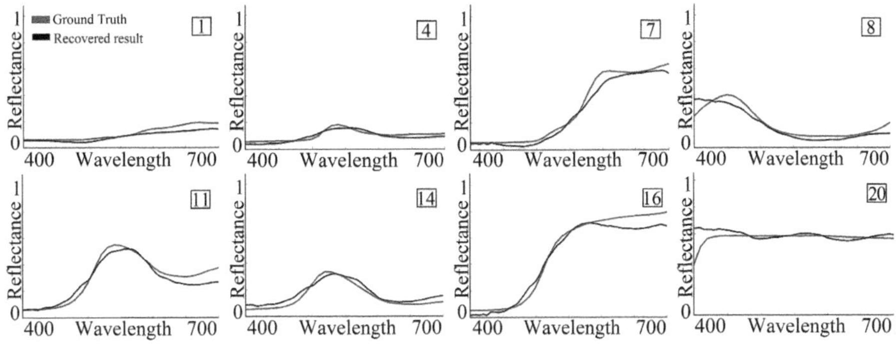


**Fig. 6.** Recovered spectral reflectance of some clips on Macbeth ColorChecker by the measurement shown in Fig. 4. Ground truth: red lines; recovered: black lines.

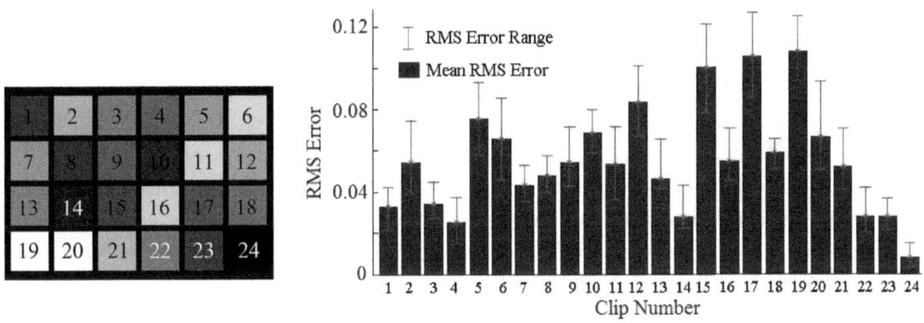


**Fig. 7.** RMS error of 24 clips of Macbeth ColorChecker for 200 measurements

500 fps camera are used as one measurement. Thus, the spectral measurements are taken at 100 Hz. But, the color wheel rotates at 120 rps. Due to the asynchronism between the DLP projector and the camera, frames captured at different times have different illumination spectra. The accuracy of the recovered results would be affected by this temporal illumination variation. Thus, we need to evaluate both the spectral accuracy and temporal accuracy of our system in this section.

We sequentially took 200 measurements (1000 frames) of a static 24-clip Macbeth ColorChecker to evaluate spectral accuracy. For each clip, we set $\gamma$ in Eq. 11 to 50 and reconstructed its spectrum based on the measurements (some results are shown in Fig. 6); then, the root mean square (RMS) error of all 200 reconstructed results was calculated. We also computed the maximum, mean, and minimum among the 200 RMS error values for every clip. The results for all 24 clips are shown in Fig. 7. We can see that, for all clips, their maximum RMS error does not deviate a lot from the minimum one. In addition, the biggest mean RMS error of all 24 clips is less than 0.11. These results demonstrate that our system can recover the spectral reflectance at a reasonable accuracy.

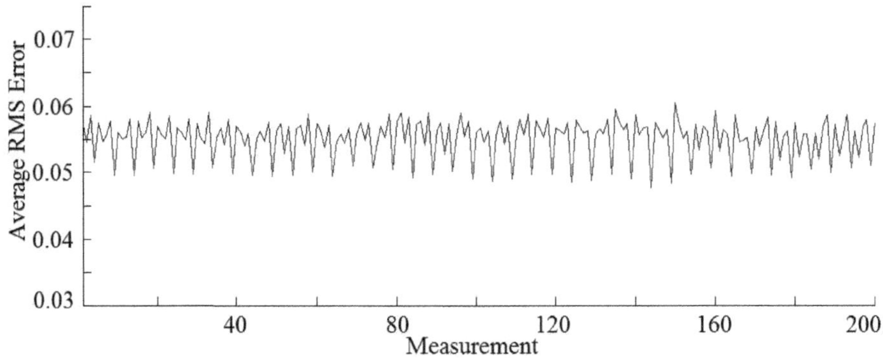


**Fig. 8.** Average RMS error for 200 measurements. Because color wheel rotates 6 rounds for every 5 measurements, a pattern of the average RMS error can be seen.

Next, we evaluated the temporal accuracy of our system. We reused the 200 measurements taken in previous test. For every measurement, we reconstructed the spectral reflectance of all 24 clips; then, the RMS error of the 24 reconstructed results was calculated; after that, we computed the average value of the 24 RMS error values, and used it as the criterion to evaluate each measurement. The results for all 200 measurements are shown in Fig. 8. The average value fluctuates in a narrow band (0.047, 0.06) which verifies the temporal accuracy of our system.

## 5 Image and Video Relighting

We used the spectral reflectance recovered by our method to do spectral relighting of a static scene as well as a moving object. To ensure there was a strong and spatial uniformly distributed light, an LCD projector (EPSON$^{TM}$ ELP-735) was used as the light source for relighting. The spectral distributions of its white, red, green, and blue were measured by a spectrometer.

### 5.1 Image Relighting

We set a static scene with fruits, vegetables, and small statues. Five consecutive frames from the scene were captured by our imaging system. Using them as one measurement, the spectral reflectance of scene points was recovered pixel by pixel. Then, the scene was spectrally relit by using Eq. 1 with the known illumination spectra of the LCD projector. A comparison between the relit results and the real captured images is shown in Fig. 9. We can see from the comparison that the computed results are very similar to the ground truth, which also reveals the accuracy of our system.

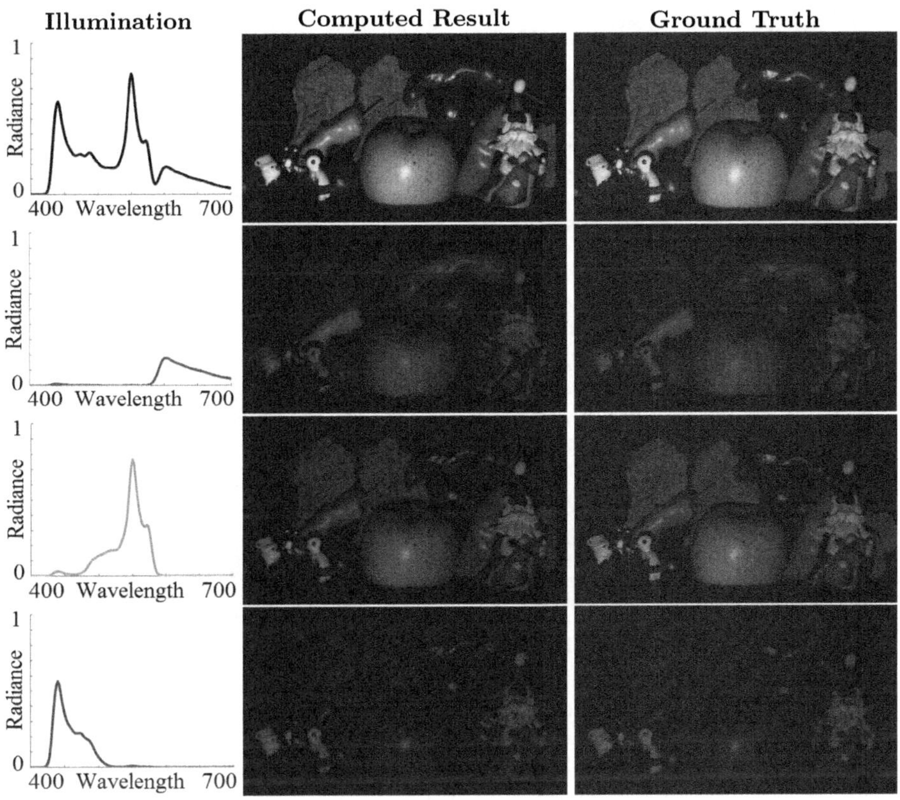


**Fig. 9.** Comparison between relit results and captured images of static scene under illuminations from a LCD projector

### 5.2 Real Video Relighting

Our system works at 100 Hz, so it is capable of measuring dynamic scenes. This capability was tested by taking spectral measurements of a manipulated toy consequently. For every measurement, the spectral reflectance of scene points was reconstructed. Based on the recovered data, the toy's movements were spectrally relit under a variety of illuminations. The results are shown in the top two rows of Fig. 10, and we can that there is a smooth movement, and the computed results look natural. In the bottom of Fig. 10, a relit result is shown in the middle. It was computed on the basis of the spectral data recovered by our system. The left is a image captured by the high-speed camera under the LCD projector's irradiation. A synthesized result to simulate captured image by a 30 fps camera is shown on the right side. Through comparisons, we can see that the relit result resembles the real captured image, and it is degraded little by the motion blur which is obvious in the synthesized result. From the comparisons, we can see that our system is robust to artifacts caused by motion. Therefore, our system is suitable for spectral reflectance recovery of dynamic scenes.

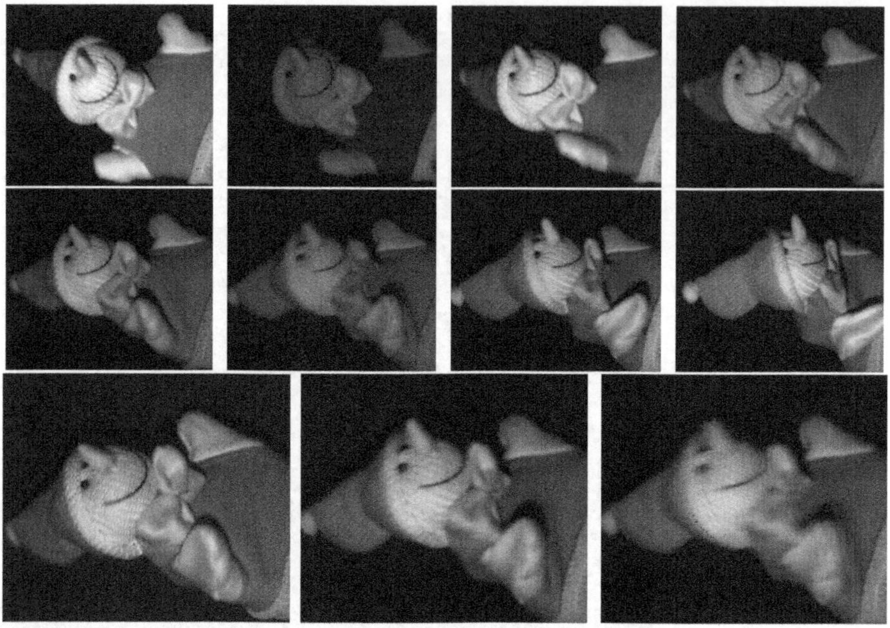

**Captured image (500 Hz) Relit result(100 Hz) Synthesized image(30 Hz)**

**Fig. 10.** Top two rows: relit results of fast-moving toy. The continuous movements through very different illuminations are shown. Bottom row: on the left is an image captured by the 500 fps camera, the middle is relit result, and the right is synthesized result to simulate captured image by a 30 fps camera. The recovered result by our system is only slightly degraded by motion blur.

## 6 Conclusion

In this work, we exploited the unique color-forming mechanism of DLP projectors. An imaging system for fast spectral reflectance recovery was built by making use of this mechanism. This system is capable of taking measurements as fast as 100 Hz. Every measurement consists of a set of sequentially captured images. For each set, the spectral reflectance of scene points can be recovered. Through intensive evaluation, the accuracy and the robustness of our system have been verified. Moreover, our system is built on easily available devices, and the excellent optical design of DLP projectors guarantees simple calibrations and a large working volume. It can be concluded that our system is practical and robust for the spectral reflectance recovery of fast-moving objects.

**Acknowledgement.** This research was supported in part by Grant-in-Aide for Scientific Research on Innovative Areas from the Ministry of Education, Culture, Sports, Science and Technology.

## References


1. Maloney, L.T., Wandell, B.A.: Color constancy: a method for recovering surface spectral reflectance. Journal of the Optical Society of America A 3, 29–33 (1986)
2. Chi, C., Yoo, H., Ben-Ezra, M.: Multi-spectral imaging by optimized wide band illumination. International Journal on Computer Vision 86, 140–151 (2010)
3. Tominaga, S.: Multichannel vision system for estimating surface and illumination functions. Journal of the Optical Society of America A 13, 2163–2173 (1996)
4. Park, J., Lee, M., Grossberg, M.D., Nayar, S.K.: Multispectral Imaging Using Multiplexed Illumination. In: Proc. IEEE International Conference on Computer Vision (2007)
5. Cohen, J.: Dependency of the spectral reflectance curves of the munsell color chips. Psychon. Science 1, 369–370 (1964)
6. Parkkinen, J.P.S., Hallikainen, J., Jaaskelainen, T.: Characteristic spectra of munsell colors. Journal of the Optical Society of America A 6, 318–322 (1989)
7. DiCarlo, J.M., Wandell, B.A.: Illuminating illumination. In: Proc. Ninth Color Imaging Conference, pp. 27–34 (2000)
8. D'Zmura, M.: Color constancy: surface color from changing illumination. Journal of the Optical Society of America A 9, 490–493 (1992)
9. Nayar, S.K., Branzoi, V., Boult, T.E.: Programmable imaging: Towards a flexible camera. International Journal on Computer Vision 70, 7–22 (2006)
10. Narasimhan, S.G., Koppal, S.J., Yamazaki, S.: Temporal dithering of illumination for fast active vision. In: Forsyth, D., Torr, P., Zisserman, A. (eds.) ECCV 2008, Part IV. LNCS, vol. 5305, pp. 830–844. Springer, Heidelberg (2008)
11. Zhang, S., Huang, P.: High-resolution, real-time 3d shape acquisition. In: Proc. IEEE Conference on Computer Vision and Pattern Recognition Workshops., vol. 3, pp. 28–37 (2004)
12. DiCarlo, J.M., Wandell, B.A.: Spectral estimation theory: beyond linear but before Bayesian. Journal of the Optical Society of America A 20, 1261–1270 (2003)
13. Tzeng, D., Berns, R.S.: A review of principal component analysis and its applications to color technology. Color Research And Application 30, 84–98 (2006)
14. Kohonen, O., Parkkinen, J., Jääskeläinen, T.: Databases for spectral color science. Color Research And Application 31, 381–390 (2006)
15. Dannemiller, J.L.: Spectral reflectance of natural objects: how many basis functions are necessary? Journal of the Optical Society of America A 9, 507–515 (1992)
16. Chiao, C., Cronin, T.W., Osorio, D.: Color signals in natural scenes: characteristics of reflectance spectra and effects of natural illuminants. Journal of the Optical Society of America A 17, 218–224 (2000)
17. Maloney, L.T.: Evaluation of linear models of surface spectral reflectance with small numbers of parameters. Journal of the Optical Society of America A 3, 1673–1683 (1986)

# Hemispherical Confocal Imaging Using Turtleback Reflector

Yasuhiro Mukaigawa[1,2], Seiichi Tagawa[1], Jaewon Kim[2], Ramesh Raskar[2], Yasuyuki Matsushita[3], and Yasushi Yagi[1]

[1]Osaka University
[2]MIT Media Lab
[3]Microsoft Research Asia

**Abstract.** We propose a new imaging method called *hemispherical confocal imaging* to clearly visualize a particular depth in a 3-D scene. The key optical component is a *turtleback reflector* which is a specially designed polyhedral mirror. By combining the turtleback reflector with a coaxial pair of a camera and a projector, many virtual cameras and projectors are produced on a hemisphere with uniform density to synthesize a hemispherical aperture. In such an optical device, high frequency illumination can be focused at a particular depth in the scene to visualize only the depth with descattering. Then, the observed views are factorized into masking, attenuation, and texture terms to enhance visualization when obstacles are present. Experiments using a prototype system show that only the particular depth is effectively illuminated and hazes by scattering and attenuation can be recovered even when obstacles exist.

## 1 Introduction

Significant effort has been made to obtain cross-sectional views of a 3-D scene. Real scenes often include obstacles such as scattering materials or opaque occluders. To clearly visualize cross-sectional views as if the scene is cut at a plane, only a particular depth has to be illuminated and haze due to scattering and attenuation should be recovered.

The simplest way to observe a particular depth is to use a large aperture lens. The large aperture makes the DOF (depth of field) shallow, and the region outside the DOF is blurred. The synthetic aperture method [24] mimics a large virtual aperture by combining many small apertures. However, obstacles are still bright and visible, while they are blurred. The confocal imaging [15] simultaneously scans two confocal pinholes over a particular depth. Since both illumination and observation are focused, clear cross-sectional views are obtained. While still visible, obstacles are darkened and blurred. Moreover, scanning requires long measuring time.

Recently, Levoy et al. [11] proposed a new imaging technique which combines synthetic aperture with the confocal imaging. Since this technique is based on light field analysis, only a particular depth can be illuminated without scanning. However, the synthesized aperture size is relatively small because rectangular

R. Kimmel, R. Klette, and A. Sugimoto (Eds.): ACCV 2010, Part I, LNCS 6492, pp. 336–349, 2011.

mirrors are aligned as a 2D array. Moreover, unwanted effects such as scattering and attenuation still remain.

In this paper, we propose a novel imaging method called *hemispherical confocal imaging*. To improve the imaging performance of the synthetic aperture confocal imaging [11], we designed *turtle reflector* which is a polyhedral mirror to approximate a hemispherical aperture with 180 degree of FOV. We introduce *focused high frequency illumination* using the turtleback reflector with a projector. This method can eliminate scattering on the focused depth, and make unfocused depth almost invisible. Moreover, we introduce *factorization* of observed views to eliminate attenuation.

**Contribution**

- By utilizing the new optical device, unfocused depth becomes almost invisible and scattering is eliminated. Moreover, the measurement is very fast, since no scanning is required.
- We have designed the *turtleback reflector* which is a polyhedral mirror circumscribed in an ellipsoid. The combination of the turtleback reflector, a projector, and a camera can synthesize a hemispherical wide aperture. The optical device can also be used for measuring the complete 8-D reflectance field on the hemisphere.
- A new imaging technique of the focused high frequency illumination is introduced. This technique enables us to separate direct and global components not on the surface but in the 3-D space because any projection pattern can be focused at the particular depth in the 3-D scene.

## 2 Related Work

### 2.1 Reflectance Field Measurement

The optical device proposed in this paper can be regarded as an 8-D reflectance field measuring device. A 4-D slice of the 8-D reflectance field under a static illumination can be recorded by scanning a camera [12] or installing multiple cameras [24]. Alternatively, a high-resolution camera is combined with a micro lens array [1], a micro mirror array [22], or masks [25]. To vary illumination, Debevec et al. [4] rotated a light source, and Sen et al. [20] used a projector as a light source. Masselus et al. [13] rotated a projector, and Matusik et al. [14] rotated both a light source and a camera to measure 6-D reflectance field. Müller et al. [17] used 151 cameras with flashes.

In principle, complete 8-D reflectance field can be measured by densely installing many projectors and cameras on a hemisphere. However, it is difficult to realize such a system due to the cost and physical interference between devices. While rotating projector and camera solves these problems, capture process is impractically long. Recently, Garg et al. [6] and Levoy et al. [11] used multiple planar mirrors and Cossairt et al. [2] used a lens array to measure a part of 8-D reflectance field, but the direction of the illumination and observation is limited to a narrow angle.

On the other hand, only our system can measure complete 8-D LF on the hemisphere covering the scene. Since our system utilizes the geometric property of an ellipsoid, many virtual cameras and projectors can be produced on a hemisphere with *uniform density*.

### 2.2 BRDF Measurement Using Mirrors

For measuring bidirectional reflectance distribution function (BRDF), mirrors are often used to replace mechanical motion. Reflected lights can be effectively measured from all directions using a hemispherical mirror [26] or a paraboloidal mirror [3]. Recently, a wide variety of mirrors such as a cylindrical mirror [10], several plane mirrors [9], an ellipsoidal mirror [16], and a combination of a paraboloidal mirror and a specially-designed dome mirror [7] have been used in conjunction with a projector and a camera.

Our turtleback reflector design was inspired by the BRDF measurement using an ellipsoidal mirror [16]. They utilized a geometric property of a rotationally symmetric ellipsoid that all rays from one focal point reflect off the ellipsoidal mirror and reach the other focal point. On the other hand, we utilized a different geometric property of an ellipsoid that the total length from one focal point to the other focal point through any surface points is constant. By utilizing this characteristic, we can produce virtual cameras and projectors at a constant distance from the target just as they are on a hemisphere. Since our purpose is not BRDF measurement but visualization of cross-sectional views, we designed a polyhedral mirror circumscribed in an ellipsoid.

### 2.3 Descattering

Incident lights to murky liquid or translucent media scatter, and the appearance becomes blurred. To obtain clear views, descattering methods have been developed. Treibitz and Schechner [21] used polarizer under water. Assuming that only single scattering is observed in optically thin media, Narasimhan et al. [18] estimated 3-D shape with descattering and Gu et al. [8] estimated 3-D distribution of inhomogeneous scattering media.

Recently, Fuchs et al. [5] combined confocal imaging with descattering which utilize the fact that scattering components have low frequency. The principle of our *hemispherical confocal imaging* is similar to this approach. However, we combine two ideas of the focused illumination proposed by Levoy et al. [11] and the high frequency illumination proposed by Nayar et al. [19]. Our approach both shortens the measuring time and eliminates scattering.

## 3 Hemispherical Confocal Imaging

Let us assume that a 3-D scene is illuminated by a light source and observed by a camera as shown in Fig.1. Even if the camera is focused on a particular depth in the scene, the captured image includes reflections from the entire scene. To

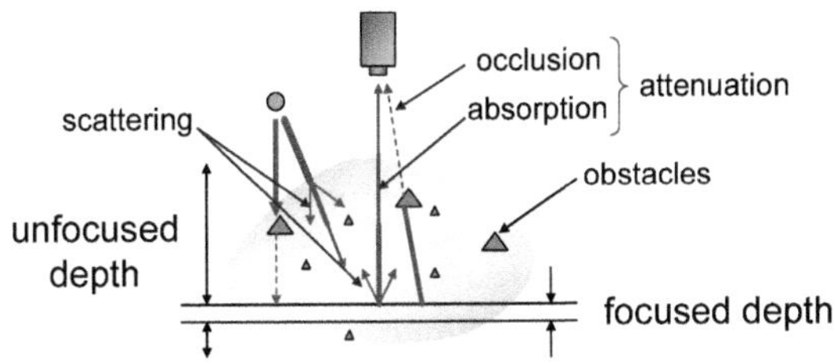


**Fig. 1.** Illumination and reflection in a 3-D scene. It is difficult to observe a particular depth due to scattering and attenuation.

**Table 1.** Comparison of several imaging methods

| | **unfocused depth** | **scanning** | **scattering** |
|---|---|---|---|
| Synthetic aperture | bright | unnecessary | remaining |
| Confocal imaging | darken | necessary | remaining |
| Synthetic aperture confocal imaging [11] | unilluminated | unnecessary | remaining |
| Confocal imaging with descattering [5] | darken | necessary | reduced |
| Our hemispherical confocal imaging | unilluminated | unnecessary | reduced |

observe the particular depth, only the depth should be illuminated. This means that both the illumination and observation should have a shallow DOF.

Even if we succeed in illuminating only the particular depth, clear views cannot be observed. The major reasons are *scattering* and *attenuation*. The scattering is caused by multi-bounce reflections in translucent media. By the scattering, the views become blurred. On the other hand, the attenuation is caused by *occlusion* due to obstacles or *absorption* due to low transparent media. By the attenuation, the illumination becomes nonuniform and the reflections are partially darkened. The following four functions are required to obtain clear views of a particular depth in a 3-D scene.

(a) The DOF should be as shallow as possible.
(b) Only the particular depth should be illuminated.
(c) Scattering should be eliminated.
(d) Attenuation should be eliminated.

To satisfy these requirements, we propose the *hemispherical confocal imaging* consisting of (1) specially designed *turtleback reflector*, and (2) focused high frequency illumination.

The turtleback reflector with coaxial camera and projector synthesizes a hemispherical aperture for both illumination and observation to solve (a). The focused high frequency illumination eliminates reflections from unfocused depth and global reflection to solve (b) and (c). Then, we factorized the observed views into masking, attenuation, and texture terms to solve (d). The merits and demerits and the number of projector and cameras of several imaging methods are summarized in Table 1 and Fig.2, respectively.

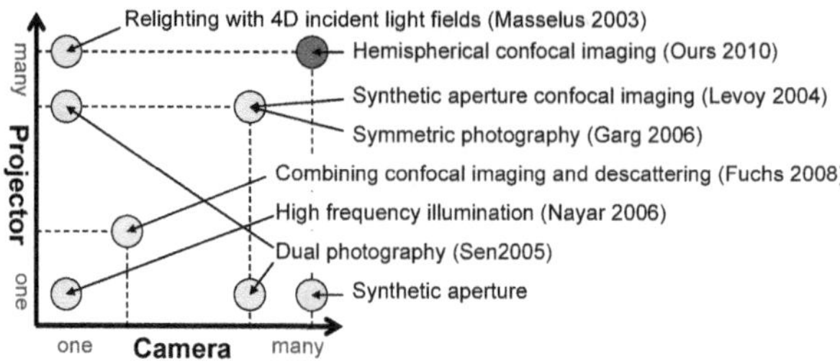


**Fig. 2.** The number of projectors and cameras of several imaging methods which use projector(s) and camera(s) for reflection analysis or reflectance field measurement

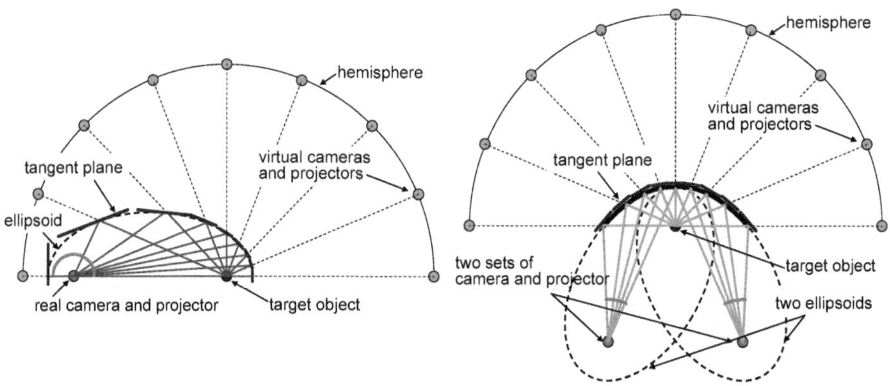


(a) design using single ellipsoid

(b) design using dual ellipsoids

**Fig. 3.** Principle of the proposed optical device. Virtual projector and cameras are distributed on a hemisphere with uniform density by using the turtleback reflector.

# 4 Optical Design of Turtleback Reflector

## 4.1 Projectors and Cameras on a Hemisphere

In our system, projectors are used as light sources to illuminate a particular depth. If the aperture size of the projector is large enough, the projected pattern is focused only within the shallow DOF and blurred elsewhere. Although it is difficult to realize a large aperture using a single lens, the combination of a number of small apertures can easily synthesize a large aperture by the synthetic aperture technique [24].

To mimic an extremely large aperture, many projectors should be placed in every direction at a constant distance and uniform density. That is, ideal locations of the projectors and cameras are on a hemisphere. If both the projectors and cameras are densely distributed on the hemisphere covering the target scene, the projectors and cameras can synthesize a hemispherical aperture with 180 degree of FOV for both illumination and observation.

### 4.2 Turtleback Reflector for Virtual Projectors and Cameras

It is difficult to place many projectors and cameras due to the cost and physical conflict. Therefore, we distribute many virtual projectors and cameras using mirrors. For this purpose we utilize the geometric property of an ellipsoid that the total length from one focal point to the other focal point through any surface points on a rotationally symmetric ellipsoid is a constant. A target object is placed at the focal point, and a projector and a camera are placed at the other focal point using beam splitter as illustrated in Fig.3(a). Planner mirrors are placed in the tangent planes to the ellipsoid. This is equivalent to hemispherical distribution of many virtual projectors and cameras with uniform density.

Actually, the design using single ellipsoid is difficult to construct because it requires real projector and camera with 180 degree of FOV. Hence, the hemisphere is evenly divided into two parts and two slanted ellipsoids are placed so that they share same focal point as shown in Fig.3(b). In this design using dual ellipsoids, a projector and a camera with a normal FOV can be used.

### 4.3 Design of Optical Device

We designed a new turtleback reflector which is a polyhedral mirror circumscribed in an ellipsoid as shown in Fig.4. The virtual projectors and cameras are placed at the nodes of a geodesic dome which is generated by dividing an icosahedron two times. While the optical device is composed of two symmetric parts, we constructed one half as a prototype to confirm the ability.

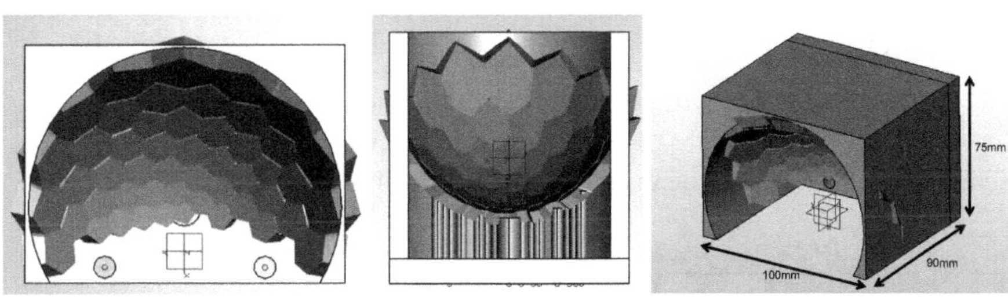


(a) front view (b) bottom view (c) dimensions

**Fig. 4.** Design of the turtleback reflector

Figure 5(a) showstheframeto fix the mirror patches made by stereo-lithography. Fifty mirror patches are attached to the frame. The turtleback reflector is combined with a high-resolution camera (PointGrey, Grass-50S5C, 2448 × 2048) and a projector (KAIREN Projector X Pro920, 640 × 480).

(a) turtleback reflector (b) total optical device

**Fig. 5.** Optical device for hemispherical confocal imaging

# 5 Focused High Frequency Illumination

## 5.1 Illumination and Reflection in a 3-D Scene

To analyze the reflectance field in a 3-D scene, we need to know how lights illuminate points in a scene, and how the reflections are observed. We divide the 3-D scene into a set of small voxels. Let $L_k$ be a set of rays which illuminate the $k$-th voxel, and $R(L_k)$ be a set of reflected rays of $L_k$ at the voxel. Since the observed view of the entire scene by a camera is expressed by a sum of the reflected rays from all voxels, the view is presented by $\sum_k R(L_k)$.

Illuminations and reflections can be regarded as a sum of direct and global components [19]. As shown in Fig.6 (a), the illumination of the $k$-th voxel can be decomposed into the direct illumination $L_k^D$ and the global one $L_k^G$. Similarly, the reflection can also be decomposed into the direct reflection $R^D(L_k)$ and global one $R^G(L_k)$. That is,

$$R(L_k) = R^D(L_k) + R^G(L_k) \quad \text{and} \quad L_k = L_k^D + L_k^G. \tag{1}$$

Hence, the observed view can be modeled as a sum of four components by

$$\sum_k R(L_k) = \sum_k (R^D(L_k^D) + R^G(L_k^D) + R^D(L_k^G) + R^G(L_k^G)) \tag{2}$$

## 5.2 Focused Illumination by Multiple Projectors

To obtain clear views of a particular depth in a 3-D scene, only the depth should be illuminated. Moreover, any global illuminations and global reflections should be eliminated to reduce scattering in the media. That is, the first term $R^D(L_k^D)$ in Eq.(2) should be measured separately.

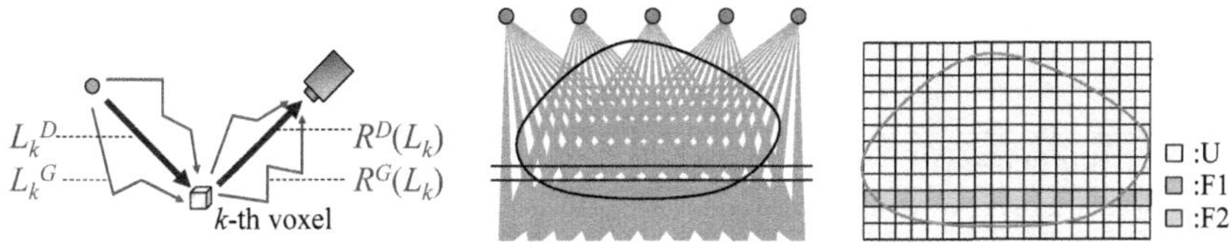


(a) four types of ray (b) focused illumination (c) classification of voxels

**Fig. 6.** Focused high frequency illumination. The high frequency patterns are focused only on the particular depth. The projection is blurred out of the DOF

By using our optical device, such special illumination and measurement can be realized. Since the virtual projectors surrounding the target scene can synthesize a large aperture, the DOF becomes very shallow. We combine the focused illumination technique proposed by Levoy et al. [11] and high frequency illumination technique proposed by Nayar et al. [19], and call the new technique *focused high frequency illumination* (**FHFI** for short). It is noted that our optical system enables the efficient combination of [11] and [19]. The former can project arbitrary focused patterns at any plane in the scene. Out of the focused plane, projected patterns become blurred. The latter can separate direct and global components on the 2-D surface. Our FHFI can separate direct and global components in the 3-D volume, because our optical device can synthesize a hemispherical aperture with 180 degree of FOV.

For the FHFI, high frequency checker board patterns are projected from each projector. The position of the white and black pixels are aligned at the depth as shown in Fig.6 (b). This means that the high frequency illumination is focused only at a particular depth. The voxels in the scene are classified into unfocused voxels $U$, and focused and illuminated voxels $F1$, and focused but unilluminated voxels $F2$ as shown in Fig.6 (c).

Compared to a white pattern, the average intensity of the high frequency illumination is darken because the half of pixels are black. Table 2 shows the relative intensities of the four reflection components for each voxel type. The global illumination to every voxel decreases by half. The direct illumination to $U$ also decreases by half because the projected patterns are blurred. The $F1$ receives full direct illumination, while the $F2$ receives no direct illumination. By combining these differences, $\sum_{k \in F1 \cup F2} R^D(L_k^D)$ which presents only direct components from voxels at the focused depth can be separated.

Let $I_P$ be a captured image when voxels of $F1$ are illuminated but voxels of $F2$ are not illuminated. Let $I_N$ be a captured image when the inverse pattern is projected. Then, these images can expressed as

$$I_P = \sum_{i \in F1} R\left(Li^D + \frac{Li^G}{2}\right) + \sum_{i \in F2} R\left(\frac{Li^G}{2}\right) + \sum_{i \in U} R\left(\frac{Li^D + Li^G}{2}\right), \quad (3)$$

$$I_N = \sum_{i \in F1} R\left(\frac{Li^G}{2}\right) + \sum_{i \in F2} R\left(Li^D + \frac{Li^G}{2}\right) + \sum_{i \in U} R\left(\frac{Li^D + Li^G}{2}\right). \quad (4)$$

**Table 2.** Relative intensities of four reflection components for each voxel type

| | $R^D(L_k^D)$ | $R^D(L_k^G)$ | $R^G(L_k^D)$ | $R^G(L_k^G)$ |
|---|---|---|---|---|
| **U** (unfocused) | 1/2 | 1/2 | 1/2 | 1/2 |
| **F1** (focused and illuminated) | 1 | 1/2 | 1 | 1/2 |
| **F2** (focused and unilluminated) | 0 | 1/2 | 0 | 1/2 |

By comparing two intensities at same position in $I_P$ and $I_N$, we can make an image $I_{max}$ which has larger intensities and an image $I_{min}$ which has smaller intensities. Since global component has only low frequency [5] [19],

$$\sum_{i \in F1} R^G(Li^D) \simeq \sum_{i \in F2} R^G(Li^D). \tag{5}$$

Therefore,

$$\begin{aligned} I_{max} - I_{min} &= \sum_{i \in F1} R^D(Li^D) + \sum_{i \in F2} R^D(Li^D) \pm (\sum_{i \in F1} R^G(Li^D) - \sum_{i \in F2} R^G(Li^D)) \\ &= \sum_{i \in F1} R^D(Li^D) + \sum_{i \in F2} R^D(Li^D) \\ &= \sum_{i \in F1 \cup F2} R^D(Li^D). \end{aligned} \tag{6}$$

This means that only the particular depth ($F1 \cup F2$) can be directly illuminated without global illuminations, and only the direct reflections can be measured without global reflections. As shown in Table 1, our method does not illuminate unfocused depth. Since no scanning is necessary, the measurement is fast. Moreover, scattering which is a major global component in translucent media is eliminated.

## 6 Factorization of the Observed Views

Although only the focused depth is illuminated and scattering is eliminated by the FHFI, the view of the focused depth is still unclear due to attenuation of the incident and reflective lights. This is due to *occlusion* and *absorption* as shown in Fig.1. Occlusion casts sharp shadows because some lights are directly interrupted by obstacles, while attenuation usually makes smooth change because lighting powers are decreased by spatially distributed low transparent media.

Fortunately, the scene is observed by several virtual cameras. Even if some lights are attenuated, other cameras may observe the scene without attenuation. Hence, we try to estimate the texture which is not affected by attenuation based on the observation from multiple cameras. We assume that there are $K$ virtual cameras and each camera has $N$ pixels. Let $O_{ij}$ be the intensity of the $i$-th pixel in the $j$-th camera. We model that the observed intensities are factorized as

$$O_{ij} = M_{ij} A_{ij} T_i. \tag{7}$$

Here, $M_{ij}$ is the *masking term* which has a value of 0 or 1. If the light is occluded by obstacles, the value becomes 0, otherwise it becomes 1. $A_{ij}$ is the *attenuation*

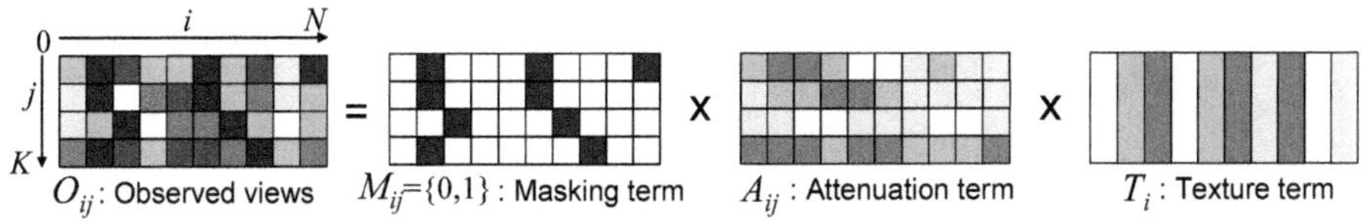


**Fig. 7.** Concept of the factorization. The observed intensities are factorized into the masking, attenuation, and texture terms to reduce attenuation.

*term* which expresses light attenuation due to absorption. $T_i$ is the *texture term* which expresses the reflectance of the particular depth. It is noted that only the texture term is independent to the viewing direction assuming Lambertian reflection. Figure 7 illustrates this relationship.

The flow of the factorization process is as follows

**STEP-1:** First, the masking term is decided. Since unfocused depths are not illuminated by the FHFI, obstacles can be easily distinguished using a simple threshold. After decision of the masking term, the following processes are done for pixels satisfying $M_{ij} = 1$.

**STEP-2:** The initial attenuation term is decided as $A_{ij} = 1$.

**STEP-3:** The texture term is calculated. Ideally, a unique reflectance should be estimated despite different camera $j$, but the observed intensities vary. This kind of problem is often seen in stereoscopy [23], so we used a median filter in a similar fashion by $T_i = \text{Median}(O_{ij}/A_{ij})$.

**STEP-4:** Update the attenuation term by $A_{ij} = O_{ij}/T_i$ to satisfy Eq.(7).

**STEP-5:** Smooth the attenuation term using a Gaussian function, because attenuation smoothly varies over the 3-D scene. Then go back to **STEP-3** until the texture term does not change.

By this factorization process, the observed views are decomposed to three terms and we can get clear texture of the particular depth without attenuation.

## 7 Experiments

### 7.1 Synthetic Aperture Using Virtual Cameras

First, we evaluated the ability of the synthetic aperture using the prototype system. A textured paper is covered by obstacle of yellow dense mesh as shown in Fig.8(a). A white uniform pattern was projected onto the scene. Figure 8(b) shows the captured image by the real camera. This image includes fifty views corresponding to fifty virtual cameras. Since all views are affected by the obstacle, it is difficult to see the texture of the sheet. Figure 8(c) shows the change of the appearance when the number of the virtual cameras increases to synthesize a large aperture. Since our optical device can synthesize a half of the hemispherical aperture, the obstacle is completely blurred and the texture becomes clear with increasing the number of virtual cameras.

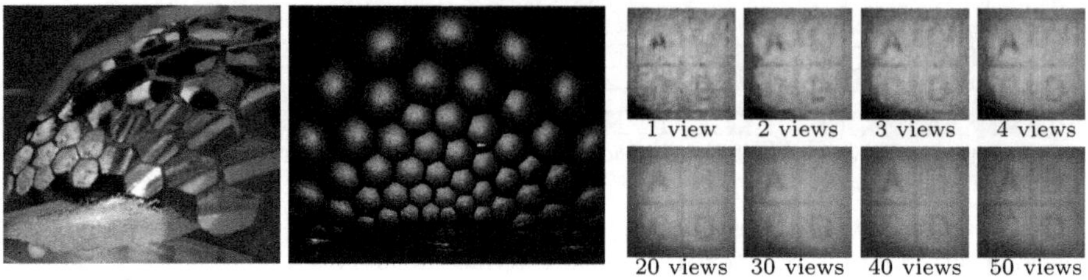


(a) target scene (b) captured image (c) change of the aperture size

**Fig. 8.** Result of synthetic aperture using our optical device

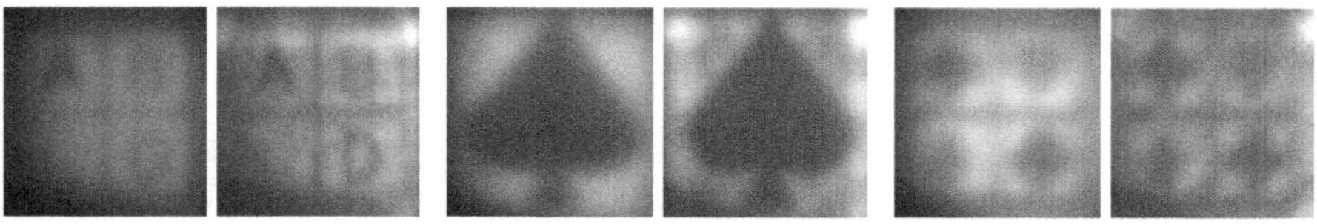

(a) (b) (c)

**Fig. 9.** Descattering by the focused high frequency illumination. Lefts: views under normal illumination. Rights: estimated direct components.

## 7.2 Descattering by Focused High Frequency Illumination

Next, we confirmed that the FHFI is effective for descattering in a 3-D volume. We covered some textured papers with a white plastic sheet. The left images of Fig.9 (a)(b)(c) show views when a white uniform pattern was projected. The appearances are blurred due to scattering. Checkered patterns in which white and black are replaced every three pixels are projected from the virtual projectors so that these patterns are aligned on the paper. Totally, eighteen images were captured by shifting the projecting pattern. The right images of Fig.9 (a)(b)(c) show the direct component when the high frequency patterns were projected. We can see that scattering in the 3-D scene is effectively reduced and the appearances become clear. While the descattering effect is not perfect, this is attributed to the low resolution of the virtual projectors in our current prototype system.

## 7.3 Factorization of the Observed Views

We confirmed the ability to visualize a particular depth in a 3-D scene by combining the FHFI and the factorization. Figure 10(a) shows the scene that an orange mesh covers a textured paper and (f) shows all views from the virtual cameras under normal illumination[1]. By simply averaging these views, a synthetic aperture image can be generated as shown in (b). Although the obstacle is blurred, the orange color of the mesh affects the paper.

[1] Although there are fifty mirror patches, only forty eight patches were used because two patches were misaligned.

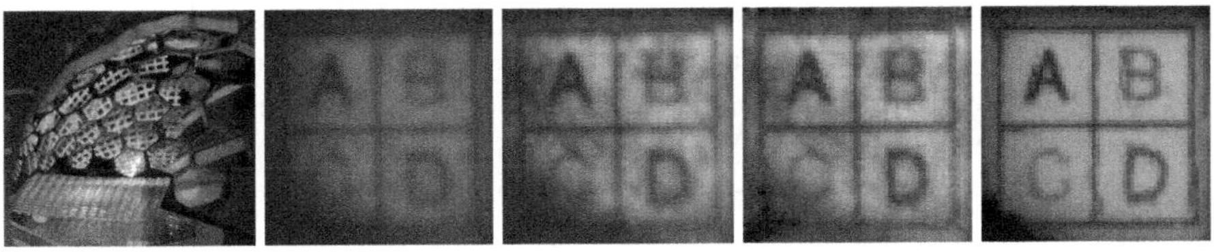


(a) scene (b) synthetic aperture (c) FHFI w/o factorization (d) texture term (e) ground truth

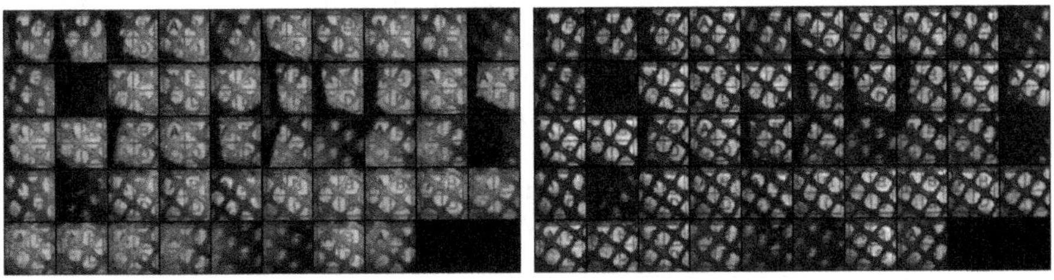

(f) views under normal illumination (g) focused high frequency illumination

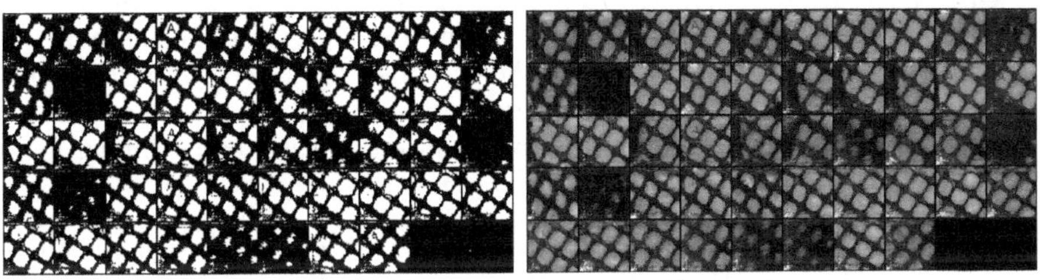

(h) masking term (i) attenuation term

**Fig. 10.** Result of the combination of the FHFI and the factorization

The mesh becomes dark by the FHFI because it is not illuminated, while the paper is bright as shown in (g). By averaging these views, the dark mesh is blurred and the orange color correctly disappears as shown in (c). However, there are uneven dark regions due to attenuation. The factorization decomposes the observed views (g) into the masking term (h), the attenuation term (i), and the texture term (d). We can see that the attenuation can be reduced especially around the letter of the black 'A' and the red 'D', since the occlusion due to the mesh is regarded as masks.

## 8 Limitations

- The resolution of the virtual projectors and cameras is low because the imaging areas of real projectors and cameras are divided for virtual ones.
- The observable area is narrow because all projectors must illuminate and all cameras must observe a common area. To enlarge the area, a large turtle reflector is necessary and it may be difficult to construct.

- The factorization is basically an ill-posed problem. For example, we can not distinguish two different scenes in which red texture is covered with colorless sheet and white texture is covered with red sheet. Some empirical constraints such as the smoothness of the attenuation are necessary.

## 9 Conclusion

We propose a new method of the hemispherical confocal imaging. This new imaging technique enables us to observe clear views of a particular depth in a 3-D scene. The originally designed turtleback reflector can divide the imaging areas so that a projector and a camera mimic a number of virtual projectors and cameras surrounding the scene. The combination of the focused high frequency illumination and the factorization can illuminate only the particular depth and eliminate scattering and attenuation. We have constructed a prototype system and confirmed the principles of the hemispherical aperture, descattering, and factorization.

One of our future works is to rebuild a more accurate optical device using a high resolution projector and evaluate the total performance, because the current prototype system only showed the principles separately. To develop some applications which can visualize any cross-sectional views of a translucent object is important. Another future work is to visualize the inside of the human body using infrared light.

**Acknowledgement.** This work is supported by a Microsoft Research CORE5 project and Grants-in-Aid for Scientific Researches (21680017 and 21650038).

## References

1. Adelson, E.H., Wang, J.Y.A.: Single Lens Stereo with a Plenoptic Camera. IEEE Tran. on PAMI, 99–106 (1992)
2. Cossairt, O., Nayar, S.K., Ramamoorthi, R.: Light Field Transfer: Global Illumination Between Real and Synthetic Objects. In: Proc. SIGGRAPH 2008, pp. 1–6 (2008)
3. Dana, K.J., Wang, J.: Device for convenient measurement of spatially varying bidirectional reflectance. J. Opt. Soc. Am. A 21(1), 1–12 (2004)
4. Debevec, P., Hawkins, T., Tchou, C., Duiker, H.P., Sarokin, W., Sagar, M.: Acquiring the Reflectance Field of a Human Face. In: Proc. SIGGRAPH 2000, pp. 145–156 (2000)
5. Fuchs, C., Heinz, M., Levoy, M., Seidel, H., Lensch, H.: Combining Confocal Imaging and Descattering. In: Proc. Computer Graphics Forum, Special Issue for the Eurographics Symposium on Rendering, vol. 27(4), pp. 1245–1253 (2008)
6. Garg, G., Talvala, E.V., Levoy, M., Lensch, H.P.A.: Symmetric Photography: Exploiting Data-sparseness in Reflectance Fields. In: Proc. EGSR 2006, pp. 251–262 (2006)
7. Ghosh, A., Achutha, S., Heidrich, W., O'Toole, M.: BRDF Acquisition with Basis Illumination. In: Proc. ICCV 2007 (2007)

8. Gu, J., Nayar, S.K., Grinspun, E., Belhumeur, P.N., Ramamoorthi, R.: Compressive Structured Light for Recovering Inhomogeneous Participating Media. In: Forsyth, D., Torr, P., Zisserman, A. (eds.) ECCV 2008, Part IV. LNCS, vol. 5305, pp. 845–858. Springer, Heidelberg (2008)
9. Han, J.Y., Perlin, K.: Measuring Bidirectional Texture Reflectance with a Kaleidoscope. ACM Transactions on Graphics 22(3), 741–748 (2003)
10. Kuthirummal, S., Nayar, S.K.: Multiview Radial Catadioptric Imaging for Scene Capture. In: Proc. SIGGRAPH 2006, pp. 916–923 (2006)
11. Levoy, M., Chen, B., Vaish, V., Horowitz, M., McDowall, I., Bolas, M.: Synthetic Aperture Confocal Imaging. In: Proc. SIGGRAPH 2004, pp. 825–834 (2004)
12. Levoy, M., Hanrahan, P.: Light field rendering. In: Proc. SIGGRAPH 1996, pp. 31–42 (1996)
13. Masselus, V., Peers, P., Dutré, P., Willems, Y.D.: Relighting with 4D incident light fields. In: Proc. SIGGRAPH 2003, pp. 613–620 (2003)
14. Matusik, W., Pfister, H., Ngan, A., Beardsley, P., Ziegler, R., McMillan, L.: Image-Based 3D Photography using Opacity Hulls. In: Proc. SIGGRAPH 2002, pp. 427–437 (2002)
15. Minsky, M.: Microscopy apparatus. US Patent 3013467 (1961)
16. Mukaigawa, Y., Sumino, K., Yagi, Y.: Multiplexed Illumination for Measuring BRDF Using an Ellipsoidal Mirror and a Projector. In: Yagi, Y., Kang, S.B., Kweon, I.S., Zha, H. (eds.) ACCV 2007, Part II. LNCS, vol. 4844, pp. 246–257. Springer, Heidelberg (2007)
17. Müller, G., Bendels, G.H., Klein, R.: Rapid Synchronous Acquisition of Geometry and Appearance of Cultural Heritage Artefacts. In: Proc. VAST 2005, pp. 13–20 (2005)
18. Narasimhan, S.G., Nayar, S.K., Sun, B., Koppal, S.J.: Structured light in scattering media. In: Proc. ICCV 2005, vol. 1, pp. 420–427 (2005)
19. Nayar, S.K., Krishnan, G., Grossberg, M.D., Raskar, R.: Fast Separation of Direct and Global Components of a Scene using High Frequency Illumination. In: Proc. SIGGRAPH 2006, pp. 935–944 (2006)
20. Sen, P., Chen, B., Garg, G., Marschner, S., Horowitz, M., Levoy, M., Lensch, H.: DualPhotography. In: Proc. SIGGRAPH 2005, pp. 745–755 (2005)
21. Treibitz, T., Schechner, Y.Y.: Active Polarization Descattering. IEEE Tran. on PAMI 31(3), 385–399 (2009)
22. Unger, J., Wenger, A., Hawkins, T., Gardner, A., Debevec, P.: Capturing and Rendering With Incident Light Fields. In: Proc. EGRW 2003, pp. 141–149 (2003)
23. Vaish, V., Szeliski, R., Zitnick, C.L., Kang, S.B., Levoy, M.: Reconstructing Occluded Surfaces using Synthetic Apertures: Stereo, Focus and Robust Measures. In: CVPR 2006, vol. II, pp. 2331–2338 (2006)
24. Vaish, V., Wilburn, B., Joshi, N., Levoy, M.: Using Plane + Parallax for Calibrate Dense Camera Arrays. In: Proc. CVPR 2004 (2004)
25. Veeraraghavan, A., Raskar, R., Agrawal, A., Mohan, A., Tumblin, J.: Dappled Photography: Mask Enhanced Cameras for Heterodyned Light Fields and Coded Aperture Refocusing. In: Proc. SIGGRAPH 2007 (2007)
26. Ward, G.J.: Measuring and Modeling anisotropic reflection. In: Proc. SIGGRAPH 1992, pp. 255–272 (1992)

# Image-Based and Sketch-Based Modeling of Plants and Trees

Sing Bing Kang

Microsoft Research, Redmond, WA, USA

**Abstract.** In this short paper, I outline representative techniques for modeling plants and trees using images and sketches. Image-based approaches have the distinct advantage that the resulting model inherits the realistic shape and complexity of a real plant or tree. Using sketches to produce tree models relies much more on prior knowledge of tree construction but makes the modeling process intuitive and easy.

## 1 Introduction

Generating models of plants and trees is highly non-trivial, since they have complex geometries. To do this, many techniques have been proposed; they can be roughly classified as primarily being rule-based, sketch-based, or image-based.

Rule-based techniques make use of small sets of generative rules or a grammar to create branches and leaves. Prusinkiewicz et al. [1] developed approaches based on the idea of the generative L-system. In [2], a collection of rules of plant growth (e.g., order of axis and phyllotaxy) is used to produce realistic-looking trees. Weber and Penn [3] used a series of geometric rules involving features such as general tree shape (including size and scale). While rule-based techniques provide some realism and editability, they generally require expertise for effective use.

Sketch-based systems were developed to provide a more intuitive way of generating plant models. For example, the system of [4] reconstructs the 3D branching pattern from 2D drawn sketches in different views by maximizing distances between branches. They use additional gesture-based editing functions to add, delete, or cut branches. The system of [5] is based on L-systems. The user draws a single free-form stroke to control the growth of a tree. Realism and amount of interactivity are issues for sketch-based systems.

There are approaches that instead use images as inputs. They range from the use of a single image and (limited) shape priors [6] to multiple images [7,8]. Since computer vision techniques tend to be imperfect, image-based techniques typically requires user feedback or correction to produce excellent results.

There are hybrid methods, such as that of [9]. Here, 3D tree models are produced from several photographs based on limited user interaction. This technique is a combination of image-based and sketch-based modeling.

I will now briefly describe three modeling methods as illustration: one image-based technique for modeling plants (with relatively large leaves), another image-based technique for modeling trees (with relatively small leaves), and a sketch-based technique for modeling trees.

R. Kimmel, R. Klette, and A. Sugimoto (Eds.): ACCV 2010, Part I, LNCS 6492, pp. 350–354, 2011.

**Fig. 1.** Overview of image-based plant modeling process

## 2 Image-Based Modeling of Plants

The system for modeling plants from images has three parts: image acquisition and structure from motion, leaf segmentation and recovery, and interactive branch recovery (Figure 1). A hand-held camera is used to capture images of the plant at different views. A standard structure from motion technique is then used to recover the camera parameters and a 3D point cloud.

Next, 3D data points and 2D images are segmented into individual leaves. An interface was designed to allow the user to easily specify the segmentation jointly using 3D data points and 2D images. The data to be partitioned are represented as a 3D undirected weighted graph that gets updated on-the-fly. For a given plant to model, the user first segments out a leaf; this is used as a deformable generic model. This generic leaf model is subsequently used to fit the other segmented data to model all the other visible leaves. This system is also designed to use the images as guides for interactive 3D reconstruction of the branches.

This approach works well with flora with relatively large and few leaves. The branches had to be interactively generated because it is typically difficult to specify their observed structure directly using plant growth parameters. Trees, on the other hand, have relatively small leaves and dense foliage. In addition, their branches tend to be more structured. Modeling a tree requires a different approach, as described next.

## 3 Image-Based Modeling of Trees

The tree modeling system also consists of three main parts: image capture and 3D point recovery, branch recovery, and leaf population (Figure 2). The recovery of the visible branches is mostly automatic, with the user given the option of

**Fig. 2.** Overview of image-based tree modeling process

refining their shapes. The subsequent recovery of the occluded branches and leaves is automatic with only a few parameters to be set by the user.

As was done by researchers in the past, the system capitalizes on the structural regularity of trees, more specifically the self-similarity of structural patterns of branches and the arrangement of leaves. The extracted local arrangement of visible branches are used as building blocks to generate the occluded ones. This is done using the recovered 3D points as hard constraints and the matte silhouettes of trees in the source images as soft constraints.

To populate the tree with leaves, the user first provides the expected average image footprint of leaves. The system then segments each source image based on color. The 3D position of each leaf segment is determined either by its closest 3D point or by its closest branch segment. The orientation of each leaf is approximated from the shape of the region relative to the leaf model or the best-fit plane of leaf points in its vicinity.

Image-based modeling techniques have the best potential of generating realistic models, but at a certain cost. First, computer vision techniques are imperfect, and thus require a certain amount of hand-holding to produce high-quality results. Second, only existing objects can be modeled. A sketch-based technique (described next) may be used as an alternative; it requires only simple sketches, and new trees can be generated.

## 4 Sketch-Based Modeling of Trees

The components of our tree sketching system are shown in Figure 3. The user needs to provide only a few connected strokes (each stroke representing a branch), and optionally the crown of the tree. The database contains typical tree exemplars and their associated global parameters. Based on the shape of the sketch, the system first selects the closest tree exemplar ("template"); the template's global parameters are subsequently used as a prior for constructing the 3D geometry of the sketch.

The sketch is assumed to be drawn under orthographic projection. This allows the problem of constructing the 3D geometry of the sketch to be reduced to estimating the depths of branch segment endpoints. The problem is formulated as an undirected graphical model (also known as Markov random field), with each branch segment as a node and its depth as a variable. In addition, rules governing the tree shape are imposed as spatial relationships between neighboring nodes. These relationships are made explicit by introducing additional nodes, producing what is called a *factor graph* (a bipartite graph with two kinds of nodes).

Solving the factor graph produces the 3D shape required. The system then propagates branches using the principle of self-similarity: it randomly selects replication blocks, scales, and reorients them, and then attaches them to open branches. If drawn, the crown constrains the overall shape of the tree during branch propagation. If the crown is not drawn, the branches are propagated by a fixed number of generations. To complete the tree model, the user can either select a leaf template from the tree database, or use the default leaf associated with the preselected template. The system then populates the tree based on botanical rules. While this is only an approximation of natural diversity, the system is capable of generating a large variety of trees.

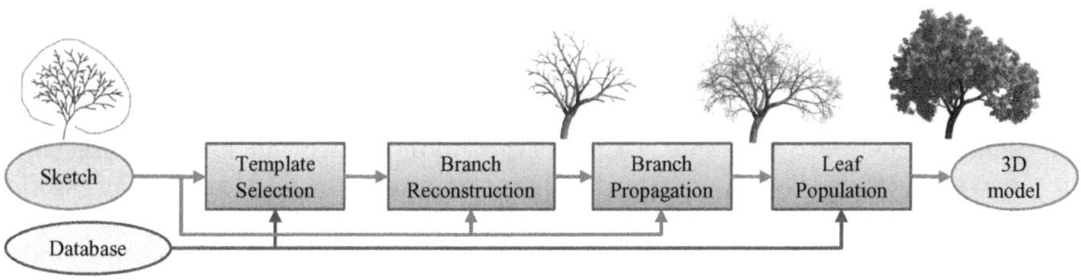


**Fig. 3.** Overview of sketch-based tree modeling process

## 5 Concluding Remarks

I have briefly described two image-based techniques (one for modeling plants and another for trees) and a sketch-based technique. These techniques are capable of generating realistic-looking models, and are attempts to investigate the tradeoffs among ease of use, degree of user interaction or feedback required, flexibility of system, and realism of output. I believe the sketch-based system scores the highest in terms of ease of use, degree of user interaction required, and flexibility. Users can generate good-looking models merely by sketching, and are able to generate new trees. Image-based techniques can only produce models of trees that exist; in addition, user feedback is required to make up for the imperfection of computer vision techniques. On the other hand, image-based techniques have the best potential for realism, since reconstruction is based on the appearance of real plants and trees. Marrying the two approaches, as was done in [9], seems to be the most promising. Unfortunately, all these approaches do not scale up well if many plants and trees need to be modeled (e.g., to generate a forest). Here, rule-based or procedural methods such as [10] are more appropriate. An interesting area of research is to derive full plant or tree growth parameters from observations.

**Acknowledgments.** This short paper is based on the SIGGRAPH publications on modeling plants [11] and trees [12], and SIGGRAPH Asia publication on sketch-based modeling [13]. I would like to thank my co-authors of those papers, namely, Yingqing Xu, Xuejin Chen, Boris Neubert, Oliver Deussen, Long Quan, Ping Tan, Gang Zeng, Jingdong Wang, and Lu Yuan.

## References

1. Prusinkiewicz, P., James, M., Mech, R.: Synthetic topiary. In: ACM SIGGRAPH, pp. 351–358 (1994)
2. de Reffye, P., Edelin, C., Francon, J., Jaeger, M., Puech, C.: Plant models faithful to botanical structure and development. In: ACM SIGGRAPH, pp. 151–158 (1988)
3. Weber, J., Penn, J.: Creation and rendering of realistic trees. In: ACM SIGGRAPH, pp. 119–127 (1995)
4. Okabe, M., Owada, S., Igarashi, T.: Interactive design of botanical trees using freehand sketches and example-based editing. Computer Graphics Forum (Eurographics) 24 (2005)
5. Ijiri, T., Owada, S., Igarashi, T.: The Sketch L-System: Global control of tree modeling using free-form strokes. In: Smart Graphics, pp. 138–146 (2006)
6. Han, F., Zhu, S.C.: Bayesian reconstruction of 3D shapes and scenes from a single image. In: IEEE Workshop on Higher-Level Knowledge in 3D Modeling and Motion Analysis, pp. 12–20 (2003)
7. Shlyakhter, I., Rozenoer, M., Dorsey, J., Teller, S.: Reconstructing 3D tree models from instrumented photographs. IEEE Computer Graphics and Applications 21, 53–61 (2001)
8. Reche-Martinez, A., Martin, I., Drettakis, G.: Volumetric reconstruction and interactive rendering of trees from photographs. ACM SIGGRAPH and ACM Transactions on Graphics 23, 720–727 (2004)
9. Neubert, B., Franken, T., Deussen, O.: Approximate image-based tree-modelling using particle flows. ACM SIGGRAPH and ACM Transactions on Graphics 26 (2007), article 88
10. Palubicki, W., Horel, K., Longay, S., Runions, A., Lane, B., Mech, R., Prusinkiewicz, P.: Self-organizing tree models for image synthesis. ACM SIGGRAPH and ACM Transactions on Graphics 28 (2009), article 58
11. Quan, L., Tan, P., Zeng, G., Yuan, L., Wang, J., Kang, S.B.: Image-based plant modeling. ACM SIGGRAPH and ACM Transactions on Graphics 25, 772–778 (2006)
12. Tan, P., Zeng, G., Wang, J., Kang, S.B., Quan, L.: Image-based tree modeling. ACM SIGGRAPH and ACM Transactions on Graphics 26 (2007), article 87
13. Chen, X., Neubert, B., Xu, Y.Q., Deussen, O., Kang, S.B.: Sketch-based tree modeling using Markov random field. ACM SIGGRAPH Asia and ACM Transactions on Graphics 27 (2008), article 109

# MOMI-Cosegmentation: Simultaneous Segmentation of Multiple Objects among Multiple Images

Wen-Sheng Chu[1], Chia-Ping Chen[2,3], and Chu-Song Chen[1,2]

[1] Research Center for Information Technology Innovation, Academia Sinica, Taipei 115, Taiwan
[2] Institute of Information Science, Academia Sinica, Taipei 115, Taiwan
[3] Dept. of CSIE, National Taiwan University, Taipei 106, Taiwan

**Abstract.** In this study, we introduce a new cosegmentation approach, *MOMI-cosegmentation*, to segment *multiple objects* that repeatedly appear among *multiple images*. The proposed approach tackles a more general problem than conventional cosegmentation methods. Each of the shared objects may even appear more than one time in one image. The key idea of MOMI-cosegmentation is to incorporate a common pattern discovery algorithm with the proposed Gibbs energy model in a Markov random field framework. Our approach builds upon an observation that the detected common patterns provide useful information for estimating foreground statistics, while background statistics can be estimated from the remaining pixels. The initialization and segmentation processes of MOMI-cosegmentation are completely automatic, while the segmentation errors can be substantially reduced at the same time. Experimental results demonstrate the effectiveness of the proposed approach over state-of-the-art cosegmentation method.

## 1 Introduction

Cosegmentation refers to simultaneous segmentation of similar objects from two or more images. While many studies [1,2,3] have shown that better segmentation from a single image could be achieved by interactive user inputs, completely automatic segmentation is possible for cosegmentation by using multiple images. The commonality across the images provides the information needed for facilitating the cosegmentation task. This idea was first introduced by Rother *et al.* [4] to segment an object of interest from an image pair, and has been applied to concurrent foreground extraction tasks, such as segmentation of image sequences [5] and several other problems [6,7,8]. Besides apparent applications in image or video editing, cosegmentation also implies several potential applications in other important areas, including biomedical imaging, video tracking, and content-based image retrieval.

The original goal of cosegmentation is to facilitate segmentation of common objects or regions by providing minimal additional information (such as just one additional image) so that better results could be obtained without user inputs.

R. Kimmel, R. Klette, and A. Sugimoto (Eds.): ACCV 2010, Part I, LNCS 6492, pp. 355–368, 2011.

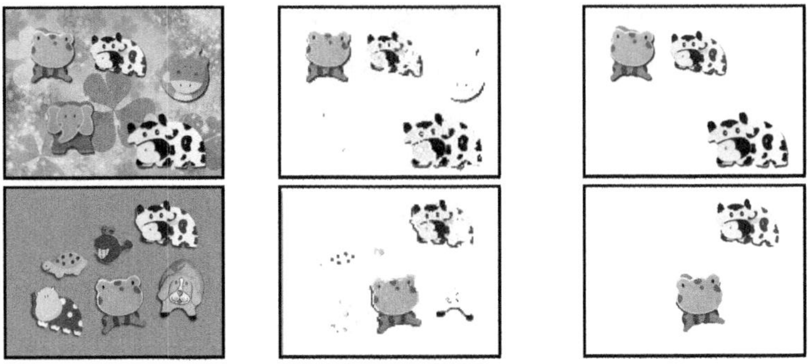

(a) Input image pair (b) Cosegmentation [9] (c) The proposed method

**Fig. 1.** Given two (or more) images (a), the objective of cosegmentation is to segment the common objects in these images. Note that the problem here is more general since multiple objects could occur multiple times in an image. (b) and (c) show the results of state-of-the-art cosegmentaion algorithm [9] and the proposed method.

It is typically designed in a class-constrained fashion, i.e., a given set of images is assumed to be of the same object class. Because each image contains only one instance of the same object, one could consider the problem as approximating the position of the common object. In practice, many images, such as our daily photos, often share more than one object in common. An object may even appear more than one time in an image. Take Fig. 1 (a) for example. Two objects, *frog* and *cow*, simultaneously appear in the image pair and *cow* appears twice in the top image. As shown in Fig. 1 (b), the cosegmentation algorithm [9] produces segmentation errors when similar colors appear both in the foreground and background regions.

In this paper, we tackle a more general problem without the assumption that only one object appears in each image. The problem becomes more difficult than conventional cosegmentation and object detection (or recognition) in several aspects. First, no prior knowledge is provided for the common objects or regions: we have no idea about what and how many the common objects are, and how many times each object appears in an image. So how can we detect common objects in an unannotated image set? An intuitive way is to exhaustively compare all sub-images at all possible positions and scales among these images. The search domain, however, is extremely huge and the computational cost increases exponentially with the number of input images. Therefore, we present a new approach, *MOMI-cosegmentation*, to address the above issues in an unsupervised framework. The novelty of MOMI-cosegmentation lies in incorporating a common pattern discovery algorithm with the proposed MRF model, which is extended from a Gibbs energy [10]. We propose to use the common pattern discovery algorithm [11] to detect coherent objects among an unannotated image set. Besides, the initialization and segmentation of the proposed approach is completely automatic, which is vital for real world applications. Fig. 1 (c) shows the results of the proposed method, where segmentation errors are significantly reduced in comparison to Fig. 1 (b) obtained by [9].

## 2 Related Work

This paper lies in the intersection between the fields of cosegmentation and common pattern discovery. In this section, we briefly review previous works in each field.

Cosegmentation belongs to the category of unsupervised techniques. Existing approaches [4,9,12] cast this problem as a minimization problem of a Markov random field (MRF), which discourages histogram dissimilarities of foreground regions between two input images. The idea proposed in [4] penalized the MRF energy by the $L_1$ histogram dissimilarities of foreground regions. However, the optimization problem regularized by the $L_1$-norm becomes more difficult to solve. Mukherjee *et al.* [12] considered the problem using the squared $L_2$ distance and showed the modified objective function leads to an optimal linear programming solution of only "half-integrality" values. Hochbaum and Singh [9] claimed that the regularization terms of histogram difference lead to difficult optimization, and proposed to replace these terms by the "carrot or stick" strategy. The optimization problem was solved more efficiently in polynomial time using only one maximum flow maximization. However, these works implicitly assumed that only one object appears in each image.

Recent approaches for common pattern discovery are [11,13,14,15]. Quack *et al.* [13] use a data mining technique to find spatial configurations of local features that frequently occur in an image set. A random partitioning approach is adopted in [14] to match all pairs of sub-images. A common pattern is then detected as the sub-image with the highest matching score. Yuan *et al.* [15] find a common pattern by gradually pruning possible candidates. Common patterns can be found by aggregating the voting maps in above methods. However, many methods implicitly assume that only one object appears in one image. In [11], common patterns are found as dense clusters in a correspondence graph represented by an incompatibility matrix. Because this work finds common patterns in a density-based clustering framework, it by nature relaxes the assumption that each image contains only one object. Nevertheless, these methods do not consider segmentation and produces undesirable segmentation artifacts.

The rest of this paper is organized as follows. In Section 3, we describe the concepts of [11] that is used to detect common objects that appears repeatedly in a set of images. The proposed segmentation model is presented in Section 4. We show the experimental results in Section 5 and give conclusions in Section 6.

## 3 Common Pattern Discovery

In this section, we review the concepts of the common pattern discovery algorithm [11]. Given a set of $N$ unannotated images, the goal is to unsupervisedly detect common objects (or regions) shared by the image set. Note that the assumption that only one common object in each image is relaxed in this approach.

**Candidate matches.** Given the $n$-th image $I_n$, we extract a set of local appearance features $\mathcal{F}_n = \{(\mathbf{p}_n^i, s_n^i, \mathbf{d}_n^i) | i = 1, \dots, |\mathcal{F}_n|\}$, where $\mathbf{p}_n^i$ and $s_n^i$ are

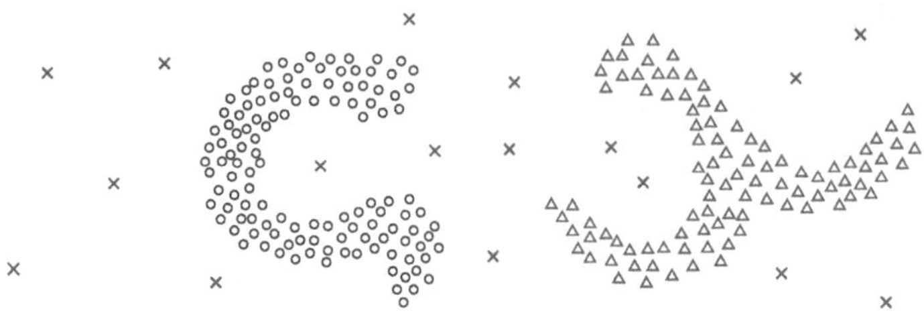

**Fig. 2.** Illustration of the correspondence graph used for density-based clustering [11]. Each node represents a candidate match $ii'$; each dense cluster can be considered as a common object (or region).

the position and the scale of the $i$-th feature in $I_n$, $\mathbf{d}_n^i$ is the corresponding feature descriptor and $|\mathcal{F}_n|$ is the number of features in $I_n$. Here, we use the Harris-Laplace corner detector and the OpponentSIFT descriptor [16] for feature extraction. Note that other options [17,18] are also applicable for computing the feature descriptor $\mathbf{d}_n^i$.

Given two images $I_m$ and $I_n$ as two sets of local features, the number of all possible correspondences across each pair of local features is enormous. It is computationally prohibitive to establish such a correspondence between each image pair. Therefore, we filter out the candidate matches $\mathcal{M}$ by

$$\mathcal{M} = \{ii' \mid \|\mathbf{d}_m^i - \mathbf{d}_n^{i'}\| < \lambda\}, \tag{1}$$

where $ii'$ stands for the match between the $i$-th feature in $I_m$ and the $i'$-th feature in $I_n$. $\lambda$ controls the maximum dissimilarity between two appearance features. Typically, $\mathcal{M}$ will contain only a small subset of all possible matches.

**Incompatibility matrix.** After the relatively small candidate matches $\mathcal{M}$ is filtered out for each image pair, the next goal is to construct an incompatibility matrix $\mathbf{D}$ for these matches between two images $I_m$ and $I_n$. The incompatibility matrix $\mathbf{D}$ measures the incoherence between a pair of "matches" in the two images. Let $i_1$ and $i_2$ denote two local features in $I_n$, $sd^n(i_1, i_2) = \|\mathbf{p}_n^{i_1} - \mathbf{p}_n^{i_2}\|$ indicates the spatial distance between $i_1$ and $i_2$. Considering each candidate match $i_2i'_2$ within the spatial $\varepsilon$-neighborhood of $i_1i'_1$, i.e., $sd^m(i_1, i_2) < \varepsilon$ and $sd^n(i'_1, i'_2) < \varepsilon$, we can compute their incompatibility as:

$$\mathbf{D}(i_1i'_1, i_2i'_2) = \alpha_1 \times \mathit{unary}(i_1i'_1, i_2i'_2) + \alpha_2 \times \mathit{binary}(i_1i'_1, i_2i'_2), \tag{2}$$

where *unary* and *binary* are the constraints used to capture the appearance dissimilarity and geometric inconsistency for each pair of candidate matches, respectively. A possible choice of the *unary* and *binary* constraints can be given as in [11]:

$$\mathit{unary}(i_1i'_1, i_2i'_2) = \frac{\|\mathbf{d}_m^{i_1} - \mathbf{d}_n^{i'_1}\| + \|\mathbf{d}_m^{i_2} - \mathbf{d}_n^{i'_2}\|}{2}, \tag{3}$$

$$\mathit{binary}(i_1i'_1, i_2i'_2) = \frac{|sd^m(i_1, i_2) - sd^n(i'_1, i'_2)|}{\sqrt{sd^m(i_1, i_2)sd^n(i'_1, i'_2)}}. \tag{4}$$

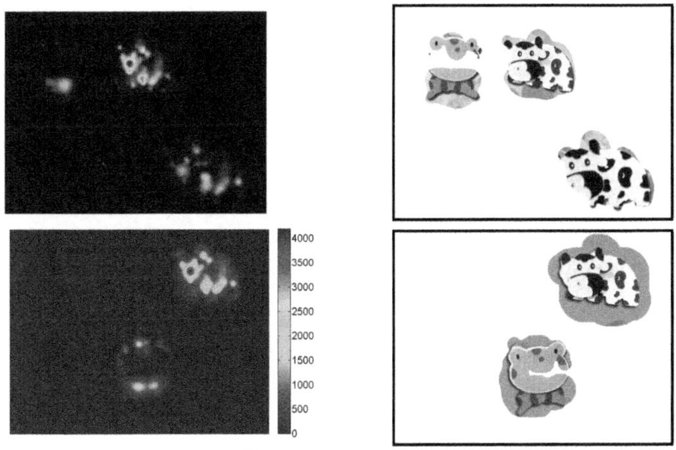


(a) Confidence maps $I_n^C$ (b) Preliminary segmentation results $I_n^P$

**Fig. 3.** The confidence maps of Fig. 1 (a). The larger value given in the confidence map implies higher possibility that the pixel is a part of a shared object.

**Correspondence graph.** Small values in $\mathbf{D}$ reflects potential correct matches of a shared object in the image pair, because appearance differences and geometric inconsistency between correct matches shall be small. Incorrect matches are likely to be inconsistent with each other with large incompatibilities. From this point of view, we can see the candidate matches $\mathcal{M}$ as nodes that forms the *correspondence graph* with corresponding linkage weights specified by $\mathbf{D}$. As illustrated in Fig. 2, correct matches tend to form dense clusters (blue circles and green triangles) with small linkage weights. The isolated nodes (red crosses) in the correspondence graph indicate the incorrect matches with large linkage weights.

**Density-based clustering.** Given the correspondence graph, the problem of finding common objects in an image set is reduced to a dense cluster discovery problem. A dense cluster, i.e., a set of nodes linked by small weights, represents a possible shared object appearing in an image pair. As we do not know in advance the shape of each cluster, clustering methods that assume each cluster has a globular shape, such as K-means and affinity propagation, are not adequate for this case. Furthermore, the number of dense clusters in the correspondence graph is not known either.

Therefore, the density-based algorithm [19] is utilized to discover clusters with arbitrary shapes in the presence of a large number of outlier matches. One of the benefits of the algorithm is that we do not have to specify the number of clusters in advance. The only parameters used are the radius $\epsilon$ of neighborhood and the density $d$ in the $\epsilon$-neighborhood. In our implementation, we fixed $\epsilon = 2000$ and $d = 20$ across all experiments.

**The confidence map.** After performing the density-based algorithm for each pair of images, we derive $(N-1)$ feature masks for each image in the unannotated image set. Each feature mask records the confidence of each local feature and indicates that how likely a local feature is a part of a common object. The confidence of the $i$-th local feature $\mathcal{F}_n^i$ in the image $I_n$ is accumulated across all $(N-1)$ feature masks. By fusing these feature masks for each image, we then obtain a confidence map of positive real values. The confidence map of Fig. 1 (a) is shown in Fig. 3. Preliminary segmentation results can be obtained by performing a simple thresholding. The preliminary segmentation results obtained from the image pair of 1 (a) are shown in Fig. 3 (b). See Fig. 4 (c) for more examples. Although this algorithm can successfully detect common objects across input images, the objects are only partly included in the preliminary segmentation results.

## 4 MOMI-Cosegmentation Incorporating Common Pattern Discovery

Conventional cosegmentaion methods [4,9,12] are restrictive in two assumptions: the input is an image pair and each image contains the same object in different backgrounds. In order to detect *multiple objects* that may appear *multiple times* in one image, we incorporate the preliminary segmentation results $I_n^P$ and the confidence maps $I_n^C$ images of $N$ images generated from the common pattern discovery algorithm [11]. We then consider the cosegmentation problem as an individual foreground/background segmentation on each image $I_n, n = 1, \ldots, N$.

The segmentation problem can be interpreted as a binary labelling problem: each pixel $p$ has to be assigned a unique label $x_p$, where $x_p$ is a binary label of 0 (background) or 1 (foreground). Let $\mathcal{V}$ be the set of all pixels in $I_n$ and $\mathcal{E}$ be the set of all adjacent pixel pairs in $I_n$. We formulate the problem of computing the optimal labels $X = \{x_p | p \in \mathcal{V}\}$ as an energy minimization of the following cost function:

$$E(X) = \lambda_{\text{color}} \sum_{p \in \mathcal{V}} E_{\text{color}}(x_p) + \lambda_{\text{smoothness}} \sum_{(p,q) \in \mathcal{E}} E_{\text{smoothness}}(x_p, x_q) + \lambda_{\text{confidence}} \sum_{p \in \mathcal{V}} E_{\text{confidence}}(x_p) + \lambda_{\text{locality}} \sum_{p \in \mathcal{V}} E_{\text{locality}}(x_p). \quad (5)$$

We introduce the different energy terms corresponding to various cues from the prior knowledge of color models, smoothness, confidence maps and locality relationship. The parameters $\lambda_{\text{color}}$, $\lambda_{\text{smoothness}}$, $\lambda_{\text{confidence}}$ and $\lambda_{\text{locality}}$ balance the contribution of each energy term. Each energy term is then described in the following subsections.

### 4.1 Color Term and Smoothness Term

The color and smoothness terms are frequently used in segmentation problems [1,20,21]. We first explain the two terms as the *fundamental model.*

**Color term.** The idea of color term is to exploit the fact that different groups of foreground or background segments tend to follow different color distributions. For an image $I_n$, we train two Gaussian mixture models (GMMs), one for the foreground and one for the background, from the given preliminary segmentation result $I_n^P$. The purpose of each GMM is to estimate the likelihood of each pixel $p$ that belongs to foreground or background based on the color cue. Each GMM is taken as full-covariance Gaussian with $K$ components (typically $K = 5$). The color term is defined as

$$E_{\text{color}}(x_p) = -\log G(p|x_p), \tag{6}$$

where each color model $G$ is given by the mixture of Gaussians:

$$G(p|x_p) = \sum_{k=1}^{K} \pi_k \frac{1}{\sqrt{\det \Sigma_k}} \exp\left(-\frac{1}{2}(p-\mu_k)^T \Sigma_k^{-1}(p-\mu_k)\right). \tag{7}$$

$G(p|x_p)$ indicates the probability that pixel $p$ belongs to the label $x_p$. Note that if the pixel $p$ is assigned to be foreground ($x_p = 1$), the summation in Eq. (7) is over the foreground GMMs for estimating the foreground likelihood of $p$; otherwise, the summation is over the background GMMs. The color term encourages the pixels to follow the labels of the most similar color model.

**Smoothness term.** The smoothness term is designed to preserve the coherence between two neighboring pixels of similar pixel values and imply a tendency to solidity of objects. This is useful in situations where matching constraints are weak, such as too sparse candidate matches or too many ambiguous colors that both occur in the foreground and the background GMMs. The smoothness term between two adjacent pixels $p$ and $q$ is defined as

$$E_{\text{smoothness}}(x_p, x_q) = [x_p \neq x_q] \exp\left(-\beta\|p-q\|^2\right), \tag{8}$$

where $[expr]$ denotes the indicator function taking value 0, 1 for the predicate $expr$ and the constant $\beta$ can be chosen to be $\left(1/2\langle\|p-q\|^2\rangle\right)$ as suggested in [10]. This term is a smoothness penalty when the neighboring pixels are labelled differently, i.e., $x_p \neq x_p$. In other words, the less similar colors of $p$ and $q$ are, the smaller cost $E_{\text{smoothness}}$ would produce, and therefore the more likely the edge between $p$ and $q$ is on the object boundary.

The minimization problem using $E_{\text{color}}$ and $E_{\text{smoothness}}$ alone is similar to that proposed in GrabCut [1]. The main distinction is that we extend the segmentation domain from the initial user-defined rectangle trimap to the entire image. The results of GrabCut used in this manner are shown in third column of Fig. 7. Although color coherence and smoothness are preserved by $E_{\text{color}}$ and $E_{\text{smoothness}}$, noticeable segmentation errors occur because of the imperfect preliminary segmentation in $I_n^P$ and the non-discriminative GMMs of the foreground and background. We then introduce two more energy terms, $E_{\text{confidence}}$ and $E_{\text{locality}}$, to recover correct foreground pixels as well as remove false "background artifacts".

## 4.2 Confidence Term

The energy functions discussed in the above section may cause the segmentation errors where correct foreground pixels are assigned to background labels. This is

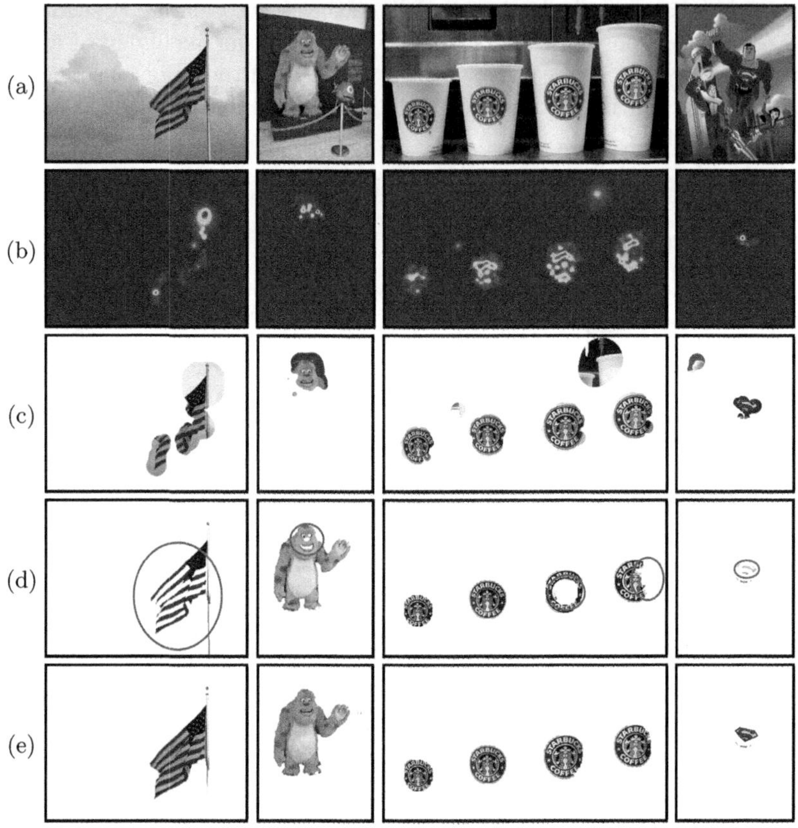


**Fig. 4.** Effects of confidence terms. (a) input images, (b) given confidence maps, (c) preliminary segmentation results, (d) GrabCut with only color terms and smoothness terms based on the preliminary results (red circles indicate the segmentation errors) and (e) segmentation with additional confidence terms.

because the similar colors between the foreground and the background models distract the labelling of foreground pixels. Fig. 4 shows examples when this type of segmentation errors occur. Take the *American flag* in Fig. 4 (d) for example, the white stripes are wrongly labelled because of its uncertain likelihood of white color between foreground and background. The goal in each column (from left to right) of this figure is to segment the *American flag*, the animation character *Sulley*, the trademark of *Starbucks* and *Superman*'s S shield, respectively.

Therefore, we resort to the cues of confidence map $I_n^C$, produced by the common pattern discovery algorithm discussed in Section 3, to resolve the color ambiguity. Specifically, we exploit the prior knowledge of confidence values to encourage good and coherent segmentations; pixels with high confidence values should be retained as foreground. Given $c(p)$ as the original confidence value of $p$ in $I_n^C$, we define the confidence term as

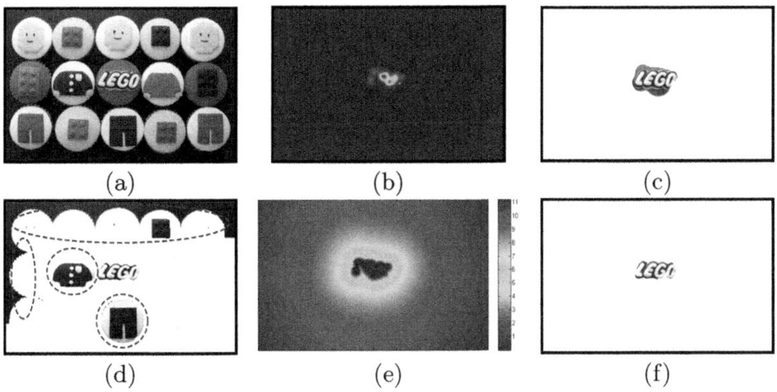


**Fig. 5.** Effects of locality terms. (a) input images, (b) given confidence maps, (c) preliminary segmentation results, (d) GrabCut based on the preliminary results (blue dashed circles indicate the background artifacts), (e) logarithmic distance map and (f) segmentation with additional confidence terms.

$$E_{\text{confidence}}(x_p) = \begin{cases} (2x_p - 1)\tilde{c}(p), & \tilde{c}(p) > 0 \\ (1 - 2x_p)\tilde{c}(p), & otherwise. \end{cases} \tag{9}$$

where $\tilde{c}(p)$ is the normalized confidence energy of pixel $p$ in $[-1, 1]$ by the sigmoid function

$$\tilde{c}(p) = 4\left(\frac{1}{1 + \exp\left(-c(p)\right)} - \frac{3}{4}\right). \tag{10}$$

Larger $\tilde{c}(p)$ refers to larger value of $c(p)$, which indicates more confidence that the pixel $p$ belongs to common objects appearing repeatedly among an image set. When $\tilde{c}(p) > 0$, $p$ has high possibility of belonging to the foreground, and thus the confidence term encourages the foreground ($x_p = 1$) likelihood by adding $\tilde{c}(p)$ and penalizes the background ($x_p = 0$) by subtracting $\tilde{c}(p)$. On the other hand, when $\tilde{c}(p) \leq 0$, we subtract $\tilde{c}(p)$ from $x_p = 1$ and add $\tilde{c}(p)$ to $x_p = 0$. As shown in Fig. 4 (d) and (e), most neglected foreground pixels could be recovered by incorporating the confidence term.

### 4.3 Locality Term

The color, smoothness and confidence cues could usually produce good results in most image sets. However, when there are color ambiguities between background and foreground GMMs, or when the number of background GMMs are not large enough to model the colors in cluttered backgrounds, incorrect segmentations in the background often occur. We call the undesirable background segments the "background artifacts".

An example is illustrated in Fig. 5. The first row displays the input image, given correspondences map and the preliminary segmentation. Segmentation based on color and smoothness cues is shown in Fig. 5 (d), where the background artifacts are marked as red circles. In order to remedy these background artifacts, we introduce the locality term

$$E_{\text{locality}}(x_p) = \log \left( \min_{q \in \mathcal{V}, c(q) > \delta} dist(p, q) \right), \tag{11}$$

where $dist(p, q) = \|\mathbf{p}_p - \mathbf{p}_q\|^2$ is the spatial distance between any pixel pairs $(p, q)$, $\delta$ controls the threshold for candidates of the reference pixel $q$ and $\sigma$ is a parameter (typically $\sigma = 20$). We use the locality term to impose the distance penalty on pixels that are away from those with confidence values higher than $\delta$. The further a pixel $p$ is away from the reference pixel $q$, the less possible $p$ belongs to the foreground. The locality term, from this perspective, is helpful to remove the background artifacts that have similar colors as foreground pixels. Fig. 5 (e) and (f) display the logarithmic distance maps and segmentation results incorporating the locality term, respectively.

## 5 Experimental Results

In this section, we discuss the experiments for evaluating the performance of the proposed method. Qualitative and quantitative analysis of the proposed approach are presented. We used the min-cut algorithm [10] to minimize the energy function $E(X)$. Throughout the following experiments, $\epsilon$ and $d$ in [11] were set to be 2000 and 20, and $K$ for the color models was fixed at 5. Parameters $\lambda_{\text{color}} = 1$ and $\lambda_{\text{smoothness}} = 40$ were set for the proposed Gibbs model, while the choices of $\lambda_{\text{confidence}}$ and $\lambda_{\text{locality}}$ were user-specified.

**Comparison with cosegmentation.** We firstly compare the proposed method with state-of-the-art cosegmentation [9]. Because the cosegmentation algorithm [9] considers only two input images, the proposed method was evaluated using only two images for fairness. In addition, [9] takes a large memory storage of additional nodes, hence the segmentation errors for [9] were reported on lower-resolution images while those were reported on full-resolution images for the proposed method. Although [9] was introduced for automatically extracting common foreground from two images, it requires manually labelling of RGB intensities for foreground and background. Our method, on the other hand, performs an automatic preliminary labelling from the results of the common pattern discovery.

The segmentation errors, i.e., the percentage of wrongly labelled pixels with respect to the whole image, were presented for five image sets as shown in Fig. 7. We also performed GrabCut [1] on each image pair as a baseline algorithm. As shown in the $3^{rd}$ and $4^{th}$ columns, GrabCut and [9] could extract the objects of interest, but suffer from the problem of color ambiguity: similar colors between foreground and background pixels. Note that in the last example, *Leaning Tower of Pisa*, the tower exhibits different colors because of different illumination. [9] fails to extract all correct foreground pixels, while our method, shown in the $5^{th}$ column, utilized the confidence term to retain correct foreground pixels and the locality term to remove mislabelled pixels in the background. Foreground misses and background artifacts can be thus considerably reduced. Our method

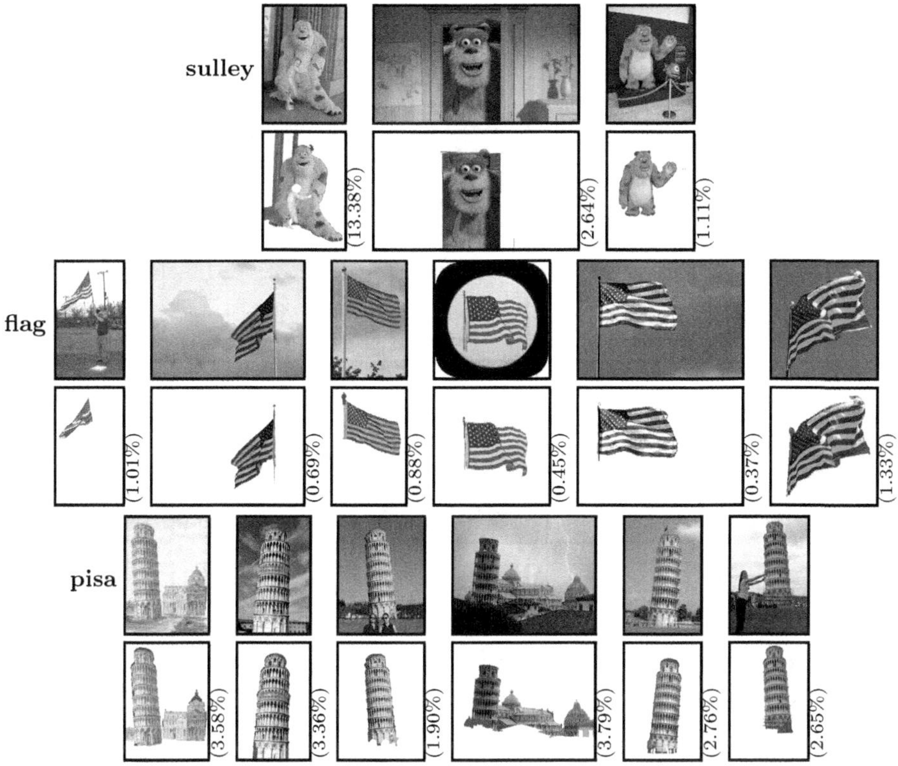


**Fig. 6.** The first row shows an input image set and the second row shows corresponding segmentation obtained by our method. Segmentation errors are shown as percentages and marked as red colors in the segmentation results.

**Table 1.** Comparison between the *fundamental model* (FM) used in [1] and the proposed *MOMI-cosegmentation* (MOMI-CS). Each method is evaluated by averaging the segmentation errors across the 12 datasets.

| set(#img) | sulley(3) | starbucks(3) | magnet(4) | flag(6) | pisa(6) | superman(7) |
|---|---|---|---|---|---|---|
| FM | 20.50 | 2.68 | 22.56 | 7.71 | 17.88 | 18.62 |
| MOMI-CS | 5.71 | 0.41 | 1.20 | 0.79 | 3.01 | 1.38 |
| set(#img) | domino(6) | heineken(8) | warcraft(6) | kfc(6) | lego(4) | pringles(8) |
| FM | 26.65 | 18.47 | 26.65 | 35.21 | 43.52 | 15.24 |
| MOMI-CS | 2.46 | 1.25 | 2.63 | 6.78 | 1.08 | 4.17 |

in these image sets produces nearly perfect results as groundtruths with very low segmentation errors. Results on more image sets will be presented shortly.

**Comparison with the *fundamental model*.** For image sets containing more than two images, we compare the performance between the proposed approach

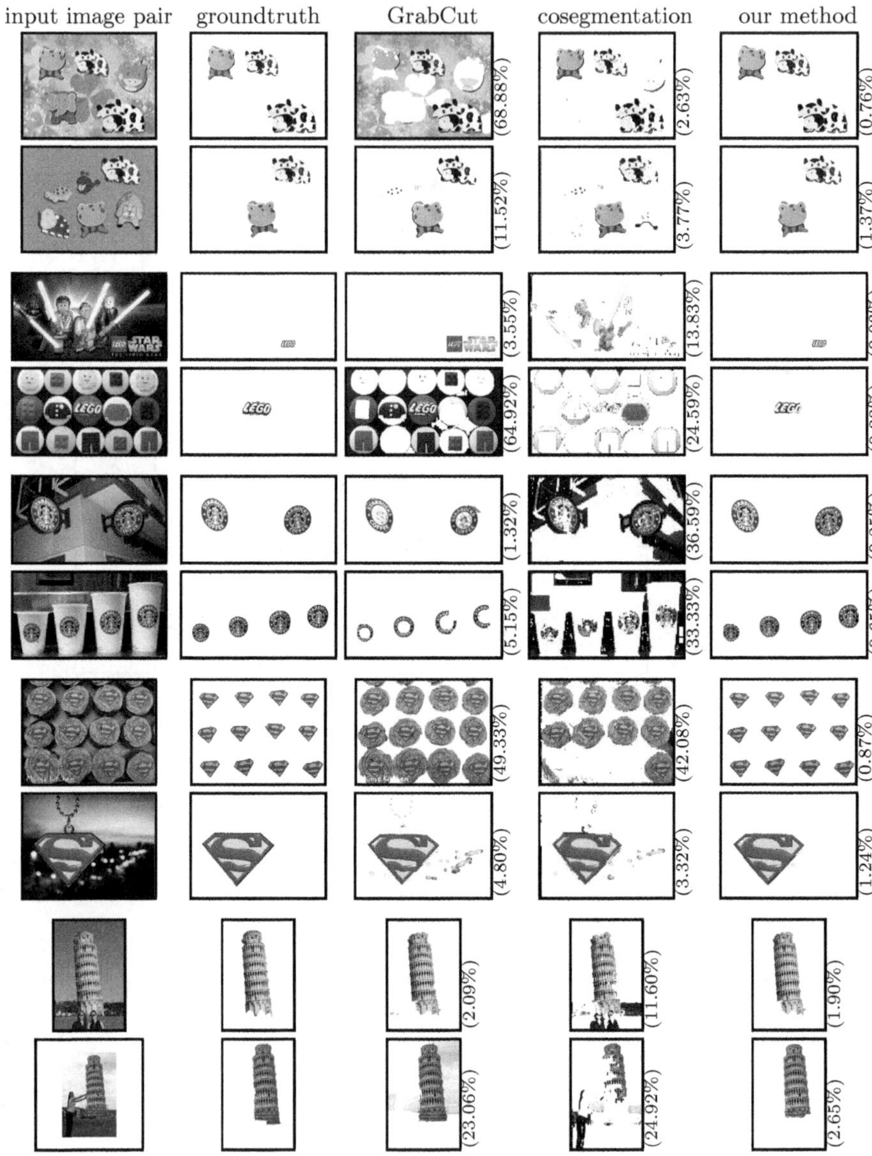


**Fig. 7.** Five examples of image pairs. Each column (from left to right) shows the input image pairs, groundtruth, GrabCut [1] results, cosegmentation [9] results and results of our method, respectively. (Errors are denoted as the percentages).

and the fundamental model used in [1]. The datasets[1] were collected from Flickr with moderate variations in illumination and scale. Groundtruths were manually labelled. Averaged segmentation errors of each dataset were presented in Table 1. The results show that good segmentation for concurrent objects can still be obtained using our method, although each dataset contains more than two images. The fundamental model used in [1] considers only color and smoothness cues, therefore produces worse results when similar colors appear in both foreground and background, as shown in the third column of Fig. 7. The proposed method achieved an average of 2.57% segmentation errors across the 12 image sets.

Besides rigid objects, we also evaluated the proposed method on some deformable objects. Both qualitative and quantitative results are shown in Fig. 6. Note that some objects of the same class may appear in heterogeneous circumstances, e.g., the second image in *sulley* is from animation while the others are real models. Similar circumstances could be found in the $4^{th}$ image (from left to right) in *flag* and the $1^{st}$ image in *pisa*. Moreover, some images are very challenging because of their cluttered backgrounds. The proposed method is capable of successfully segment the shared objects, and produced satisfactory results with less than 6% averaged segmentation errors in these mega-pixel images.

## 6 Conclusion

In this paper, we proposed a new cosegmentation approach called MOMI-cosegmentation, which is more general and scalable in many aspects. Compared to conventional cosegmentation methods, the proposed approach can deal with more than two input images, and allow multiple objects to appear more than one time in an image. Although the domain of searching the common objects in multiple images is computationally prohibitive, we combined color, smoothness, confidence and locality cues and incorporated a common pattern discovery algorithm to achieve satisfactory segmentation. Foreground misses and background artifacts can be efficiently reduced using our method. In addition, label initialization and segmentation process are automatic in MOMI-cosegmentation. The experiments have demonstrated that the performance of the proposed method outperforms state-of-the-art cosegmentation method [9].

**Acknowledgement.** This work was supported in part by the National Science Council, Taiwan, under the grants NSC99-2631-H-001-020 and NSC98-2221-E-001-012-MY3. We would also like to thank Professor Vikas Singh for providing the source code of [9].

## References

1. Rother, C., Kolmogorov, V., Blake, A.: Grabcut: Interactive foreground extraction using iterated graph cuts. In: ACM SIGGRAPH, p. 314. ACM, New York (2004)
2. Li, Y., Sun, J., Tang, C.K., Shum, H.Y.: Lazy snapping. ACM Trans. on Graphics 23, 303–308 (2004)

[1] The dataset is available at
`http://imp.iis.sinica.edu.tw/ivclab/research/coseg/`

3. Riklin-Raviv, T., Sochen, N., Kiryati, N.: Shape-based mutual segmentation. IJCV 79, 231–245 (2008)
4. Rother, C., Minka, T., Blake, A., Kolmogorov, V.: Cosegmentation of image pairs by histogram matching-incorporating a global constraint into MRFs. In: CVPR, vol. 1 (2006)
5. Cheng, D.S., Figueiredo, M.: Cosegmentation for image sequences. In: International Conference on Image Analysis and Processing, pp. 635–640 (2007)
6. Sun, J., Kang, S.B., Xu, Z.B., Tang, X., Shum, H.Y.: Flash Cut: Foreground Extraction With Flash And No-flash Image Pairs. In: CVPR, pp. 1–8 (2007)
7. Cao, L., Fei-Fei, L.: Spatially coherent latent topic model for concurrent object segmentation and classification. In: ICCV (2007)
8. Gallagher, A.C., Chen, T.H.: Clothing cosegmentation for recognizing people. In: CVPR, pp. 1–8 (2008)
9. Hochbaum, D.S., Singh, V.: An efficient algorithm for co-segmentation. In: ICCV (2009)
10. Boykov, Y., Jolly, M.P.: Interactive graph cuts for optimal boundary and region segmentation of objects in ND images. In: ICCV, vol. 1, pp. 105–112 (2001)
11. Chen, C.P., Chu, W.S., Chen, C.S.: Common pattern discovery with high-order constraints by density-based cluster discovery (2010) (submitted for publication)
12. Mukherjee, L., Singh, V., Dyer, C.R.: Half-integrality based algorithms for cosegmentation of images. In: CVPR (2009)
13. Quack, T., Ferrari, V., Leibe, B., Gool, V.L.: Efficient mining of frequent and distinctive feature configurations. In: ICCV, pp. 1–8 (2007)
14. Yuan, J., Wu, Y.: Spatial random partition for common visual pattern discovery. In: ICCV, pp. 1–8 (2007)
15. Yuan, J., Li, Z., Fu, Y., Wu, Y., Huang, T.S.: Common spatial pattern discovery by efficient candidate pruning. In: International Conference on Image Processing (2007)
16. van de Sande, K., Gevers, T., Snoek, C.: Evaluating color descriptors for object and scene recognition. IEEE Trans. on PAMI (2010) (in press)
17. Lowe, D.G.: Distinctive image features from scale-invariant keypoints. IJCV 60, 91–110 (2004)
18. Mikolajczyk, K., Schmid, C.: Scale and affine invariant interest point detectors. IJCV 60, 63–86 (2004)
19. Ester, M., Kriegel, H.P., Sander, J., Xu, X.: A density-based algorithm for discovering clusters in large spatial databases with noise. In: Knowledge Discovery and Datamining, pp. 226–231 (1996)
20. Sun, J., Zhang, W., Tang, X., Shum, H.-Y.: Background cut. In: Leonardis, A., Bischof, H., Pinz, A. (eds.) ECCV 2006. LNCS, vol. 3952, pp. 628–641. Springer, Heidelberg (2006)
21. Guillemaut, J.Y., Kilner, J., Hilton, A.: Robust graphcut scene segmentation and reconstruction for free-viewpoint video of complex dynamic scenes. In: ICCV (2009)

# Spatiotemporal Closure

Alex Levinshtein[1], Cristian Sminchisescu[2], and Sven Dickinson[1]

[1] University of Toronto
{babalex,sven}@cs.toronto.edu
[2] University of Bonn
cristian.sminchisescu@ins.uni-bonn.de

**Abstract.** Spatiotemporal segmentation is an essential task for video analysis. The strong interconnection between finding an object's spatial support and finding its motion characteristics makes the problem particularly challenging. Motivated by closure detection techniques in 2D images, this paper introduces the concept of *spatiotemporal closure*. Treating the spatiotemporal volume as a single entity, we extract contiguous "tubes" whose overall surface is supported by strong appearance and motion discontinuities. Formulating our closure cost over a graph of spatiotemporal superpixels, we show how it can be globally minimized using the parametric maxflow framework in an efficient manner. The resulting approach automatically recovers coherent spatiotemporal components, corresponding to objects, object parts, and object unions, providing a good set of multiscale spatiotemporal hypotheses for high-level video analysis.

## 1 Introduction

Spatiotemporal segmentation refers to the task of partitioning a video sequence into coherently moving objects. While such partitioning does not correspond to a full video interpretation, it can prove to be an essential component for higher-level tasks, including tracking, object recognition, video retrieval, or activity recognition. What makes spatiotemporal segmentation challenging is the strong coupling that exists between the estimation of an object's spatial support and the estimation of its motion parameters. On one hand, local motion estimates may be unreliable, especially in untextured regions, and larger spatial support is needed for accurate motion estimation. On the other hand, appearance alone may not be enough to recover the object's spatial support in cases of heterogeneous object appearance or low contrast with the background, and we may need to rely on motion to define the correct spatial support for objects. This chicken and egg problem forces most spatiotemporal segmentation techniques to resort to restrictive modeling assumptions or suboptimal solutions to the problem.

This paper introduces a novel spatiotemporal grouping approach with minimal modeling assumptions and a globally optimal algorithm for segmentation. Similar to prior methods, we represent the whole video stack using a graph with node affinities encoding appearance and motion similarity. In this manner, our

R. Kimmel, R. Klette, and A. Sugimoto (Eds.): ACCV 2010, Part I, LNCS 6492, pp. 369–382, 2011.

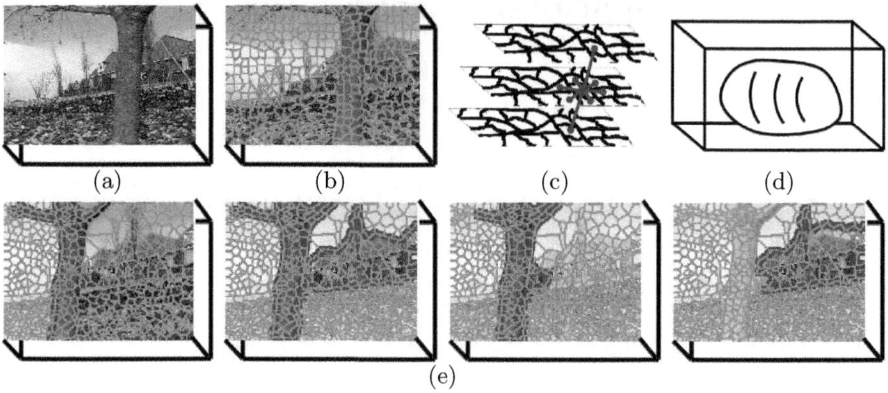


**Fig. 1.** Overview of our approach illustrated on the flower garden sequence. (a) Spatiotemporal volume; (b) Spatiotemporal superpixels; (c) Superpixel graph with edges encoding appearance and motion affinity; (d) Optimizing our spatiotemporal closure corresponds to finding a closed surface cutting low affinity graph edges; (e) Our optimization framework results in multiple multiscale hypotheses, corresponding to objects, object unions, and object parts.

segmentation approach encodes global information without making overly restrictive modeling assumptions. A number of methods approach the problem in a similar manner. However, they commonly employ greedy clustering algorithms [1,2,3,4,5], resort to approximate methods for optimizing global NP-hard costs [6,7,8], assume a known number of objects [6,7,8,9], or work with pixels whose small spatial support results in unreliable motion or appearance features [5,9].

We propose to solve the video segmentation problem by extending the concept of 2D image closure into the spatiotemporal domain, where the perception of closure would correspond to 3D "tubes" whose overall boundary is strongly supported by appearance and motion discontinuities. Fig. 1 illustrates the main steps of our approach. Building on our 2D closure detection framework [10], we formulate spatiotemporal closure detection inside a spatiotemporal volume (Fig. 1a) as selecting a subset of spatiotemporal superpixels whose collective boundary falls on such discontinuities (Fig. 1b). Our spatiotemporal superpixels, based on the framework of [11], provide good spatiotemporal support regions for the extraction of appearance and motion features, while limiting the undersegmentation effects that plague other superpixel extraction techniques due to their lack of compactness and temporal stability. We proceed by forming a superpixel graph whose edges encode appearance and motion similarity of adjacent superpixels (Fig. 1c). Closure detection is posed as the optimization of a global, unbalanced normalized cuts (Ncuts) cost over the superpixel graph (Fig. 1d). Similar to [12], we optimize our unbalanced Ncuts cost with the parametric maxflow approach [13] that is not only able to efficiently find a globally optimal closure solution, but returns multiple closure hypotheses (Fig. 1e). This not only eliminates the need for estimating the number of objects in a video sequence, as all objects exhibiting sufficient closure are extracted, but can result in hypotheses that oversegment objects into parts or merge adjacent objects. The use of such multiscale hypotheses was shown to facilitate state-of-the-art object recognition in images [14]. Similarly, multiple spatiotemporal segmentation hypotheses can serve tasks such as action recognition, video synopsis and indexing [15].

In the following sections, we begin by reviewing related work on spatiotemporal segmentation (Section 2). Next, in Section 3, we introduce our problem formulation. It is here that our cost function is described. Section 4 details all the steps of our algorithm. In Section 5, we evaluate our framework, comparing different superpixel affinities and evaluating against an alternative optimization framework. Finally, in Section 6, we draw conclusions and outline our plans for future work.

## 2 Related Work

A full interpretation of a dynamic scene is a great challenge in computer vision. Tracking methods often adopt a high-level probabilistic scene representation, where objects are modeled with low-dimensional state vectors whose probability at any given instance is a function of the observed data and the temporal dynamics. Inferring object states in real world motion sequences is a difficult task in the face of occlusion, camera motion, and variability in object appearance, dynamics, and shape. As a result, tracking techniques are forced to restrict their models of observed data likelihood and motion [16,17], or resort to approximate techniques to infer object states [18,19]. In contrast, our focus in this paper is on spatiotemporal segmentation. Unlike tracking, where objects are represented at a high level, spatiotemporal segmentation is a low-level task that aims to automatically extract precise object boundaries given generic perceptual grouping regularities, such as similarity, proximity and common fate.

Spatiotemporal segmentation methods can be divided into two categories, layer-based approaches and generic segmentation techniques (a good review is provided in Megret and Dementhon [20]). In the first category, a scene is represented using overlapping layers, with each layer capturing a coherently moving object or part [21,22,23,24,25]. Most such approaches are limited by either assuming a fixed number of layers, assuming a restricted motion model per layer, or resorting to suboptimal techniques that iteratively estimate the spatial extent and the motion of each layer. Nevertheless, this strong global model of a scene enables layer-based methods to successfully segment objects in video sequences in the presence of occlusion, appearance changes, and other effects. In this work, however, we will focus on more generic, less restrictive models for spatiotemporal segmentation.

The second category of approaches does not enforce strong models and attempts to segment a video based on generic spatiotemporal information. Methods mainly differ in their segmentation algorithms and their treatment of the spatiotemporal volume, with some methods analyzing the volume in a framewise manner and others treating it as a single 3D entity. One set of techniques models moving objects with active contours. In Bascle and Deriche [26], motion is modeled with a global warp which is found by correlating internal region appearance in successive frames. After the warp, however, only appearance information is used to update the region's contour. Paragios and Deriche [27] propose a more elegant geodesic active contour formulation. Unlike [26], both motion and appearance information are used in active contour evolution and their level-set framework enables them to easily handle the splitting and merging of contours. However, they assume a static background model to facilitate automatic contour initialization and better tracking. A similar method is proposed by Chung et al. [28], who employ the EM framework to iterate between region motion estimation and segmentation using active contours, but unlike [27] do not rely on a static background. Cremers and Soatto [29] propose a more holistic approach and treat the spatiotemporal volume as a single entity instead of working with pairs of frames. However, their approach provides no automatic initialization and does not estimate the number of objects in a scene.

A different set of techniques opts for a more bottom-up approach, and finds spatially and temporally coherent clusters. Similar to methods based on active contours, some of these approaches handle the spatiotemporal volume in a frame-wise fashion [1,2,3,4,30]. While such techniques are more applicable to realtime segmentation, some opt to treat the video stack as a single entity facilitating more global constraints. Dementhon [5] and Greenspan et al. [9] are examples of two techniques that represent videos with distributions in a low-dimensional feature space (7D in [5] and 6D in [9]). While it enables them to efficiently segment videos by employing non-parametric (a mean-shift-based technique in [5]) or parametric (GMM in [9]) clustering, such low-dimensional models may prove too restrictive for many motion sequences. Instead of explicitly modeling video sequences in some Euclidean space, segmentation can be formulated as an optimization of a global cost that is based on pairwise similarities between neighboring points in the spatiotemporal stack. For example, in [6,7], video segmentation is formulated as a normalized cuts problem, further extended by Huang et al. [8] to handle more global interactions. Our approach falls under this category and is closely related to Ncuts as it also defines a global cost function. However, unlike Ncuts-based techniques that are forced to resort to approximate solutions, we are able to find an exact global optimum of our cost. Moreover, the number of clusters does not have to be specified a priori, as we automatically detect a multiscale set of spatiotemporal clusters.

## 3 Problem Formulation

We formulate the detection of spatiotemporal segments as a superpixel selection problem. To that end, we define our closure cost to be the unbalanced normalized cuts cost over a superpixel graph. Out of an exponential number of superpixel subsets we will efficiently select subsets corresponding to coherent spatiotemporal segments.

Given a superpixel segmentation of every frame in a video, we start by building a superpixel graph with spatial and temporal connections. Let $\boldsymbol{X}$ be an indicator vector for all the superpixels across all frames, with each element being in the set $\{0,1\}$. We connect each superpixel to its spatial and temporal neighbors and define an affinity $W_{ij}$ for each pair of neighboring superpixels $i$ and $j$, encoding the similarity of the two superpixels. Setting $D_i = \sum_j W_{ij}$, we optimize the following closure cost:

$$C(\boldsymbol{X}) = \frac{cut(\boldsymbol{X})}{volume(\boldsymbol{X})} = \frac{\sum_{ij} X_i(1-X_j)W_{ij}}{\sum_i D_i X_i} = \frac{\sum_i D_i X_i - 2\sum_{i<j} X_i X_j W_{ij}}{\sum_i D_i X_i} \tag{1}$$

where $cut(\boldsymbol{X})$ is the sum of the affinities of all the edges between selected and unselected superpixels, and $volume(\boldsymbol{X})$ is the sum of all the affinities originating from the selected superpixels. Minimizing the ratio $C(\boldsymbol{X})$ is equivalent to minimizing the numerator $cut(\boldsymbol{X})$ while maximizing the denominator $volume(\boldsymbol{X})$. The cut between selected and unselected superpixels is small when selected superpixels are strongly separated from the rest in terms of their appearance and motion. Normalization by volume pushes the solution towards large and compact subsets of superpixels that are homogeneous in terms of appearance and motion.

The above is called the unbalanced normalized cuts cost. It is similar to our 2D closure cost in [10], with the exception that the numerator measures the cut instead of the gap and is normalized by affinity volume instead of area. That said, we will show that the affinities $W_{ij}$ can also include the length of the boundary between superpixels or their area to give larger superpixels a greater influence.

Unlike the standard normalized cuts cost, which is NP-hard to optimize, our closure cost can be minimized efficiently using parametric maxflow [13]. In parametric maxflow, the problem of ratio minimization is converted to minimizing a parametrized difference of the numerator and the denominator. For the cost in Eqn. 1, the parametric maxflow cost is:

$$\begin{aligned} C(\boldsymbol{X},\lambda) &= cut(\boldsymbol{X}) - \lambda \cdot volume(\boldsymbol{X}) \\ &= \sum_i D_i X_i - 2\sum_{i<j} X_i X_j W_{ij} - \lambda \sum_i D_i X_i \end{aligned} \tag{2}$$

Different $\lambda$'s correspond to different weights of the cut against the affinity volume. Parametric maxflow can optimize the above parametrized cost, efficiently finding all the different *breakpoints* (interval boundaries) of $\lambda$ between which the optimal solution $X$ is fixed, resulting in an increasing sequence of breakpoints $\lambda_0, \lambda_1, \lambda_2, \ldots$ . Kolmogorov et al. [13] show that while the solution $\boldsymbol{X}^*$ in range $0 \le \lambda \le \lambda_0$ corresponds to the global minimum of $C(\boldsymbol{X})$, consecutively larger breakpoints $\lambda_1, \lambda_2, \ldots$ are also related to ratio optimization. In fact, the optimal solution $\boldsymbol{X}^i$ of $C(\boldsymbol{X},\lambda)$ in the interval $[\lambda_i, \lambda_{i+1}]$, is also an optimal solution of $\min_{volume(\boldsymbol{X}) \ge T} C(\boldsymbol{X})$, where $T = volume(\boldsymbol{X}^i)$. Therefore, employing parametric maxflow results in several solutions where optimal cuts are found with increasing affinity volume constraints. We refer the reader to [13] for more details on the parametric maxflow method.

# 4 Algorithm Details

Our algorithm consists of several stages. We start by extracting the superpixels for each frame of the video. Subsequently, we construct a superpixel graph where each superpixel is connected to its spatial and temporal neighbors. Each superpixel edge is assigned an affinity that measures the degree of superpixel similarity. Once the graph is built, optimal cuts are found using parametric maxflow. Finally, we post-process the solutions to detect connected components, remove similar or spurious results, and generate other potentially good solutions. The following subsections describe each of these stages[1].

## 4.1 Superpixel Extraction

We begin by extracting superpixels from every frame using the TurboPixels approach of Levinshtein et al. [11]. Instead of using the algorithm in its raw form, we modify it to obtain more temporally coherent superpixels. We start by extracting superpixels in the first frame using the original form of the superpixel algorithm in [11]. Instead of reseeding the superpixels in the next frame on a regular grid, we use the current frame's superpixel to drive the seeding procedure. To that end, we first compute the optical flow using the Lucas-Kanade (LK) algorithm. The LK algorithm returns the flow for every pixel in every frame, together with a measure of reliability for each pixel flow. For every superpixel, we compute a weighted average of the flow over all the reliable pixels, where pixels that are closer to the superpixel centroid have larger weights. Superpixels with an insufficient number of reliably flowing pixels get a flow of $(0, 0)$. The result is a *superpixel flow*, with motion flow vector $\boldsymbol{V_i}$ for every superpixel $i$ (Fig. 2).

**Fig. 2.** Superpixel flow. The arrow within each superpixel indicates the motion flow vector of this superpixel. Yellow arrows indicate reliable flows, while red arrows correspond to unreliable flows.

Taking the superpixel flow for every superpixel, we project the center of each superpixel to the next frame according to the computed flow. These projected

[1] See the Approach Overview section at `http://www.cs.toronto.edu/~babalex/SpatiotemporalClosure/supplementary_material.html` for a graphical overview of the method.

centers serve as the initial seeds for the superpixel evolution in the next frame. We repeat this process for all the frames in the video, giving us a much more temporally stable superpixel segmentation. In addition, we also modify the superpixel algorithm to use a Pb-based [31] affinity rather than the original grayscale gradient-based affinity proposed in [11][2].

## 4.2 Superpixel Affinity

Once the superpixels are extracted, we form spatial and temporal edges in the superpixel graph. Every edge is assigned an affinity $W_{ij}$ that measures the similarity of the two superpixels (Fig. 1c).

To form spatial connections, we find the immediate spatial neighbors of each superpixel in each frame. Spatial neighbors of superpixel $i$ are defined as superpixels in the same frame that share some boundary with superpixel $i$. The formation of temporal connections follows the same approach as was used in the superpixel extraction technique. Each superpixel in frame $f$ (except the superpixels in the last frame) is connected to one superpixel in frame $f+1$. The correspondence is determined based on the superpixel flow vectors. The center of superpixel $i$ from frame $f$ is projected to frame $f+1$ according to the superpixel flow $\boldsymbol{V_i}$. We form an edge between superpixels $i$ and $j$, where superpixel $j$ is the superpixel in frame $f+1$ that contains the projected center of superpixel $i$.

Motivated by [8], our superpixel affinity $W_{ij}$ for a spatial edge $(i,j)$ is defined as the combination of appearance ($W^a_{ij}$) and motion ($W^m_{ij}$) affinities. Appearance affinity is obtained by computing the histogram intersection of the grayscale (or color, if available) histograms of the two superpixel regions (we use 30 bin histograms for grayscale and $4 \times 4 \times 4$ histograms for RGB). Motion affinity is computed by comparing the flow vectors of the two superpixels, $\boldsymbol{V_i}$ and $\boldsymbol{V_j}$, and is equal to $W^m_{ij} = 1 - \frac{\|\boldsymbol{V_i}-\boldsymbol{V_j}\|}{\max\{\|\boldsymbol{V_i}\|,\|\boldsymbol{V_j}\|\}}$ capped to the range $(0,1)$. Since our superpixel graph construction incorporates superpixel flow already, we include the motion affinity only for spatial edges. Finally, to give larger superpixels more influence, we augment the affinity by weighting it with the product of areas of the two superpixels ($A_i$ and $A_j$). Combining that with the goal of not grouping two superpixels if either their appearance or motion is dissimilar results in the following superpixel affinity:

$$W_{ij} = \begin{cases} A_i A_j \min\left(W^a_{ij}, W^m_{ij}\right), & (i,j) \text{ are in the same frame} \\ A_i A_j W^a_{ij}, & (i,j) \text{ are in different frames} \end{cases} \tag{3}$$

Since our graph has edges for only a small spatial neighborhood of superpixels with edge affinities encoding both appearance and motion, we will refer to it as S-AM. In Section 5 we will compare this graph construction to other graphs with modified spatial connectivity and different superpixel affinities.

[2] See the Superpixel Extraction section at `http://www.cs.toronto.edu/~babalex/SpatiotemporalClosure/supplementary_material.html` for a better visualization of superpixel extraction.

### 4.3 Optimal Cuts for Each Shot

At this point, we have a superpixel graph and thus can apply the parametric maxflow framework to optimize the cost in Eqn. 1. However, prior to running the optimization framework, we first detect the shot boundaries in the video with the goal of independently finding closures for each shot.

Temporal superpixel edges across shot boundaries are unreliable. Thus if a video is composed of multiple shots, running the optimization on the whole video results in undesirable solutions. Since this is not the focus of this work, we take a very simplistic approach to shot boundary detection. Similar to the appearance affinity between superpixels, we compute an appearance affinity between consecutive frames by comparing the grayscale histograms of whole frames using the histogram intersection kernel. This results in a $F-1$ dimensional vector of consecutive frame affinities (where $F$ is the number of frames). The shot boundaries correspond to the detected minima in this vector (Fig. 3). Given the detected shots, we build a subgraph for every shot by selecting the superpixels and the edges that are contained in the shot. We optimize the cost in Eqn. 1 for all the subgraphs and concatenate the results.

Note that optimizing the cost in Eqn. 1 directly results in a trivial solution where all the superpixels are selected for which $cut(\boldsymbol{X}) = 0$ and $volume(\boldsymbol{X}) > 0$, resulting in $C(\boldsymbol{X}) = 0$. Moreover, we want to be able to weaken affinities in order to handle the cases of potential bleeding between foreground and background due to appearance or motion similarity. We solve the first problem by introducing infinite penalties for a subset of superpixels in the graph, preventing the trivial solution. Specifically, we run the optimization six times for each shot. In the first four runs, all the superpixels on the left, right, top, and bottom frame boundary respectively, are assigned an infinite penalty. In the two additional runs, we assign infinite penalties first to all top and bottom superpixels, and then to all

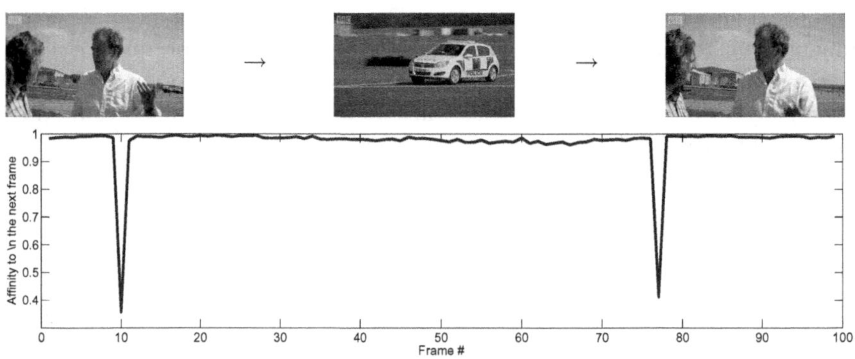


**Fig. 3.** Shot detection by finding minima in consecutive frame affinities. The top row shows a video containing 3 shots. The shot changes from people to car at frame 11 and back to people at frame 78. The bottom row shows a corresponding drop in consecutive frame affinity for these frames. These minima are detected in order to find the shot boundaries.

left and right superpixels. To handle the second issue, we augment the closure affinity in Eqn. 3 to :

$$W'_{ij} = \begin{cases} A_i A_j \left(\min\left(W^a_{ij}, W^m_{ij}\right)\right)^\alpha, & (i,j) \text{ are in the same frame} \\ A_i A_j \left(W^a_{ij}\right)^\alpha, & (i,j) \text{ are in different frames} \end{cases} \quad (4)$$

The exponent $\alpha$ controls the contribution of weak affinities. Increasing the exponent effectively lowers all the affinities towards 0, thereby preventing bleeding, but also increases the relative difference between weak and strong affinities. In the results section, we will analyze the effect of changing $\alpha$ on performance and suggest an optimal value for $\alpha$.

### 4.4 Post-Processing

Running parametric maxflow on the spatiotemporal superpixel graph results in hundreds and sometimes thousands of breakpoints. Some of the solutions differ by a very minor increase in area, while others contain multiple connected components. Furthermore, some desirable solutions are missed. Since our goal is to yield a small number of spatiotemporal hypotheses that capture coherently moving objects in the scene, we post-process the results to narrow down the number of solutions to a more manageable number and in the process generate additional good solutions. While such post-processing no longer guarantees the optimality of the solutions according to Eqn. 1, the resulting solutions still have a low closure cost and empirically yield a better set of hypotheses than the original solutions from parametric maxflow. Post-processing consists of the following 3 stages:

1. **Filtering solutions and generating new ones by analyzing the area change:** As previously stated, parametric maxflow results in solutions that minimize the cut with increasing area constraints. Some solutions corresponding to consecutive breakpoints $(\lambda_i, \lambda_{i+1})$ are almost equivalent in their superpixel selections and differ by a very small increase in area. We filter out the solutions where such an increase is insignificant (less then 1% of relative area increase). Conversely, for all other solutions we detect consecutive solution pairs where the relative area increase is above a threshold (more than 5%) and generate a new solution subtracting one superpixel subset from another.
2. **Selecting connected components and removing small solutions:** Some solutions up to this point contain only a few superpixels or select superpixels in a very small number of frames. We filter out these solutions by keeping only the solutions with at least 2 superpixels, with total area that is at least 1% of the frame area, and that participate in at least 5 frames. We run a connected component analysis for all the remaining solutions. Each solution that contains multiple connected components in space-time is split, generating one solution for each connected component.
3. **Removing duplicate solutions:** The above post-processing steps can result in the generation of duplicate solutions. In this final step we remove duplicate solutions.

For our test videos, this post-processing step reduces the number of solutions of a single run of parametric maxflow from several hundreds to an average of $20 - 80$ solutions.

## 5 Evaluation

We first perform a qualitative analysis of our approach on several short video sequences. Some sequences (such as the flower garden sequence) are grayscale, while others contain color. In the case of color sequences, we make use of this additional information, comparing color histograms instead of grayscale when computing superpixel affinities. The frame size for each video is on the order of $300 \times 300$ pixels, with the length of a video ranging from around 10 frames to 250 frames (hippo sequence). Based on quantitative evaluation (described in later paragraphs), we set $\alpha = 6$ for our qualitative experiments. We also perform a quantitative evaluation on a test dataset [32], comparing different graph constructions and affinity variations, as well as evaluating our approach against standard normalized cuts on the same graphs. The computational bottlenecks of the approach are the preprocessing steps: Pb edge detection, superpixel extraction, and optical flow computation, each taking several seconds per frame. Once a superpixel graph is built, each run of the optimization using parametric maxflow finishes in less than 5 seconds on the whole video, followed by all the postprocessing steps taking approximately 1 second.

Fig. 4 shows our qualitative results. For each sequence we show a frame from the original video and visualize several interesting solutions[3]. In the car sequence, several objects of interest were successfully recovered, such as the car and the heads of the people. Moreover, a part of the car (windshield) is also recovered in one of the solutions, indicating that our method can be used for part-based object recognition in videos or for action recognition that requires the tracking of parts. In the galloping horse sequence, the horse was correctly recovered in the middle of the sequence. A fence is also discovered as one of the solutions. However, in the beginning of the sequence it is partially merged with the horse due to poor superpixel boundaries and affinities between the horse and the background, which is also the reason for the incomplete solution in the Pepsi sequence. The horse example also illustrates that our framework works best when with large objects, as small objects usually have higher closure cost and tend to be undersegmented by superpixels. The table sequence illustrates that our framework can detect most objects in the scene. Finally, the hippo sequence illustrates how an additional solution (dog) can be generated by subtracting one solution (hippo) from another (hippo and dog).

For quantitative evaluation of our method we use 27 sequences from the dataset of Stein et al. [32]. Each sequence has a ground truth video segmentation mask, marking one foreground object. Given a set of detected spatiotemporal figures for a sequence, we choose the solution with the maximal F measure

[3] See the Results at `http://www.cs.toronto.edu/~babalex/SpatiotemporalClosure/supplementary_material.html` for a video visualization of the results.

**Fig. 4.** Qualitative video figure/ground segmentation results. We display one sample frame from a sequence, followed by several interesting solutions.

$(\frac{2 \cdot Precision \cdot Recall}{Precision + Recall})$ relative to the ground truth. We report the average F measure across all sequences.

We compare different variations of our algorithm, as well as replace our parametric maxflow minimization of the unbalanced normalized cuts cost with standard normalized cuts. Unlike our method, normalized cuts requires a user specified number of clusters. Therefore, to compare with our approach we run normalized cuts with 5, 10, 15, 20, and 25 clusters and concatenate all the results. Recall that our previously described graph construction (S-AM) includes only the immediate spatial neighbors and adds the motion affinity $W_{ij}^m$ for spatial edges. We define additional variations over this construction:

- S-A - Same graph as S-AM, but with affinity only including **appearance** $W_{ij} = A_i A_j \left(W_{ij}^a\right)^\alpha$
- L-AM - Same as S-AM but with **larger** spatial connectivity. In addition to the edges in S-AM we add edges between all superpixels in the same frame whose centroids are less than $R$ apart, where $R$ is five times the radius of an average superpixel.
- L-A - Same as L-AM, but with affinity only including appearance $W_{ij} = A_i A_j \left(W_{ij}^a\right)^\alpha$

We compare our method (SC) to normalized cuts (NCuts) for all the above graph constructions. While we are able to solve the unbalanced normalized cuts problem in a globally optimal fashion, normalized cuts cost is NP-hard to optimize and therefore only an approximation is provided. Despite that, the cut balancing in NCuts further constrains the solutions to be balanced and compact and helps to avoid bleeding, while our closure cost pushes the solutions to contain more superpixels which may result in undersegmentation. Fig. 5 illustrates the performance as we vary $\alpha$. We also observe that our method achieves comparable results using S-AM and L-AM, indicating that our increase of spatial connectivity has only a marginal effect on the results. Note that the video sequences

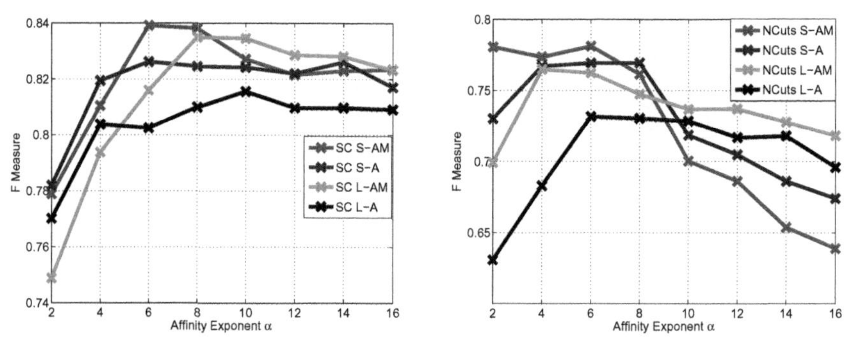


SC NCuts

**Fig. 5.** Quantitative evaluation of spatiotemporal closure detection. We compare the performance of each method (SC on the left and NCuts on the right) on four different graph constructions.

in the test dataset mostly contain large objects. Thus undersegmentation as a result of incorrect superpixels or our unbalanced normalized cuts closure cost is less of a concern, resulting in SC outperfoming the standard NCuts.

## 6 Conclusions

We began by motivating the problem of bottom-up spatiotemporal segmentation. We proceeded by extending work in bottom-up 2D closure detection to spatiotemporal closure detection in videos. Defining our closure cost over spatiotemporal superpixels was shown to facilitate better affinity computation and lead to more stable solutions. Finally, we employ parametric maxflow not only to efficiently find a global optimum of our spatiotemporal closure cost, but recover several multiscale segmentations giving a full hierarchical description of a dynamic scene. The limitations of our framework are particularly apparent when small, low-contrast objects are present, occasionaly leading to object undersegmentation. Therefore, in future work we will improve our spatiotemporal superpixel approach to recover larger, more meaningful superpixels, without sacrificing speed or accuracy. In addition, we will also explore other graph constructions and will design a better superpixel affinity by learning the best composition of motion and appearance cues in a supervised manner.

## References


1. Wang, D.: Unsupervised video segmentation based on watersheds and temporal tracking. CirSysVideo 8, 539–546 (1998)
2. Moscheni, F., Bhattacharjee, S., Kunt, M.: Spatiotemporal segmentation based on region merging. PAMI 20, 897–915 (1998)
3. Gelgon, M., Bouthemy, P.: A region-level motion-based graph representation and labeling for tracking a spatial image partition. Pattern Recognition 33, 725–740 (2000)
4. Deng, Y., Manjunath, B.: Unsupervised segmentation of color-texture regions in images and video. PAMI 23, 800–810 (2001)
5. DeMenthon, D.: Spatio-temporal segmentation of video by hierarchical mean shift analysis. In: SMVP (2002)
6. Shi, J., Malik, J.: Motion segmentation and tracking using normalized cuts. In: ICCV, pp. 1154–1160 (1998)
7. Fowlkes, C., Belongie, S., Malik, J.: Efficient spatiotemporal grouping using the nyström method. In: CVPR, pp. 231–238 (2001)
8. Huang, Y., Liu, Q., Metaxas, D.: Video object segmentation by hypergraph cut. In: CVPR, pp. 1738–1745 (2009)
9. Greenspan, H., Goldberger, J., Mayer, A.: Probabilistic space-time video modeling via piecewise gmm. PAMI 26, 384–396 (2004)
10. Levinshtein, A., Sminchisescu, C., Dickinson, S.: Optimal Contour Closure by Superpixel Grouping. In: Daniilidis, K., Maragos, P., Paragios, N. (eds.) ECCV 2010. LNCS, vol. 6312, pp. 480–493. Springer, Heidelberg (2010)
11. Levinshtein, A., Stere, A., Kutulakos, K.N., Fleet, D.J., Dickinson, S.J., Siddiqi, K.: Turbopixels: Fast superpixels using geometric flows. PAMI 31, 2290–2297 (2009)

12. Carreira, J., Sminchisescu, C.: Constrained parametric min-cuts for automatic object segmentation. In: CVPR (2010)
13. Kolmogorov, V., Boykov, Y., Rother, C.: Applications of parametric maxflow in computer vision. In: ICCV (2007)
14. Li, F., Carreira, J., Sminchisescu, C.: Object Recognition as Ranking Holistic Figure-Ground Hypotheses. In: CVPR (2010)
15. Pritch, Y., Rav-Acha, A., Peleg, S.: Nonchronological video synopsis and indexing. PAMI 30, 1971–1984 (2008)
16. Welch, G., Bishop, G.: An introduction to the kalman filter. Technical report (1995)
17. Black, M., Jepson, A.: Eigentracking: Robust matching and tracking of articulated objects using a view-based representation. IJCV 26, 63–84 (1998)
18. Isard, M., Blake, A.: Condensation - conditional density propagation for visual tracking. IJCV 29, 5–28 (1998)
19. Comaniciu, D., Ramesh, V., Meer, P.: Kernel-based object tracking. PAMI 25, 564–577 (2003)
20. Megret, R., DeMenthon, D.: A survey of spatio-temporal grouping techniques. Technical report, University of Maryland, College Park (2002)
21. Wang, J., Adelson, E.: Representing moving images with layers. TIP 3, 625–638 (1994)
22. Weiss, Y., Adelson, E.H.: A unified mixture framework for motion segmentation: Incorporating spatial coherence and estimating the number of models. In: CVPR, p. 321 (1996)
23. Weiss, Y.: Smoothness in layers: Motion segmentation using nonparametric mixture estimation. In: CVPR, p. 520 (1997)
24. Jojic, N., Frey, B.J.: Learning flexible sprites in video layers. In: CVPR, vol. 1, p. 199 (2001)
25. Jepson, A.D., Fleet, D.J., Black, M.J.: A layered motion representation with occlusion and compact spatial support. In: Heyden, A., Sparr, G., Nielsen, M., Johansen, P. (eds.) ECCV 2002. LNCS, vol. 2350, pp. 692–706. Springer, Heidelberg (2002)
26. Bascle, B., Deriche, R.: Region tracking through image sequences. In: ICCV, p. 302 (1995)
27. Paragios, N., Deriche, R.: Geodesic active contours and level sets for the detection and tracking of moving objects. PAMI 22, 266–280 (2000)
28. Chung, D., MacLean, W., Dickinson, S.: Integrating region and boundary information for spatially coherent object tracking. IVC 24, 680–692 (2006)
29. Cremers, D., Soatto, S.: Motion competition: A variational approach to piecewise parametric motion segmentation. IJCV 62, 249–265 (2005)
30. Patras, I., Lagendijk, R.L., Hendriks, E.A.: Video segmentation by map labeling of watershed segments. PAMI 23, 326–332 (2001)
31. Martin, D.R., Fowlkes, C.C., Malik, J.: Learning to detect natural image boundaries using local brightness, color, and texture cues. PAMI 26, 530–549 (2004)
32. Stein, A., Hoiem, D., Hebert, M.: Learning to find object boundaries using motion cues. In: ICCV (2007)

# Compressed Sensing for Robust Texture Classification

Li Liu[1], Paul Fieguth[2], and Gangyao Kuang[1]

[1] School of Electronic Science and Engineering, National University of Defense Technology, Changsha, Hunan, China 410043
dreamliu2010@gmail.com, kuangyeats@vip.sina.com
[2] Department of System Design Engineering, University of Waterloo, Waterloo, Ontario, Canada N2L 3G1
pfieguth@uwaterloo.ca

**Abstract.** This paper presents a simple, novel, yet very powerful approach for texture classification based on compressed sensing. At the feature extraction stage, a small set of random features is extracted from local image patches. The random features are embedded into a bag-of-words model to perform texture classification, thus learning and classification are carried out in the compressed domain. The proposed unconventional random feature extraction is simple, yet by leveraging the sparse nature of texture images, our approach outperforms traditional feature extraction methods which involve careful design and complex steps. We report extensive experiments comparing the proposed method to the state-of-the-art in texture classification on four databases: CUReT, Brodatz, UIUC and KTH-TIPS. Our approach leads to significant improvements in classification accuracy and reductions in feature dimensionality, exceeding the best reported results on CUReT, Brodatz and KTH-TIPS.

## 1 Introduction

The classification of texture is a key problem in computer vision and pattern recognition, especially for real-world texture images with great intra-class variability due to illumination variations, rotations, viewpoint changes and nonrigid deformations.

By extracting features from a local patch, most feature extraction methods focus on local texture information, characterized by the gray level patterns surrounding a given pixel; however texture is also characterized by its global appearance, representing the repetition of and the relationship among local patterns. Recently, a bag-of-words (BoW) model, borrowed from the text literature, has opened up new prospects for texture classification [1][2][3][4][5][6]. The BoW model encodes both the local texture information, by using features from local patches to form textons, and the global texture appearance, by statistically computing an orderless histogram.

Very popular is the use of large support filter banks to extract texture features at multiple scales and orientations [1][2][3]. However, more recently, in [4] the

R. Kimmel, R. Klette, and A. Sugimoto (Eds.): ACCV 2010, Part I, LNCS 6492, pp. 383–396, 2011.

**Table 1.** Summary of texture datasets used in the experiments. Example textures are provided in the supplemental material.

| Texture Dataset | Dataset Notation | Image Rotation | Controlled Illumination | Scale Variation | Texture Classes | Sample Size | Samples per Class | Samples in Total |
|---|---|---|---|---|---|---|---|---|
| CUReT | $\mathcal{D}^{\text{CUReT}}$ | √ | √ | | 61 | $200 \times 200$ | 92 | 5612 |
| UIUC | $\mathcal{D}^{\text{UIUC}}$ | √ | | √ | 25 | $640 \times 480$ | 40 | 1000 |
| BrodatzFull | $\mathcal{D}^{\text{BFull}}$ | | | | 111 | $215 \times 215$ | 9 | 999 |
| Brodatz90 | $\mathcal{D}^{B90}$ | | | | 90 | $128 \times 128$ | 25 | 2250 |
| KTH-TIPS | $\mathcal{D}^{\text{KT}}$ | | √ | √ | 10 | $200 \times 200$ | 81 | 810 |

authors challenge the dominant role that filter banks have been playing in texture classification, claiming that classification based on textons directly learned from the raw image patches outperforms textons based on filter bank responses.

The key parameter in patch-based classification is patch size. Small sizes cannot capture large-scale structures that may be the dominant features of some textures, and are highly sensitive to noise and other variations, whereas large patch sizes lead to high storage and computational complexity. Therefore, it is natural to ask whether high-dimensional patch vectors can be projected into a lower dimensional subspace without suffering great information loss.

The compressed sensing (CS) approach [7][8], which has been the motivation for this research, is therefore appealing because of its surprising result that high-dimensional sparse data can be accurately reconstructed from just a few nonadaptive linear random projections. When applying CS to texture classification, the key question is therefore how much information about high-dimensional sparse texture signals in local image patches can be preserved by random projections, and whether this leads to any advantages in classification.

The proposed method is computationally simple, yet very powerful. Instead of performing texture classification in the original high-dimensional patch space or making efforts to figure out a suitable feature extraction method, by using random projections of local patches we perform texture classification in a much lower-dimensional compressed patch space. The theory of CS implies that the precise choice of the number of features is no longer critical: a small number of random features, above some threshold, contains enough information to preserve the underlying local texture structure.

Finally, since textures often appear on undulating real world surfaces, the invariances to illumination, rotation, viewpoint, and scale must also necessarily be local rather than global [9]. To avoid the complexity of local invariance in [5] [6], in this paper, we develop simple, yet very powerful rotation-invariant descriptors by *sorting* local patches.

Section 2 reviews the CS background. With the development of the CS approach in Section 3 and the rotation invariance descriptors in Section 4, Section 5 provides extensive experimental results for the CS and sorted CS classifiers and comparative evaluation with the current state-of-the-art on the databases listed in Table 1.

## 2 CS Background

The theory of compressed sensing has recently been brought to the forefront by the work of Candès and Tao [7] and Donoho [8], who have shown the advantages of random projections for capturing information about sparse or compressible signals. The premise of CS is that a small number of nonadaptive linear measurements of a compressible signal or image contain enough information for near perfect reconstruction and processing. This emerging theory has generated enormous amounts of research with applications to high-dimensional geometry [10], image reconstruction [11], and machine learning [12] etc.

CS exploits the fact that many signal classes have a low-dimensional structure compared to the high-dimensional ambient space. Therefore, a small number of nonadaptive measurements in the form of randomized projections can capture most of the salient information in a signal and can approximate the signal well and allow it to be reconstructed.

- The key assumption in CS is that of *sparsity* or *compressibility*. Let $\underline{\boldsymbol{x}} \in \mathbb{R}^{n\times 1}$ be an unknown signal of length $n$ and $\boldsymbol{\Psi} = [\underline{\psi}_1 \ ... \ \underline{\psi}_n]$ an orthonormal basis, where $\underline{\psi}_i \in \mathbb{R}^{n\times 1}$, such that

$$\underline{\boldsymbol{x}} = \sum_{i=1}^{n} \theta_i \underline{\psi}_i = \boldsymbol{\Psi}\underline{\boldsymbol{\theta}} \tag{1}$$

  where $\underline{\boldsymbol{\theta}} = [\theta_1 \ ... \ \theta_n]^T$ denotes the vector of coefficients that represents $\underline{\boldsymbol{x}}$ in the basis $\boldsymbol{\Psi}$, as illustrated in Fig. 1. Signal $\underline{\boldsymbol{x}}$ is said to be *sparse* or *compressible* if most of the coefficients in $\underline{\boldsymbol{\theta}}$ are zero or can be discarded without much loss of information.
- Let $\boldsymbol{\Phi}$ be an $m \times n$ sampling matrix, with $m \ll n$, such that

$$\underline{\boldsymbol{y}} = \boldsymbol{\Phi}\underline{\boldsymbol{x}} = \boldsymbol{\Phi}\boldsymbol{\Psi}\underline{\boldsymbol{\theta}} \tag{2}$$

  where $\underline{\boldsymbol{y}}$ is an $m \times 1$ vector of measurements. The sampling matrix $\boldsymbol{\Phi}$ must allow the reconstruction of length-$n$ signal $\underline{\boldsymbol{x}}$ from $m$ measurements in $\underline{\boldsymbol{y}}$.

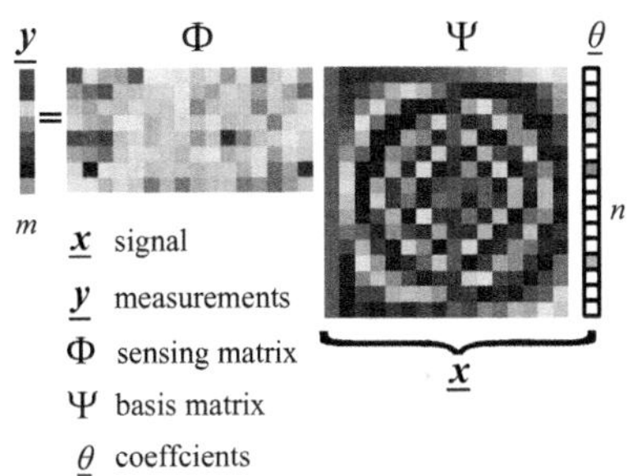


**Fig. 1.** Compressed Sensing measurement process

Since the transformation from $\underline{\boldsymbol{x}}$ to $\underline{\boldsymbol{y}}$ is a *dimensionality reduction*, in general there is an information loss, however the measurement matrix $\boldsymbol{\Phi}$ can be shown to preserve the information in sparse and compressible signals if it satisfies the so-called *restricted isometry property* (RIP) [13]. Intriguingly, a large class of random matrices have the RIP with high probability [7][8][13].

- Signal reconstruction takes the $m$ measurements in $\underline{\boldsymbol{y}}$, the random measurement matrix $\boldsymbol{\Phi}$, and the basis $\boldsymbol{\Psi}$ to reconstruct $\underline{\boldsymbol{\theta}}$. A large number of approaches have been proposed to solve the reconstruction problem, however the algorithms tend to be computationally burdensome.

CS theory has been used for classification in the SRC algorithm [12] for face recognition. In contrast to the texture problem, however, the SRC algorithm is based on global features, whereas texture classification almost certainly depends on the relationship between a pixel and its neighborhood. Furthermore SRC is reconstruction based, explicitly reconstructing the sparse $\underline{\boldsymbol{\theta}}$, a computationally intensive step we wish to avoid.

## 3 The CS Classifier

Let us begin by formulating a basic CS classifier, with a robust extension development in Section 4. The premise underlying CS is one of signal sparsity or compressibility, and the compressibility of textures is certainly well established. Certainly most natural images are compressible, as extensive experience with the wavelet transform has demonstrated. Textures, being roughly stationary/periodic, are all the more sparse. Furthermore from the large literature on texture classification via feature extraction, the degrees of freedom underlying a texture are clearly few in number.

The local patch vector $\underline{\boldsymbol{x}} \in \mathbb{R}^{n\times 1}$ is embedded into a lower-dimensional space $\underline{\boldsymbol{y}} \in \mathbb{R}^{m\times 1}$:

$$\underline{\boldsymbol{y}} = \boldsymbol{\Phi}\underline{\boldsymbol{x}} \tag{3}$$

ideally where $m \ll n$. Clearly $\boldsymbol{\Phi} \in \mathbb{R}^{m\times n}$, $m < n$ loses information in general, since $\boldsymbol{\Phi}$ has a null space, implying the indistinguishability between $\underline{\boldsymbol{x}}$ and $\underline{\boldsymbol{x}} + \underline{\boldsymbol{z}}$, for $\underline{\boldsymbol{z}} \in \mathcal{N}(\boldsymbol{\Phi})$. The challenge in identifying an effective feature extractor $\boldsymbol{\Phi}$ is to have the null-space of $\boldsymbol{\Phi}$ orthogonal to the low-dimensional subspace of sparse signal $\underline{\boldsymbol{x}}$.

Ideally, we wish to ensure that $\boldsymbol{\Phi}$ is information-preserving, by which we mean that $\boldsymbol{\Phi}$ provides a stable embedding that approximately preserves distances between all pairs of signals, such that for any two patches, $\underline{\boldsymbol{x}}_1$ and $\underline{\boldsymbol{x}}_2$

$$1 - \epsilon \leq \frac{\|\boldsymbol{\Phi}(\underline{\boldsymbol{x}}_1 - \underline{\boldsymbol{x}}_2)\|_2}{\|\underline{\boldsymbol{x}}_1 - \underline{\boldsymbol{x}}_2\|_2} \leq 1 + \epsilon \tag{4}$$

for small $\epsilon > 0$. One of the key results in [13] from CS theory is the Restricted Isometry Property, which states that (4) is indeed satisfied by certain random

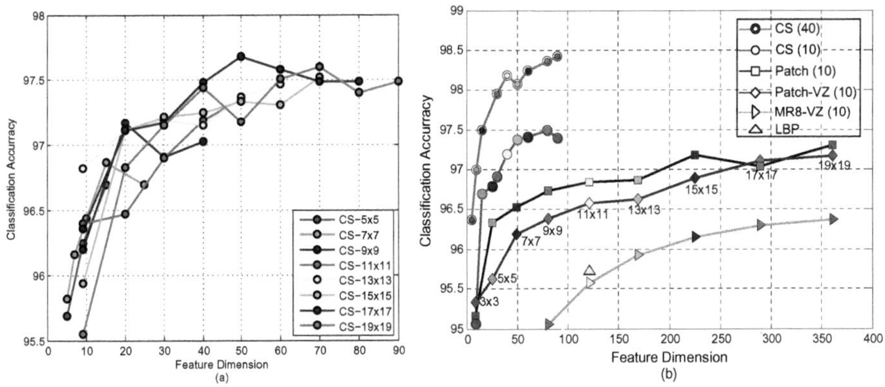


**Fig. 2.** Classification results on $\mathcal{D}^{\text{CUReT}}$ as a function of feature dimensionality (Except for LBP, whose results are shown as a function of patch size). The bracketed values denote the number of textons $K$ per class. "Patch-VZ" and "MR8-VZ" results are quoted directly from the paper of Varma and Zisserman [4]. Classification rates obtained based on the same patch size are shown in the same color.

matrices, including a Gaussian random matrix $\boldsymbol{\Phi}$. It is on this basis that we propose to use the emerging theory of compressed sensing to rethink texture classification.

We wish to preserve both local texture information, contained in a local image patch, and global texture appearance, representing the repetition of the local textures and the relationship among them. We choose a texton-based approach, an effective local-global representation [1][2][3][4], trained by adaptively partitioning the feature space into clusters using $K$-means. For an input data set $\mathcal{Y} = \{\underline{\boldsymbol{y}}_1, ..., \underline{\boldsymbol{y}}_{|\mathcal{Y}|}\}$, $\underline{\boldsymbol{y}}_i \in \mathbb{R}^{d\times 1}$, and an output texton set $\mathcal{W} = \{\underline{\boldsymbol{w}}_1, ..., \underline{\boldsymbol{w}}_K\}$, $\underline{\boldsymbol{w}}_i \in \mathbb{R}^{d\times 1}$, the quality of a clustering solution is measured by the average quantization error

$$Q(\mathcal{Y}, \mathcal{W}) = \frac{1}{|\mathcal{Y}|} \sum_{j=1}^{|\mathcal{Y}|} \min_{1\leq k\leq K} \|\underline{\boldsymbol{y}}_j - \underline{\boldsymbol{w}}_k\|_2^2 \tag{5}$$

However, $Q(\mathcal{Y}, \mathcal{W})$ goes as $K^{-2/d}$ for large $K$ [14], a problem when $d$ is large, since $K$ is then required to be extremely large to obtain satisfactory cluster centers, with computational and storage complexity consequences. On the other hand, Varma and Zisserman [4] have shown that image patches contain sufficient information for texture classification, arguing that the inherent loss of information in the dimensionality reduction of feature extraction leads to inferior classification performance. CS addresses the dilemma between these two perspectives very neatly. The high-dimensional texture patch space has an intrinsic dimensionality that is much lower, therefore CS is able to perform feature extraction without information loss. On the basis of the above analysis, we claim that the CS and BoW approaches are complementary, and will together lead to superior performance for texture classification.

**Table 2.** Experimental results for the CS classifier on $\mathcal{D}^{\mathrm{B90}}$, 13 samples per class for training and 12 for testing. Means and standard deviations have been computed over 20 runs. Ten textons used per class.

| Method | Patch Size 3 × 3 | 5 × 5 | 7 × 7 |
|---|---|---|---|
| Dim | 5 | 10 | 25 |
| CS | 94.3% ± 0.30% | **95.4% ± 0.39%** | 95.0% ± 0.34% |
| Dim | 9 | 25 | 49 |
| Patch | 94.0% ± 0.11% | 94.7% ± 0.21% | 94.8% ± 0.30% |
| Scale | 1 | 2 | 3 |
| $\mathrm{LBP}^{riu2}$ | 87.7% ± 0.63% | 93.6% ± 0.34% | 94.8% ± 0.37% |

The actual classification algorithm is the texton-based method of [4] except that instead of using image patch vector $\underline{\boldsymbol{x}}$, the compressed sensing measurements $\underline{\boldsymbol{y}} = \boldsymbol{\Phi}\underline{\boldsymbol{x}}$ derived from $\underline{\boldsymbol{x}}$ are used as features, where we choose $\boldsymbol{\Phi}$ to be a Gaussian random matrix, i.e., the entries of $\boldsymbol{\Phi}$ are independent zero-mean, unit-variance normal.

Figure 2 plots the classification accuracy for the **CUReT dataset** $\mathcal{D}^{\mathrm{CUReT}}$ (see Table 1), using the same subset of images and the same experimental setup as Varma and Zisserman [3] [4]. Figure 2 (b), in particular, presents a comparison of CS, Patch, MR8, and LBP, with the results averaged over tens of random partitions of training and testing sets. The CS method significantly outperforms all other methods, a clear indication that the CS matrix preserves the salient information contained in the local patch (as predicted by CS theory) and that performing classification in the compressed patch space is not a disadvantage. In contrast to the Patch method, not only does CS offer higher classification accuracy, but also at a much lower-dimensional feature space, reducing storage requirements and computation time, and allowing more textons per class.

Table 2 shows the classification accuracy on **Brodatz90** dataset $\mathcal{D}^{\mathrm{B90}}$. Due to the impressive diversity and perceptual similarity of some textures in the Brodatz database, some of which essentially belong to the same class but at different scales, while others are so inhomogeneous that a human observer would arguably be unable to group their samples correctly, 90 texture classes from the Brodatz album were kept. The proposed method performs better than the Patch method, although by a relatively small margin. The example demonstrates that the proposed CS method can successfully classify 90 texture classes from the Brodatz dataset, despite the large number of classes contained in $\mathcal{D}^{B}$ which can cause a high risk of mis-classification.

**Table 3.** Comparison of highest classification performance on $\mathcal{D}^{\mathrm{CUReT}}$ with a common experimental setup, except for Zhang [6] who used EMD/SVM classifier

| Method | LBP | MR8 | Patch | Patch-MRF | Zhang et al. | **CS** |
|---|---|---|---|---|---|---|
| Accuracy | 95.72% | 97.43% | 97.17% | 98.03% | 95.5% | **98.43%** |

Table 3 presents the overall best classification performance achieved for any parameter setting. The proposed CS method gives the highest classification accuracy of 98.43%, even higher than the best of Patch-MRF in [4], despite the fact that the model dimensionality of the Patch-MRF method is far larger than that of the CS.

# 4 The SCS Classifiers

Motivated by the striking classification results in Figure 2 and Table 3, we would like to further capitalize on the CS approach by proposing a robust variant. Existing schemes to achieve rotation invariance in the patch vector representation include estimating the dominant gradient orientation of the local patch [3] [4], marginalizing the intensities weighted by the orientation distribution over angle, and adding rotated patches to the training set. The dominant orientation estimates tend to be unreliable, especially for blob regions which lack strong edges at the center, and finding the dominant orientation for each local patch is computationally expensive. To avoid the ambiguity of identifying a dominant direction, and the clustering challenge with learning over all rotated patches, instead we just use

$$\underline{\boldsymbol{y}} = \boldsymbol{\Phi}\ sort(\underline{\boldsymbol{x}}) \tag{6}$$

where we sort over all (or parts) of $\underline{\boldsymbol{x}}$. Since sorting ignores the ordering of elements in $\underline{\boldsymbol{x}}$, the $sort(\underline{\boldsymbol{x}})$ is clearly rotation invariant (excepting the effects of pixellation). Classification using (6) will be referred to as Sorted CS (SCS).

## 4.1 Sorted Pixel Values

For example, suppose we reorder the patch vector by taking the center pixel value $x_{0,0}$ of the patch of size $(2a+1) \times (2a+1)$ as its first entry and simply sort the other $n-1$ pixels:

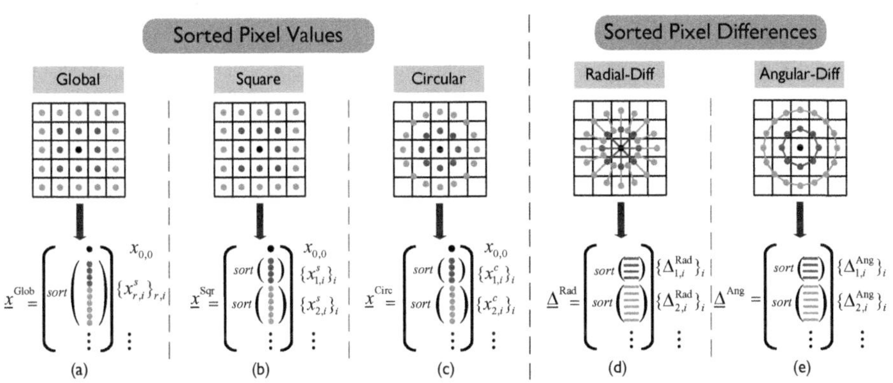


**Fig. 3.** Sorting schemes on an example patch of size 5 × 5-pixels: sorting pixels (a, b, c) or sorting pixel differences (d, e). The pixels may be taken natively on a square grid (a, b) or interpolated to lie on rings of constant radius (c, d, e).

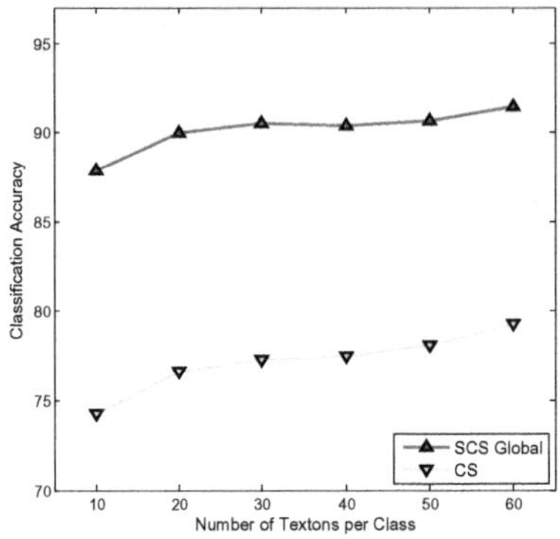


**Fig. 4.** Comparison of the simplest sorted CS against basic CS: classification accuracy as a function of number of textons per class on $\mathcal{D}^{\text{UIUC}}$ with 20 samples per class for training, using a patch of size $9 \times 9$ and a dimensionality of 30. Results are averaged over tens of random partitionings of the training and testing set.

$$\underline{\boldsymbol{x}}^{\text{Glob}} = [x_{0,0}, sort([x_{1,0}, ..., x_{1,p_1-1}, ..., x_{a,0}, ..., x_{a,p_a-1}])]^T \tag{7}$$

where $x_{r,i}$, $0 \leq i < p_r$ refers to the $r$th concentric square of pixels (see Figure 3 (a)) and $p_r = 8r$, $1 \leq r \leq a$. Since sorting deletes all location information, clearly $\underline{\boldsymbol{x}}^{\text{Glob}}$ is invariant to rotation. The advantages of our sorting approach include simplicity and noise robustness.

Figure 4 motivates this idea, showing a jump (from below 80% to above 90%) in classification performance, compared to basic CS in classifying the challenging UIUC database. This surprising experimental result, despite the quality of the basic CS classifier in Table 3, confirms the effectiveness of the sorting strategy.

Clearly, global sorting provides a poor discriminative ability, since crude sorting over the whole patch (the center pixel excluded) leads to an ambiguity of the relationship among pixels from different scales. A natural extension of global sorting is to sort pixels of the same scale. We propose two kinds of sorting schemes, illustrated in Figure 3 (b) and (c). Our schemes follow a strategy similar to some recently developed descriptors like SIFT [15], SPIN and RIFT [5]: They subdivide the region of support, and instead of sorting, they adopt histogramming strategy and compute a histogram of appearance attributes (pixel values or gradient orientations) inside each subregion.

In our proposed approach, sorting provides stability against rotation, while sorting at each scale individually preserves some spatial information. In this way, a compromise is achieved between the conflicting requirements of greater geometric invariance on the one hand and greater discriminative power on the other. As can be seen from Figure 5, sorting over concentric squares or circular rings both offer an improvement over global sorting.

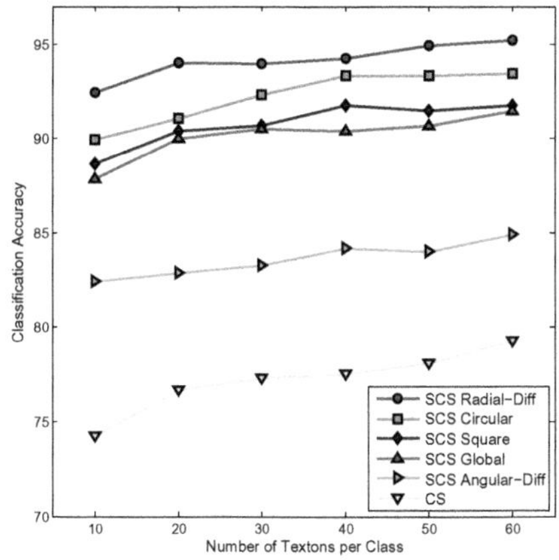


**Fig. 5.** Like Figure 4, but comparing all sorting schemes with basic CS

## 4.2 Sorted Pixel Differences

Sorting each ring of pixels loses any sense of spatial coupling, whereas textures clearly possess a great many spatial relationships. Therefore we propose sorting radial or angular differences, illustrated in Figure 3 (d) and (e). It is worth noting that gray-level differences have been successfully used in a large number of texture analysis studies [16][17][18][19].

We propose pixel differences in radial and angular directions on a circular grid, different from the traditional pixel differences which are computed in horizontal and vertical directions on a regular grid. In particular, radial differences encode the inter-ring structure, thus sorted radial differences can achieve rotation invariance while preserving the relationship between pixels of different rings, which has not been explored by many rotation invariant methods such as LBP. The sorted radial and angular difference descriptors are computed as:

$$\begin{aligned} \underline{\Delta}^{\text{Rad}} &= [sort(\Delta^{\text{Rad}}_{1,0}, ..., \Delta^{\text{Rad}}_{1,p_1-1}),\ ...\ sort(\Delta^{\text{Rad}}_{a,0}, ..., \Delta^{\text{Rad}}_{a,p_a-1})]^T \\ \underline{\Delta}^{\text{Ang}} &= [sort(\Delta^{\text{Ang}}_{1,0}, ..., \Delta^{\text{Ang}}_{1,p_1-1}),\ ...\ sort(\Delta^{\text{Ang}}_{a,0}, ..., \Delta^{\text{Ang}}_{a,p_a-1})]^T \end{aligned} \tag{8}$$

where

$$\begin{aligned} \Delta^{\text{Rad}}_{r,i} &= x^{\delta_r}_{r,i} - x^{\delta_r}_{r-1,i}, \quad \delta_r = 2\pi/p_r \\ \Delta^{\text{Ang}}_{r,i} &= x_{r,i} - x_{r,i-1} \end{aligned} \tag{9}$$

Figure 5 plots the classification results of CS and sorted CS on $\mathcal{D}^{\text{UIUC}}$. The results show that all sorted CS classifiers perform significantly better than the CS classifier, where sorted radial differences performed the best, and the sorted interpolated-circular the best among the pixel-value methods. In general, the performance increases with an increasing number of textons used per class.

# 5 Experimental Evaluation

## 5.1 Image Data and Experimental Setup

Since near-perfect overall performance has been shown in Section 3 for the CUReT database, for our comprehensive experimental evaluation we have used another three datasets, summarized in Table 1, derived from the three most commonly used texture sources: the UIUC database [5], the Brodatz database [20] and the KTH-TIPS [21] database.

The **UIUC** dataset $\mathcal{D}^{\text{UIUC}}$ [5] has been designed to require local invariance. Textures are acquired under significant scale and viewpoint changes, and uncontrolled illumination conditions. Furthermore, the database includes textures with varying local affine deformations, and nonrigid deformation of the imaged texture surfaces. This makes the database very challenging for classification. For our classification experiments on UIUC, we replicate as closely as possible the experiments described by Lazebnik et al [5].

The Brodatz database [20] is perhaps the best known benchmark for evaluating texture classification algorithms. For the **BrodatzFull** dataset $\mathcal{D}^{\text{BFull}}$, we keep all 111 classes. To the best of our knowledge, there are relatively few publications actually reporting classification results on the whole Brodatz database. Performing classification on the entire database is challenging due to the relatively large number of texture classes, the small number of examples for each

**Table 4.** Experimental results for $\mathcal{D}^{\text{BFull}}$ and $\mathcal{D}^{\text{UIUC}}$: all results for our proposed approach are obtained by number of textons used per class $K = 10$ for $\mathcal{D}^{\text{BFull}}$ and $K = 40$ for $\mathcal{D}^{\text{UIUC}}$, except SCS Radial-Diff (Best), which are the best obtained by varying $K$ up to 40 for $\mathcal{D}^{\text{BFull}}$ and $K$ up to 80 for $\mathcal{D}^{\text{UIUC}}$. Results of Lazebnik et al. and Zhang et al. are quoted directly from [5] and [6].

| | $\mathcal{D}^{\text{BFull}}$ Patch Size | | | | $\mathcal{D}^{\text{UIUC}}$ Patch Size | | |
|---|---|---|---|---|---|---|---|
| Method | $5 \times 5$ | $9 \times 9$ | $11 \times 11$ | $15 \times 15$ | $5 \times 5$ | $9 \times 9$ | $13 \times 13$ |
| Dimensionality | 10 | 30 | 40 | 60 | 10 | 30 | 50 |
| CS | 89.1% | 90.8% | 89.2% | 88.4% | 79.6% | 77.5% | 76.3% |
| SCS Global | 83.6% | 81.9% | 82.0% | 79.8% | 90.3% | 90.4% | 89.1% |
| SCS Square | 84.0% | 87.2% | 85.8% | 86.8% | 90.8% | 91.8% | 91.2% |
| SCS Circular | 86.7% | 88.4% | 87.4% | 85.7% | 90.8% | 93.3% | 92.1% |
| SCS Radial-Diff | 93.1% | 94.7% | 95.1% | 94.7% | 91.4% | 94.3% | 95.4% |
| SCS Radial-Diff (Best) | 94.7% | **96.2**% | 95.8% | 95.5% | 91.5% | 95.2% | **96.3**% |
| SCS Angular-Diff | 88.9% | 89.8% | 92.4% | 90.5% | 77.1% | 84.2% | 86.5% |
| Scale | 2 | 3 | 4 | 5 | 2 | 3 | 5 |
| LBP | 87.5% | 88.9% | 89.7% | 89.9% | 75.6% | 81.5% | 86.1% |
| Lazebnik Best [5] | 88.2% | | | | 96.1% | | |
| Best from Zhang [6] | 95.9% | | | | **98.7**% | | |

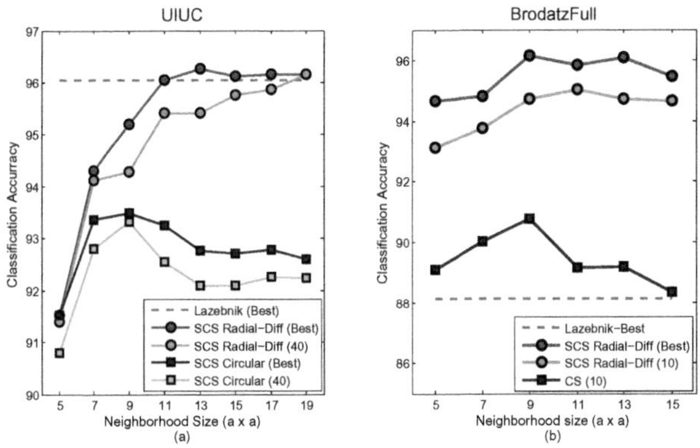


**Fig. 6.** Classification accuracy as a function of patch size: (a) Results on $\mathcal{D}^{\mathrm{UIUC}}$. The SCS Radial-Diff Best curve shows the best results obtained by varying $K$ up to 80. Similarly, SCS Circular Best curve is obtained by varying $K$ up to 60. (b) Results on $\mathcal{D}^{\mathrm{BFull}}$. The SCS Radial-Diff Best curve is obtained in the same way as in (a), but with $K$ up to 40 being tried. In both (a) and (b), the results for Lazebnik Best are the highest classification accuracies, directly quoted from the original paper [5].

class, and the lack of intra-class variation. In order to obtain results comparable with Lazebnik et al.[5] and Zhang et al. [6], we used the same dataset as them, dividing each texture image into nine non-overlapping subimages.

The **KTH-TIPS** dataset $\mathcal{D}^{\mathrm{KT}}$ [21] contains 10 texture classes with each class having 81 images, captured at 9 lighting and rotation setups and 9 different scales.

Implementation details: Each sample is intensity normalized to have zero mean and unit standard deviation. All results are reported over tens of random partitions of training and testing sets. Each extracted CS vector is normalized via Weber's law. Histograms/$\chi^2$ and nearest neighbor (NN) classifier are used. Half of the samples per class are randomly selected for training and the remaining half for testing, except for $\mathcal{D}^{\mathrm{BFull}}$, where three samples are randomly selected as training and the remaining six as testing.

### 5.2 Experimental Results and Performance Analysis

The results for datasets $\mathcal{D}^{\mathrm{BFull}}$ and $\mathcal{D}^{\mathrm{UIUC}}$, the same datasets used by Lazebnik et al. [5] and Zhang et al. [6], are shown in Table 4 and Figure 6. As expected, among the CS methods, SCS Radial-Diff performs best on both of these two datasets. The radial-difference method significantly outperforms LBP, outperforms the method of Lazebnik et al. [5] for both BrodatzFull and UIUC, and outperforms the method of Zhang et al. [6] on BrodatzFull. This latter result should be interpreted in light of the fact that Zhang et al. use scale invariant and affine invariant channels and a more advanced classifer (EMD/SVM), which is important for $\mathcal{D}^{\mathrm{UIUC}}$ where some textures have significant scale changes and affine variations.

**Table 5.** Comparison results for $\mathcal{D}^{\mathrm{B90}}$: all results for our proposed approach are obtained by $K = 10$ except SCS Radial-Diff (Best), which are the best obtained by varying $K = 10, 20, 30, 40$

| Method | Patch Size $5 \times 5$ | $7 \times 7$ | $9 \times 9$ | $11 \times 11$ |
|---|---|---|---|---|
| Dim | 10 | 25 | 30 | 40 |
| CS | 95.4% | 95.0% | 94.5% | 93.6% |
| SCS Radial-Diff | 97.2% | **97.6%** | 97.5% | 97.4% |
| SCS Radial-Diff (Best) | 97.5% | **98.2%** | 98.1% | 97.6% |
| Dim | 25 | 49 | 81 | 121 |
| Patch | 94.7% | 94.8% | 94.0% | 93.2% |
| Scale | 2 | 3 | 4 | 5 |
| $\mathrm{LBP}^{riu2}$ | 93.6% | 94.8% | 95.1% | 94.7% |

**Table 6.** Comparisons of classification results of the Basic CS and the SCS Radial-Diff on $\mathcal{D}^{\mathrm{CUReT}}$ with $K = 10$

| Patch Size | CS | SCS Radial-Diff |
|---|---|---|
| $7 \times 7$ | 96.80% | 96.33% |
| $9 \times 9$ | 96.91% | 96.61% |

**Table 7.** Experimental results for $\mathcal{D}^{KT}$: all results for our proposed approach are obtained by $K = 20$ except SCS Radial-Diff (Best), which are the best obtained by varying $K = 10, 20, 30, 40$. Results of Zhang et al are quoted directly from [6].

| Method | Patch Size $7 \times 7$ | $9 \times 9$ | $11 \times 11$ | $13 \times 13$ | $15 \times 15$ |
|---|---|---|---|---|---|
| CS | 95.6% | 95.2% | 94.6% | 94.1% | 94.2% |
| SCS Global | 94.0% | 93.3% | 93.7% | 92.6% | 92.0% |
| SCS Square | 96.1% | 94.7% | 95.4% | 95.5% | 95.6% |
| SCS Circular | 95.1% | 95.4% | 95.1% | 94.7% | 94.0% |
| SCS Radial-Diff | 96.0% | 96.6% | 96.8% | 96.9% | **97.1%** |
| SCS Radial-Diff (Best) | 96.5% | 97.2% | 97.3% | 97.4% | **97.4%** |
| SCS Angular-Diff | 91.0% | 91.1% | 90.9% | N/A | N/A |
| Best from Zhang [6] | 96.1% | | | | |

Motivated by the strong performance of SCS Radial-Diff, we return to the comparison of Section 3. Table 5 shows the classification results on $\mathcal{D}^{\mathrm{B90}}$ with homogeneous or near homogeneous textures, in comparison to the state-of-the-art. We can see that near perfect performance can be achieved by the proposed SCS Radial-Diff approach. In terms of $\mathcal{D}^{\mathrm{CUReT}}$, [4] showed that the incorporating of rotation invariance is not so helpful, nevertheless the performance penalty in Table 6 for incorporating rotation invariance is very modest.

Finally, Table 7 lists the results for the KTH-TIPS database $\mathcal{D}^{\mathrm{KT}}$. Note that $\mathcal{D}^{\mathrm{KT}}$ has controlled imaging and a small number of texture classes. Textures in this dataset have no obvious rotation, though they do have controlled scale variations. From Table 7, we can see that SCS Radial-Diff again performs the best, outperforming all methods in the extensive comparative survey of Zhang et al. [6].

## 6 Conclusions

In this paper, we have described a classification method based on representing textures as a small set of compressed sensing measurements of local texture patches. We have shown that CS measurements of local patches can be effectively used in texture classification. The proposed method has been shown to match or surpass the state-of-the-art in texture classification, but with significant reductions in time and storage complexity. The main contributions of our paper are the proposed CS classifier, and the novel sorting scheme for rotation invariant texture classification. Among the sorted descriptors evaluated in this paper, the sorted radial difference descriptor is simple, yet it yields excellent performance across all databases.

The proposed CS approach outperform all known classifiers on the CUReT databases, and the proposed SCS Radial Difference approach outperforms all known classifiers on the Brodatz90, BrodatzFull, and KTH-TIPS databases.

## References


1. Leung, T., Malik, J.: Representing and Recognizing the Visual Appearance of Materials Using Three-Dimensional Textons. International Journal of Computer Vision 43, 29–44 (2001)
2. Cula, O.G., Dana, K.J.: 3D Texture Recognition Using Bidirectional Feature Histograms. International Journal of Computer Vision 59, 33–60 (2004)
3. Varma, M., Zisserman, A.: A Statistical Approach to Texture Classification from Single Images. International Journal of Computer Vision 62, 61–81 (2005)
4. Varma, M., Zisserman, A.: A Statistical Approach to Material Classification Using Image Patches. IEEE Trans. Pattern Analysis and Machine Intelligence 31, 2032–2047 (2009)
5. Lazebnik, S., Schmid, C., Ponce, J.: A Sparse Texture Representation Using Local Affine Regions. IEEE Trans. Pattern Analysis and Machine Intelligence 27, 1265–1278 (2005)

6. Zhang, J., Marszalek, M., Lazebnik, S., Schmid, C.: Local Features and Kernels for Classification of Texture and Object Categories: A Comprehensive Study. International Journal of Computer Vision 73, 213–238 (2007)
7. Candès, E., Tao, T.: Near-Optimal Signal Recovery from Random Projections: Universal Encoding Stratigies? IEEE Trans. Information Theory 52, 5406–5425 (2006)
8. Donoho, D.L.: Compressed Sensing. IEEE Trans. Information Theory 52, 1289–1306 (2006)
9. Mellor, M., Hong, B.-W., Brady, M.: Locally Rotation, Contrast, and Scale Invariant Descriptors for Texture Analysis. IEEE Trans. Pattern Analysis and Machine Intelligence 30, 52–61 (2008)
10. Donoho, D., Tanner, J.: Observed Universality of Phase Transitions in High-Dimensional Geometry, with Implications for Modern Data Analysis and Signal Processing. Phil. Trans. R. Soc. A 367, 4273–4293 (2009)
11. Romberg, J.: Imaging via Compressive Sampling. IEEE Signal Processing Magazine 25, 14–20 (2008)
12. Wright, J., Yang, A., Ganesh, A., Sastry, S.S., Ma, Y.: Robust Face Recognition via Sparse Representation. IEEE Trans. Pattern Analysis and Machine Intelligence 31, 210–217 (2009)
13. Candès, E., Tao, T.: Decoding by Linear Programming. IEEE Trans. Information Theory 51, 4203–4215 (2005)
14. Graf, S., Luschgy, H.: Foundations of Quantization for Probability Distributions. Springer, New York (2000)
15. Lowe, D.: Distinctive Image Features from Scale-Invariant Keypoints. International Journal of Computer Vision 60, 91–110 (2004)
16. Conners, R.W., Harlow, C.A.: A Theoretical Comparison of Texture Algorithms. IEEE Trans. Pattern Analysis and Machine Intelligence 2, 204–222 (1980)
17. Unser, M.: Sum and Difference Histograms for Texture Classification. IEEE Trans. Pattern Analysis and Machine Intelligence 8, 118–125 (1986)
18. Ojala, T., Valkealahti, K., Oja, E., Pietikäinen, M.: Texture Discrimination with Multidimensional Distributions of Signed Gray-Level Differences. Pattern Recognition 34, 727–739 (2001)
19. Ojala, T., Pietikäinen, M., Mäenpää, T.: Multiresolution Gray-Scale and Rotation Invariant Texture Classification with Local Binary Patterns. IEEE Trans. Pattern Analysis and Machine Intelligence 24, 971–987 (2002)
20. Brodatz, P.: Textures: A Photographic Album for Artists and Designers. Dover Publications, New York (1966)
21. Hayman, E., Caputo, B., Fritz, M., Eklundh, J.-O.: On the Significance of Real-World Conditions for Material Classification. In: Pajdla, T., Matas, J(G.) (eds.) ECCV 2004. LNCS, vol. 3024, pp. 253–266. Springer, Heidelberg (2004)

# Interactive Multi-label Segmentation

Jakob Santner, Thomas Pock, and Horst Bischof

Institute of Computer Graphics and Vision
Graz University of Technology, Austria

**Abstract.** This paper addresses the problem of interactive multi-label segmentation. We propose a powerful new framework using several color models and texture descriptors, Random Forest likelihood estimation as well as a multi-label Potts-model segmentation. We perform most of the calculations on the GPU and reach runtimes of less than two seconds, allowing for convenient user interaction. Due to the lack of an interactive multi-label segmentation benchmark, we also introduce a large publicly available dataset. We demonstrate the quality of our framework with many examples and experiments using this benchmark dataset.

## 1 Introduction

Image segmentation is one of the elementary fields in computer vision and has been studied intensively over the last decades. It describes the task of dividing an image into a finite number of non-overlapping regions. In contrast to unsupervised segmentation, where algorithms try to find consistent image regions autonomously, interactive segmentation deals with partitioning an image based on user-provided input. These concepts are widely used, *i.e.* every major image editing software package (*Adobe Photoshop*, *Corel Draw*, *GIMP etc.*) features an interactive segmentation algorithm.

As the name suggests, interactive segmentation needs human interaction: The user has to provide information on what he wants to segment, usually by drawing brush strokes, rectangles or contours (Fig. 1). Based on this input the algorithm produces a segmentation result, which is typically not exactly what the user intended to get. Therefore, the user can manipulate the input (*e.g.* by adding seed points) or change parameters and re-segment the image until the desired result is obtained. This leads to two key properties for a good interactive segmentation method:

- It has to be computationally efficient. Interactive tools needing more than a few seconds to compute are not used, no matter how good their results are.
- The method should produce the desired results with as little interaction as possible. Therefore, it has to quickly 'understand' what the user wants to segment.

These requirements are mutually exclusive: To increase a method's capability to 'understand' what the user wants to segment, it needs increasingly more

R. Kimmel, R. Klette, and A. Sugimoto (Eds.): ACCV 2010, Part I, LNCS 6492, pp. 397–410, 2011.

(a) (b) (c)

**Fig. 1.** Different ways to provide user input for an interactive segmentation algorithm: Besides bounding rectangles (a) as used *e.g.* in [1], a very common concept are brush-stroke tools like in [2,3,4] (b) or like the *Quick Selection Tool* in *Adobe Photoshop* (c)

sophisticated models and image features, which in turn results in increased computational complexity.

One of the early approaches to interactive segmentation is the method of Mortensen and Barret [5] called *Intelligent Scissors* or *Magnetic Lasso*. They try to find the optimal boundary between user-provided seed nodes based on the gradient magnitude. This method, which is implemented in *GIMP* and *Photoshop*, works well on images where the gradient magnitude distinctively describes the segment boundaries, however, on highly textured areas lots of user interaction is needed. To obtain smooth segment boundaries, many recent methods perform regularization by penalizing the boundary lengths, *e.g.* by minimizing the Geodesic Active Contour (GAC) energy introduced by Caselles et al. [6]. This minimization can be performed in the discrete setting using graph based approaches ([1,7]) as well as in the continuous domain with weighted Total Variation ([2,8]). Such algorithms yield impressive results on natural, artificial as well as medical images. However, they rely on simple intensity or color features modeled with basic learning algorithms such as Gaussian mixtures models or histograms, which limits their applicability to problems where color provides sufficient information for a successful segmentation.

Applying more sophisticated features increases the complexity of the modeling stage, but in turn allows to segment images where color or intensity alone is not descriptive enough. Recently, two methods demonstrated the integration of such texture descriptors into an interactive segmentation framework: Han et al. [9] applied a multi-scale nonlinear structure tensor in a graph-cut framework. In previous work [3], we segmented texture by using HoG descriptors learned with Random Forests [10] in a weighted TV environment.

All of the methods referenced so far can only solve binary segmentation problems. To perform multi-label segmentation, some approaches simply combined such binary segmentation results: Donoser et al. [11] showed excellent results on the Berkeley Segmentation Database [12] with an unsupervised method based on covariance features and affinity propagation. They combined binary segmentations to obtain their multi-label results. This approach leads to unassigned

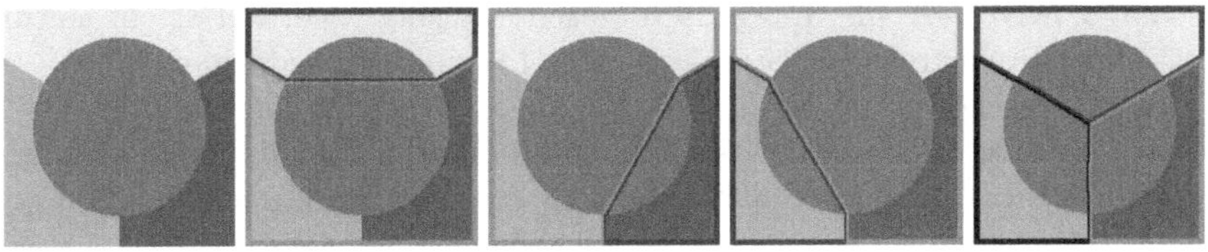


**Fig. 2.** The gray circle in the center of (a) has the same distance in the RGB space to the three border regions, thus the same likelihood for all three labels (b-d). By partitioning this image by applying sequential binary segmentations, unlabeled regions occur which require additional post-processing. Segmentation methods capable of solving multi-label problems find the optimal partition (e).

pixels, which they eliminated in an additional post-processing step. There are several drawbacks of such combination approaches (Fig. 2):

- For every label in the image at least one binary segmentation problem has to be solved.
- Ambiguities depending on the order of combination of the binary solutions might occur.
- Additional post-processing steps are required in case of unassigned regions.

To address these issues, we need an algorithm capable of partitioning an image into several labels simultaneously. Recently very promising multi-label segmentation algorithms based on solving the Potts model have been proposed [13].

The contribution of this paper is two-fold: First, we demonstrate a powerful interactive multi-label segmentation framework consisting of several different color- and texture-based image features, Random Forest classification and a multi-class regularization based on the Potts model. By computing the core parts of our framework on the GPU, we are able to perform all steps fast enough for convenient user interaction. Second, we introduce a novel benchmark database for interactive segmentation, which we employ to assess the performance of our framework. This benchmark, as well as an in-depth description of our framework [24], is publicly available on our website[1].

The remainder of the paper is organized as follows: In Section 2, we describe all parts of our framework thoroughly. In Section 3 we introduce a novel benchmark dataset and demonstrate the performance of our method until we conclude this work and given an outlook in Section 4.

## 2 Method

In this section, we describe the core parts of our framework: Like many other interactive segmentation methods, our algorithm consists of three major stages:

[1] www.gpu4vision.org

- **Feature Stage:** The image is represented using a certain set of features, starting from grayscale or color values to sophisticated interest point and texture descriptors.
- **Model Stage:** Based on user input, a model for the different segments is estimated and evaluated. This amounts to a supervised classification problem, where the user input is used as training set and the rest of the image as test set. The output of this stage is the likelihood for every pixel to belong to a certain image segment.
- **Segmentation Stage:** Depending on the quality of the model, the likelihood is typically very noisy. Therefore, spatial regularization is applied to the likelihood to produce a smooth segmentation with distinctive boundaries.

### 2.1 Features

The image description is an essential part for interactive segmentation as it gives a bias towards the type of images which can be segmented (Fig. 3). While color models work well for many natural images, the segmentation of *e.g.* medical images strongly benefits from texture descriptors. In previous work [3], we showed implementation strategies to reach interactive performance when using several different image descriptions simultaneously for segmentation. Following this approach, we employ several color models (Grayscale, RGB, HSV and CIELAB) as well as features to describe textural properties (entire image patches, Haralick [15] features as well as Local Binary Patterns [16]), all implemented on GPUs. If two or more features are selected, their feature vectors are simply concatenated to form one single feature vector per pixel.

The Haralick features employed in our framework offer two parameters: $N$ represents the number of discrete grayvalues for the construction of the graylevel-cooccurence matrix, which is sampled from an $s \times s$ square image patch. The Local Binary Patterns implemented are uniform and rotationally invariant, with $P$ points sampled on a circle with radius $R$. For the generation of the histograms, a square patch of $3 \cdot R \times 3 \cdot R$ is employed.

(a)

(b)

**Fig. 3.** The employed image description determines which type of images can be segmented: While (a) can easily be partitioned using color models, (b) will need some kind of texture description for good results

### 2.2 Model

In interactive segmentation, the user has to provide seeds for every label he wants to segment, which form the training set of a supervised classification task. Then the feature vectors for every pixel are evaluated, yielding the likelihood that the pixel belongs to a certain label (Fig. 4). To fulfill the speed requirements of the interactive setting, the classification algorithm needs to tackle multi-class problems in high-dimensional feature spaces with a large number of data samples efficiently. Random Forests have shown to be well suited for such problems [3] due to their training and evaluation speed as well as their feature selection capability. The training of the Random Forests employed in this work is optimized for multiple CPU cores, the evaluation is computed on the GPU.

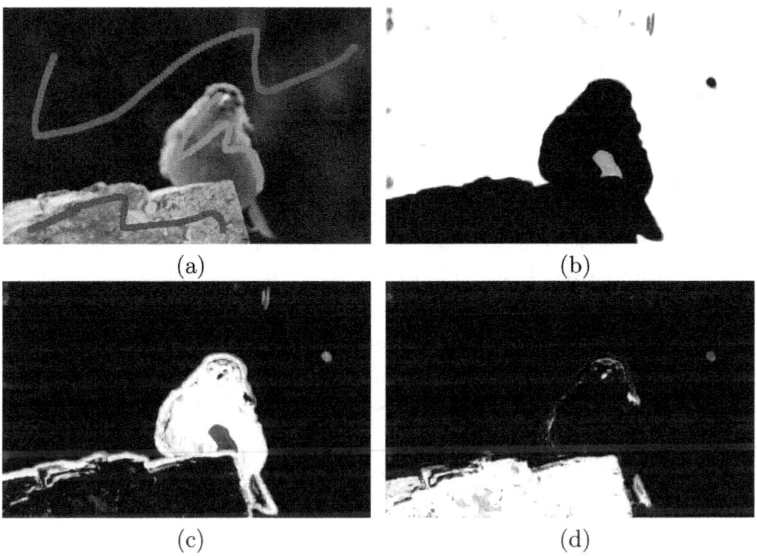


**Fig. 4.** After the user has marked seeds for every label (a), the seeds form the training set in a supervised classification problem. The obtained model is evaluated on every pixel: (b-d) show the resulting probabilities for each pixel to belong to a certain label (*i.e.* the light pixels in b are very likely to belong to label 0 (red seeds) *etc.*).

### 2.3 Multi-label Segmentation

The label likelihood obtained from the learning algorithm is typically very noisy (Fig. 4). Therefore, a common approach is to employ some kind of regularization to obtain spatially coherent labels. Among all the different image segmentation models, the Potts model appears to be the most appropriate since it is simple and does not assume any ordering of the label space.

The Potts model [17] was originally proposed to model phenomena of solid state physics. It reappeared in a computer vision context as a special case of the famous Mumford-Shah functional [18]. The aim of the Potts model is to partition the image domain $\Omega \subset \mathbb{R}^2$ into $N$ pairwise disjoint sets $E_i$. Although originally formulated in a discrete setting, the continuous setting turned out to be more appropriate for computer vision applications [13]. In the continuous setting the Potts model is written as

$$\min_{E_i,\, i=1,\dots,N} \left\{ \frac{1}{2} \sum_{i=1}^{N} \mathrm{Per}_D(E_i; \Omega) + \lambda \sum_{i=1}^{N} \int_{E_i} f_i(x)\, dx \right\} ,$$
$$\text{such that } \bigcup_{i=1}^{N} E_i = \Omega, \quad E_i \cap E_j = \emptyset,\ \forall i \neq j , \tag{1}$$

The first term penalizes the length of the partition (measured in the space induced by the metric tensor $D(x)$) and hence leads to spatially coherent solutions. The second term is the data term, which takes a point-wise defined weighting function as input, *e.g.* in our case the likelihood output of the learning algorithm. The parameter $\lambda \geq 0$ is used to control the trade-off between regularity and data fidelity.

In the last years, different algorithms have been proposed to minimize the Potts model. The most widely used algorithm is the $\alpha$-expansion algorithm of Boykov and Kolmogorov [19], which tries to approximately minimize the Potts model in a discrete setting by solving a sequence of globally optimal binary problems. In a continuous setting, level set based algorithms (*e.g.* [20]) have been used for a long time but suffer from non-optimality problems. Ignited by the work of Chan, Esedoglu and Nikolova [21], several globally optimal algorithms were proposed that minimize the Potts model in a continuous setting. For example, a continuous version of the $\alpha$-expansion algorithm has been recently studied in [14]. In this work we make use of the convex relaxation approach of [13], since it provides a unified framework and has been shown to deliver excellent results.

The starting point of [13] is to rewrite the abstract Potts model (1) by means of a convex total variation functional

$$\min_{u_i,\, i=1,\dots,N} \left\{ \frac{1}{2} \int_{\Omega} \sqrt{\nabla u_i(x)^T D(x) \nabla u_i(x)} dx + \lambda \int_{\Omega} u_i f_i dx \right\},$$
$$u_i(x) \geq 0 , \quad \sum_{i=1}^{N} u_i(x) = 1 , \quad \forall x \in \Omega , \tag{2}$$

where $u_i : \Omega \to \{0, 1\}$ are the labeling functions, *i.e.* $u_i(x) = 1$, if $x \in E_i$ and $u_i(x) = 0$ else. The first term is the anisotropic total variation of the binary function $u_i$ and coincides with the anisotropic boundary length of the set $E_i$.

Unfortunately, this minimization problem is non-convex, since the space of binary functions form a non-convex set. The idea of [13] is therefore to relax the set of binary functions to the set of functions that can vary between 0 and 1, *i.e.* $u_i : \Omega \to [0, 1]$. This convex relaxation turns the problem into a convex optimization problem, which can then be solved using a first order primal-dual algorithm. Although the relaxation does not guarantee that the original binary problem is solved optimally, it was shown in [13] that it very often delivers globally optimal solutions for practical problems. We have implemented the primal-dual algorithm on the GPU, which leads to interactive computing times.

The metric tensor $D(x)$ can be used to locally affect the length metric. It is therefore reasonable to locally adapt the metric tensor to intensity gradients of the input image $I$. That is, segmentation boundaries will be attracted by strong edges of the input image and hence lead to more precise segmentation boundaries. For simplicity, we approximate the metric tensor by means of a diagonal matrix

$$D(x) = diag(g(x)) ,$$

where the function $g(x)$ is an edge detector function given by

$$g(x) = \exp^{-\alpha|\nabla I|} .$$

The parameter $\alpha$ is used to control the influence of the edge detector function to the segmentation boundaries.

# 3 Experiments

In this section we introduce a novel benchmark dataset for interactive segmentation, which we use to assess the quality of our framework. We furthermore evaluate the computational performance of the building blocks of our method.

## 3.1 Benchmark

Benchmarking the quality of an interactive segmentation method is not straightforward because of the human interaction in the loop. Unsupervised segmentation benchmarks such as the Berkeley dataset [12] cannot be used, as they provide only hand-labeled ground-truth data, but no seeds. Arbelaez and Cohen [22] used the center point of ground-truth labels of the Berkley dataset as seeds for their segmentation tool. They showed impressive results, however, this evaluation does not reflect the circumstances in interactive segmentation at all. Hence, most interactive segmentation algorithms lack quantitative evaluation.

To address this issue, we created a benchmark consisting of 262 seed - groundtruth pairs on 158 different natural images. We wanted to know what users expect when they draw seeds into an image. Therefore, we let them draw seeds as well as a ground-truth labeling corresponding to the segmentation result they would like to obtain (Fig. 5). The ground-truth labeling is a partition of the image into pairwise disjoint regions with free topology. The seed pixels are stored as the path the user took with his mouse cursor, seeds for the

(a) (b) (c)

**Fig. 5.** A seed-ground-truth pair: For the given image (a), a user provided seeds for 8 different labels (b) as well as the ground-truth segmentation he would expect to get from those seeds (c)

background could optionally be sampled randomly within the ground-truth background region. As different frameworks usually employ different types of tooltips for seed generation, everybody using this benchmark dataset may apply his own tooltip upon the stored mouse path. See Fig. 5b as an example: Here, a solid brush with a radius of 5 pixels is applied to generate the seeds from the stored mouse path. In our experiments, we use an airbrush with a opacity of 5 percent and a radius of 7 pixels (*i.e.* a solid brush where only a random 5 percent of the pixels are taken into account).

Many of the quality measures of established segmentation benchmarks describe the accuracy of the segment boundaries only. We want that the resulting segmentations are close to the ground-truth labeling of the user, such that the amount of further interaction to yield the desired segmentation is as small as possible. Therefore, we chose the arithmetic mean of the Dice evaluation score [23] over all segments as evaluation score for our benchmark. This score relates the area of two segments $|E_1|$ and $|E_2|$ with the area of their mutual overlap $|E_1 \cap E_2|$ such that

$$dice(E_1, E_2) = \frac{2|E_1 \cap E_2|}{|E_1| + |E_2|}, \quad (3)$$

where $|\cdot|$ denotes the area of a segment. Given $GT_i$ the ground-truth labeling for the $i$-th of $N$ segments, the evaluation score for one image amounts to

$$score = \sum_{i=1}^{N} dice(E_i, GT_i) = \sum_{i=1}^{N} \frac{2|E_i \cap GT_i|}{|E_i| + |GT_i|} \quad (4)$$

### 3.2 Image Features

In this experiment, we compare the average score of different features as well as different combinations of features. We applied Random Forests with 200 trees, the segmentation was computed for 750 iterations with $\lambda = 0.2$ and $\alpha = 15$. The average score over the whole dataset looks as follows:

| Type | Feature (Dimension) | Benchmark Score |
|---|---|---|
| Color Models | Grayscale (1) | 0.728 |
| | RGB (3) | 0.877 |
| | HSV (3) | 0.897 |
| | CIELAB (3) | 0.898 |
| Textural Features | Image Patches 17 × 17 (289) | 0.814 |
| | Haralick $N = 32, s = 13$ (13) | 0.855 |
| | LBP $P = 20, s = 8$ (22) | 0.819 |

The parameter settings of the textural features were optimized *w.r.t.* the benchmark score exhaustively. The color models show a better average performance than the intensity value and the grayscale-based textural features. This leads to the assumption, that most of the images in the benchmark are better separable by color than by texture. However, the observation that the average performance of the textural features is higher than the performance of the intensity alone shows that local texture is descriptive in some of the images too (Fig. 6).

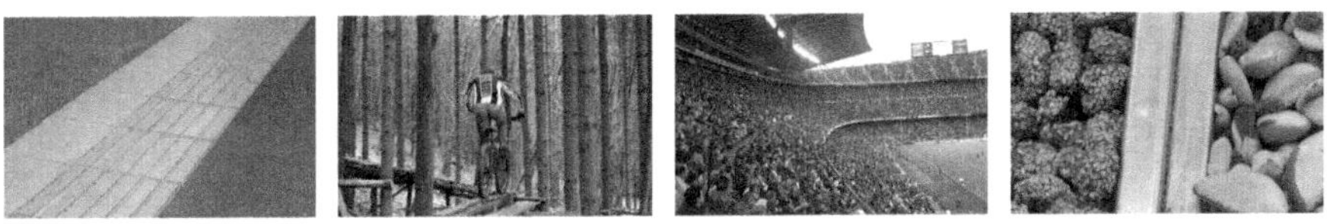

(a) (b) (c) (d)

**Fig. 6.** Images of the dataset exhibit different performance depending on their representation: (a) and (b) yield a high score when represented with RGB values. In (c) the crowd and the roof of the stadium have similar colors, just as the nuts in (d). The two latter images yield a higher score when represented using grayscale image patches.

Now we combine different image features by simply concatenating their feature vectors, to find out whether the perfomance can be increased:

| Feature | Dimension | Benchmark Score |
|---|---|---|
| Gray RGB HSV CIELAB | 10 | 0.896 |
| CIELAB Haralick $N = 32, s = 5$ | 16 | 0.916 |
| CIELAB LBP $P = 16, R = 3$ | 21 | 0.920 |

While the combination of the color features leads to no improvement, combining a color model with a texture descriptor yields a higher score (Fig. 7): The result of the CIELAB model improves from 0.898 to 0.916 when combined with Haralick features, using Local Binary Patterns increases the performance to 0.920.

### 3.3 Tooltip

In this experiment, we assess the influence of different tooltips to the benchmark performance. Based on the previous results, we employ a combination of CIELAB color vectors and LBP features with $P = 16, R = 3$ in the upcoming experiments.

| Tooltip | Radius | Benchmark Score |
|---|---|---|
| Solid Brush | 5 | 0.917 |
| | 9 | 0.926 |
| | 13 | 0.925 |
| Airbrush | 5 | 0.908 |
| | 9 | 0.919 |
| | 13 | 0.927 |

These results show, that the airbrush yields comparable results to the solid brush, however, the smaller number of seed pixels leads to a faster model stage. Therefore, in the following experiments, we employ an airbrush tooltip with a radius of 13 pixels.

**Fig. 7.** Image (a) is the segmentation result using the HSV color model as feature. In (b), we additionally employ Haralick texture features to describe local structure, which leads to a significantly improved result.

### 3.4 Random Forest / Repeatability

Our framework has two sources of randomness: The Random Forests as well as the airbrush tooltip. In this experiment, we want to evaluate the influence of this randomness to the benchmark score. Furthermore, in order to improve the runtime of the framework, we want to find out whether similar benchmark scores can be achieved with a smaller number of trees in the Random Forest. We conducted 30 identically parametrized runs of our benchmark with Random Forests with 30 trees. The average benchmark score from these runs was 0.926, with a standard deviation of 0.0013.

### 3.5 Runtime

Finally, we want to give a detailed overview of the computational performance of our framework. The runtimes stated in this section are the average runtimes of the algorithm stages over one benchmark run, conducted on a desktop PC featuring a 2.6 GHz quad-core processor and an NVIDIA Geforce GTX 280 GPU.

Computing the image features is typically done only once before the segmentation is performed. We implemented all image features on the GPU allowing for dense feature calculation in about a third of a second. The time needed to

**Fig. 8.** Results of our framework on the benchmark dataset proposed in this paper

train a Random Forest mainly depends on the number of trees, the dimension of the feature space as well as the complexity of the learning tasks. While the training of the classifiers is optimized for multi-core CPUs, we perform the evaluation for every image pixel on the GPU. For a typical benchmark segmentation problem, the training of 30 trees takes $\approx$ 750 milliseconds, the evaluation takes $\approx$ 100 milliseconds. The time spent for computing the segmentation model depends on the number of labels as well as the number of computed iterations. The massive parallel computation power of the GPU makes this algorithm suited for

**Fig. 9.** Segmentation of images taken from the Berkeley Dataset [12]

interactive segmentation: 750 iterations on a four-label problem are solved in about 1000 milliseconds.

| Algorithm Stage | Operation | Runtime [ms] |
|---|---|---|
| Features | CIELAB + $LBP_{16,3}$ | 340 |
| Model | Training | 750 |
| | Evaluation | 100 |
| Segmentation | | 1000 |

For the interactivity, only the runtime of the model and segmentation stage is important (as the features need to be computed only once). The runtime of these stages amounts to less than two seconds.

### 3.6 Qualitative Examples

Fig. 8 shows 24 images of our benchmark dataset taken from a run using all features, a Random Forest with 100 trees and a lambda value of 0.2. Fig. 9 shows interactive segmentations of images taken from the Berkeley Segmentation Dataset.

## 4 Conclusion

In this paper we proposed a powerful interactive multi-label texture segmentation framework. We showed that by using GPUs and multi-core implementations, the extraction of color and texture features, the training and evaluation of random forests as well as the minimization of a multi-label Potts model can be performed fast enough for convenient user interaction.

We additionally presented a large novel benchmark dataset for interactive multi-label segmentation and evaluated the single building blocks of our framework. We demonstrated the performance of our method in numerous images, both from own as well as other datasets.

In future work, we are interested in extending our framework towards three-dimensional data and videos in spatial-temporal representation.

**Acknowledgement.** This work was supported by the Austrian Research Promotion Agency (FFG) within the project VM-GPU (813396) as well as the Austrian Science Fund (FWF) under the doctoral program Confluence of Vision and Graphics W1209. We also greatly acknowledge NVIDIA for their valuable support.

## References

1. Rother, C., Kolmogorov, V., Blake, A.: "GrabCut": Interactive foreground extraction using iterated graph cuts. ACM Trans. Graph. 23, 309–314 (2004)
2. Unger, M., Pock, T., Trobin, W., Cremers, D., Bischof, H.: TVSeg - Interactive total variation based image segmentation. In: BMVC 2008, Leeds, UK (2008)

3. Santner, J., Unger, M., Pock, T., Leistner, C., Saffari, A., Bischof, H.: Interactive texture segmentation using random forests and total variation. In: BMVC 2009, London, UK (2009)
4. Vezhnevets, V., Konouchine, V.: "Grow-Cut" - Interactive multi-label n-d image segmentation. In: Proc. Graphicon, pp. 150–156 (2005)
5. Mortensen, E.N., Barrett, W.A.: Intelligent scissors for image composition. In: SIGGRAPH 1995, pp. 191–198. ACM, New York (1995)
6. Caselles, V., Kimmel, R., Sapiro, G.: Geodesic active contours. IJCV 22, 61–79 (1995)
7. Boykov, Y.Y., Jolly, M.P.: Interactive graph cuts for optimal boundary & region segmentation of objects in n-d images. In: ICCV 2001, vol. 1, pp. 105–112 (2001)
8. Bresson, X., Esedoglu, S., Vandergheynst, P., Thiran, J., Osher, S.: Global Minimizers of The Active Contour/Snake Model. Technical report, EPFL (2005)
9. Han, S., Tao, W., Wang, D., Tai, X.C., Wu, X.: Image segmentation based on grabcut framework integrating multiscale nonlinear structure tensor. Trans. Img. Proc. 18, 2289–2302 (2009)
10. Breiman, L.: Random forests. Machine Learning 45, 5–32 (2001)
11. Donoser, M., Urschler, M., Hirzer, M., Bischof, H.: Saliency driven total variation segmentation. In: ICCV (2009)
12. Martin, D., Fowlkes, C., Tal, D., Malik, J.: A database of human segmented natural images and its application to evaluating segmentation algorithms and measuring ecological statistics. In: ICCV (2001)
13. Pock, T., Chambolle, A., Cremers, D., Bischof, H.: A convex relaxation approach for computing minimal partitions. In: CVPR (2009)
14. Olsson, C., Byrd, M., Overgaard, N.C., Kahl, F.: Extending continuous cuts: Anisotropic metrics and expansion moves. In: ICCV (2009)
15. Haralick, R.M., Shanmugam, K., Dinstein, I.: Textural features for image classification. IEEE Transactions on Systems, Man and Cybernetics 3, 610–621 (1973)
16. Ojala, T., Pietikainen, M., Maenpaa, T.: Multiresolution gray-scale and rotation invariant texture classification with local binary patterns. PAMI 24, 971–987 (2002)
17. Potts, R.B.: Some generalized order-disorder transformations. Proc. Camb. Phil. Soc. 48, 106–109 (1952)
18. Mumford, D., Shah, J.: Optimal approximation by piecewise smooth functions and associated variational problems. Comm. Pure Appl. Math. 42, 577–685 (1989)
19. Boykov, Y., Veksler, O., Zabih, R.: Fast approximate energy minimization via graph cuts. PAMI 23, 1222–1239 (2001)
20. Chan, T., Vese, L.: Active contours without edges. IEEE Trans. Image Processing 10, 266–277 (2001)
21. Chan, T., Esedoglu, S., Nikolova, M.: Algorithms for finding global minimizers of image segmentation and denoising models. SIAM Journal of Applied Mathematics 66, 1632–1648 (2006)
22. Arbelaez, P., Cohen, L.: Constrained image segmentation from hierarchical boundaries. In: CVPR (2008)
23. Dice, L.R.: Measures of the amount of ecologic association between species. Ecology 26, 297–302 (1945)
24. Santner, J.: Interactive Multi-label Segmentation. PhD thesis, Graz University of Technology (2010)

# Four Color Theorem for Fast Early Vision

Radu Timofte and Luc Van Gool

ESAT-PSI / IBBT, Katholieke Universiteit Leuven, Belgium
{Radu.Timofte,Luc.VanGool}@esat.kuleuven.be

**Abstract.** Recent work on early vision such as image segmentation, image restoration, stereo matching, and optical flow models these problems using Markov Random Fields. Although this formulation yields an NP-hard energy minimization problem, good heuristics have been developed based on graph cuts and belief propagation. Nevertheless both approaches still require tens of seconds to solve stereo problems on recent PCs. Such running times are impractical for optical flow and many image segmentation and restoration problems. We show how to reduce the computational complexity of belief propagation by applying the Four Color Theorem to limit the maximum number of labels in the underlying image segmentation to at most four. We show that this provides substantial speed improvements for large inputs to a variety of vision problems, while maintaining competitive result quality.

## 1 Introduction

Much recent work on early vision algorithms such as image segmentation, image restoration, stereo matching, and optical flow models these problems using Markov Random Fields (MRF). Although this formulation yields an NP-hard energy minimization problem, good heuristics have been developed based on graph cuts [2] and belief propagation [12,8]. A comparison between the two different approaches for the case of stereo matching is described in [9]. Nevertheless both approaches still require tens of seconds to solve stereo problems on recent PCs. Such running times are impractical for optical flow and many image segmentation and restoration problems. Alternative, faster methods generally give inferior results.

In the case of Belief Propagation (BP), a key reason for its slow performance is that the algorithm complexity is proportional to both the number of pixels in the image, and the number of *labels* in the underlying image segmentation, which is typically high. If we could limit the number of labels, its speed performance should improve greatly. Our key observation is that by modifying the propagation algorithms we can reuse labels for non-adjacent segments. Since image segments form a planar graph, they therefore require at most four labels by virtue of the Four Color Theorem (FCT).

In this paper we use the Four Color Theorem [6] from planar maps in early vision. FCT states that for any 2D map there is a four-color covering such that contiguous regions sharing a common boundary (with more than a single point)

R. Kimmel, R. Klette, and A. Sugimoto (Eds.): ACCV 2010, Part I, LNCS 6492, pp. 411–424, 2011.

do not have the same color. F. Guthrie first conjectured the theorem in 1852 (*Guthrie's problem*). The consequence of this theorem is that when an image, seen as a planar graph, is segmented into contiguous regions, there are only four colors to be assigned for each pixel/node. Algorithmically speaking, for each pixel/node there is only one of four decisions that can be taken.

This paper aims to exploit this result from graph theory by presenting new algorithmic techniques that substantially improve the running time of BP, as well as providing alternatives to fast local methods for early vision problems. In the case of image segmentation, we obtain results comparable to traditional fast loglinear Minimum Spanning Tree (MST)-based methods [3]. Our implementation has linear time complexity in the number of pixels. The same principle is demonstrated for image restoration, independent of the number of labels. Stereo matching can benefit by replacing the color segmentation step from existing methods with our segmentation. All the previously mentioned cases and the optical flow estimation are also addressed by using a proposed multi-scale loopy belief propagation which works with only four labels. In practice, we obtain faster local and global techniques that are as accurate as other well-known standard methods.

To our knowledge, Vese and Chan [11] are the first to use the Four-Color Theorem in computer vision for their multiphase level set framework, in the piecewise smooth case. The Four-Color Theorem also provides a hard upper bound of two bits to the amount of space required to store a pixel in color-coded image partitions, which is still sub-optimal according to information theory [1].

The general framework for the problems we consider here can be defined as follows (we use the notation and formulation from [4]). Let $P$ be the set of pixels $p$ in an image and $L$ be a set of labels. The labels correspond to the quantities that we want to estimate at each pixel, such as disparities, intensities, or classes. A labeling $f$ then assigns a label $f_p \in L$ to each pixel $p$. Typically, a label varies smoothly except for a limited number of places, where it changes discontinuously, i.e. at segment edges. A labeling is evaluated through an energy function,

$$E(f) = \sum_{(p,q)\in N} V(f_p, f_q) + \sum_{p\in P} D_p(f_p) \tag{1}$$

where $N$ are the edges in the four-connected image grid. $V(f_p, f_q)$ is the pairwise cost or 'discontinuity cost' of assigning labels $f_p$ and $f_q$ to two neighboring pixels. $D_p(f_p)$ is the unary cost or 'data cost' of assigning label $f_p$ to pixel $p$. Finding a labeling with minimum energy corresponds to the MAP estimation problem for an appropriately defined MRF.

The remainder of the paper is structured as follows. Section 2 reviews Loopy Belief Propagation (LBP). In Section 3, the Four Color Theorem-based techniques are incorporated in both the LBP framework and a fast forward BP approximation. Section 4 describes the experiments that were conducted, and how the proposed techniques can be used for tasks like image segmentation, image restoration, stereo matching, and optical flow. The conclusions are drawn in Section 5.

## 2 Loopy Belief Propagation

For inference on MRFs, Loopy Belief Propagation can be used [12]. In particular, the max-product approach finds an approximate minimum cost labeling of energy functions in the form of Eq. (1). As an alternative to a formulation in terms of probability distributions, an equivalent formulation uses negative log probabilities, where the max-product becomes a min-sum. Following Felzenszwalb and Huttenlocher [4], we use this formulation as it is numerically more robust and makes more direct use of the energy function.

The max-product BP algorithm passes messages around on a graph defined by the 4-connected image grid. Each message is a vector, with the number of possible labels as dimension. Let $m^t_{p \to q}$ be the message that node $p$ passes on to a neighboring node $q$ at time $t$, $0 < t \leq T$. Consistent with the negative log formulation, all entries in $m^0_{p \to q}$ are initialized to zero. At each iteration new messages are computed as follows:

$$m^t_{p \to q}(f_q) = \min_{f_p}(V(f_p, f_q) + D_p(f_p) + \sum_{s \in N(p) \setminus q} m^{t-1}_{s \to p}(f_p)) \qquad (2)$$

where $N(p) \setminus q$ denotes all of $p$'s neighbors, except $q$. After $T$ iterations, a belief vector is computed for each node,

$$b_q(f_q) = D_q(f_q) + \sum_{p \in N(q)} m^T_{p \to q}(f_q) \qquad (3)$$

Finally, for each node $q$, the label $f^*_q$ that minimizes $b_q(f_q)$, is selected. The standard implementation of this message passing algorithm runs in $O(nk^2T)$ time, with $n$ the number of pixels, $k$ the number of possible labels, and $T$ the number of iterations. Essentially it takes $O(k^2)$ time to compute each message and there are $O(n)$ messages per iteration. In [4] the time of computing each message is reduced to $O(k)$ for particular classes of data cost functions such as truncated linear and quadratic models combined with a Potts model for the discontinuity cost.

### 2.1 Multiscale BP on the Grid Graph

As in [4], we can color the 2D grid in a checkerboard pattern where every edge connects nodes of different colors, so the grid graph is bipartite. By using this property, we compute only half of the messages corresponding to the same color nodes with the color alternating at each iteration. This scheme has the advantage of using the same memory space for storing the updated messages. Another technique explained in [4] and used here addresses the problem of BP information flow that implies the use of many iterations to cover large distances. One solution is to perform BP in a coarse-to-fine manner, so that the long-range interactions between nodes are captured by shorter ones in coarser graphs. The minimization function does not change. In this hierarchical approach, BP runs at one resolution level in order to get estimates for the next finer level. Better initial estimates help to speed up the convergence. The pyramidal structure works as follows. The zero-th level is the original image. The $i$-th level corresponds to a lower-resolution version with blocks of $2^i \times 2^i$ pixels grouped together. The resulting blocks are still connected in a grid structure (see Fig.1). For a block $b$, the adapted data cost of assigning label $f_b$ is

$$D_b(f_b) = \sum_{p \in b} D_p(f_p) \tag{4}$$

where the sum runs over all pixels in the block. The multiscale algorithm first solves the problem at the coarsest level, where the messages are initialized to zero. Subsequent finer levels take the previous, coarser level as initialization. In the 4-connected grid each node $p$ sends messages in all 4 directions, *right*, *left*, *up*, *down*. Let $r_p^t$ be the message sent by node $p$ to the *right* at iteration $t$, and similarly $l_p^t$, $u_p^t$, $d_p^t$ for the other directions. This is just a renaming. If $q$ is the right neighbor of $p$ then $r_p^t = m_{p \to q}^t$ and $l_q^t = m_{q \to p}^t$. For a node $p$ at level $i-1$, the messages will be initialized with the ones obtained by solving the level $i$ for the containing block $p'$:

$$\begin{array}{ll} r_{p,i-1}^0 \leftarrow r_{p',i}^T & \quad l_{p,i-1}^0 \leftarrow l_{p',i}^T \\ u_{p,i-1}^0 \leftarrow u_{p',i}^T & \quad d_{p,i-1}^0 \leftarrow d_{p',i}^T \end{array} \tag{5}$$

Note that the total number of nodes in a quad-tree is 4/3 the number of nodes at the finest level, so the overhead introduced by the multiscale approach is as much as 1/3 of the original standard single scale approach, but it results in a greatly reduced number of iterations (between five and ten iterations instead of hundreds) for convergence. In general, the number of levels needed is proportional to $\log_2$ of the image diameter.

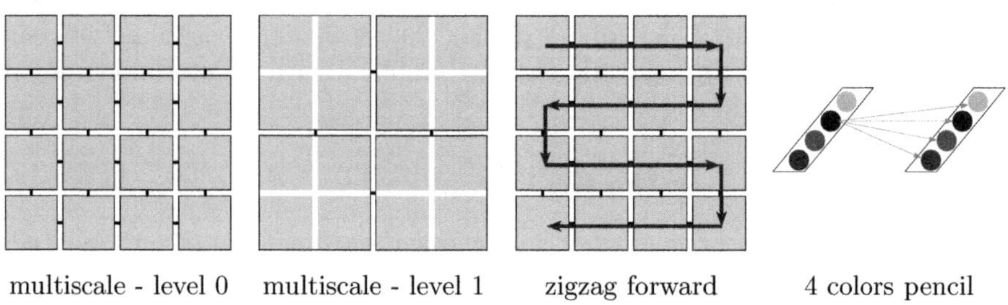


**Fig. 1.** Levels in multiscale and zigzag forward traversal for a 2D grid. A 4 color pencil (potential connections) between two neighbors.

### 2.2 Forward Belief Propagation

We propose a traversing order for the nodes in the 2D grid as another way to speed up the propagation of beliefs. Applying sequentially the message updates in the imposed order allows a belief from one node to influence all the following nodes in the traversing order considered.

Following the traversal order while updating the messages corresponds to a Forward Belief Propagation (FBP) variant, while Backward Belief Propagation (BBP) corresponds to the reverse order of the traversal.

Here we consider the traversing order to be in a zigzag/square wave pattern (see Fig. 1), thus by starting from top-left we go to the top-right, then we go down one line and take the path back to left, where we move to the next line and repeat the procedure until no line remains unvisited. This traversing order has some advantages in canceling/reducing the incorrectly propagated beliefs.

The message costs have to be normalized. One way of doing it is to simply divide by the number of messages used for computing them. In our 4 -connected grid graph, we divide the costs of the newly updated messages by 4.

Note that the FBP variant can be used in the multiscale framework as described in Section 2.1. In our implementations we take the forward traversal order (Fig. 1) in the odd iterations and the backward traversal order in the even iterations. Replacing the checkerboard scheme with FBP traversing in the multiscale BP framework (BP-FH) from [4] provides very similar performance in the stereo and restoration tasks. We refer to this method as Forward Belief Propagation - Multi Scale (FBP-MS). Without multiscale guidance, FBP usually gets stuck in a poor solution.

## 3 Four-Color Theorem in BP

In this paper, inspired by the Four-Color Theorem, we reduce the number of working labels to at most *four*. This is possible by using an extra parameter $\mu$ for the quantity estimated in each pixel, which encapsulates the role of the old labels (data anchors). By keeping only *four* labels, one for each possible color of one pixel/node in the 2D grid graph, we obtain an algorithm that is linear in the number of pixels and very fast in practice. Moreover, our results are comparable to those obtained when using standard max-product BP, efficient multiscale BP [4], or graph cut algorithms to minimize energy functions of the form in Eq. (1). In the case of stereo matching, we quantify this using the benchmark in [7].

On the right side of Fig.1 we show two neighboring nodes/pixels with 4 colors/states and a *pencil*. The pencil shows the possible links for one state to connect to the states of the neighboring node/pixel. Here, as usual, only the best connection is used, the one that minimizes a certain energy.

The Four-Color Theorem says that we can cover an image with disjoint segments, where each segment has one of no more than four colors, and no two segments sharing a border of more than one pixel have the same color. The total number of segments for the problems considered here is a priori unknown but upper bounded by the number of pixels. Let $S$ be the set of segments. Thus, locally each pixel can belong to one of four colors and the color itself represents the segment of pixels connected through the same color. Equation (1) can be rewritten as

$$E(s) = \sum_{(p,q)\in N} V(s_p, s_q) + \sum_{p\in P} D_p(s_p) \tag{6}$$

where a labeling $s$ assigns a segment $s_p$ from the set of segments $S$ to each pixel $p$ from $P$.

### 3.1 BP with Four-Colors

In order to work with four colors, the labeling $f$, in our new formulation, assigns $f_p \in C$ to each pixel $p \in P$, where the cardinality of $C$ is 4 ($|C| = 4$). In the left part of Fig. 2 a 4-connected grid neighborhood is depicted for the Four-Colors case. The edges/connections to the neighboring nodes are picked to minimize the costs in the BP formulation. For each pixel $p$, each possible color/state $c$, and each neighbor $q$, we have a data parameter that is continuously updated through message passing, $\mu^t_{p\to q}(c)$ at iteration $t$. Also, $\mu^t_q(c)$ is updated by using the incoming messages (including $\mu^t_{p\to q}(c)$). The data parameter, which represents a quantity to be estimated in each pixel, is the equivalent of the labels in the original BP formulation, where the labels are in a bijection to the set of quantities to be estimated. The initial values for $\mu^0_p(c)$ and $\mu^0_{p\to q}(c)$ depend on the data; we set them to the best observation we have. For example, in stereo matching, these will be the best scoring disparities in each pixel (in a *winner-take-all* sense) and in image restoration, the original pixel values. When we compute the new messages (Eq. (2)), we also store the color $a^t_{p\to q}$ for which we have the minimum message energy at iteration $t$:

$$a^t_{p\to q}(f_q) = \arg\min_{f_p}(V(f_p, f_q) + D_p(f_p) + \sum_{s\in N(p)\backslash q} m^{t-1}_{s\to p}(f_p)) \tag{7}$$

Also, at each iteration, part of the message is the data parameter estimation:

$$\mu^t_{p\to q}(f_q) = \mu^{t-1}_p(a^t_{p\to q}(f_q)) \tag{8}$$

We call this formulation BP-FCT, standing for Belief Propagation with Four-Color Theorem principle.

### 3.2 FBP with Four-Colors

FBP (see Section 2.2) does not produce good quality results on its own (e.g., in the absence of multiscale guidance). A solution is to consider a different updating scheme at each step which employs local consistency. Instead of computing standard message updating, where it is assumed that the history and neighborhood belief are stored/propagated in all the messages reaching the current node and taking simply the best updates, we are exploiting the ordering introduced in FBP, and keep track of the links/connections/edges which provided the best costs for each node and state/color in the grid. Thus the nodes are processed sequentially, following the imposed order. We compute the current best costs only based on the connections to the previously processed nodes. These are (for

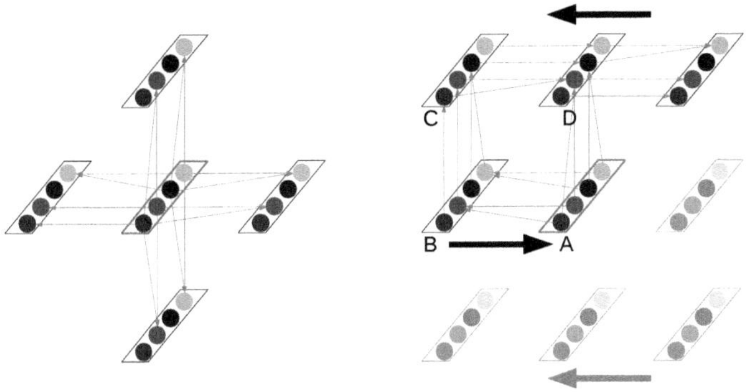


**Fig. 2.** Four Colors pixel neighborhood used in computing message updates, the case for BP-FCT (*left*) and FBP-FCT(*right*)

an inner node) the previously processed node (in the traversal order) and the neighboring node from the previously processed line.

For each pixel $x$ and each possible color/state $c_x \in C$, we keep the minimum obtained energy $E(x, c_x)$, the average image intensity/color $\mu_{x,c_x}$ of the segment $s_{x,c_x}$ where $x$ belongs, and the number of pixels in the current segment $ns_{x,c_x}$. We use the following notations:

$cp(x, c_x)$ - **c**olor/state from the **p**receding pixel in the FBP traversing order that contributed to the energy $E(x, c_x)$ (i.e: $cp(A, blue) = yellow$ in Fig. 2).

$lp(x)$ - **l**ink to the **p**receding pixel in the FBP traversing order for $x$.

$cu(x, c_x)$ - **c**olor/state from the **u**pper pixel in the FBP traversing order that contributed to the energy $E(x, c_x)$ (i.e: $cu(A, red) = green$ in Fig. 2).

$lu(x)$ - **l**ink to the **u**pper pixel in the FBP traversing order for $x$ (i.e: $lu(A) = D$, $lp(A) = B$ in Fig. 2).

The right part of Fig. 2 shows the general case where for the current node $A$ we are using only what we know in the already traversed nodes $(B, C, D)$. To enforce local consistency, we calculate the energy of each potential state of a pixel $A$ not only from the previous pixel $B$, but also from the state of the pixel $D$ above induced by the state in $B$. For example, in Fig. 2, to compute the energy to transition from the *green* state in $B$ to the *red* state in $A$ we observe that when $B$ is green, $D$ is also green. Therefore, the potential energy in $A$'s red state, if $B$'s green state is considered as connection, is the sum of the transition energy from $B$'s green state and $D$'s green state. Computing the best costs for a node $A$'s state requires first considering all of previously processed node $B$'s states each paired with the state in $D$ (the neighboring node in the previously processed line) that propagates to it, then taking the minium computed energy among these alternatives.

Thus, the energy $E(x, c_x)$ is computed consistently if $lu(x) = lp(lu(lp(x)))$ and $cu(x, c_x) = cp(lu(lp(x)), cu(lp(x), cp(x, c_x)))$, which enforces that the connecting color of the previously processed pixel to be the result/propagation of the connecting color of the neighboring pixel from the previously processed line:

$$E(x, c_x) = \min_{c_p \in C} \{E(lp(x), c_p) + h((x, c_x), (lp(x), c_p)) + h((x, c_x), (lu(x), c_{up}))\} \tag{9}$$

where $c_{up} = cp(lu(lp(x)), cu(lp(x), c_p))$, and

$$h((x, c_x), (y, c_y)) = \begin{cases} dist(pxl_x, pxl_y, \sigma_c) + dist(\mu_{x,c_x}, \mu_{y,c_y}, \sigma_m) & \text{if } c_x = c_y \\ 2 - [dist(pxl_x, pxl_y, \sigma_c) + dist(\mu_{x,c_x}, \mu_{y,c_y}, \sigma_m)] & \text{if } c_x \neq c_y \end{cases} \tag{10}$$

where $dist(\mu_{x,c_x}, \mu_{y,c_y}, \sigma_m)$ is the data cost, and

$$dist(u, v, \sigma) = 1 - \exp\left(-\frac{\|u - v\|_1}{2\sigma^2}\right) \tag{11}$$

The distance/penalty $dist(u, v, \sigma)$ between pixel values penalizes a for segment discontinuities between similar values ($\|u - v\|_1 < 2\sigma^2$), while if the pixels are very different, the penalty is reasonable small. The function models the distribution of noise among neighboring pixels, as $\sigma$ is an estimation of the camera noise. The argument $c_p$ that gives the minimum energy $E(x, c_x)$ is stored in $cp(x, c_x) = c_p$. Also the link to the upper pixel state is stored in $cu(x, c_x) = c_{up}$.

Given the previous notations, the segment color $\mu_{x,c_x}$ and the approximated number of pixels which belong to the segment, $ns_{x,c_x}$, are obtained as follows:

$$ns_{x,c_x} = \begin{cases} ns_{lp(x),cp(x,c_x)} + 1 & \text{if } c_x = cp(x, c_x) \\ ns_{lu(x),cu(x,c_x)} + 1 & \text{if } c_x = cu(x, c_x) \wedge c_x \neq cp(x, c_x) \\ 1 & \text{otherwise} \end{cases} \tag{12}$$

$$\mu_{x,c_x} = \begin{cases} (\mu_{lp(x),cp(x,c_x)} ns_{lp(x),cp(x,c_x)} + pxl_x)/ns_{x,c_x} & \text{if } c_x = cp(x, c_x) \\ (\mu_{lu(x),cu(x,c_x)} ns_{lu(x),cu(x,c_x)} + pxl_x)/ns_{x,c_x} & \text{if } c_x = cu(x, c_x) \wedge c_x \neq cp(x, c_x) \\ pxl_x & \text{otherwise} \end{cases} \tag{13}$$

## 4 Experiments

We now demonstrate how the proposed methods can be used in different early vision applications that can be formulated as energy minimization problems through a MRF model.

For all experiments we provide details about discontinuity costs ($V(f_p, f_q)$), data costs ($D_p(f_p)$), and message updates/computations ($\mu_p^t(f_p)$, $m_{p\to q}^t(f_q)$). Also, we provide a comparison with well-known standard methods. The images used are from the Berkeley Segmentation Dataset [5], and the Middlebury Stereo Datasets [7], and [4].

All the provided running times were obtained on an Intel Core 2 Duo T7250 (2.0GHz/800Mhz FSB, 2MB Cache) notebook with 2GB RAM. More results are available at: http://homes.esat.kuleuven.be/~rtimofte/

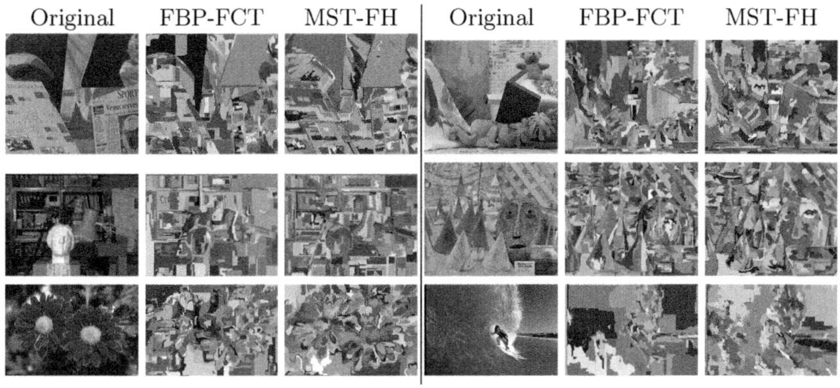


**Fig. 3.** Segmentation results

### 4.1 Image Segmentation

For comparison, we use our proposed method based on the Four-Color Theorem - FBP-FCT (see Section 3.2) and a standard Minimum Spanning Tree based method - MST-FH from [3].

The oversegmentation in our FBP-FCT is addressed in a similar fashion to MST-FH in [3]. The smaller (than an imposed minimum size) segments are merged with other segments until no segment with a size less than the minimum imposed remains. We use the original MST-FH implementation with the original parameters ($\sigma = 0.8$, $k = 300$). Our FBP-FCT method uses an initial smoothing of the image with $\sigma = 0.7$, $\sigma_c = 3.2$, and $\sigma_m = 4.2$. We set $k = 300$ and the minimum size for both methods is 50 pixels. Fig. 3 depicts the image segmentation results for several cases.

MST-FH is an $O(n \log n)$ algorithm, while FBP-FCT is $O(n|C|^2)$. This means (and our tests show) that FBP-FCT and MST-FH have comparable running times on low resolution images while for high resolutions images ($\log n > |C|^2$), the FBP-FCT is faster. For example, *Venus* RGB image with $434 \times 383$ pixels (top-left in Fig. 3) is processed by MST-FH in about 200ms, while our approach takes 250ms.

### 4.2 Image Restoration

The image restoration problem is a case where usually the number of labels (in an MRF/BP formulation) is very large since it is equal to the number of intensity levels/colors used. We argue that this way of seeing the problem, besides being computationally demanding, does not take advantage of the relation that exists between labels as intensities. The intensity values are obtained through uniform sampling of a continuous signal, therefore carry direct information on their relative closeness. Instead of updating labels through selection from neighboring labels, such update can result from a mathematical operation on such neighboring labels, e.g. by averaging.

In the BP-FCT formulation (see Section 3.1) we take the following updating function for the intensity data at each message:

$$\mu_q^t(f_q) = \frac{k_r I(q) + \sum_{p \in N(q) \setminus q} (\mu_{p \to q}^t(f_q)[a_{p \to q}^t(f_q) = f_q])}{k_r + \sum_{p \in N(q) \setminus q} [a_{p \to q}^t(f_q) = f_q]} \tag{14}$$

where $[\cdot = \cdot]$ is the Iverson bracket for Kronecker's delta, i.e. it returns 1 for equality, 0 otherwise. $k_r$ is the weight/contribution of the observed value $I(q)$ in the updated data term. Here we use $k_r = 0.25$.

The data cost for a pixel $p$, at iteration $t$ is:

$$D_p(f_p) = \min(\|I(p) - \mu_p^{t-1}(f_p)\|_2, \tau_1) \tag{15}$$

and counts for the difference between the estimated intensity value $\mu_p^t(f_p)$ at iteration $t$ and the observed one $I(p)$ for the pixel $p$ under labeling $f_p$. The discontinuity cost is given by:

$$V(f_p, f_q) = s \min(\|\mu_p^{t-1}(f_p) - \mu_q^{t-1}(f_q)\|_2, \tau_2) \tag{16}$$

We compare BP-FCT (Section 3.1) and FBP-FCT (Section 3.2) with BP-FH[4]. For this purpose the gray-scale images are corrupted by adding an independent and identically-distributed Gaussian noise with zero-mean and variance 30. We use the original available implementation for multiscale BP-FH. For fair comparison we apply, as in BP-FH, a Gaussian smoothing with standard deviation 1.5 before processing the corrupted images. We set $\sigma_c = \sigma_m = 3.2$ for FBP-FCT, while BP-FCT uses for discontinuity truncation $\tau_2 = 200$, for data truncation $\tau_1 = 10000$ and for the rate of increase in the cost $s = 1$.

Reducing the number of labels when $|L| \gg |C|^2$ assures a considerable speed-up of our proposed methods (BP-FCT - $O(n|C|^2T)$, FBP-FCT - $O(n|C|^2)$), over the standard multi-scale BP-FH which has a time complexity of $O(n|L|T)$. Image restoration results are shown in Fig. 4. For all the images tested BP-FCT and FBP-FCT had better or similar restoration performance (in terms of PSNR) than the BP-FH method, while being up to 100 times faster. For example, the *Boat* gray-scale corrupted image with $321 \times 481$ pixels (top row in Fig. 4) is processed by FBP-FCT in about $200ms$, by BP-FCT in 4 seconds, while BP-FH takes more than 30 seconds. The computed PSNR (in dB) for the *Penguin* images are: $PSNR_{Corrupted} = 18.87$, $PSNR_{FBP-FCT} = 25.76$, $PSNR_{BP-FCT} = 27.21$, $PSNR_{BP-FH} = 26.02$. Having a restoration method that works for a single channel image (gray levels), usually the multi-channel images (e.g: RGB) are processed for each channel individually and the restored image is the union of the restored channels.

### 4.3 Stereo Matching

For stereo matching, in the BP-FCT (see Section 3.1) framework we define the following cost and update functions.

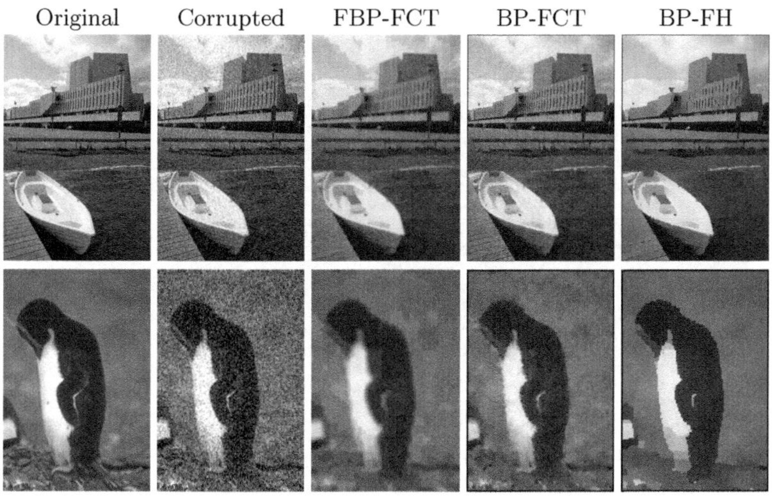


**Fig. 4.** Restoration results. BP-FCT and FBP-FCT perform better than BP-FH (in terms of PSNR) while being 1 up to 2 orders of magnitude faster.

For a pixel $p = (i, j)$ whose intensity is $I_l(i, j)$ in the left camera image, the cost for a disparity of $d$ is:

$$DSI(p, d) = s \min(\|I_l(i, j) - I_r(i - d, j)\|^2, \tau_s) \tag{17}$$

where $s = 0.1$ and $\tau_s = 10000$. The data cost at iteration $t$ is:

$$D_p(f_p) = DSI(p, \mu_p^{t-1}(f_p)) \tag{18}$$

The data update at iteration $t$ is:

$$\mu_q^t(f_q) = \begin{cases} \arg\min_d DSI(p, d) & \text{if } W(q) = \emptyset \\ \arg\min_{\mu_{p\to q}^t(f_q), p \in W(q)} (DSI(p, \mu_{p\to q}^t(f_q))) & \text{otherwise} \end{cases} \tag{19}$$

where $W(q) = \{p \in N(q) | a_{p\to q}^t(f_q) = f_q\}$.

The discontinuity cost is given by:

$$V(f_p, f_q) = \begin{cases} 0 & \text{if } f_p = f_q \\ \tau_v & \text{otherwise} \end{cases} \tag{20}$$

where $\tau_v = 40$ in our experiments. Figure 5 depicts results of the BP-FCT method in comparison with BP-FH. Here we see a case where our approach does not improve upon the full BP-FH [4] formulation. The main reason for the poorer performance is the discrete nature of the cost function. There is no smooth transition in costs from one disparity to a neighboring one with respect to difference in absolute values. This makes it difficult to define an updating function for the data estimation when we work with four colors. Our intuition is to pick the best disparity from the same color segment and in the absence of connections

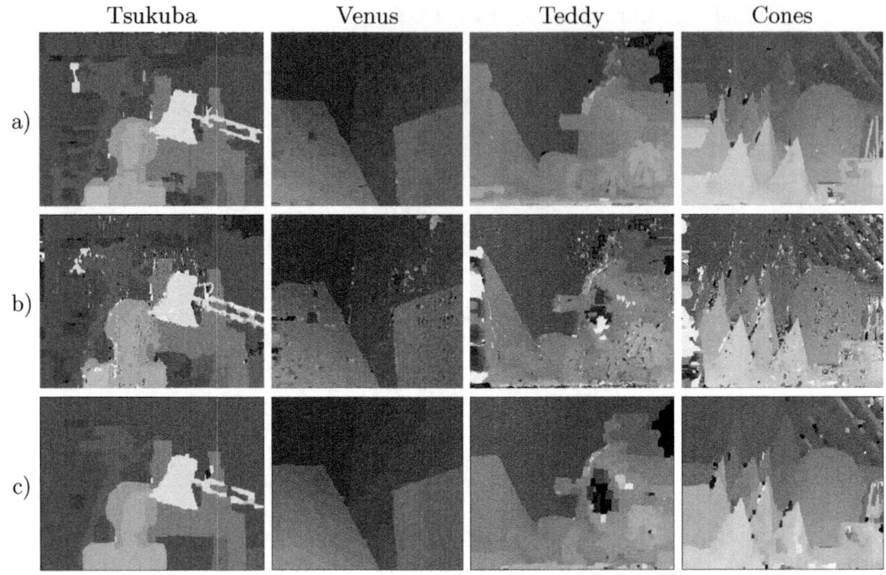


**Fig. 5.** Stereo matching results. *a*) Our implementation of [10] where for image segmentation we use our FBP-FCT, *b*) BP-FCT, *c*) BP-FH from [4].

to neighbors with the same color, to return to the best observed disparity (see Eq. (19)). The drawback is that our proposed method needs more iterations per level to achieve a similar performance to BP-FH. In our experiments, it takes between five and ten times more iterations than BP-FH to achieve similar performance, however this is not directly seen in the running time for large inputs, since our method has $O(n|C|^2T)$ complexity and BP-FH $O(n|L|T)$. Starting from the case where $|L| > 4|C|^2$ our proposed method (BP-FCT) is similar or considerably faster ($|L| \gg 4|C|^2$).

For stereo matching based on image segmentation, our proposed segmentation method (FBP-FCT) can be integrated as a fast oversegmentation step. Figure 5 depicts results where we used our segmentation and our implementation of the pipeline from [10]. This is a *winner-take-all* method that combines fast aggregation of costs in a window around each pixel with costs from the segment support to which the pixel belongs. According to the Middlebury benchmark [7], this implementation ranks $76^{th}$ out of 88 current methods.

### 4.4 Optical Flow

In motion flow estimation, the labels correspond to different displacement vectors. While in stereo matching the disparities were evaluated along the scanline, here the displaced/corresponding pixels are to be found in a surrounding window in the paired images. Thus, the set of labels goes quadratic when compared with stereo. We are using the cost functions as defined for stereo matching in the BP-FCT framework (see Section 4.3), where $d$ is a mapping for displacements.

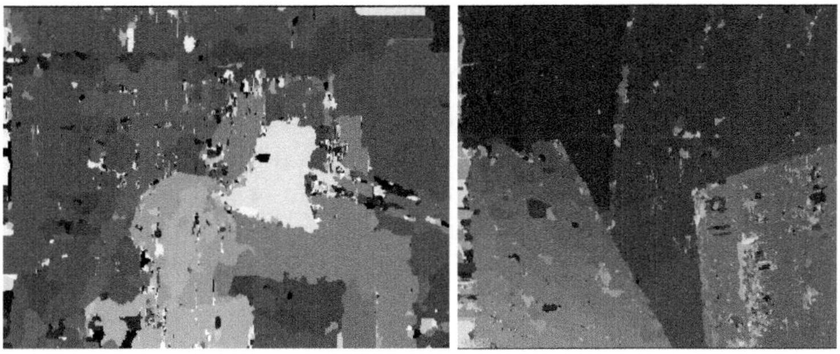

**Fig. 6.** Optical flow results for the *Tsukuba* and *Venus* image pairs

For a pixel $p = (i, j)$ whose intensity is $I_l(i, j)$ in the left image, the cost for a displacement of $d(i, j) = (d_i, d_j)$ is:

$$DSI(p, d) = s \min(\|I_l(i, j) - I_r(i - d_i, j - d_j)\|^2, \tau_s) \tag{21}$$

where $s = 0.1$ and $\tau_s = 10000$. The other cost functions are given by Eqs (18), (19), and (20). We keep the same values for parameters as in the stereo case, Section 4.3.

Note that increasing the set/space of displacement values will not increase the computational time of our Four-Color Theorem based BP approach, since the set is decoupled from the working set of labels, the four colors. However, the *DSI* still needs to be computed for having good initial estimates. In our case, the optical flow case takes as much running time as the stereo case (see Section 4.3) for the same size images, when the *DSI* computation is neglected. Figure 6 shows the results on standard stereo image pairs by using the optical flow formulation. For this we are considering as disparity the distance from the left pixel to the corresponding pixel in the right image. We see that the results are worse but very close to the ones obtained in the specific stereo formulation (Fig. 5). Increasing the number of displacements from 16(20) in the *Tsukuba*(*Venus*) pair in the stereo case to as much as 1024(1600) (about two orders magnitude bigger) causes a drop of only 4% in the quality of the results, but does not increase the computational time of the BP-FCT algorithm. However, the *DSI* computation time taken individually increases linearly with the number of possible displacements.

## 5 Conclusions

We have presented how the Four-Color Theorem based on the max-product belief propagation technique can be used in early computer vision for solving MRF problems where an energy is to be minimized. Our proposed methods yield comparable results with other methods, but improve either the speed for large images and/or large label sets (the case of image segmentation, stereo matching and optical flow), or both the performance and speed (the case of image restoration).

The Four Color Theorem principle is difficult to apply in cases where the label set is discrete and no natural order/relation between them can be inferred. This is the case for stereo matching and optical flow, where the disparity cost function takes discrete, unrelated values. This causes slower convergence, but is compensated by the low time complexity of the methods, independent of the number of labels. Thus, the proposed methods perform faster than the standard methods considered here, at least for large inputs.

**Acknowledgement.** This work was supported by the Flemish IBBT-iCOCOON project.

## References

1. Agarwal, S., Belongie, S.: On the non-optimality of four color coding of image partitions. In: IEEE International Conference on Image Processing (2002)
2. Boykov, Y., Veksler, O., Zabih, R.: Fast approximate energy minimization via graph cuts. IEEE Transactions on Pattern Analysis and Machine Intelligence 23, 1222–1239 (2001)
3. Felzenszwalb, P.F., Huttenlocher, D.P.: Efficient graph-based image segmentation. International Journal of Computer Vision 59 (2004)
4. Felzenszwalb, P.F., Huttenlocher, D.P.: Efficient belief propagation for early vision. International Journal of Computer Vision 70 (2006)
5. Martin, D., Fowlkes, C., Tal, D., Malik, J.: A database of human segmented natural images and its application to evaluating segmentation algorithms and measuring ecological statistics. In: International Conference on Computer Vision (2001)
6. Robertson, N., Sanders, D.P., Seymour, P.D., Thomas, R.: A new proof of the four colour theorem. Electron. Res. Announc. Amer. Math. Soc. 2, 17–25 (1996)
7. Scharstein, D., Szeliski, R.: A taxonomy and evaluation of dense two-frame stereo correspondence algorithms. International Journal of Computer Vision 47, 7–42 (2002)
8. Sun, J., Zheng, N.N., Shum, H.Y.: Stereo matching using belief propagation. IEEE Transactions on Pattern Analysis and Machine Intelligence 25, 787–800 (2003)
9. Tappen, M.F., Freeman, W.T.: Comparison of graph cuts with belief propagation for stereo, using identical MRF parameters. In: International Conference on Computer Vision (2003)
10. Tombari, F., Mattoccia, S., Di Stefano, L., Addimanda, E.: Near real-time stereo based on effective cost aggregation. In: International Conference on Pattern Recognition (2008)
11. Vese, L.A., Chan, T.F.: A multiphase level set framework for image segmentation using the Mumford and Shah model. International Journal of Computer Vision 50, 271–293 (2002)
12. Weiss, Y., Freeman, W.T.: On the optimality of solutions of the max-product belief propagation algorithm in arbitrary graphs. IEEE Transactions on Information Theory 47, 723–735 (2001)

# A Unified Approach to Segmentation and Categorization of Dynamic Textures

Avinash Ravichandran[1], Paolo Favaro[2], and René Vidal[1]

[1] Center for Imaging Science, Johns Hopkins University, Baltimore, MD, USA
[2] Dept. of Electrical Engineering and Physics, Heriot-Watt University, Edinburgh, UK

**Abstract.** Dynamic textures (DT) are videos of non-rigid dynamical objects, such as fire and waves, which constantly change their shape and appearance over time. Most of the prior work on DT analysis dealt with the classification of videos of a single DT or the segmentation of videos containing multiple DTs. In this paper, we consider the problem of joint segmentation and categorization of videos of multiple DTs under varying viewpoint, scale, and illumination conditions. We formulate this problem of assigning a class label to each pixel in the video as the minimization of an energy functional composed of two terms. The first term measures the cost of assigning a DT category to each pixel. For this purpose, we introduce a *bag of dynamic appearance features* (BoDAF) approach, in which we fit each video with a linear dynamical system (LDS) and use features extracted from the parameters of the LDS for classification. This BoDAF approach can be applied to the whole video, thus providing a framework for classifying videos of a single DT, or to image patches (superpixels), thus providing the cost of assigning a DT category to each pixel. The second term is a spatial regularization cost that encourages nearby pixels to have the same label. The minimization of this energy functional is carried out using the random walker algorithm. Experiments on existing databases of a single DT demonstrate the superiority of our BoDAF approach with respect to state-of-the art methods. To the best of our knowledge, the problem of joint segmentation and categorization of videos of multiple DTs has not been addressed before, hence there is no standard database to test our method. We therefore introduce a new database of videos annotated at the pixel level and evaluate our approach on this database with promising results.

## 1 Introduction

Dynamic textures (DT) are video sequences of non-rigid dynamical objects that constantly change their shape and appearance over time. Some examples of dynamic textures are video sequences of fire, smoke, crowds, and traffic. This class of video sequences is especially interesting since they are ubiquitous in our natural environment. For most practical applications involving DTs, such as surveillance, video retrieval, and video editing, one is interested in finding the types of DTs in the video sequences (categorization) and the regions that these DTs occupy (segmentation). For example, in the case of surveillance one wants to know

R. Kimmel, R. Klette, and A. Sugimoto (Eds.): ACCV 2010, Part I, LNCS 6492, pp. 425–438, 2011.

whether a fire has started or not, and if so, its location and volume. Specifically, given a video sequence that contains multiple instances of multiple DTs, in this paper we are interested in obtaining, for each pixel in the image, a label for its category. This task is referred to as *joint segmentation and categorization.*

Most of the existing works on DT analysis have addressed either categorization alone [1,2,3,4,5] or segmentation alone [6,7,8,9,10]. Existing categorization algorithms proceed by first modeling the entire video sequence using a single linear dynamical system (LDS). The parameters of these LDSs are then compared using several metrics such as subspace angles [6], KL-Divergence [7], Binet-Cauchy kernels [8], etc. The classification is then done by combining these metrics with classification techniques such as $k$-nearest neighbors ($k$-NN) or support vector machines (SVMs). One of the biggest shortcoming of these approaches is that the metrics used for comparing DTS are not designed to be invariant to changes in viewpoint, scale, etc. As a consequence, these methods perform poorly when a single class contains DTs with such variabilities. In [10], we addressed this issue by using multiple local models instead of a single global model for the video sequence. This bag-of-systems (BoS) approach was shown to be better at handling variability in imaging conditions of the data. Nonetheless, although there exists a large body of work addressing the problem of categorizing DTs, these methods suffer from several drawbacks. Firstly, all the approaches mentioned above assume that the video sequence contains a single DT. Hence, these methods cannot directly be used for video sequences with multiple DTs. Secondly, the choice of the metric used in these approaches requires that the training and testing data have the same number of pixels. This poses a challenge when one wants to compare local regions of a video sequence and adds additional overhead for normalizing all video sequences to the same spatial size.

On the other hand, existing DT segmentation algorithms model the video sequence locally at a patch level [3] or at a pixel level [1,2,4,5] as the output of an LDS. The parameters of these LDSs are used as cues for segmentation as opposed to using traditional features (e.g., intensity and texture). Such cues are then clustered with additional spatial regularization that ensures contiguous segmentations. The segmentation is then obtained by iterating between clustering the LDS parameters and extracting new LDS parameters. Such an iterative procedure is solved using an expectation maximization approach [3,4,5] or a level set based approach [1,2]. As a consequence, these algorithms require an initialization and can be sensitive to it.

While none of the above methods addresses the joint segmentation and categorization of DTs, one could leverage such techniques by first segmenting the video into multiple regions and then categorizing each region. In practice, however, this incurs in several problems. First, a manual initialization for the segmentation algorithm may not be available (especially for large databases). Second, since the segmentation results have been obtained without any relation to semantic categories, a single segment may contain multiple categories of interest. Finally, even if we were given an oracle segmentation algorithm, current categorization methods for DTs are all designed for classifying whole videos only.

In order to jointly segment and categorize video sequences containing multiple DTs, one would like an algorithm that requires no initialization. Clearly, the brute force alternative of searching through all possible segmentations is also unfeasible, especially when considering several categories. Furthermore, the algorithm should be able to explicitly handle intra class variability, i.e., changes in viewpoint, scale, illumination, etc.

**Paper Contributions.** In this paper, we propose a scheme that possesses all the above characteristics. The following are our paper contributions.

1. We introduce a new classification framework that provides either *global classifiers* for categorizing an entire video sequence or *local classifiers* for categorizing a region in the video sequence. This framework explicitly accounts for the invariance to changes in imaging conditions such as viewpoint, scale, etc. We propose a bag of dynamic appearance features (BoDAF) representation that uses SIFT features extracted from the parameters of the LDS. Unlike the case of image categorization, where the bag-of-features (BoF) approach results in one histogram per image, in our case we have multiple histograms per video. As we will show later, this is due to the fact that we are comparing the parameters of the LDSs using the BoDAF approach. Since there are several parameters for a single LDS, this gives rise to multiple histograms. This poses challenges in adopting the BoF approach directly. Hence, we propose a method to that extends the BoF approach to multiple histograms.
2. We then propose a joint categorization and segmentation algorithm for DTs. To the best of our knowledge, there does not exists any other work that handles both the segmentation and categorization simultaneously for DTs. Our joint segmentation and categorization approach proceeds by dividing the video sequences into superpixels by adapting existing techniques. In doing so, we assume that the superpixels do not change as a function of time. Each superpixel is then classified using local classifiers, which we will introduce in this paper. Finally, since objects are spatially contiguous, we introduce a spatial regularization framework. The joint segmentation and categorization problem is then posed as a multi-label optimization problem and solved using the random walker framework. The weights of the random walker algorithm are defined based on two terms: the output of the classifier for each superpixel and the similarity between adjacent superpixels.
3. Our final contribution is to provide a large database of dynamic textures annotated with accurate spatial labeling. We hope that this database can be used in the future as a benchmark for other algorithms.

We wish to point out that the approach described above is completely different from [10] where we use a bag-of-systems approach to categorize video sequences of DTs. In our approach, we use a single global LDS and extract features from the parameters of this model as opposed to using multiple local LDSs to model a video sequence as in [10]. Additionally, we propose a new method to compare DTs, while [10] uses the traditional subspace angles as the metric of comparison.

# 2 Categorization of Videos of a Single Dynamic Texture Using a Bag of Dynamic Appearance Features

In this section, we review the linear dynamical system (LDS) model that we use for video categorization. Then, we propose a classification scheme which is based on features from the LDS parameters. As we will see in §3, these classifiers will be the building blocks for our joint categorization and recognition framework.

## 2.1 Dynamic Texture Model

We denote a video sequence of a DT as $V = \{I(t)\}_{t=1}^{F}$, where $F$ is the number of frames. This video sequence is modeled as the output of a LDS given by

$$z(t+1) = Az(t) + Bv(t) \quad \text{and} \quad I(t) = C^0 + Cz(t) + w(t), \tag{1}$$

where $z(t) \in \mathbb{R}^n$ is the hidden state at time $t$, $A \in \mathbb{R}^{n \times n}$ is the matrix representing the dynamics of the hidden state, $B \in \mathbb{R}^{n \times m}$ is the input-to-state matrix, $C \in \mathbb{R}^{p \times n}$ maps the hidden state to the output of the system, $C^0$ represents the mean image of the video sequence and $w(t)$ and $v(t)$ are the zero-mean measurement noise and process noise, respectively. The order of the system is given by $n$, and the number of pixels in each frame of the sequence is given by $p$ . Notice that the $i^{th}$ column of $C$, $C^i$ is the vectorization of an image with $p$ pixels. Hence, the matrix $C$ can be also considered as a basis for the video frames consisting of $n$ images. These images along with the mean image are called *dynamic appearance images* (DAI) i.e., $\mathcal{C} = \{C^0, C^1, \dots C^n\}$ .

The parameters of the LDS that model a given video sequence can be identified using a suboptimal approach as outlined in [11]. However, the identified system parameters are unique only up to a change of basis. This basis ambiguity can be overcome by using a canonical form. In this paper, we use the Jordan Canonical form (JCF) outlined in [12]. In what follows, we show how we compare video sequences using the parameters of their corresponding LDS.

## 2.2 Representation and Comparison of LDSs

We assume that we are given $N$ video sequences $\{V_j\}_{j=1}^{N}$, each of which belongs to one of $L$ classes. Most categorization algorithms for DTs rely on analyzing the parameters of the LDS that model the video sequences. Traditional methods [6,7,8] compare the parameters $(A_j, C_j)$ of the LDS by using several existing metrics as outlined in §1. However, one major drawback of such metrics is that they do not necessarily capture any semantic information and hence they might not be meaningful in the context of categorization. For instance, two videos from two different classes might be judged similar because their parameters are close in the chosen metric. However, two videos from the same class under different viewpoint, scale, and illumination conditions might be judged different. Previous methods [10] have addressed the latter problem by using multiple local models and by assuming that viewpoint changes are negligible at small scales. In this paper we follow an alternative approach. We show that variability in viewpoint, scale, etc., can be accounted for with a mix of global and local representations.

In our approach we extract features at each pixel of each DAI. To achieve invariance to scale, orientation, and partially to illumination changes, we consider SIFT features [13]. When extracting such features we do not use any interest point detector to prune them, because we do not know in advance which regions are discriminative. While these features account for the local characteristics of a DT, we also need to capture the global statistics of the DT and provide a common representation for comparing different DTs. One such representation is the bag-of-features (BoF) approach [14,15]. In a BoF framework, features extracted from an image are first quantized using a *visual codebook* and then used to compute a histogram of codeword occurrences. Since the histograms are assumed to follow a characteristic distribution for each class, they are used as features for classification. This approach is very popular for object recognition due to its simplicity and fairly good performance.

In our case, rather than comparing two images, we need to compare two $C$ matrices, which are two sets of images. This introduces several challenges in adopting the BoF scheme, especially in the formation of the codebook and the quantization of features. The simplest way to form a codebook is to cluster all the features extracted from all the DAIs. This would result in a single codebook for all the DAIs. However, each of the DAIs captures a different piece of information. The mean $C^0$, for example, contains the low frequency texture information. In contrast, as the order $n$ of the LDS increases, the corresponding DAI, $C^n$, describes higher frequency details of the textures. Hence, a single codebook would dilute the differences between the details captured by each DAI. To prevent this from happening, we build separate codebooks for each of the DAIs. We use $K$-means to cluster the features from a DAI to form the codebook for that DAI.

Let the codebooks have size $K$ for each DAI. We represent each DAI using histograms. Such histograms can be calculated in several ways from the features in the DAI and the codebook. The most common choice is called the *term frequency* (TF) [14]. This approach essentially counts the number of codewords that occur in each DAI. Another approach is called the *term frequency - inverse document frequency* (TF-IDF) [15]. This approach discounts the codewords that occur in all the images and concentrates on the ones that occur less frequently. We refer the reader to [14,15] for details about these approaches. Thus, for every video sequence $V_j$ and its associated set of appearance images $\mathcal{C}_j = \{C_j^i\}_{i=0}^n$, we represent the appearance parameters using the set of histograms $\mathcal{H}_j = \{h_j^i \in \mathbb{R}_+^K\}_{i=0}^n$. Therefore, instead of $(A_j, \mathcal{C}_j)$, which was traditionally used for categorization, we now have $(A_j, \mathcal{H}_j)$.

### 2.3 Classifiers

The first step towards building a classifier for DTs is to define a distance to compare the representations $M_j = (A_j, \mathcal{H}_j)$ of each pair of video sequences. Among the several existing methods for comparing histograms, popular choices include the $\chi^2$ distance and the square root distance, which are given by

$$d_{\chi^2}(h_1,h_2) = \frac{1}{2}\sum_{k=1}^{K}\frac{|h_{1k}-h_{2k}|^2}{h_{1k}+h_{2k}} \quad \text{and} \quad d_{\surd}(h_1,h_2) = \cos^{-1}(\sum_{i=1}^{K}\sqrt{h_{1k}h_{2k}}), \quad (2)$$

respectively, where $h_{jk}$ denotes the $k^{th}$ element of the histogram vector $h_j \in \mathbb{R}^K$. As for comparing the $A$ matrices, since they are already in their canonical form, one could use any preferred choice of the matrix norm. Also, one can form a kernel from these distances by using a radial basis function (RBF). Such a kernel is given by $\mathcal{K}(x,y) = e^{-\gamma d(x,y)}$, where $\gamma$ is a free parameter.

In order to build a single classifier for multiple histograms, we first explored a multiple kernel learning (MKL) framework [16]. In this framework, one is given multiple kernels $\{\mathcal{K}_i\}_{i=1}^{n_b}$, which are referred to as basis kernels. These kernels can be, for example, the kernels between each DAI histogram or the product of the kernels between all the DAI histograms, or the kernel between the $A$ parameters. Given these kernels, a single kernel is built as $\mathcal{K} = \sum_{k=1}^{n_b} w_k\mathcal{K}_k$, by choosing a set of weights such that $\sum w_k = 1$. The weights can be automatically learned from the data by training an SVM classifier. We refer the reader to [16] for more details on this method.

When we used the MKL algorithm to obtain the weights, we noticed that, in practice, the weight corresponding to the kernel based on the matrix $A$ and the weight corresponding to the kernels based on individual DAI histograms were nearly zero, while the weight corresponding to the product kernel was nearly 1. For this reason we decided to disregard the kernel on the $A$ matrix and the kernels on individual DAI histograms, and adopted the product kernel on the DAI histograms. One can show that for RBF kernels, using a product kernel is equivalent to concatenating the histograms into one single vector $H_j \in \mathbb{R}^{(n+1)K}$. Hence, we choose the concatenated version of the histogram as our representation for $\mathcal{H}_j$.

As for the choice of the classifiers for the histograms, we can use discriminative classifiers such as $k$-NNs or SVMs for the classification. The kernel of choice for SVMs is the RBF kernel, which we introduced earlier on. Alternatively, one could use a generative approach for the classifier. One such classification algorithm is the naïve Bayes classifier [14]. This algorithm works directly with the codeword occurrence as opposed to using the histogram representation. We refer the readers to [14] for details on implementing such a classifier.

## 3 Joint Segmentation and Categorization of Videos of Multiple Dynamic Textures

The classifiers described so far have been defined as *global classifiers* for the purpose of categorizing videos containing a single DT. However, in this paper we are interested in categorizing video sequences containing multiple DTs. In this section, we formulate a solution to this problem under the simplifying assumption that the class of a pixel does not change as a function of time, so that the segmentation of the video into multiple DTs is common across all the frames.

When this segmentation is known, one can easily adapt the method described in §2 so that it can be applied to classifying each segmented region. More specifically, we can fit a single LDS to the video sequence as before, and then form multiple concatenated histograms of DAIs, one per region, by restricting the feature counts to the features within each region. Afterwards, each region can be classified independently based on its histograms. In general, however, the segmentation is not known and one needs to address both problems simultaneously.

In what follows, we describe the proposed framework for joint segmentation and categorization of DTs, which consists of three main ingredients: dividing the video sequence into superpixels, constructing a local classifier for each superpixels, and finding the segmentation and categorization of the video by minimizing an energy functional defined on the graph of superpixels.

### 3.1 Superpixels

The first step in our approach is to divide each video into superpixels. This step is required as the decision of whether a single pixel belongs to one class or another cannot be made based on the information from a single pixel. This calls for exploring supports over larger regions. However, searching through the space of all admissible supports drastically increases the complexity of the algorithm. One solution is to use a fixed window around each pixel. However, a fixed window could contain pixels from two different DTs. A better solution is to choose regions that are unlikely to contain two different DTs. This issue can be resolved by using superpixels, i.e., image patches, that are naturally defined by the data rather than by the user. Also, since superpixels contain pixels that are similar, we can categorize the whole superpixel as opposed to classifying each pixel contained in it. This reduces the computational complexity of the method.

Our algorithm for computing superpixels builds a feature vector at each pixel by concatenating the values of the normalized DAIs with the 2D pixel location. This results in an $n+3$ dimensional vector, where $n+1$ components are from the DAIs. We then use the quick shift algorithm [17] to extract the superpixels from such features.

### 3.2 Classifiers

Given a training video sequence, we first obtain its LDS parameters. We then cluster the dense SIFT features extracted from the DAIs of all the training videos in order to form the codebooks. We then divide the DAIs into superpixels as describe above. By restricting the feature counts to the features within each superpixel, we can build *local histograms*. All the local histograms from all the DAIs of all the training videos are then used to train *local classifiers*.

Given a test video sequence, we compute its LDS parameters and DAIs as before. We then quantize the features within each superpixel to form a local histogram. We then use the local classifiers to assign, to each pixel inside a superpixel, the cost of assigning it to each of the DT categories. This procedure already provides us with a segmentation and a categorization of the video. However, this segmentation may not be smooth, since each superpixel is classified independently. Therefore, we find the joint segmentation and categorization of the video by minimizing an energy functional, which is the sum of the above classification cost plus a regularization cost, as described next.

### 3.3 Joint Segmentation and Categorization

We solve the joint segmentation and categorization problem using a graph theoretic framework. We define a weighted graph $G = (\mathcal{V}, \mathcal{E})$, where $\mathcal{V}$ denotes the set of nodes and $\mathcal{E}$ denotes the set of edges. The set of nodes $\mathcal{V} = \{s_i\}_{i=1}^{N_s} \cup \{g_l\}_{i=1}^{L}$ is given by the $N_s$ superpixels obtained from the DAIs, $\{s_j\}_{j=1}^{N_s}$, and by $L$ auxiliary nodes, $\{g_l\}_{l=1}^{L}$, one per each DT category. The set of edges $\mathcal{E} = \mathcal{E}_s \cup \mathcal{E}_c$ is given by the edges in $\mathcal{E}_s = \{(i,j) | s_i \sim s_j\}$, which connect neighboring superpixels $s_i \sim s_j$, and by the edges in $\mathcal{E}_c = \{(i,l) | s_i \sim g_l\}$, which connect each $s_i$ to every $g_l$.

We also associate a weight to every edge, which we denote by $w_{ij}$ if $(i,j) \in \mathcal{E}_s$ and by $\tilde{w}_{ij}$ if $(i,j) \in \mathcal{E}_c$. We set $\tilde{w}_{il}$ to be the probability that superpixel $s_i$ belongs to class $l$, $P(c_l|s_j)$, as obtained from the trained classifiers. We restrict our attention to the SVM and the naïve Bayes classifiers for this case. For weights between $s_i$ and $s_j$, we set $w_{ij} = \mathcal{K}_\chi(H_{s_i}, H_{s_j})$, where $\mathcal{K}_\chi$ is defined as

$$\mathcal{K}_\chi(H_{s_i}, H_{s_j}) = 2 \sum_{m=1}^{K(n+1)} \frac{H_{s_i}^m \cdot H_{s_j}^m}{H_{s_i}^m + H_{s_j}^m}, \tag{3}$$

where $H_{s_i}$ is the concatenated histogram from superpixel $s_i$. This is a kernel derived from the $\chi^2$ distance. This weight measures the similarity of the concatenated histograms between the neighboring superpixels in order to enforce the spatial continuity. In order to have the flexibility to control the influence of the prior and the smoothing terms, we introduce two other parameters $\alpha_1 > 0$ and $\alpha_2 > 0$ and redefine the weights as $\tilde{w}_{ij} = \alpha_1 P(c_j|s_j)$ and $w_{ij} = \alpha_2 \mathcal{K}_\chi(H_{s_i}, H_{s_j})$. Varying these parameters will determine the contribution of these weights.

In order to solve the joint segmentation and categorization problem, we associate with each node $s_i$, a vector $\mathbf{x}_i = \left[x_i^1, \cdots, x_i^L\right]$, where $x_i^l$ is the probability that $s_i$ belongs to class $l$. For the auxiliary nodes $g_l$, this probability is denoted by $\tilde{x}_l^i = \delta(l-i)$, where $\delta(\cdot)$ is the discrete Dirac delta. A label $y_i$ for each node is then obtained via $y_i = \arg\max_{y \in \{1,\dots,L\}} x_i^y$. These labels can be estimated using several techniques such as graph cuts [18] and the random walker [19]. In this work, we propose to solve this problem using a random walker (RW) formulation as this formulation generalizes more easily to the multi-label case.

In the RW framework, the nodes $g_l$ are considered as terminal nodes. A RW starting at the node $s_i$ can move to one of its neighbors $s_j$ with a probability $w_{ij} / \sum_{k \in \mathcal{N}_i} w_{ik}$, where $\mathcal{N}_i$ is the set of neighbors of $s_i$. We are interested in the probability that a RW starting from $s_i$ reaches first the node $g_l$. This probability is precisely the value $x_i^l$. It can be shown that the value $x_i^l$ can be obtained by minimizing the following cost function

$$E(\mathbf{x}) = \sum_{l=1}^{L} \left[ \sum_{(i,j) \in \mathcal{E}_c} \tilde{w}_{il} (x_i^l - \tilde{x}_j^l)^2 + \sum_{(i,j) \in \mathcal{E}_s} w_{ij} (x_i^l - x_j^l)^2 \right], \text{ s.t. } \forall i \sum_{l=1}^{L} x_i^l = 1. \tag{4}$$

As shown in [19], the minimizer of this cost function can be obtained by solving $L-1$ linear systems. Since by construction $w_{ij} \geq 0$ and $\tilde{w}_{il} \geq 0$, it can be shown that our cost function is convex and has a unique global minimum. We refer the reader to [19] for details on how this function can be minimized.

# 4 Experiments

This section presents two sets of experiments, which evaluate the proposed framework on the categorization of video sequences of a single DT, and on the joint segmentation and categorization of video sequences of multiple DTs. For the first task we use existing datasets and for the second one we propose a new dataset.

**UCLA-8 [10]:** This data set is a reorganized version of the UCLA-50 dataset [6] and consists of 88 video sequences spanning 8 classes. Here each class contains the same semantic texture (e.g., water), but under different conditions such as viewpoint and scale. Although this dataset consists of fewer classes than UCLA-50, this dataset is much more challenging than the UCLA-50.

**Traffic Database [7]:** This dataset consists of 254 videos spanning 3 classes (heavy, medium and light traffic) taken from a single fixed camera. The videos contain traffic conditions on a highway at different times of the day and under different weather conditions. However, the clips share the same viewpoint and area on the highway. We use this dataset to highlight the fact that our classification algorithm works even when the variations between the classes are subtle with respect to the appearance and large with respect to the temporal process.

**Dyntex [20]:** Existing datasets, i.e., UCLA-8 and Traffic, are not well-suited for testing our joint segmentation and categorization algorithm. This is because most of the video sequences in UCLA-8 contain only a single texture and seldom any background. As for the traffic dataset, since every video sequence shares the same view point, all video sequences have almost the same annotation. However, the Dyntex dataset is well-suited for our purposes. This dataset consists of 656 video sequences, without annotation either at the pixel level or at the video level. Even though there is a large amount of video sequences, there are several issues with adopting the dataset for testing. The first one is that this dataset contains video sequences that are taken by a moving camera. Such video sequences are beyond the scope of this paper. It also contains video sequences that are not DTs. Moreover, several of the classes in this dataset have very few samples, e.g., 2-3 video sequences, which is too little to be statistically significant. Hence, we used the 3 largest classes we could obtain from this dataset: waves, flags and fountains. This gave us 119 video sequences, which we manually annotated at the pixel level. Sample annotation from the dataset can be found in Figure 1.

## 4.1 Whole Video Categorization

We now show experimental results on the UCLA-8 and traffic databases, which contain video sequences of a single DT. As mentioned earlier, we used SIFT

**Fig. 1.** Sample annotations for the subset of the Dyntex dataset we use for the joint segmentation and categorization. (Color convention: Black for background, red for flags, green for fountain and blue for waves).

features extracted from the DAIs. We extracted these features on a dense regular grid with a patch size of $12\times12$ as opposed to using interest points. We also tested our algorithm on descriptors extracted using interest points and we observed that the performance was significantly lower than with dense SIFT features.

We used both the TF and the TF-IDF representations to obtain the histogram of each DAI from its features and the codebook. However, our experimental results showed that the TF representation performs better than the TF-IDF representation. Hence, all results presented in this section are obtained using the TF approach. For the classifiers, we used $k$-NN with $k=1$ and $k=3$ using $d_{\chi^2}$ and $d_{\surd}$ as the distance metric. We denote these classifiers as 1NN- TF-$\chi^2$, 1NN-TF- $\surd$, 3NN-TF-$\chi^2$ and 3NN-TF-$\surd$. We also used the SVM and naïve Bayes (NB) classifier. The kernel used for the SVM was the RBF kernel based on the $\chi^2$ distance. The free parameters for the SVM classifier were obtained using a 2-fold cross validation on the training dataset.

**UCLA-8:** We split the data into 50% training and 50% testing as in [10]. We modeled the video sequences using LDSs of order $n=3$ and did not optimize this parameter. All the results reported here are averaged over 20 runs of our algorithm. This is because in every run of our algorithms the codewords can change. This is attributed to the convergence properties of the $K$-Means algorithms.

Table 1 shows the results of the different classifiers on UCLA-8. This table shows the maximum categorization performance over all choices of cluster sizes. From this table we observe that the NB algorithm outperforms all the other classifiers. We compared our method with the bag-of-systems (BoS) approach outlined in [10] and the method of using a single global model and the Martin distance as outlined in [6]. The performance of our method is at 100% as opposed to 80% using the BoS and 52% using a global LDS and the Martin distance.

Also shown in Table 1 is the performance of just using an individual vector of the DAI, i.e., we use only one $h_j^i$ as opposed to using the concatenated $H_j$. Table 1 shows these results under the columns titled $C^i$, $i=0\ldots3$. From this table we see that, although there exists a basis that performs well, its performance does not match that of our approach. This shows that our approach is able to successfully combine the information from all the DAIs. One could argue that since these results are based on the DAI, the dataset could be classified only using

**Table 1.** Categorization performance on UCLA-8 using different approaches

| Classifier | $C^0$ | $C^1$ | $C^2$ | $C^3$ | Our Approach | Concat Frames |
|---|---|---|---|---|---|---|
| 1NN-TF-$\sqrt{}$ | 70 | 82 | 90 | 85 | **97** | 80 |
| 1NN-TF-$\chi^2$ | 70 | 81 | 90 | 83 | **94** | 79 |
| 3NN-TF-$\sqrt{}$ | 71 | 77 | 93 | 83 | **97** | 80 |
| 3NN-TF-$\chi^2$ | 71 | 77 | 91 | 82 | **93** | 81 |
| NB | 72 | 80 | 91 | 88 | **100** | 82 |
| SVM | 64 | 76 | 90 | 78 | **95** | 80 |

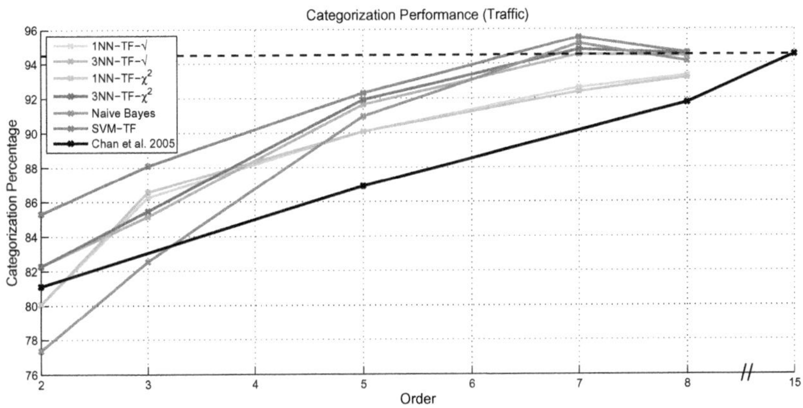


**Fig. 2.** Categorization performance of different classifiers on the Traffic Dataset of [7]

the texture information, i.e., using the frames of the video sequence themselves. To address this issue, we substituted $n+1$ video frames in lieu of the $n+1$ DAIs in our classification framework. These frames were chosen at regular time intervals from the video sequence. The results of using such an approach are summarized in Table 1 under the column Concat Frames. We see that the performance of this approach is about 80% in all the cases. This accuracy is comparable to that of the BoS approach. This shows that comparing the parameters as opposed to comparing only the frames of the video sequences increases the categorization performance.

**Traffic Dataset:** For the Traffic dataset, we used the same experimental conditions as in [7], where the data was split into 4 different sets of 75% training and 25% testing. The maximum performance reported in [7] for this dataset is 94.5% for an order of $n = 15$. We report our results averaged over 10 runs of our algorithm using a codebook size of $K = 42$ for each of the DAI. Figure 2 shows the performance of our algorithm for different classifiers and different orders of the LDSs. It can be seen that for this dataset, we obtain a categorization performance of 95.6% for $n = 7$. Although the increase with respect to the state of the art is not as large as in the case of the UCLA-8 dataset, we notice that:

**Table 2.** Performance of Joint Segmentation and Categorization Algorithm

| Method | Percentage Correctly Classified | | | | | Intersection/Union metric | | | | | APV |
|---|---|---|---|---|---|---|---|---|---|---|---|
| | Background | Waves | Flags | Fountain | Average | Background | Waves | Flags | Fountain | Average | |
| NB+RW | 41.9 | **99.6** | **77.1** | **71.3** | **72.5** | 0.41 | 0.81 | 0.55 | **0.30** | 0.51 | 70.4 |
| NB | 47.5 | 97.2 | 70.8 | 59.5 | 68.8 | 0.46 | 0.79 | 0.48 | 0.25 | 0.50 | 69.6 |
| SVM+RW | **87.5** | 97.1 | 66.9 | 30.2 | 70.4 | **0.69** | **0.87** | **0.60** | 0.25 | **0.60** | **81.7** |
| SVM | 80.6 | 94.2 | 62.0 | 41.9 | 69.7 | 0.66 | 0.82 | 0.50 | 0.29 | 0.57 | 78.4 |

(a) For any given order, our classifiers outperform [7]; (b) Our maximum classification performance is obtained at almost half the order as that of [7]. This clearly exhibits the fact that our classification framework is more powerful than traditional methods of comparison and works well across different datasets.

### 4.2 Joint Segmentation and Categorization

We first describe some of our implementation details. We use approximately 50% of the data for training and testing. We identify a LDS of order $n = 3$ for each video sequence. We use the quick-shift algorithm for extracting superpixels. This algorithm has three parameters: $\lambda$ which controls the tradeoff between the spatial importance and the importance of the DAI, $\sigma$ which is the scale of the kernel density estimation and $\tau$ which determines the maximum distance between the features of the members in the same region. We use the following parameters: $\lambda = 0.25$, $\tau = 3$, $\sigma = 3$ for all our experiments. These values were chosen so that there were no extremely small superpixels. Using the manual annotations, we assign to each superpixel the label of the majority of its pixels. Concatenated histograms are then extracted from the DAIs by restricting the codeword bin counts to the support described by each superpixel. We then train both as SVM and a naïve Bayes classifier using these histograms. We used $K = 80$ as the size for the codebook for each DAI and $\alpha_1/\alpha_2 = 3/4$ for the RW) formulation. We report our results before and after applying the RW.

In order to compare our results with the ground truth segmentation, we use 3 metrics. For each class, we report the percentage of pixels across all the videos that are correctly assigned this class label. We also report our performance using the intersection/union metric. This metric is defined as the number of correctly labeled pixels of that class, divided by the number of pixels labeled with that class in either the ground truth labeling or the inferred labeling. The last metric is the average of the percentage of pixels that are correctly classified per video sequences (APV). The results of our algorithm are shown in Table 2. From this table we observe that the NB classifier does not perform well on the background class while the SVM classifier does not perform well on the fountain class.

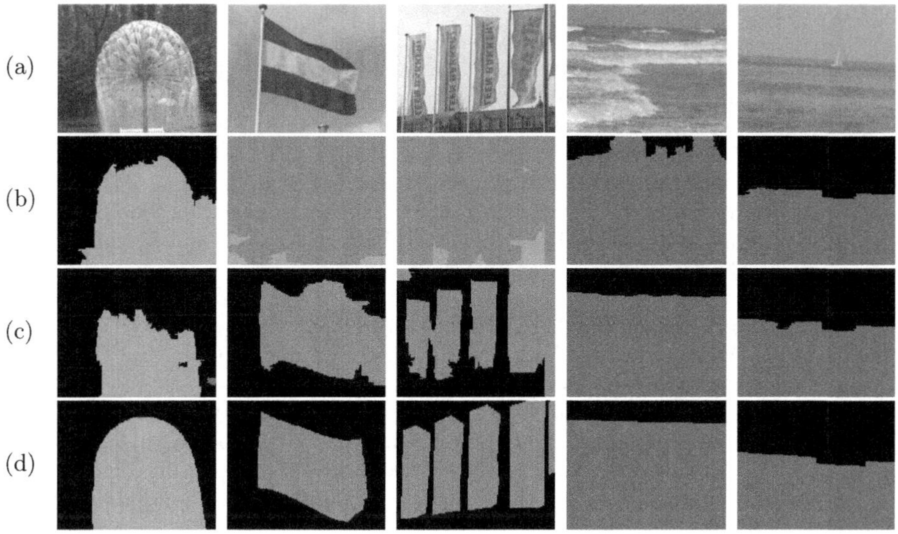


**Fig. 3.** Sample segmentations. (a) sample frame from the video sequence; (b) segmentation using Bayes + RW; (c) segmentation using SVM + RW; (d) ground truth segmentation (annotation).

The SVM classifier overall is better over two metrics when compared to the NB classifier. Figure 3 shows a few sample segmentations from our dataset.

## 5 Conclusions

In this paper, we have proposed a method to address the problem of joint segmentation and categorization of DTs. To the best of our knowledge, this problem has not been addressed before. Instead of first segmenting the video sequence and then categorizing each segment, we provided a unified framework that can simultaneously perform both tasks. A key component in this framework is the new classification scheme that we have proposed. We have shown that this scheme performs very well for classifying whole videos and can also be used locally for our joint segmentation and categorization framework. Our results showed that by using this classifier, we outperformed existing methods for DT categorization. Since there are no available datasets to test our framework, we provided what is, to the best of our knowledge, the first dynamic texture database with annotations at the pixel level. We hope that this database can be used as a benchmark for future research.

**Acknowledgements.** The authors would like to thank Daniele Perone for his help with the annotations and Lucas Theis for his help with the experiments. This work was partially supported by grants ONR N00014-09-1-0084 and ONR N00014-09-1-1067.

## References

1. Doretto, G., Cremers, D., Favaro, P., Soatto, S.: Dynamic texture segmentation. In: IEEE Int. Conf. on Computer Vision, pp. 44–49 (2003)
2. Ghoreyshi, A., Vidal, R.: Segmenting dynamic textures with ising descriptors, ARX models and level sets. In: Vidal, R., Heyden, A., Ma, Y. (eds.) WDV 2005/2006. LNCS, vol. 4358, pp. 127–141. Springer, Heidelberg (2007)
3. Chan, A., Vasconcelos, N.: Modeling, clustering, and segmenting video with mixtures of dynamic textures. IEEE Trans. on Pattern Analysis and Machine Intelligence 30, 909–926 (2008)
4. Chan, A., Vasconcelos, N.: Layered dynamic textures. IEEE Trans. on Pattern Analysis and Machine Intelligence, 1862–1879 (2009)
5. Chan, A., Vasconcelos, N.: Variational layered dynamic textures. In: IEEE Conf. on Computer Vision and Pattern Recognition (2009)
6. Saisan, P., Doretto, G., Wu, Y.N., Soatto, S.: Dynamic texture recognition. In: IEEE Conf. on Computer Vision and Pattern Recognition, vol. II, pp. 58–63 (2001)
7. Chan, A., Vasconcelos, N.: Probabilistic kernels for the classification of autoregressive visual processes. In: IEEE Conf. on Computer Vision and Pattern Recognition, pp. 846–851 (2005)
8. Vishwanathan, S., Smola, A., Vidal, R.: Binet-Cauchy kernels on dynamical systems and its application to the analysis of dynamic scenes. Int. Journal of Computer Vision 73, 95–119 (2007)
9. Vidal, R., Favaro, P.: Dynamicboost: Boosting time series generated by dynamical systems. In: IEEE Int. Conf. on Computer Vision (2007)
10. Ravichandran, A., Chaudhry, R., Vidal, R.: View-invariant dynamic texture recognition using a bag of dynamical systems. In: IEEE Conf. on Computer Vision and Pattern Recognition (2009)
11. Doretto, G., Chiuso, A., Wu, Y., Soatto, S.: Dynamic textures. Int. Journal of Computer Vision 51, 91–109 (2003)
12. Ravichandran, A., Vidal, R.: Video registration using dynamic textures. In: Forsyth, D., Torr, P., Zisserman, A. (eds.) ECCV 2008, Part II. LNCS, vol. 5303, pp. 514–526. Springer, Heidelberg (2008)
13. Lowe, D.: Distinctive image features from scale-invariant keypoints. Int. Journal of Computer Vision 20, 91–110 (2003)
14. Dance, C., Willamowski, J., Fan, L., Bray, C., Csurka, G.: Visual categorization with bags of keypoints. In: European Conf. on Computer Vision (2004)
15. Sivic, J., Zisserman, A.: Video google: A text retrieval approach to object matching in videos. In: IEEE Int. Conf. on Computer Vision, pp. 1470–1477 (2003)
16. Rakotomamonjy, A., Bach, F., Canu, S., Grandvalet, Y.: Simplemkl. Journal of Machine Learning Research 9, 2491–2521 (2008)
17. Vedaldi, A., Soatto, S.: Quick shift and kernel methods for mode seeking. In: Forsyth, D., Torr, P., Zisserman, A. (eds.) ECCV 2008, Part IV. LNCS, vol. 5305, pp. 705–718. Springer, Heidelberg (2008)
18. Boykov, Y., Jolly, M.: Interactive graph cuts for optimal boundary & region segmentation of objects in N-D images. In: IEEE Int. Conf. on Computer Vision, pp. 105–112 (2001)
19. Grady, L.: Multilabel random walker image segmentation using prior models. In: IEEE Conf. on Computer Vision and Pattern Recognition, pp. 763–770 (2005)
20. Péteri, R., Huskies, M., Fazekas, S.: Dyntex: A comprehensive database of dynamic textures (Online Dynamic Texture Database)

# Learning Video Manifold for Segmenting Crowd Events and Abnormality Detection

Myo Thida[1,2], How-Lung Eng[1], Monekosso Dorothy[2], and Paolo Remagnino[2]

[1] Institute for Infocomm Research, Singapore
Tel.: +065-6408 2505
mthida@i2r.a-star.edu.sg
[2] Digital Image Research Centre
Faculty of Computing Information Systems and Mathematics
Kingston University, UK

**Abstract.** This paper addresses the problem of analyzing video events in crowded scenes. A novel manifold learning method is proposed to achieve visualization and modeling of video events in a low dimensional space. In the proposed approach, a video is considered as a trajectory of frames in a low-dimensional space. This low-dimensional representation of a video preserves the spatio-temporal property of a video as well as the characteristic of the video. Different tasks of video content analysis such as visualization, video event segmentation and abnormality detection are achieved by analyzing these video trajectories based on the Hausdorff distance similarity measure. We evaluate our proposed method on the state-of-the-art public data-sets containing different crowd events. Qualitative and quantitative results show the promising performance of the proposed method.

## 1 Introduction

Visual analysis of a crowded scene has recently became an active research area due to a rapid increase in the demand for public safety and security. Several attempts have been made to analyze videos of crowded scenes and provide a semantic interpretation of the monitored scene. This higher level interpretation of a scene can lead to several applications such as understanding crowd behaviours, and abnormality detection. In this paper, we propose a method to address the problem of abnormality detection in a crowded scene based on an analysis of spatial temporal motion information in a manifold space. This provides an advantage of visualizing and identifying different crowd events in a low dimensional space and detect abnormality. The main contributions of this paper are summarized as follows:

- We propose a framework to segment different events and detect abnormality in crowded scenes using a novel manifold learning method, called Spatio-Temporal Laplacian Eigenmap (ST-LE). The manifold learning method drives a compact trajectory description that reflects the characteristics of crowd behaviour.

R. Kimmel, R. Klette, and A. Sugimoto (Eds.): ACCV 2010, Part I, LNCS 6492, pp. 439–449, 2011.

– We propose a trajectory-based analysis method for understanding of video content in a crowded scene. We address two issues related to crowd dynamic analysis: segmenting a video into different crowd events and detection of abnormal events.
– We provide both qualitative and quantitative results to validate the proposed method. Performance evaluation is carried out recently published data-sets [1] and [2].
– The proposed method has several desirable properties: i) it provides a mean to visualize and identify different crowd events, ii) it provides a tool to break down a video sequence into different clusters where each cluster represents an event, and iii) it detects abnormal crowd behaviour.

This paper is organized as follows. Section 2 reviews related work on video analysis of crowded scene and manifold learning methods in video content analysis. Section 3 explains our proposed method, where each sub-section provides a detail explanation on representation of video inputs, manifold learning and analysis of videos in the manifold space. Qualitative and quantitative experimental results are presented and discussed in Section 4 and conclusion is given in Section 5.

## 2 Related Work

### 2.1 Crowd Video Analysis

The existing methods on crowd video analysis can be divided into two main approaches. The first approach analyzes crowd behaviours based on individual movements. Hence, to understand crowd behaviours, it is necessary to detect individuals in a crowd, track these individuals from frame to frame and analyze trajectories to model normal crowd behaviours. Wang *et al.* [3] used statistical models of trajectories to distinguish between normal and abnormal crowd behaviours. Hu *et al.* [4] used trajectories of moving vehicles to model typical motion patterns of a traffic scene. The model later was used to detect abnormal activities and predict traffic paths. The use of neural networks for modelling motion patterns from trajectories was proposed by Johnson and David [5]. Later, the model of motion patterns was employed for traffic predictions. However, tracking individuals in a crowd remains a challenging and ill-posed problem.

To address this limitation, in recent years, researchers have proposed to analyze crowd behaviours based on the holistic properties of the scene. Instead of modelling trajectories of a crowd, the second approach mainly focus on building a crowd motion model using the instantaneous motions of the video, for example, optical flow of the whole frame. Andrade *et al.* [6,7,8,9] combined principle component analysis of optical flow vectors, Hidden Markov Models and spectral clustering for detecting crowd emergency events. Hu and Shah [10,11] modelled the motion patterns of a scene by merging optical flow vectors from all video frames based on a sink-seeking process which finds entries, exits and paths of moving particles. Recently, particle advection method becomes popular for crowd video analysis. In [12], an interaction force between pedestrians is estimated based on optical flow and particle advection method. The normal pattern of this force is later used to model the dynamics of a crowded scene and detect abnormal behaviours in crowds. Similarly, Wu et al. [13] used particle trajectories for modelling crowded scenes and abnormality detection. However, the research focus so far has been mainly on the tasks of segmenting and modelling crowd motion patterns in a scene. The lack of research effort on crowd events understanding is justified by the complexity of the problem. This suggests that the development of automated video event segmentation should be pursued for crowded scenes. Initial research reported in [14,15] motivates us to work on visualization and segmentation of crowd events. In this paper, we exploit temporal coherence between video frames and use the manifold learning algorithm for understanding of crowd events. The proposed manifold learning method provides us a tool to visualize and segment crowd events in a low-dimensional space.

### 2.2 Manifold Learning in Video Content Analysis

Manifold learning algorithms or non-linear dimensionality reduction methods have been used as an initial step for many computer vision tasks. These methods generate a low-dimensional space so that high dimensional data can be better visualized and processed efficiently in this low-dimensional space. One of the earliest applications of manifold learning algorithms is to visualize image sets and classify images based on the embedded coordinates. In this context, face recognition [16,17,18] and pose estimation [19] are popular applications where manifold learning has shown promising results. Most commonly, these applications are performed in three basic steps. First, a manifold learning algorithm is applied to a training set of images to find a low-dimensional embedding space. Second, the new image is projected into the manifold space spanned by the trained data set. Third, the new image is classified using a nearest neighbor classifier or a more complex classifier such as Bayesian or Support vector machine.

The remarkable achievements in visualization and classification of face image sets have led researchers to explore the use of manifold learning algorithms for semantic interpretation of temporal sequences. A few research work in this direction includes visualization of a video [14,15], facial expression recognition [20] and human activity recognition [21,22,23]. The work closely related to this paper is the visualization of videos [14,15]. The authors have proved that a video corresponds to a trajectory in an embedded space and different appearances on manifolds indicate different video events. However, both work [14,15] demonstrated their methods only on videos containing a single object.

## 3 Proposed Method

### 3.1 Representations of Video Frames

We propose an approach to represent each frame of a video using a Histogram of Optical Flow, which is defined as follows. We first compute optical flow over a $n$ by $n$ grid between two successive frames using the technique in [24]. The distribution of optical flow in each grid is then represented using a weighted histogram of B bins, where weight in each bin corresponds to the magnitude of optical flow in one particular direction. Then, a representation of an image is obtained by concatenating all histograms of grids into a long vector. Mathematically, it is defined as:

$$\mathbf{x} = [\mathbf{h}_k], \quad k = \{1, 2, 3, \ldots, K\} \tag{1}$$

where $k$ is an index of each grid and $K$ is total number of grids in each frame. $\mathbf{h}_k$ is a weighted histogram of B bins for a particular grid, $k$.

### 3.2 Learning Video Trajectories: Using Spatio-temporal Laplacian Eigenmap (ST-LE)

Given a set of video frames, our objective here is to find a compact representation of the video. Recently, manifold learning algorithms such as Isomap[25] and Laplacian Eigenmaps(LE)[26] were used to represent a high-dimensional video data into a trajectory in the manifold space [14,15]. However, the traditional manifold learning algorithms such as Isomap and LE ignore the temporal coherence between frames which is a useful information for video data. In this paper, we propose a novel manifold learning algorithm, called Spatio-Temporal Laplacian Eigenmap (ST-LE) for representation of videos in the low-dimensional space.

First, we construct a weighted neighborhood graph to group the video frames based on spatio-temporal relations. In this graph, each data-point is connected to its nearest neighbors in a weighted manner, i.e., the weight of the edge connecting between a data-point and its first nearest neighbor is higher than the weight of the edge between the data-point and its second nearest neighbor. The weight of the edge connecting two video frames, $i$ and $j$ are computed using a kernel function defined as follows:

$$\omega_{ij} = \exp(-d_t \times d_s), \tag{2}$$

where $d_t$ gives the temporal relationship between two frames and $d_s$ is the dis-similarity measure between two frames in the feature space. The temporal distance, $d_t$ is considered based on the consecutive frames in the time-ordered video sequence and defined as:

$$d_t = \frac{\|t_i - t_j\|}{h}, \tag{3}$$

where $t_i$ and $t_j$ are temporal index of frames and $h$ is the width of the temporal window. Parameter $h$ depends on the temporal consistency of video frames (smaller temporal consistency means that the width of the temporal window should be smaller). In our experiments, the parameter $h$ is fixed to length of the video by 4.

The spatial distance between two frames is defined as the weighted summation of distance measure between corresponding grids. Mathematically, our dis-similarity measure is given as:

$$d_s(\mathbf{x}_i, \mathbf{x}_j) = \sum_{k=1}^{K} \alpha_k \times d(\mathbf{h}_k^i, \mathbf{h}_k^j), \tag{4}$$

where K is total number of grids in an image and $\alpha_k$ is the weight for each position. Parameter $\alpha$ depends on the prior knowledge of the scene. For example, $\alpha$ should be zero for the background position. In this paper, we assume that there is no prior information about the scene, and the value is fixed to 1 for all grids. $d(\mathbf{h}_k^i, \mathbf{h}_k^j)$ can be any distance measure between two histograms of corresponding locations in frames, $i$ and $j$. In our proposed method, we define distance measure between two histograms as follows:

$$d(\mathbf{h}_k^i, \mathbf{h}_k^j) = 1 - \frac{\mathbf{h}_k^i \cdot \mathbf{h}_k^j}{\|\mathbf{h}_k^i\|\|\mathbf{h}_k^j\|}, \tag{5}$$

where $\mathbf{h}_k^i$ refers to the vector of weighted histogram for grid k from frame $i$ and $\mathbf{h}_k^j$ refers to the vector of corresponding location from frame $j$.

In the second step, we find a low-dimensional embedding space by minimizing the following cost function:

$$\phi(\mathbf{Y}) = \sum_{ij} \omega_{ij} \|\mathbf{y}_i - \mathbf{y}_j\|^2, \tag{6}$$

where $\omega_{ij}$ is given by equation (2) and $\mathbf{Y}$ is the low-dimensional embedding of the entire video where each entry, $\mathbf{y}_i$ is the low-dimensional representation of a video frame. The above equation can be expanded as:

$$\phi = \sum_{ij} \omega_{ij} \left[\|\mathbf{y}_i\|^2 + \|\mathbf{y}_j\|^2 - 2\mathbf{y}_i\mathbf{y}_j\right], \tag{7}$$

$$= \sum_{ij} \omega_{ij} \|\mathbf{y}_i\|^2 + \sum_{ij} \omega_{ij} \|\mathbf{y}_j\|^2 - 2\sum_{ij} \omega_{ij}\mathbf{y}_i\mathbf{y}_j, \tag{8}$$

$$= \sum_{i} m_{ii} \|\mathbf{y}_i\|^2 + \sum_{j} m_{jj} \|\mathbf{y}_j\|^2 - 2\sum_{ij} \omega_{ij}\mathbf{y}_i\mathbf{y}_j, \tag{9}$$

$$= 2\mathbf{Y}^{\mathbf{T}}\mathbf{M}\mathbf{Y} - 2\mathbf{Y}^{\mathbf{T}}\mathbf{W}\mathbf{Y}, \tag{10}$$

$$= 2\mathbf{Y}^{\mathbf{T}}\mathbf{L}\mathbf{Y}, \tag{11}$$

where $\mathbf{W}$ is the weighted neighborhood matrix and its entry, $\omega_{ij}$ is given by equation (2). $\mathbf{M}$ is the diagonal weight matrix in which each entry is a total sum of each row of weight matrix, $\mathbf{W}$, and computed as

$$m_{ii} = \sum_{j} \omega_{ij}. \tag{12}$$

The graph Laplacian $\mathbf{L}$ of the weighted neighbor graph $\mathbf{W}$ is computed by $\mathbf{L} = \mathbf{M} - \mathbf{W}$. Hence, minimizing $\phi(\mathbf{Y})$ reduces to finding optimum $\mathbf{Y}$:

$$\mathbf{Y}_{\text{opt}} = \arg\min(\mathbf{Y}^T L\mathbf{Y}) \quad \text{subject to } \mathbf{Y}^T M\mathbf{Y}, \tag{13}$$

and this is equal to solving the generalized eigenvalue problem,

$$\mathbf{Ly} = \lambda\mathbf{My}. \tag{14}$$

for the smallest $k_s$ non-zero eigenvalue problems.

The above embedding process maps each video frame into a low-dimensional point, $\mathbf{y}_i$. The number of dimensions for the subspace is selected based on the relative difference between two adjacent eigen-values, $\lambda$. The temporal order of video frames defines a path in the embedded space. As a result, the above step transforms a video segment of $T$ frames into a set of $T$ points in the low-dimensional space, where each point corresponds to the successive frames in the original sequence. Figure 1 (first row) shows a video trajectory in $3D$ space generated using Isomap [27], LE [27] and our proposed ST-LE method. This video contains a crowd whose individuals start running in the middle of the sequence. It is observed that our proposed method generates a smooth trajectory where the transition between two different events is clearly shown.

### 3.3 Clustering Video Trajectories: Similarity Measure

Our proposed ST-LE discovers the internal structure of the video and produces a smooth trajectory for each video sequence. To analyze these video trajectories for different problems of video understanding, we first need to define a similarity measure between these trajectories. In this paper, we propose to use a Hausdorff distance to measure the similarity between trajectories. Given two short trajectories, $\mathbf{S}_1 = \{\mathbf{y}_1, \mathbf{y}_2, \mathbf{y}_3, \cdots \mathbf{y}_{T1}\}$ and $\mathbf{S}_2 = \{\mathbf{y}_1, \mathbf{y}_2, \mathbf{y}_3, \cdots \mathbf{y}_{T2}\}$, where $\mathbf{y}_i$ is the low-dimensional representation of the video frame, and $T_1$ and $T_2$ are the durations of these two short segments, we measure the similarity as follows:

$$d_H(\mathbf{S}_1, \mathbf{S}_2) = \min(d(\mathbf{S}_1, \mathbf{S}_2), d(\mathbf{S}_2, \mathbf{S}_1)), \quad \text{where} \tag{15}$$

$$d(\mathbf{S}_1, \mathbf{S}_2) = \frac{1}{T_1} \sum_{i=1}^{T_1} \min_i(|\mathbf{y}_i^1 - \mathbf{y}_j^2|), \quad \text{where} \quad j = \{1, 2, \ldots, T_2\}. \tag{16}$$

$$d(\mathbf{S}_2, \mathbf{S}_1) = \frac{1}{T_2} \sum_{j=1}^{T_2} \min_j(|\mathbf{y}_i^1 - \mathbf{y}_j^2|), \quad \text{where} \quad i = \{1, 2, \ldots, T_1\}. \tag{17}$$

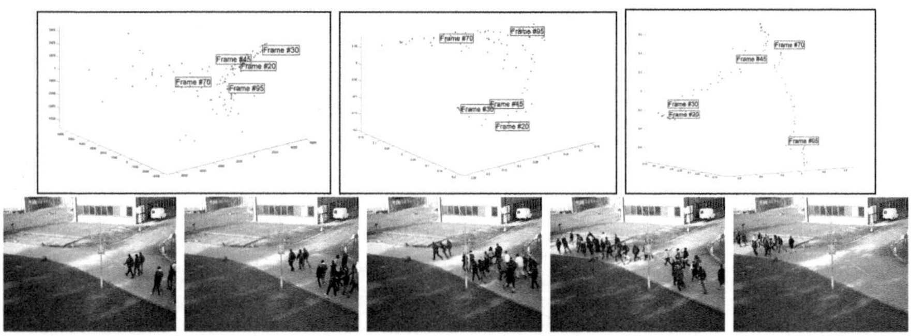

**Fig. 1.** Trajectory of a video sequence with two Crowd events (left: Isomap [27], center: LE [27], right: our proposed ST-LE method) and representative frames (Frame #20: a crowd enters the scene, Frame #30: crowd is walking, Frame #45: crowd starts to run, Frame #70: crowd is running} and Frame #95: crowd left the scene)

# 4 Video Understanding Applications and Experimental Results

## 4.1 Segmenting a Video Trajectory

In this section, we explain how to analyze video trajectories extracted by our proposed manifold learning algorithm for video segmentation. Our objective here is to segment a video sequence into different clusters where each cluster represents an event.

In Figure 2, we show the video trajectories of six different crowd sequences generated by our proposed method. These sequences from PETS 2009 [1] contains one or more of the following events: walking, running, evacuation (rapid dispersion), local dispersion, crowd formation and splitting. To segment each video sequence into different crowd events, we first divide a long trajectory into a set of non-overlapping short segments of width $T$. Parameter $T$ is selected based on the combination of the frame-rate and the expected minimum duration of an event. In this paper, $T$ is fixed to $2\times$ frame-rate. Then, we perform an unsupervised spectral clustering based on the trajectory similarity measure introduced in the previous section.

To evaluate the segmentation performance, we create a ground truth by manually segmenting each video sequence into different crowd events based on the definition provided in [28]. Table 1 shows the comparison between our segmentation result and the ground truth.

## 4.2 Abnormality Detection

Detecting abnormal events can be solved through measuring similarity between representatives of normal sequences and a test sequence in the low dimensional space. To validate this, we test our method using the crowd activity dataset from University of Minnesota [2]. This data-set includes eleven video sequences

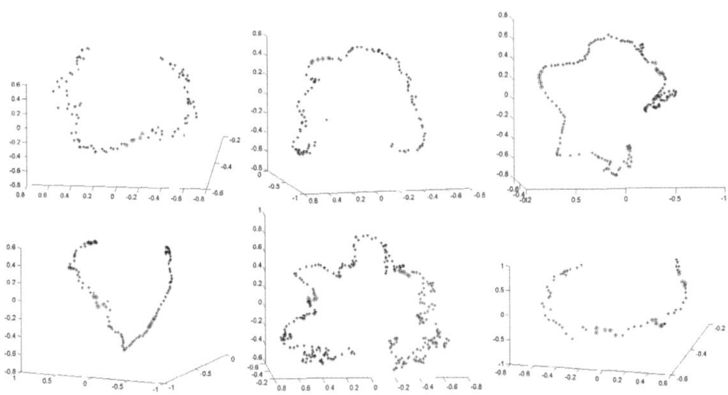

**Fig. 2.** Video trajectories of Six different video sequences. Each sequence contains two or more crowd-events. Different events are shown in different colors while circles show transition between events

**Table 1.** Time intervals for different crowd events. Each bracket, [start frame: end frame], gives time interval for each individual event.

| Seq Name | Ground Truth | Proposed Method |
|---|---|---|
| 14:16-A | [0-40][41-107] | [0-42][43-107] |
| 14:16-B | [108-165][166-222] | [108-162][163-222] |
| 14:27 | [1-95][96-124][125-184]] | [1-90][91-125][126-184] |
| 14:31 | [1-55][56-105][106-130] | [1-50][51-90][91-130] |
| 14:33-A | [1-100][101-190][191-310]] | [1-95][96-185][186-310] |
| 14:33-B | [310-340][341-361][362-377] | [310-340][341-351][352-365][366-377] |

of three different scenarios: 2 sequences for $1^{st}$ scenario, 6 sequences for $2^{nd}$ scenario and 3 sequences for $3^{rd}$ scenario. Each sequence contains a normal starting section and abnormal ending section.

In our framework, we first generate a video trajectory for each sequence and divide the long trajectory into a set of short trajectories. Each trajectory is represented by a $(k_s \times T)$ trajectory where $k_s$ is the reduced dimensionality and $T$ is the temporal duration for each segment. Then, we extract representative trajectories for the normality of each sequence using the spectral clustering and trajectory similarity measure introduced in section 3.3. We use 3/4 of the normal segments for training and the remaining part of the video sequence including both normal and abnormal segments for testing. The normality score for each test segment is computed as follow:

$$P_{test} = \exp(-\min_r(d_H(\mathbf{S}_{test}, \mathbf{S}_r))), \quad \text{where} \quad r = \{1, 2, \ldots, C\}. \tag{18}$$

$d_H(\mathbf{S}_{test}, \mathbf{S}_r)$ is the similarity measure between the test trajectory, $\mathbf{S}_{test}$ and the representative trajectory, $\mathbf{S}_r$, computed using equation (15). $C$ is the total number of representative trajectories. Based on a fixed threshold on the normality score, we label each segment as normal or abnormal. In the experiments, we set the range of the threshold value to be between $[0.1, 0.9]$.

Figure 3 shows results of abnormal event detection by our proposed method when it is evaluated on all the 11 video sequences of the Minnesota's crowd activity data-set [2]. In each row, the ROC curve and qualitative results for a selective

**Table 2.** Comparison of our proposed method and the state-of-the-art methods. (Please note that in the first row, we report the best, the worst and the average performance of our proposed method. In second to fourth rows, we list the results from [13] and [12] as reported in their respective papers. For [13], to our best understanding, the result is obtained from selected 6 sequences).

| Method | Area under ROC |
|---|---|
| Our proposed method | 0.99(max) 0.97(average) 0.95(min) |
| Lagrangian [13] | 0.99 |
| Social Force Model [12] | 0.96 |
| Optical flow [12] | 0.84 |

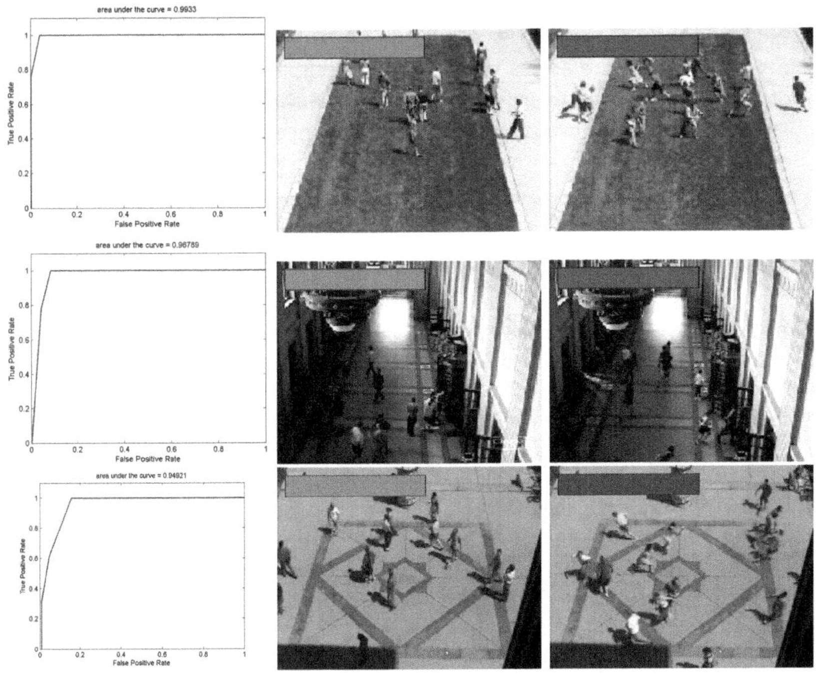


**Fig. 3.** Experimental results for abnormality detection for three videos of different crowded scenes. Each row represents the results for a video sequence of different scenarios. The left column shows the ROC curve for individual sequence, while the center column and the right column show the normal frame and abnormal frame of the video. Green bar and red bar on the left corner of the frame indicate normality and abnormality respectively. (Please note that the figures in this paper are best viewed in color and the reader is requested to refer to online version for best view).

sequence of each scenario is shown. The results demonstrate the promising performance of our proposed method. Table 2 compares the performance of our proposed method and the state-of-the-art methods in terms of area under the ROC curve. In the table, we report maximum (the best), minimum (the worst) and average area under the ROC curves for the tested 11 video sequences. The table also lists the results as reported in [12] and [13]. For [13], to our best understanding, the result is obtained from 6 out of the 11 sequences. As can be seen, the performance of our method is promising, having higher detection accuracy than social force model [12] and pure optical flow in general [12].

## 5 Conclusions

In this paper, we have presented a manifold-learning based method for video content analysis in a crowded scene. The proposed method is based on a spatio-temporal modeling of a video sequence in a low-dimensional embedded space.

We model each video sequence as a video trajectory in the embedded space. The generated video trajectories serve as compact, yet information, representations for analyzing time series data. Experiments have been performed on the state-of-the-art public data-sets on crowded scenes. The results show that our proposed method provides a promising performance in video event segmentation and visualization as well as abnormality detection.

## References

1. PETS 2009 benchmark data (2009), http://www.cvg.rdg.ac.uk/PETS2009
2. Unusual crowd activity dataset, http://mha.cs.umn.edu
3. Wang, X., Tieu, K., Grimson, E.: Learning semantic scene models by trajectory analysis. In: Leonardis, A., Bischof, H., Pinz, A. (eds.) ECCV 2006. LNCS, vol. 3953, pp. 110–123. Springer, Heidelberg (2006)
4. Hu, W., Xiao, X., Fu, Z., Dan, X., Tan, T., Steve, M.: A System for Learning Statistical Motion Patterns. IEEE Transactions on Pattern Analysis and Machine Intelligence 28, 1450–1464 (2006)
5. Johnson, N., Hogg, D.: Learning the Distribution of Object Trajectories for Event Recognition. Image and Vision Computing 14, 583–592 (1996)
6. Andrade, E., Fisher, R.: Simulation of Crowd problems for Computer Vision. In: Proceedings of 19th International Conference on Pattern Recognition (ICPR 2005), vol. 3, pp. 71–80 (2005)
7. Andrade, E., Fisher, R., Blunsden, S.: Modelling Crowd Scenes for Event Detection. In: Proceedings of 19th International Conference on Pattern Recognition (ICPR 2006), vol. 1, pp. 175–178 (2006)
8. Andrade, E., Blunsden, S., Fisher, R.: Performance Analysis of Event Detection Models in Crowded Scenes. In: Proceedings of Workshop on 'Towards Robust Visual Surveillance Techniques and Systems' at Visual Information Engineering, pp. 427–432 (2006)
9. Andrade, E., Fisher, R., Blunsden, S.: Detection of Emergency Events in Crowded Scenes. In: Proceedings of IEE International Symposium on Imaging for Crime Detection and Prevention (ICDP 2006), pp. 528–533 (2006)
10. Hu, M., Ali, S., Shah, M.: Detecting Global Motion Patterns in Complex Videos. In: Proceedings of International Conference on Pattern Recognition (ICPR 2008), Tempa, Florida, pp. 1–5. IEEE, Los Alamitos (2008)
11. Min Hu, S.A., Shah, M.: Learning Motion Patterns in Crowded Scenes Using Motion Flow Field. In: Proceedings of International Conference on Pattern Recognition (ICPR 2008), Tempa, Florida, pp. 1–5. IEEE, Los Alamitos (2008)
12. Mehran, R., Oyama, A., Shah, M.: Abnormal Crowd Behavior Detection using Social Force Model. In: Proceedings of the Conference on Computer Vision and Pattern Recognition (CVPR 2009), Maimi Beach, Florida, pp. 935–942. IEEE, Los Alamitos (2009)
13. Wu, S., Moore, B.E., Shah, M.: Chaotic invariants of lagrangian particle trajectories for anomaly detection in crowded scenes. In: Proceedings of the Conference on Computer Vision and Pattern Recognition (CVPR 2010), San Francisco, CA, USA. IEEE, Los Alamitos (2010)
14. Pless, R.: Image spaces and video trajectories: Using isomap to explore video sequences. In: Proceedings of IEEE International Conferecne on Computer Vision, vol. 2, pp. 1433–1440. IEEE, Los Alamitos (2003)

15. Tziakos, I., Cavallaro, A., Xu, L.Q.: Video event segmentation and visualisation in non-linear subspace. Pattern Recognition Letter 30, 123–131 (2009)
16. Wu, Y., Chan, K.L., Wang, L.: Face recognition based on discriminative manifold learning. In: Proceedings of IEEE International Conferecne on Image Processing, vol. 4, pp. 171–174 (2004)
17. He, X., Yan, S., Hu, Y., Niyogi, P., Zhang, H.J.: Face recognition using laplacianfaces. IEEE Transactions on Pattern Analysis and Machine Intelligence 27, 328–340 (2005)
18. Wang, R., Shan, S., Chen, X., Gao, W.: Manifold-manifold distance with application to face recognition based on image set. In: Proceedings of IEEE International Conferecne on Computer Vision and Pattern Recognition, pp. 1–8 (2008)
19. Elgammal, A., Lee, C.S.: Inferring 3d body pose from silhouettes using activity manifold learning. In: Proceedings of IEEE International Conferecne on Computer Vision and Pattern Recognition, vol. 2, pp. 681–688 (2004)
20. Chang, Y., Hu, C., Turk, M.: Probabilistic expression analysis on manifolds. In: Proceedings of IEEE International Conferecne on Computer Vision and Pattern Recognition, vol. 2, pp. 520–527 (2004)
21. Schuldt, C., Laptev, I., Caputo, B.: Recognizing human actions: A local svm approach. In: Proceedings of IEEE International Conferecne on Pattern Recognition, vol. 3, pp. 32–36. IEEE, Los Alamitos (2004)
22. Wang, L., Suter, D.: Analyzing human movements from silhouettes using manifold learning. In: Proceedings of IEEE International Conference on Video and Signal Based Surveillance, pp. 1–7 (2006)
23. Wang, L., Suter, D.: Learning and matching of dynamic shape manifolds for human action recognition. IEEE Transactions on Image Processing 16, 1646–1661 (2007)
24. Brox, T., Bruhn, A., Papenberg, N., Weickert, J.: High accuracy optical flow estimation based on a theory for warping. In: Pajdla, T., Matas, J(G.) (eds.) ECCV 2004. LNCS, vol. 3024, pp. 25–36. Springer, Heidelberg (2004)
25. Tenenbaum, J.B., de Silva, V., Langford, J.C.: A global geometric framework for nonlinear dimensionality reduction. Sciene 290, 2319–2323 (2000)
26. Belkin, M., Niyogi, P.: Laplacian eigenmaps for dimensionality reduction and data representation. Neural Computation 15, 1373–1396 (2003)
27. MANIfold learning matlab demo, `http://www.math.ucla.edu/~wittman/mani/index.html`
28. Garate, C., Bilinski, P., Bremond, F.: Crowd Event Recognition Using HOG Tracker. In: Proceedings of Eleventh International Workshoop on Performance Evaluation of Tracking and Surveillance (Winter-PETS 2009). IEEE, Los Alamitos (2009)

# A Weak Structure Model for Regular Pattern Recognition Applied to Facade Images

Radim Tyleček and Radim Šára

Center for Machine Perception
Faculty of Electrical Engineering,
Czech Technical University,
Prague, Czech Republic

**Abstract.** We propose a novel method for recognition of structured images and demonstrate it on detection of windows in facade images. Given an ability to obtain local low-level data evidence on primitive elements of a structure (like window in a facade image), we determine their most probable number, attribute values (location, size) and neighborhood relation. The embedded structure is weakly modeled by pair-wise attribute constraints, which allow structure and attribute constraints to mutually support each other. We use a very general framework of reversible jump MCMC, which allows simple implementation of a specific structure model and plug-in of almost arbitrary element classifiers. The MC controls the classifier by prescribing it "where to look", without wasting too much time on unpromising locations.

We have chosen the domain of window recognition in facade images to demonstrate that the result is an efficient algorithm achieving performance of other strongly informed methods for regular structures like grids, while our general model covers loosely regular configurations as well.

## 1 Introduction

Recent development in construction of virtual worlds like Google Earth or Bing Maps 3D heads toward higher level of detail and fidelity. Popularity of application such as Street View shows that reconstruction of urban environments plays an important role in this area. While acquisition of extensive data in high resolution for this purpose is feasible today, their automated processing is now the limiting factor for delivering more realistic experience and it is a task for computer vision at the same time. In urban settings, typical acquired data are images of buildings' facades and their interpretation can help discover 3D structure and reduce the complexity of the resulting model; for example, it would allow going beyond planar assumptions in dense street view reconstruction presented by [1]. Complexity is particularly important when the representation has to scale with the size of cities in applications such as [2] who plan to combine range data with images. The work of [3] dealing directly with structural regularity in 3D data also supports our ideas.

While facades as man-made scenes exhibit intensive regularity and structure when compared to arbitrary natural scenes, they still present a great variety of

R. Kimmel, R. Klette, and A. Sugimoto (Eds.): ACCV 2010, Part I, LNCS 6492, pp. 450–463, 2011.

styles, configurations and appearance. The design of a general facade model that is able to cover their range is thus a challenging problem, and several approaches have been proposed to deal with it.

Shape grammars, as introduced in [4] and later picked up by [5], are the basic essence for all recent methods based on procedural modeling to overcome the limitations of traditional segmentation techniques. The idea of shape grammars is that image can be explained by combining rules and symbols.

Some aspects of probabilistic approach were first discussed in [6], including the use of Reversible Jump Markov Chain Monte Carlo (RJMCMC). The proposed grammar is simple, based on splitting and the results are demonstrated for highly regular facades only. In a similar fashion [7] determines the structure by splitting facade to a regular grid of individual tiles and subdividing them. Meyer and Reznik [8] presented a pipeline for multi-view interpretation, where heuristics based on interest points were designed to detect positions of windows, and subsequently used MCMC to localize their borders. Ripperda [9] has designed a comprehensive dictionary of rules, on which the proposed method substantially depends; the results presented on simple facades show this approach has difficulty to achieve good localization.

The most recent method of [10] combines trained randomized forest classifiers with shape grammar to segment Haussmannian facades into eight classes. Their model assumes windows form a grid while allowing different intervals. In the second step, positions of rows and columns are stochastically estimated by a specific random walk algorithm that does not propose dimension changes. They evaluated their results quantitatively on a limited dataset of Haussmannian facades in Paris which is available online.

The majority of the mentioned algorithms for single-view facade interpretation work with hard constraint on grid configurations of windows and employ strong domain-specific heuristics. Additionally, they require user design of specific grammar or training, while both processes are prone to overfitting. Our contribution is in the design of segmentation framework with the following properties:

- a general model allows a simple implementation avoiding strong domain specific heuristics,
- structure is not modeled by a global grid, but softly by local pair-wise constraints, allowing loosely regular configurations,
- different element classifiers can be conveniently plugged in,
- efficient interpretation is achieved as the classifier is guided by the sampler and need not even visit all image pixels in practice,
- the number, spacing and exact size of facade elements need not to be known in advance and does not rely on preprocessing that can fail i.e. in irregular cases like in Fig. 4.

Since windows are the most prominent elements of a facade, we choose detection of window-like image elements to be the target of this paper.

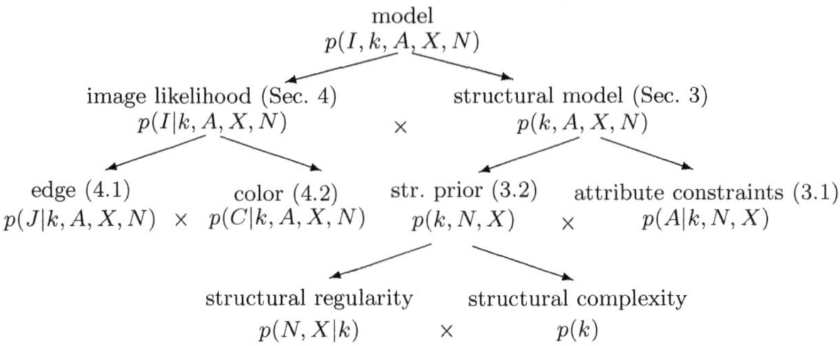


**Fig. 1.** Hierarchy in probability model, numbers in brackets are section references

## 2 Structural Recognition Framework

We consider the problem of recognizing elements in an image, like windows in a facade. Our model parameters (variables) consist of complexity $k$ (the number of windows), shape attributes $A$ (i.e. size, aspect), location attributes $X$ (window center locations) and element neighborhood relation $N$. The recognition task can then be formulated as follows: Given image data $I$, we search for model parameters $\theta = (k, A, X, N)$ by finding the mode of the following joint distribution $p(I, \theta)$

$$\theta^* = \arg\max_{\theta} p(I|\theta)p(\theta), \tag{1}$$

which is computed with Bayes theorem from data likelihood $p(I|\theta)$ and structural model prior $p(\theta)$. We will decompose our probability model hierarchically as shown in Fig. 1 and propose pdfs specific for the task of window detection in facade images. Then we can apply stochastic RJMCMC framework to find the optimal value $\theta^*$ by effectively sampling from the space of possible combinations of parameters $\theta$. More details on its implementation will be given in the following sections.

## 3 Structural Model

The structural model is based on pair-wise element neighborhood and attribute constraints, yielding bottom-up approach. We are given a set of $k \in \mathrm{N}$ element locations $X = \{x_i \in \mathrm{R}^2; i = 1, \ldots, k\}$. Our neighborhood representation is based on a planar graph $G(X) = \{V(X), D(X)\}$, where vertices $V(X) = \{v_i;\ i = 1, \ldots, k\}$ correspond to elements and edges $D(X) = \{(u, v);\ u, v \in V(X)\}$ to relative neighborhood relationship between them.

Since we are dealing with image elements attributed by their locations $X$ in image plane, we can limit the edge set $D(X)$ to a reasonable planar subgraph and *Relative Neighborhood Graph* (RNG) turns out to be a natural choice [11].

It is defined by the following condition: Two points $u$ and $v$ are connected by an edge whenever there does not exist a third point $r$ that is closer to both $u$ and $v$ than they are to each other (in Euclidean metric). It is known that RNG is a unique subgraph of *Delaunay Triangulation* (DT), and can be computed from it efficiently, in $O(n)$ time. This choice defines a function $X \mapsto G(X)$, where the graph is uniquely constructed from a set of element locations $X$.

We define neighbors as elements that are in immediate proximity of each other and such that they share some attributes. This neighborhood $N$ is to be recovered as a part of the solution, and we represent it by binary labels $N = \{l_{uv} \in \{0, 1\}\,;\, (u, v) \in D(X)\}$ for edges indicating mutual neighborhood of two elements when $l_{uv} = 1$. Such two elements are then members of the same structural component, where all connected elements are related by attribute similarity constraints. Labels $l_{uv} = 0$ allow the existence of dissimilar elements in proximity of each other.

An edge $(u, v)$ has an orientation attribute $o_{uv} \in \{\mathtt{h}, \mathtt{v}\}$, which is a function of locations $x_u, x_v$ of elements on its endpoints. It is given by the angle $\psi$ between vertical direction and line connecting element locations. The case of $|\psi| < \frac{\pi}{4}$ determines vertical orientation ($\mathtt{h}$), the other case is horizontal ($\mathtt{v}$). This choice defines a function $D(X) \mapsto \{\mathtt{h}, \mathtt{v}\}$.

The prior probability model $p(k, N, X, A) = p(A|k, N, X)p(k, N, X)$ splits into attribute constraints $p(A|k, N, X)$ and structure prior $p(k, N, X)$. The parameters of the underlying distributions were chosen empirically.

### 3.1 Attribute Constraints

The attribute constraints evaluate the similarity of two neighboring elements (in terms of $N$); such attributes can be shape or appearance.

For facades, we assume our elements can be represented by a rectangular shape template with its borders parallel to image borders. The shape attributes $A = \{W, H, T\} = \{(w_i, h_i, t_i)\,; i = 1, \ldots, k\}$ are described in Fig. 2 and the column width $t_i = t$ is given and fixed. Our attribute constraints will then reflect the fact neighboring windows most probably have the same dimensions. We start by decomposition

**Fig. 2.** *Left:* Window shape template is parametrized by its width $w_i \in (0, 1)$, height $h_i \in (0, 1)$, both relative to image height $I_h$, and the width of the central column $t_i \in (0, 1)$ relative to the window width. *Right:* Shape template (red) is matched with image edges (blue).

$$p(A|k, N, X) = p(W|H, k, N, X)p(H|k, N, X)1(A|X), \tag{2}$$

where $p(W|H, k, N, X) = \prod_{i=1}^{k} p(w_i|h_i)$ is the aspect ratio with distribution $p(w_i|h_i) = \beta(\frac{w_i}{w_i+h_i}, \alpha_r, \beta_r)$. When any of the windows overlap with another, we set unit function $1(A|X) = 0$, effectively avoiding such window configuration.

To model constraints on heights $H$, we introduce a set of latent variables $h_c$, one for each component $c$ of graph $G(X)$ with neighborhood $N$. The height similarity within components is enforced in

$$p(H|k, N, X) = \prod_{c} \left( p(h_c) \prod_{i \in V_c} p(h_i|h_c) \right), \tag{3}$$

where $c$ is from the set of all components, $V_c$ is the set of windows in the component $c$ and $p(h_c) = \beta(h_c, \alpha_h, \beta_h)$ is the common height prior. Each height in a component $c$ should be most probably equal to $h_c$, which is expressed by $p(h_i|h_c) = \mathcal{N}(h_i - h_c, 0, \sigma_h)$.

### 3.2 Structural Prior

The structure prior $p(k, N, X) = p(N, X|k)p(k)$ combines structural regularity $p(N, X|k)$ and complexity $p(k)$.

**Structural Regularity.** In order to model multiple assumptions on $p(N, X|k)$, we express it as a probability mixture [12]:

$$p(N, X|k) = \omega_1 p_a(X|N)p(N) + \omega_2 p_s(X|N)p(N) + \omega_3 p_c(N|X)p(X), \tag{4}$$

where $\sum_{i=1}^{k} \omega_i = 1, \omega_{123} = \frac{1}{3}$ and $k$ was omitted in $p(\cdot)$ for simplicity. We assume element locations in $p(X)$ are mutually independent and uniformly distributed in image. The neighborhood prior $p(N) = \prod_{(u,v)} p(l_{uv})$ takes into account the possibility of suppressing an edge where $p(l_{uv} = 0) = p_{sup}$, $p(l_{uv} = 1) = 1 - p_{sup}$ and $p_{sup} = 0.01$ is the probability of a suppressed edge.

*Alignment.* The first assumption on the position of elements is that neighboring elements should be horizontally or vertically aligned. We model this by measuring angles $\varphi(x_u, x_v) \in (-\frac{\pi}{4}, \frac{\pi}{4})$ between the line connecting element locations $x_u x_v$ and horizontal ($o_{uv} = \mathtt{h}$) resp. vertical ($o_{uv} = \mathtt{v}$) direction, and express them in

$$p_a(X|N) = \prod_{(u,v) \in D(X)} p(x_u, x_v|l_{uv}), \tag{5}$$

where $p(x_u, x_v|l_{uv} = 1) = \beta(\varphi'(x_u, x_v), \beta_\varphi, \beta_\varphi)$, $\beta_\varphi = 50$ and $\varphi'(x_u, x_v) = \frac{2}{\pi}(\varphi_{uv} + \frac{\pi}{4}) \in (0, 1)$ is the angle normalized to unit interval. The probability in the case of a suppressed edge is $p(x_u, x_v|l_{uv} = 0) = p_{a0}$.

*Spacing.* The second assumption is that the distance between elements in a horizontal or vertical neighborhood should most probably be equal. We model this by comparing distances to horizontal and vertical neighbors in

$$p_s(X|N) = \prod_{(u,v,z)\in D^2(X)} p(x_u, x_v, x_z|l_{uv}, l_{vz}) \tag{6}$$

where $(u, v, z)$ denotes a pair of edges $(u, v), (v, z), u \neq z$ with the common vertex $v$ and the same orientation. The distance term is expressed by $p(x_u, x_v, x_z|l_{uv} = l_{vz} = 1) = \beta(\frac{\Delta_{uv}}{\Delta_{uv}+\Delta_{vz}}, \beta_\Delta, \beta_\Delta)$, where $\beta_\Delta = 50$ and $\Delta_{uv} = |x_u - x_v|$ are distances to the neighbors. As in the previous case, the probability in the cases with any suppressed edge is $p(x_u, x_v, x_z|l_{uv} \neq 1 \vee l_{vz} \neq 1) = p_{s0}$.

*Configurations.* We model higher-order dependencies in the structure configurations with

$$p_c(N|X) = \prod_{i=1}^{k} p(l_{ij}|(i,j) \in D(X)), \tag{7}$$

where the probabilities $p(l_{ij}|(i,j) \in D(X))$ model the expected degree of a given vertex $i$, including orientation of edges $(i, j)$ connected to it, i.e. the typical grid configuration is to have two vertical and two horizontal edges incident with vertex $i$.

**Table 1.** Neighborhood configuration prior $p(l_{ij}|(i,j) \in D(X))$, where $deg_h(i), deg_v(i)$ are functions of neighboring labels $l_{ij}$. The $p_{c0} = 10^{-4}$ is the probability of a single (unstructured) window, $p_{c1} = 0.099$ is the probability of a single row or column of windows, $p_{c2} = 0.9$ is the probability of a window grid, $p_{c3} = 10^{-5}$ is the probability of more dense configurations.

| $deg_h(i), deg_v(i)$ | 0h | 1h | 2h | 3h |
|---|---|---|---|---|
| 0v | $p_{c0}$ | $\frac{1}{2}p_{c1}$ | $\frac{1}{(m-2)}p_{c1}$ | $p_{c3}$ |
| 1v | $\frac{1}{2}p_{c1}$ | $\frac{1}{4}p_{c2}$ | $\frac{2}{(m-2)}p_{c2}$ | $p_{c3}$ |
| 2v | $\frac{1}{(n-2)}p_{c1}$ | $\frac{2}{(n-2)}p_{c2}$ | $\frac{1}{(m-2)(n-2)}p_{c2}$ | $p_{c3}$ |
| 3v | $p_{c3}$ | $p_{c3}$ | $p_{c3}$ | $p_{c3}$ |

With the grid assumption and the window size prior, we can estimate the number of rows $m = \frac{1}{2\mu_h}$ and columns $n = \frac{1}{2\mu_h r_h}$, assuming the space between the windows to be equal to the window size. This heuristic plays only a minor role in our model and helps us to derive the vertex configuration probability $p(l_{ij}|(i,j) \in D(X))$. It is given in Table 1, where rows and columns correspond to the number of horizontal and vertical edges connected to the window vertex. The maximum degree of a vertex in RNG is six with at most three horizontal and three vertical edges.

**Structural Complexity.** The prior for number of elements can be modeled with Poisson distribution $p(k) = \text{Pois}(k, mn)$ based on the estimation of number of rows $m$ and columns $n$ given above.

# 4 Data Likelihood

The data likelihood $p(I|K, N, A, X)$ is solely task-specific and can be chosen arbitrarily as long as it can be evaluated by means of probability density or likelihood ratio.

In the task of window detection in facade images, the input is image $I = \{i;\, i = 1, \ldots, I_w \cdot I_h\}$ defined as a set of pixels and we assume it is rectified, i.e. the windows borders are parallel to the image borders, and $I_w, I_h$ are image width and height.

We want to express the probability of observing image $I$ if window parameters and structure are given. We combine two features: image edges $J$ and color $C$ in $p(I|k, A, X, N) = p(J|k, A, X, N)p(C|k, A, X, N)$. We use color to detect regions of interest and edge features for localization of the windows' borders.

## 4.1 Edge Likelihood

We assume that window borders correspond to edges, and use Canny detector to find them. However, this model will not fully hold in real world situations, when we obtain the input by detecting edges in a picture—there can be windows which do not have all pixels with underlying edges and vice versa, some edges do not belong to any windows at all. The latter case will typically prevail.

We use binary imaging model for window edges represented by oriented edge image $J = \{J_i \in \{0, 1, 2\}\,; i \in I\}$, where $J_i = 1$ if pixel $i$ belongs to an horizontal edge detected in $I$ (foreground), resp. $J_i = 2$ for vertical edge; otherwise $J_i = 0$ (background). We define $d(J) \in (0, 1)$ as a distance transform of the edge image $J$ normalized by $\max(I_h, I_w)$. We use the gradient of $d(J)$ to distinguish between horizontal and vertical edges. Similarly, we introduce edge image $R(A, X)$ rendered from the current configuration specified by attributes $A, X$ and the shape template in Fig. 2 with nearest neighbor discretization. Assuming pixel independence, we can write $p(J|A, X) = \prod_{i \in I} p(J_i|R_i(A, X))$ where the probability of observing a pixel $i$ in the edge image $J$ given the rendered configuration $R$ is

$$
\begin{aligned}
p(J_i = 0|R_i = 0) &= p_{\mathrm{TN}} = 1 - 2p_{\mathrm{FN}}, \\
p(J_i \in \{1, 2\}\,|R_i = 0) &= p_{\mathrm{FN}} = 0.1, \\
p(J_i = 0|R_i \in \{1, 2\}) &= p_{\mathrm{FP}}(d(i))(1 - p_{\mathrm{FX}}),\ d(i) > 0, \\
p(J_i = 1|R_i = 1) = p(J_i = 2|R_i = 2) &= p_{\mathrm{TP}} = p_{\mathrm{FP}}(0), \\
p(J_i = 2|R_i = 1) = p(J_i = 1|R_i = 2) &= p_{\mathrm{FX}},
\end{aligned}
\tag{8}
$$

where $p_{\mathrm{FP}}(d(i)) = \beta(d(i), \beta_{\mathrm{FP}} = 500, 1)$ makes rectangles close to edges more probable and acts as a guide for directing the random walk. The $p_{\mathrm{FX}} = 10^{-9}$ is the probability assigned when the edge specified by the configuration crosses an image edge with opposite direction.

The edge likelihood can be efficiently evaluated from pre-computed integral edge images, one for each orientation, yielding constant computational complexity $O(1)$ per edge; this speed-up is possible thanks to rectified images and helps make random sampling (described in Sect. 5) very efficient.

### 4.2 Color Likelihood

A pixel color classifier matches the input RGB color image $C = \{c_i \in (0,1)^3; i = 1,\dots,k\}$ with a unimodal Gaussian distribution $\mathcal{N}(\bar{C}, \Sigma_C)$ for window pixels. Its mean $\bar{C} = (0.33, 0.36, 0.38) \in (0,1)^3$ and covariance $\Sigma_C$ of window color were trained on a single representative facade image and correspond to dark colors; higher mean in blue channel is related to the reflection of sky in window glass. We use the classifier to segment pixels either to foreground (window) or background (non-window) sets $C_f \cup C_b = I$. Assuming pixel independence, the probability of observing segmented image is

$$p(C|A,X) = \prod_{i \in C_f} p_f(c_i|A,X) \prod_{j \in C_b} p_b(c_j|A,X), \tag{9}$$

where the foreground color model is expressed by $p_f(C_i|A,X) = \mathcal{N}(\bar{C}, \Sigma_C)$, the background probability $p_b(c_j|A,X) = p_b$ is constant and we evaluate foreground pixels only. Similarly to edge likelihood, color likelihood can be evaluated using pre-computed integral images in linear time.

## 5 Recognition Algorithm

We have chosen reversible jump Markov Chain Monte Carlo (RJMCMC) framework [13] that fits our task of finding the most probable interpretation of the input image in the terms of target probability $p(\theta, I)$ in (1), which has a very complex pdf as it is a joint probability of both attributes and structure. Our solution $\theta^*$ is found as the most probable parameter value the chain visits in a given number of samples.

While the MCMC algorithm is simple, we need to carefully design proposal distribution $q$ that should approximate target distribution $p(\theta, I)$ well while it is easy to sample from it. We should point out that the quality of the resulting interpretation is determined by the probability model and the time necessary to reach the solution is influenced by the proposal distributions. It turns out that by exploiting the estimated structure we can efficiently guide the random walk of our chain by repeatedly sampling the new state $\theta'$ from the vicinity of the current state from conditional probability $q(\theta'|\theta)$.

We use an independent sampler $q(\theta|I)$ to initialize the Markov chain, which samples the initial state $\theta_0$ either from the prior distribution $\theta \sim q(\theta)$ or exploits some image information in $\theta \sim q(\theta|I)$. This involves sampling the number of elements $k \sim q(k)$ first and then their attribute values $(X, A) \sim q(X, A)$ independently. In practice we choose sampler to start with $k_0 = 1$.

The conditional sampler $q(\theta'|\theta, I) \to \theta'$ is a mixture of individual samplers such that each modifies a subset of parameters $\theta$ based on a specific proposal distribution $q_m(\theta'|\theta, I)$. The main sampler only chooses from $q(m)$ which of the individual samplers $m$ will be used to propose the next move. We will now propose the set of samplers that will explore the space of parameters $\theta$. Their design must fulfill Markov Chain properties of detailed balance and reversibility of all moves, i.e. given a move there must always exist a reverse move $m'$, and their probability ratio must be reflected in the acceptance of Metropolis-Hastings (MH) algorithm:

$$A = \min\left\{1, \frac{p(\theta', I)}{p(\theta, I)} \cdot \frac{q(m'|\theta')}{q(m|\theta)}\right\}. \quad (10)$$

### 5.1 Metropolis-Hastings Moves

Moves introduced in this section do not modify the model complexity $k$ and can be thus evaluated by a classical MH algorithm (10).

*Attribute modification.* This move picks up an element $i \sim \mathcal{U}(\{1, \ldots, k\})$ from discrete uniform distribution and perturbs some of its attributes values randomly. Additionally, attribute samplers can be designed to exploit image likelihood to increase the acceptance rate. In the window detection scenario, we have implemented three variants for this type of proposals:

- *Drift* - random variation of position $x'_i = x_i + \Delta$, $\Delta \sim \mathcal{N}(0, \sigma_\Delta)$ without changing the size,
- *Resize* - randomly pick up one of four window sides (left/right/top/bottom) and move it by $\Delta$,
- *Flip* - fix one of the window sides and flip the window around it.

*Element resampling.* This move is a more radical variant of the previous one, we pick up an element $i$ and change of all its attributes by sampling from the prior distribution $a'_i, x'_i \sim q(a_i, x_i)$ or $a'_i, x'_i \sim q(a_i, x_i|I)$ if possible.

*Attribute constraint enforcement.* This move proposes changes to the attributes according to the current neighborhood, $a'_i, x'_i \sim q(a_i, x_i|A, X, N)$. We pick up a random edge $(u, v) \sim \mathcal{U}(D(X))$ and direction ($u \Rightarrow v$ or $v \Rightarrow u$) and transfer attribute values over the edge from one element to another according to the specific constraints, i.e. $a'_u = a_v$. For facades, we transfer both position and size from one element to the other in dimension given by orientation of the connected edge, i.e. height and vertical position for horizontal edge.

*Structure modification.* We include move to allow changes to the neighborhood structure: it picks up a random edge $q_d \to (u, v)$ and changes its label $l'_{uv} = 1 - l_{uv}$, effectively suppressing or recovering the edge.

Proposals for latent heights $h_c$ are performed similarly by choosing uniformly component $c$ and then sampling $h_c \sim \mathcal{N}(\bar{h}_c, \sigma_h)$, where $\bar{h}_c = \frac{1}{|V_c|} \sum_{i \in V_c} h_i$ is the mean height in the component.

### 5.2 Reversible Jump Moves

We also need to find the number of elements $k$, that controls the dimension of parameters $A, X$. In order to compare the models in different dimensions, we need to define dimension matching functions $q_{\rightarrow}, q_{\leftarrow}$ for both direct and reverse moves. Then the acceptance ratio can be calculated as $A = \min\{1, \alpha\}$, where

$$\alpha = \frac{p(\theta', I)}{p(I)} \cdot \frac{q(m|\theta')}{q(m'|\theta)} \cdot \frac{q_{\leftarrow}(u_{\leftarrow}|\theta')}{q_{\rightarrow}(u_{\rightarrow}|\theta)} \cdot J_{\rightarrow}, \tag{11}$$

where $\rightarrow$ refers to direct move, $\leftarrow$ to reverse move, $u$ are dimension matching variables and $J_{\rightarrow} = \left|\frac{\partial f_{\rightarrow}(\theta, u_{\rightarrow})}{\partial(\theta, u_{\rightarrow})}\right|$ is the Jacobian of the transformation, following the notation given in [13]. There are three moves:

*Birth.* By inserting a new element into our model we propose an increase of dimension $k \rightarrow k' = k + 1$. We choose the communication variables to be $u_{\rightarrow} = [a_*, x_*]$, where we sample the attributes of the new element $a_*, x_* \sim q(a, x)$ and obtain a new state where $A' = \{A, a_*\}$ and $X' = \{X, x_*\}$. The corresponding dimension matching function is $f_{\rightarrow}(A, X, u_{\rightarrow}) = f_{\rightarrow}(\{A, X\}, [a_*, x_*])$, which inserts $a_*$ into the set, and its Jacobian $J_{\rightarrow} = 1$. We will use the following notation within this paper: terms in [ ] refer to communication variables and terms in { } to parameters. The reverse move is *death*, for which we have no communication variable $u_{\leftarrow} = [\ ]$, only choose an element $i$ to be removed from the set. To establish reversibility, we define inverse matching function as $f_{\leftarrow}(A', X', u_{\leftarrow}) = f_{\leftarrow}(\{A', X'\}, [\,])$, where $a_i, x_i$ are the removed attributes and $A = A' \setminus a_i$, $X = X' \setminus x_i$. The corresponding birth move acceptance is then

$$\alpha_{birth} = \frac{p(\theta', I)}{p(I)} \frac{q(m|\theta')}{q(m'|\theta)} \cdot \frac{q(i|k')}{q(*|k)} \cdot \frac{1}{q_{\rightarrow}(a^*|A)} \cdot 1, \tag{12}$$

where $q_{\rightarrow}(a_*|A) = p(a)$ is the prior probability of the new window, $q(i|k') = \frac{1}{k'}$ and $q(*|k) = \frac{1}{k}$ are the probabilities of selecting the windows $a_*, a_i$.

*Death.* By removing an existing element from the set we propose a decrease of dimension $k \rightarrow k' = k - 1$, and choose a window $i \sim \mathcal{U}(1, k)$ to be removed. With an appropriate change of labeling, the derivation of death move will be the same as for birth, except for the inversion of ratios in (12).

*Replicate.* This is a special case of the *birth* jump that exploits the structure for predicting values for the new elements according to attribute constraints, which can be generally described as sampling from $a_*, x_* \sim q(a, x|N)$. For facades, we uniformly sample an edge $(u, v) \sim \mathcal{U}(D(X))$ and place the new window to the position according to $x_* = x_u + \alpha(x_v - x_u)$, where we choose $\alpha \sim \mathcal{U}\left(\{\frac{1}{2}, \frac{1}{3}, \frac{2}{3}, 2, -1\}\right)$ and calculate the new height by $h_* = \frac{1}{2}(h_u + h_v)$ and the width $w_*$ analogically.

### 5.3 Convergence and Complexity

We have found that the typical necessary number of MCMC samples (classifier calls) is proportional to image size in pixels $|I|$ (from 30% for easy instances to 200% for difficult ones). This is a good news, we expected that the number will grow exponentially with scene complexity. As a result, we fixed the number of samples in our current method to a pessimistic estimate, but our experiments suggest that significantly shorter sampling time could be achieved with suitably designed stopping condition.

## 6 Experimental Results

We have performed a number of experiments with the implementation of window detection in facades of various styles to demonstrate the universality of our approach. We have run the Markov Chain for $5 \cdot 10^5$ iterations in our experiments, which roughly equals to visiting all pixels in the analyzed images.

Because of a very recent appearance of a first public dataset known to us with quantitative results in [10], we are among the first to compare with them. The test part of the dataset consists of 10 rectified and annotated images of facades from a street in Paris, which share attributes of Haussmannian style but differs in lightning conditions. Direct comparison is not possible, because they segment facade pixels into eight different classes of elements and our window detector defines only two (window/non-window). To deal with this issue, we have merged the columns of confusion matrix given in [10] into two, and the results are given in Table 2. All parameters of our model were fixed for this experiment, specifically the size prior was set such that the most probable relative window height is $h = 0.1$ and aspect ratio $r = 0.5$.

**Table 2.** Quantitative results on Haussmannian dataset [10] shown in percentage of pixels from class specified in a row. Second column displays the percentage of pixels of given class in the whole test set. RF stands for Randomized Forest, PS for Procedural Segmentation. Our window detection rate of 83% is comparable to 81% rate for PS (in bold face).

| ground truth[10] | | RF [10] | | PS [10] | | proposed | | mapping of our classes | |
|---|---|---|---|---|---|---|---|---|---|
| class | area | hit | miss | hit | miss | hit | miss | *window* | *non-window* |
| *window* | 11 | 30 | 70 | **81** | 19 | **83** | 17 | • | |
| *wall* | 48 | 38 | 62 | 83 | 17 | 84 | 16 | | • |

The numbers in Table 2 for *window* and *wall* classes show that our weak structure model slightly outperforms Procedural Segmentation (PS) framework [10]. This is clearly a success, because PS benefits from a randomized forest combining 8 classifiers, trained on $15 \times 15$ pixel patches in 20 images from the same street as the test data, and a grammar specifically designed for Haussmannian style. In contrast, our method is guided by far weaker cues: color of individual pixels, rectangular shape matching with image edges and size prior. In our case the dominant role plays the weak structural model that emerges from the data: it is able to select among objects of interest proposed by local classifiers and, at the same time, support windows completing the structure even where the classifier response is low. This allows us to achieve good results even when illumination varies and partial occlusion of windows is present, as shown in Fig. 3. Poor

a) Monge No. 13  b) Monge No. 43  c) Monge No. 50

**Fig. 3.** Visualization of results on part of Parisian dataset [10], facade a) is occluded by plants, in facade b) cast shadow is present. False positive windows in c) are also window-like regions: They have good response from both classifiers and match with the neighbors. Detected windows are shown in red, neighborhood edges in green and image edges are emphasized in blue. Results on the complete test set are available as supplemental material.

results of Randomized Forest (RF) segmentation from [10] included in Table 2 give an idea how entirely unstructured approaches perform on this data.

For classes different than *window* and *wall* the results cannot be directly compared with the other methods, but allow us to analyze the behavior of our method in such classes. Balconies are typically overlapping windows in Haussmannian style, but such overlaps are somehow randomly annotated as *window* or *balcony* in the ground truth [10], even when the appearance is the same, introducing some amount of ambiguity in the results. The *shop* class areas are actually formed by shop-windows and the wall around them, and the visualized results show that our detector follows this interpretation. The roof area was difficult for our approach, since the color classifier considers them window-like.

While the authors in [10] claim their segmentation framework generalizes on some mild variants of Haussmannian facades, we can say our framework is not limited to any particular style at all. To prove this, we demonstrate results on modern buildings in Fig. 5 and 4 a).

Finally, we have made experiments with loosely regular facade of Frank Gehry's *Dancing House* shown in Fig. 4 b), where window alignment shows significant deviation from grid structure. We were successful in correctly locating all windows lying on the major plane as well as their neighborhood. The ability to handle sparse regular structures is presented on the right in Fig. 4 c).

a) Modern facade b) Irregular facade c) Sparse structure

**Fig. 4.** Results on facade images from Prague

**Fig. 5.** Interpreted facades of a modern building. *Left:* Simple shape template with $t = 1$ fails to detect light windows. *Right:* Change to $t = 0.33$ improves the result significantly as the response from edge likelihood is stronger.

# 7 Conclusion and Future Work

We have presented a recognition framework that uses a weak structure model to locate elements in images, and demonstrated its potential in the task of window detection in facades. Our experiments have demonstrated that structural regularity given by pair-wise attribute constraints can efficiently guide a stochastic process that estimates element locations and neighborhood at the same time. We have shown that the conjunction of a weak non-specific classifier and a weak structural model can lead to performance that would be hardly achievable by a well-trained specific classifier. Despite the seemingly complex description of the model, the ideas are simple and the implementation is straightforward.

In our future we would like to endow our recognition framework with more powerful classifiers and an ability to handle relations on multiple levels that would i.e. allow two different structural components to overlap.

**Acknowledgment.** This work has been supported by Google Research Award, by the Czech Ministry of Education under project MSM6840770012 and by Grant Agency of the CTU Prague under project SGS10/278/OHK3/3T/13.

## References

1. Micusik, B., Kosecka, J.: Piecewise planar city 3D modeling from street view panoramic sequences. In: Proc. CVPR (2009)
2. Hohmann, B., Krispel, U., Havemann, S., Fellner, D.: CITYFIT: High-quality urban reconstructions by fitting shape grammars to images and derived textured point cloud. In: Proc. of the International Workshop 3D-ARCH (2009)
3. Pauly, M., Mitra, N., Wallner, J., Pottmann, H., Guibas, L.: Discovering structural regularity in 3D geometry. Transactions on Graphics 27, 43 (2008)
4. Gips, J.: Shape grammars and their uses. Birkhäuser, Basel (1975)
5. Zhu, S., Mumford, D.: A stochastic grammar of images. Foundations and Trends in Computer Graphics and Vision 2, 362 (2006)
6. Alegre, F., Dellaert, F.: A probabilistic approach to the semantic interpretation of building facades. In: International Workshop on Vision Techniques Applied to the Rehabilitation of City Centres (2004)
7. Müller, P., Zeng, G., Wonka, P., Van Gool, L.: Image-based procedural modeling of facades. Transactions on Graphics 26, 85 (2007)
8. Mayer, H., Reznik, S.: Building facade interpretation from uncalibrated wide-baseline image sequences. ISPRS Journal of Photogrammetry and Remote Sensing 61, 371–380 (2007)
9. Ripperda, N., Brenner, C.: Data driven rule proposal for grammar based facade reconstruction. Photogrammetric Image Analysis 36, 1–6 (2007)
10. Teboul, O., Simon, L., Koutsourakis, P., Paragios, N.: Segmentation of building facades using procedural shape prior. In: Proc. CVPR (2010)
11. Toussaint, G.T.: The relative neighbourhood graph of a finite planar set. Pattern Recognition 12, 261–268 (1980)
12. McLaughlan, G.J.: Finite Mixture Models. Wiley, Chichester (2000)
13. Green, P.J.: Reversible jump Markov chain Monte Carlo computation and Bayesian model determination. Biometrika 82, 711–732 (1995)

# Multiple Viewpoint Recognition and Localization

Scott Helmer, David Meger, Marius Muja, James J. Little, and David G. Lowe

University of British Columbia

**Abstract.** This paper presents a novel approach for labeling objects based on multiple spatially-registered images of a scene. We argue that such a multi-view labeling approach is a better fit for applications such as robotics and surveillance than traditional object recognition where only a single image of each scene is available. To encourage further study in the area, we have collected a data set of well-registered imagery for many indoor scenes and have made this data publicly available. Our multi-view labeling approach is capable of improving the results of a wide variety of image-based classifiers, and we demonstrate this by producing scene labelings based on the output of both the Deformable Parts Model of [1] as well as a method for recognizing object contours which is similar to chamfer matching. Our experimental results show that labeling objects based on multiple viewpoints leads to a significant improvement in performance when compared with single image labeling.

## 1 Introduction

Object recognition is one of the fundamental challenges in Computer Vision. However, the framework in which it is typically evaluated, by labeling bounding boxes within a single image of each scene, is quite different from the scenario present in many applications. Instead, in domains ranging from robotics, to recognition of objects in surveillance videos, to analysis of community photo collections, spatially registered imagery from multiple viewpoints is available. Spatial information can be aggregated across viewpoints in order to label objects in three dimensions, or simply to further verify the uncertain inference performed in each individual image.

This paper proposes such a scene labeling approach, by which we refer to labeling the objects in a scene. We do not choose a particular target application nor tailor the approach to a specific classification function. Instead we present a method that takes multiple well-registered images of a scene and image-space classification results in those images as input and determine an improved set of 3D object locations that are consistent across the images. Our method for locating consistent regions consists of two steps. The first step is a sampling procedure that draws a finite set of candidate 3D locations in order to avoid the high computational cost of considering every potential location. The second step scores these potential locations based on how well they explain the outputs of the image-based classifier in all available viewpoints. Experimental analysis

R. Kimmel, R. Klette, and A. Sugimoto (Eds.): ACCV 2010, Part I, LNCS 6492, pp. 464–477, 2011.

shows that this method produces significant increases in labeling accuracy when compared against the image-based classifiers upon which it is based.

Figure 1 illustrates a scenario for which a top-performing method on the Pascal Visual Object Categories (VOC) challenge mis-labels several objects in a scene. We have observed that, for such scenes, occlusion and appearance similarity between categories are the most significant challenges for correct recognition. In the left image of Figure 1, two bowls are not detected because their contours are broken by occlusion. Also, the alignment of a bottle and bowl forms a mug-like contour, which causes a false positive for the single-image appearance model in another case. In contrast, labeling from multiple viewpoints achieves correct inference because such accidental alignments occur in only a fraction of the views, and the relatively larger number of views without occlusion support one another to give confident detections. The correct labelings for both scenarios are shown in the right image of Figure 1.

The contribution of this paper is a novel scene labeling strategy based on imagery from multiple viewpoints as well as a new data set suitable for evaluation of such an approach. Our data set contains spatially registered imagery from many viewpoints of a number of realistic indoor scenes. We have made this data set publicly available, as part of the UBC Visual Robot Survey (UBC VRS[1]) as we hope the availability of such data will encourage other authors to consider the problem of recognizing objects from a number of viewpoints, rather than in single still images.

The next section of this paper describes related work in multi-view scene labeling. This is followed by a technical description of our method in Section 3. Next we describe the data set that we have collected, and provide results for our approach evaluated on this data. The paper is concludes with a discussion of future work and outstanding problems.

## 2 Related Work

View-point independent category recognition is currently an active area of research, with a number of new approaches being advanced, [2,3,4]. These approaches attempt to perform viewpoint-independent inference, which would, in principle, remove the requirement to have images from multiple viewpoints to annotate a scene. However, these methods typically require annotated training data from a semi-dense sampling of viewing directions and in some cases require additional information such as a video sequence [2]. While viewpoint invariant category recognition is a promising direction, we argue that for certain categories and scenes, multiple viewpoint recognition is advantageous, as in Figure 1.

Integrating information across many images has been a major focus of active vision. Several authors have described Bayesian strategies to combine uncertain information between views, [5,6]. In particular [6] have previously suggested the use of a generative model of object appearance conditional on the object label and other confounding variables such as pose and lighting, along with a

[1] http://www.cs.ubc.ca/labs/lci/vrs/index.html

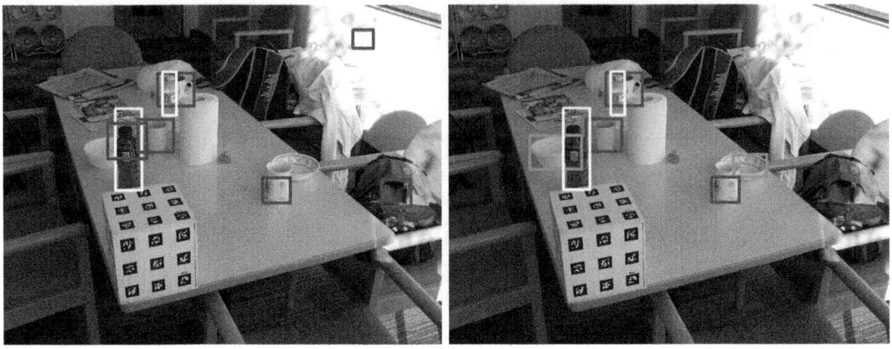

**Fig. 1.** On the left is an image labeled with the Deformable Parts Model from [1], a state-of-the-art approach to recognize objects in single images. On the right is a result from our multi-view approach. The challenges presented by occlusion and clutter are overcome by fusing information across images. Bowls are labelled in green, mugs in red and bottles in white. A threshold equivalent to 0.65 recall has been used for both methods.

sequential update strategy in order to solve this problem. However, active vision has typically been used to recognize object instances or categories for which accurate pose estimation is possible. We extend some of these ideas for more general object categories where accurate pose estimation is still a challenge.

Several authors have also recently considered fusing information across the frames of a video sequence to improve over single-frame performance. For example, Andriluka *et al.* [7] use a bank of viewpoint-dependent human appearance models and combine these with a learned motion prior to gather consistent information across motion tracks. Also, Wojek *et al.* [8] infer the location of pedestrians simultaneously with the motion of a vehicle to achieve localization in 3D from an on-board camera. The probabilistic scoring function described in Section 3 is similar to the approaches used in each of these methods.

The recent work by Coates and Ng [9] most closely resembles our own, although developed independently. Here, they first use multiple images and rough registration information to determine possible corresponding detections, using similar techniques to us. The posterior probability for each set of corresponding detections is computed, where non-maximum suppression is used to discard errant correspondences. Their work differs from ours most significantly in that the posterior probability for a correspondence is based solely on appearance, where as our work includes geometric information as well. In addition, their experimental validation is limited, presenting multi-view results on a single category, where the difference in viewpoint is not particularly significant. Our work presents a formulation that is more general with more extensive experiments to demonstrate the utility of multi-viewpoint recognition.

Numerous robotic systems with object recognition capabilities have also been previously presented. Many systems are targeted for outdoor navigation such as intelligent driving systems (see several recent vision based methods among far too many others too mention [10,11]), however in most cases these systems have attempted to recognize just a few of the most relevant object categories for the task of safe navigation: typically pedestrians, cars, in some cases stop signs. In contrast, indoor robots attempting to perform human-centric tasks that require a broader sampling of visual categories include [12,13,14,15]. Many of these systems have provided us with inspiration and share aspects of our own approach. However, we are not aware of any platform that reasons so explicitly about the integration of information across viewpoints. cases combined with imagery (e.g. [14,18]). Notably the approach of [18] is capable of recognizing objects from publicly available 3D models. These approaches also have not to our knowledge considered multiple viewpoints. NOTE: The last work isn't really relevant to our work

Another contribution of our work is the publication of a new dataset, called the UBC VRS. There are several existing data sets that contain imagery of objects from multiple viewpoints, similar to the one described in this paper. Savarese *et al.* [16] is perhaps the most similar since it focuses on challenging object categories and each image shows an object instance in a realistic scene. However, this data set has only a single instance in each image, the objects occupy nearly all of the image and there is little clutter. In each of these aspects, the data presented in this paper is more challenging. Several other datasets also feature imagery from multiple viewpoints, which is intentionally either less realistic, less cluttered or both [17,19]. To our knowledge, the data presented in this paper represents the largest available set of *spatially-registered* imagery for realistic scenes that is both publicly available and annotated with the presence of challenging object categories.

## 3 Method

We define multiple-viewpoint object localization as the task of inferring the 3D location, scale and pose of a set of objects from image-space detections in well-registered views. Each image-space hypothesis represents a ray in 3D, so objects observed from several views with matching detections will produce many nearly intersecting rays. This set of rays should mutually agree upon position, location and scale of a single 3D object. Our method involves locating a set of objects, $C$, that maximizes the the conditional likelihood of the set of image-space detections, $F$: $p(F|C)$.

Section 3.1 describes our likelihood function in more detail. Then, Section 3.2 grounds our discussion by describing 2 image-space detectors that we have used as inputs. In Section 3.3, a method for proposing candidate object locations based on single factors within the full model is developed. This technique proposes a larger number of candidate objects than is optimal, and so a final assignment and scoring procedure returns a final set of inferred objects, as is described in Section 3.4.

### 3.1 Likelihood Model

In order to describe the likelihood model in detail, we expand upon the definitions of 3D objects $c$ and image-space detections $f$. Each $c \epsilon C$ has a 3D position $X$, and azimuth orientation $\theta$. As all of the objects considered have a single up-direction and our registration process allowed us to directly observe the gravity direction, elevation angle is neglected here. Each detection $f$ (also referred to as a feature) consists of a category label, a bounding box $b$, a response $v$, and (depending on the detector) an azimuth orientation $\theta$. We seek to optimize $p(F|C)$, and this requires a mapping $h$, such that $h(c) = F_c$ where $F_c$ is the set of detections accounted for by object $c$. We assume that every detection is generated by at most one 3D object, and we enforce that all detections not assigned to a cluster as assigned to a null cluster. Briefly, valid assignments are those for which the 3D object projects near to the detected bounding box. We will expand upon our search for $h$ shortly. For now, we express the goal of our inference as maximizing:

$$p(F|C) = \sum_h p(F, h|C) \tag{1}$$

$$= \sum_h p(F|h, C)p(h|C) \tag{2}$$

$$\simeq \sum_h \prod_{c \in C} \prod_{f \in h(c)} p(f|h, c)p(h|C) \tag{3}$$

In Equation (3) we assume detections $f$ are conditionally independent given an assignment to generating objects. We approximate the above with q(C),

$$q(C) = \max_h \prod_{c \in C} \prod_{f \in h(c)} p(f|h, c)p(h|C) \tag{4}$$

Therefore, the target for maximization is our selection of potential objects in the scene $C$.

The above approach is not uncommon in object classification. It is similar in spirit to the constellation models of Fergus *et al.* [20], with the exception that our features are detector responses from multiple viewpoints.

We continue to decompose Equation (4) in order to express it in terms of the geometric priors available to our system. Define $d_f$ as the distance from the camera centre to the projection of $X$ onto the $Z$ axis of the camera for which detection $f$ occurred, $z_f$ as the focal length of that camera and $X_f$ as the reprojection of $X$ into the image. Given a mapping, we define the score for an object $c$ as similar to the first term in Equation (4), that is:

$$score(c) = \prod_{f \in h(c)} p(f|h, c) \tag{5}$$

$$= \prod_{f \in h(c)} p(v_f, \theta_f, b|c) \tag{6}$$

$$= \prod_{f \in h(c)} p(v_f|c)p(\theta_f|\theta_c)p(|b_{centre} - X_f||X)p(b_{scale}|cat) \quad (7)$$

The first term in Equation (7) represents the generative appearance model of the detector, discussed above. The second term represents agreement in pose. Here we utilize a Gaussian distribution on the angular difference, with $\mu = 0$ and $\sigma = pi/8$. In the case where the detector does not provide a pose estimate (as is the case with DPM), we omit this term. The third term penalizes distance between the reprojected object centre $X_f$ and the bounding box center for that detection. In this case, $|b_{centre} - X_f|$ is scored using a Gaussian with $\mu = 0$ and $\sigma = \sqrt{b_{area}}/4$, truncated to 0 when $X_f$ lies outside the bounding box. The final term is a scale prior represented as a Gaussian centred about the expected size of each object category. Using $z_f$, $d_f$, and the scale prior, the last term is a Gaussian with parameters $\{z_f\mu/d_f, z_f\sigma/d_f\}$.

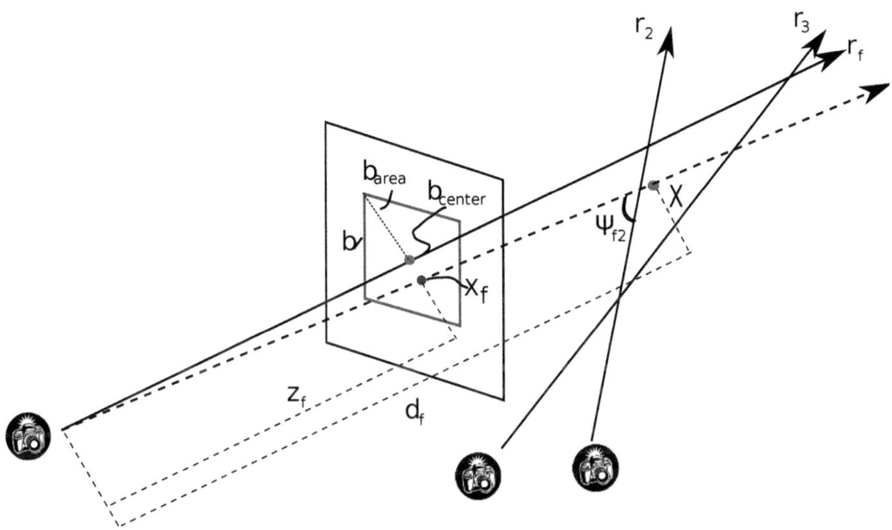


**Fig. 2.** The geometry of the 3D scene labeling problem. Bounding boxes identified by image-based classifiers project to rays ($r_f$,$r_2$, $r_3$) in 3D. Near intersections of rays suggest an object's 3D location, $X$, if the suggested scale (using $d_f$, $z_f$, $b_{center}$) agrees with the scale prior and the reprojection of $X$ onto the image plane, $X_f$, is close to $b_{center}$. If an azimuth pose is suggested by the detector, then we can utilize $\psi_{f2}$ as well to determine if the detections agree on an object pose $\theta$.

### 3.2 Object Detectors

Our approach is not dependent upon a particular object recognition technique. Instead, we can produce scene labelings based on any object detector that produces a finite number of responses, $f$, each detailing a bounding box $b$, a score $v$, and possibly a pose estimate for the object $\theta$. Ideally, we have a generative model for the classifier. That is, we know the probability of each score value $v$ given the presence of the object class. We can utilize validation data to build an empirical distribution for $v$. In our implementation, we have utilized two different object classifiers to demonstrate the applicability of our approach.

**Deformable Parts Model.** The Discriminatively Trained Deformable Part Model (DPM) [1] was one of the best performing methods in the 2009 Pascal VOC. DPM is based on a complex image feature that extends the Histogram of Oriented Gradients (HOG) descriptor [21]. HOG describes an object as a rigid grid of oriented gradient histograms. The more recent DPM approach extends HOG by breaking the single edge template into a number of parts with probabilistic (deformable) locations relative to the object's centre. DPM detections are weighted based on the margin of the SVM to produce a real-valued score. We translate this score into an observation model, $v$ from above, with a validation step. The classifier is originally trained using a set of hand collected imagery from the internet, along with other publicly available datasets. We have employed DPM classifiers for 3 of our studied object categories: Mugs, Bottles, and Bowls.

**Boundary Contour Models.** We wanted to explore the possibility of using object pose in our method, so we implemented a simple classifier that not only outputs a probability $v$, but also a pose. We have discretized the azimuth into 8 viewpoints, and represent each viewpoint as a separate classifier. The classifier for one viewpoint has as its model a single boundary contour. For a particular window in the image, the edges are compared to a scaled version of the boundary contour using a version of oriented chamfer matching [22], and this distance is represented as $v$. Using a validation set we have empirically modeled, $p(v|cat, \theta)$, the distribution of $v$ when the object with pose $\theta$ is present in the image. This classifier is used in the sliding window paradigm, using non-maximum suppression to return a finite set, $F$, of responses $f = (v, b, \theta)$, as required. The training and validation data we used for the shoe classifier came primarily from Savarese *et al.* [16], with a few additional images acquired from internet imagery.

### 3.3 Identifying Potential 3D Objects

We seek an efficient way to find a set of objects $C$ that will maximize Equation (4). This is accomplished by casting rays passing from the camera's centre through the centre of the bounding box. The size of the bounding box, along with the scale prior, suggests where along the ray that a potential object could be located. With a multitude of rays in the scene, locations of near-intersection for an object category suggest the presence of an object. See Figure 2 for an example.

Determining a reasonable set of near-intersections can be challenging, depending on the nature of the scene and the false positive rate of the object detector. For all pairs $(i, j)$ of rays for a particular category, we use four properties to construct a boolean adjacency matrix, $A$, that indicates rays that might correspond to the same 3D object. First, $i$ and $j$ cannot come from the same image. Second, the minimum distance between the rays must satisfy: $d_{i,j} < 0.5\mu$, where $\mu$ and $\sigma$ are the scale priors for the category. Third, the scale of the object, $s_i$, (suggested by the bounding box size and distance along the ray $i$) must satisfy: $\|s_i - \mu\| < 2\sigma$. Finally, for a classifier that supplies pose information, the rays must agree somewhat on the azimuth angle of the object. More precisely, the angle between the two rays, $\psi$, must be within $\pi/4$ of the expected angle between the two detections. We apply these hard thresholds in order to produce a boolean adjacency matrix, and significantly reduce the potential near-intersections that must be considered in later stages.

Using $A$, for all valid pairs of rays $(i, j)$ we compute the 3D point $X$ that minimizes the reprojection error between $X$ and the bounding box centres in the both image planes. This $X$ becomes a potential object $c$. Then, we again utilize $A$ to determine potential agreeing rays (i.e. constructing $h$) to compute $score(c)$, including only those rays for which $c$ explains their geometric quantities (bounding box size, and position) better than a uniform prior. In the case of detections that also return pose, we also infer the best object pose using the rays that are assigned to $c$. The result of this process is a much larger set of objects than are likely.

### 3.4 Maximum Likelihood Object Localization

The final step in our approach is to determine a set of candidate objects that approximately optimizes the likelihood function. We use a greedy strategy to select final objects from the large set of candidates proposed previously, and construct a matching $h$. That is, we score each object $c$, and iteratively add the one achieving the highest score to $C$, afterwards assigning its supporting detections, $F_c$ to $c$, ie $h(c) = F_c$. We then remove these detections and their rays from all remaining potential objects. Following this, we recompute the scores, and repeat this process until all detections $f$ are assigned to an object $c \in C$, or to the null object. We will finally end up with a matching $h$, the objects $C$, and a score for each of the objects $c \in C$.

At this stage, we have an assignment of each 2D detection to a 3D potential object in the case that multi-view consensus was found, or to nothing if it was not. We attempt to use this matching to re-score the input image-space detections such that they reflect the inferred geometry as well as possible. If a detection has been mapped to an object, we assign the score of the object to the 2D detection. If the detection has mapped to a null object, the score remains what it would have been in the single view case since we could bring no additional information to explain the geometric quantities.

## 4 Experiments

We have evaluated the scene labeling performance of our technique using the UBC VRS dataset, a collection of well-registered imagery of numerous real-world scenes. The next Section will describe this data set in detail. Section 4.2 will subsequently describe an experimental technique for fair evaluation of scene labeling approaches. It also contains the results generated from these approaches, which illustrate the performance of our technique.

### 4.1 Collecting Images from Multiple Registered Viewpoints

Numerous interacting factors affect the performance of a multi-view registration system. Many are scene characteristics, such as the density of present objects, the appearance of each instance and the environment lighting. Others are artifacts of the image collection process, such as the number of images in which each object instance is visible at all and whether its appearance is occluded. Ideally, we would like to evaluate our technique on imagery with similar properties to likely test scenarios. As discussed in Section 2, existing datasets are not suitable to evaluate realistic scene labeling because they either lack significant clutter or are generated synthetically.

Therefore, we have collected a new dataset, the UBC VRS, containing a variety of realistic indoor scenes imaged from a variety of viewpoints. Each scene contained many of our evaluation object categories without our intervention. In a few cases, we have added additional instances in order to increase the volume of evaluation data, but we have been careful to preserve a realistic object distribution. The physical settings present in the dataset include 11 desks, 8 kitchens and 2 lounges. In addition, we have augmented the highly realistic scenes with several "hand-crafted" scenarios, where a larger than usual number of objects were placed in a simple setting. We have assembled 7 shoe-specific, and 1 bottle-specific scene of this nature.

As mentioned, each scene has been imaged from a variety of viewpoints, and each image has been automatically registered into a common coordinate frame using a fiducial target of known geometry. Fiducial markers are a common tool for tasks ranging from motion capture for the movie industry to 3D reconstruction. Our target environment involves highly cluttered, realistic backgrounds, and so simple coloured markers or uniform backgrounds (i.e. green screens) are not desirable. Instead, we have constructed a 3D target from highly unique visual patterns similar to those described in [23,24,25]. This target can be robustly detected with image processing techniques, and image points corresponding to known 3D positions (marker corners) can be extracted to sub-pixel accuracy. For the experiments in this paper, we have estimated a pinhole model for our cameras offline, so these 2D-3D correspondences allow the 3D pose of the camera to be recovered.

When evaluating practical inference techniques aimed at realistic scenarios, repeatability and control of experiments is of highest importance. In order to allow other researchers to repeat our experiments, we have released the entire set of imagery used for to generate all of the following results as part of the UBC VRS dataset at the address http://www.cs.ubc.ca/labs/lci/vrs/index.html

### 4.2 Evaluation

To measure localization performance, we compare the output of our automated labeling procedure with ground truth annotations produced by a human labeler. Our labeling procedure follows the the Pascal VOC format, which is a well-accepted current standard in Computer Vision. Specifically, each object is labelled using a tight bounding box and 2 additional boolean flags indicate whether the object is truncated (e.g. due to occlusion) and/or difficult (e.g. visible, but at an extremely small scale). Instances flagged as difficult or truncated are not counted in numerical scoring.

We also employ the evaluation criterion used for the VOC localization task. That is, each object label output by a method is labeled as a true or false positive based the ratio of area of intersection vs area of union between the output

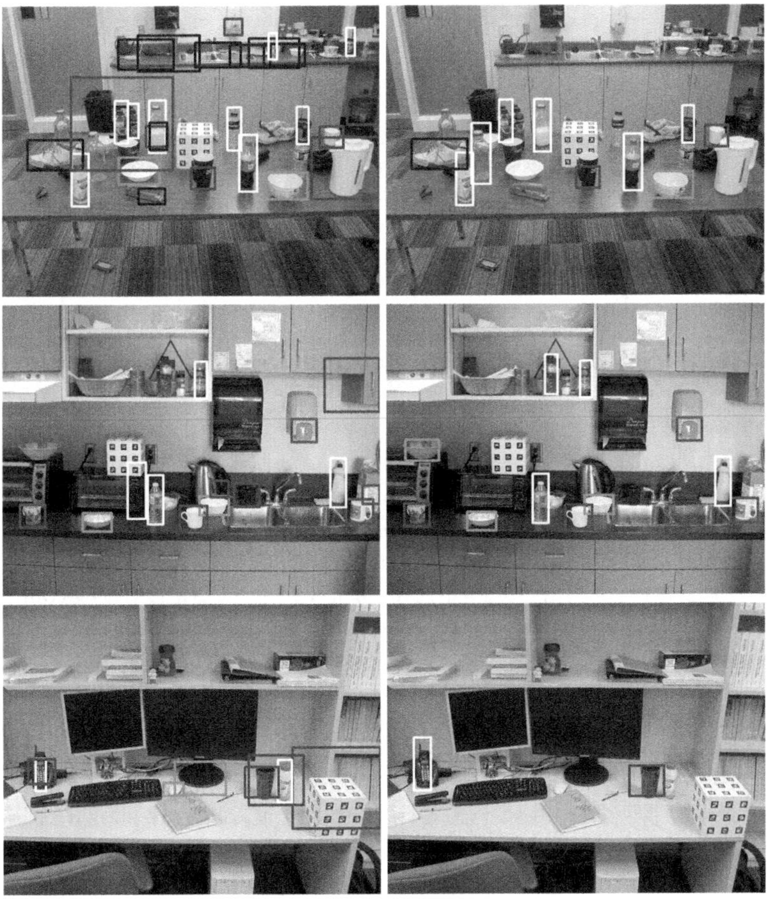

**Fig. 3.** Image-based detections (left) and multi-viewpoint detections from our method (right). Mugs are shown in red, shoes in blue, bowls in green, and bottles in white. A 0.65 recall threshold of used for all categories but shoes which use recall of 0.25.

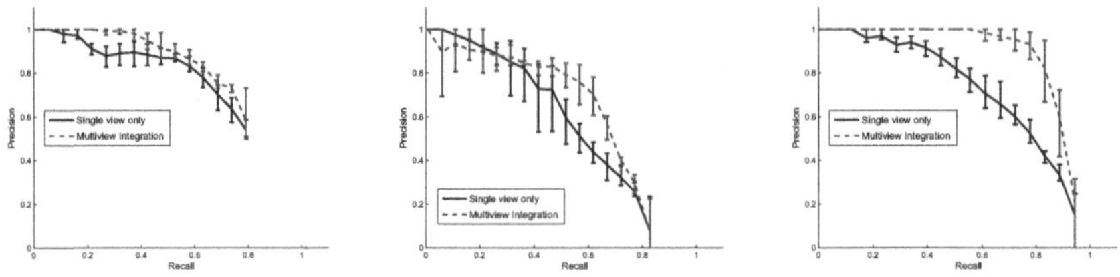


Bottles Mugs Bowls

**Fig. 4.** The above Recall-Precision curves show the multi-view approach as compared to the single view approach when the number of viewpoints available is fixed to 3

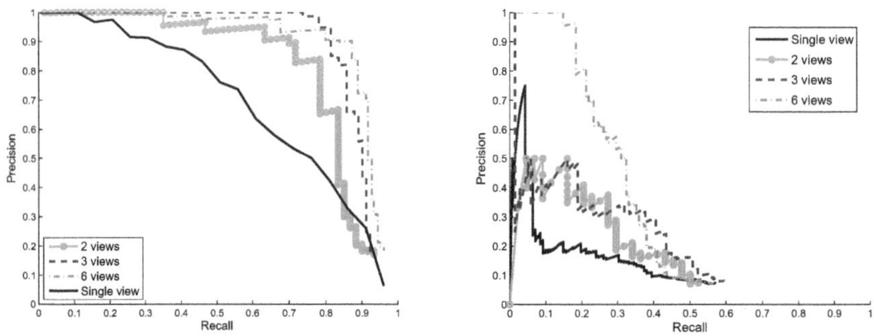


Bowls Shoes

**Fig. 5.** The performance of our system generally increases as the number of views for each scene is increased

bounding box and the closest ground truth annotation of the same category. A precision-recall curve is used to summarize detection results over a variety of possible thresholds, and this curve can be summarized into a single value by summing the area under the curve (AUC).

Our first experiment utilizes the evaluation criteria described to compare the scene labeling produced by our method with the labeling produced by image-space classification methods. For each scene, we perform numerous trials of labeling, to achieve statistical significance. In each trial we select a sub-set of 3 images obtained from well-separated viewpoints. Trials are made independent by randomizing the starting location for this viewpoint selection, such that the labeling procedure sees mostly non-overlapping sets of images between trials. The results of all trials over all scenes in the testing database are shown in Figure 4.

The multi-view approach significantly outperforms labeling based on single images. This is somewhat expected given that a multi-view approach can utilize more information. We have analyzed the situations where the multi-view procedure is not able to infer the correct labeling, and comment on these briefly here. First, we note that there are situations where the appearance based detector simply fails, suggesting further work on the object detectors. Second, there are

**Table 1.** A summary of results generated when evaluating our approach for a variety of object categories. Each value in the table summarizes precision and recall over all possible thresholds with the area under the curve (AUC).

| Number of Views | 1 | 2 | 3 | 6 |
|---|---|---|---|---|
| Mugs | 0.57 | 0.60 | 0.65 | 0.67 |
| Bottle | 0.67 | 0.75 | 0.76 | 0.75 |
| Bowl | 0.71 | 0.79 | 0.86 | 0.90 |
| Shoe | 0.1 | 0.13 | 0.18 | 0.28 |

| Scale Prior | Disabled | Enabled |
|---|---|---|
| Mugs | 0.60 | 0.65 |
| Bottle | 0.69 | 0.76 |
| Bowl | 0.84 | 0.86 |

a number of objects that cause inter-category confusion, even a low recall. For example, the top of a mug or a plate look similar to a bowl in most viewpoints. This could be remedied by including structure information or priors that preclude different objects occupying the same space. We leave this for future work.

We have also studied the contribution that several system components make to our labeling accuracy, and here we describe the effect of each. First, we varied the number of viewpoints considered for each scene. For brevity, only the results for the category *bowl* are shown in Figure 5 and the results for the remaining categories are displayed in a more compact form in Table 1. The general trend is that additional viewpoints lead to more accurate labeling. There is however, a notable difference in the behaviour between classes identified with the DPM detector (mug, bowl, bottle) and those identified with the contour detector (shoe).

For the mug, bowl and bottle, the addition of a second view of the scene yields a significant increase in performance, a third view gives strong, but slightly lessening improvement, and further additional views begin to yield less and less improvement. Our analysis of this trend is that the DPM detector gives sufficiently strong single-view performance, that after aggregating information across only a small number of images, nearly all the object instances with reasonably recognizable appearances are identified correctly. Adding additional viewpoints beyond the third does increase the confidence with which these instances are scored, but it can no longer change the labels such that an instance moves from incorrect to correct, and thus only modest improvement in the curve is possible.

On the contrary, the result from the shoe detector is interesting in that the performance for two viewpoints is little better than for a single image, but the performance does increase significantly between the third and sixth image considered. Our analysis of these results shows that this trend results from the relatively lower recall of the shoe detector for single images. In this case considering more viewpoints increases the chance of detecting the shoe in at least a portion of the images. Moreover, since the shoe detector is sensitive to pose, accidental agreement between hypotheses is unlikely.

Finally, we examined the effect of the scale prior on labeling performance. Table 1 demonstrates that the AUC score improves for each of the classes considered when the scale prior is applied. The use of scale improves the set of clusters that are proposed by improving the adjacency matrix, and it also improves the accuracy of MAP inference for cluster scoring.

## 5 Conclusions

This paper has presented a multi-view scene labeling technique that aggregates information across images in order to produce more accurate labels than the state-of-the-art single-image-classifiers upon which it is based. Labelling scenes from many viewpoints is a natural choice for applications such as the analysis of community photo collections and semantic mapping with a mobile platform. Our method is directly applicable to applications where accurate geometry has been recovered, and as our results demonstrate, the use of information from many views can yield a significant improvement in performance.

## References

1. Felzenszwalb, P., McAllester, D., Ramanan, D.: A discriminatively trained, multi-scale, deformable part model. In: Proc. IEEE CVPR (2008)
2. Su, H., Sun, M., Fei-Fei, L., Savarese, S.: Learning a dense multi-view representation for detection, viewpoint classification and synthesis of object categories. In: Proc. IEEE ICCV (2009)
3. Thomas, A., Ferrari, V., Leibe, B., Tuytelaars, T., Gool, L.V.: Using multi-view recognition and meta-data annotation to guide a robot's attention. Int. J. Robotics Research (2009)
4. Liebelt, J., Schmid, C., Schertler, K.: Viewpoint-independent object class detection using 3d feature maps. In: Proc. IEEE CVPR (2008)
5. Whaite, P., Ferrie, F.: Autonomous exploration: Driven by uncertainty. Technical Report TR-CIM-93-17, McGill U. CIM (1994)
6. Laporte, C., Arbel, T.: Efficient discriminant viewpoint selection for active bayesian recognition. Int. J. Computer Vision 68, 1573–1405 (2006)
7. Andriluka, M., Roth, S., Schiele, B.: Monocular 3d pose estimation and tracking by detection. In: Proc. IEEE CVPR (2010)
8. Wojek, C., Roth, S., Schindler, K., Schiele, B.: Monocular 3D scene modeling and inference: Understanding multi-object traffic scenes. In: Daniilidis, K., Maragos, P., Paragios, N. (eds.) ECCV 2010. LNCS, vol. 6314, pp. 467–481. Springer, Heidelberg (2010)
9. Coates, A., Ng, A.Y.: Multi-camera object detection for robotics. In: Proc. IEEE Int. Conf. Robotics and Automation (2010)
10. Leibe, B., Schindler, K., Cornelis, N., Gool, L.V.: Coupled object detection and tracking from static cameras and moving vehicles. IEEE Trans. Pattern Analysis Machine Intelligence (2008)
11. Wojek, C., Walk, S., Schiele, B.: Multi-cue onboard pedestrian detection. In: CVPR, pp. 1–8 (2009)
12. Kragic, D., Björkman, M.: Strategies for object manipulation using foveal and peripheral vision. In: Proc. IEEE ICVS (2006)
13. Gould, S., Arfvidsson, J., Kaehler, A., Sapp, B., Meissner, M., Bradski, G., Baumstarck, P., Chung, S., Ng, A.: Peripheral-foveal vision for real-time object recognition and tracking in video. In: Proc. IJCAI (2007)
14. Rusu, R.B., Holzbach, A., Beetz, M., Bradski, G.: Detecting and segmenting objects for mobile manipulation. In: Proc. ICCV, S3DV Workshop (2009)
15. Ye, Y., Tsotsos, J.K.: Sensor planning for 3d object search. Computer Vision and Image Understanding 73, 145–168 (1999)

16. Savarese, S., Fei-Fei, L.: 3d generic object categorization, localization and pose estimation. In: Proc. IEEE ICCV (2007)
17. Viksten, F., Forssen, P.E., Johansson, B., Moe, A.: Comparison of local image descriptors for full 6 degree-of-freedom pose estimation. In: Proceedings of the IEEE International Conference on Robotics and Automation, ICRA (2009)
18. Forssen, P.E., Meger, D., Lai, K., Helmer, S., Little, J.J., Lowe, D.G.: Informed visual search: Combining attention and object recognition. In: ICRA, pp. 935–942 (2008)
19. LeCun, Y., Huang, F., Bottou, L.: Learning methods for generic object recognition with invariance to pose and lighting. In: Proceedings of the IEEE Conference on Computer Vision and Pattern Recognition (2004)
20. Fergus, R., Fei-Fei, L., Perona, P., Zisserman, A.: Learning object categories from google's image search. In: Proc. of the 10th IEEE International Conference on Computer Vision, ICCV (2005)
21. Dalal, N., Triggs, B.: Histograms of oriented gradients for human detection. In: Proc. of the IEEE Conference on Computer Vision and Pattern Recognition, San Diego, USA, vol. 2, pp. 886–893 (2005)
22. Shotton, J., Blake, A., Cipolla, R.: Multiscale categorical object recognition using contour fragments. IEEE Transactions on Pattern Analysis and Machine Intelligence 30, 1270–1281 (2008)
23. Fiala, M.: Artag, a fiducial marker system using digital techniques. In: CVPR 2005, vol. 1, pp. 590–596 (2005)
24. Poupyrev, I., Kato, H., Billinghurst, M.: Artoolkit user manual, version 2.33. Human Interface Technology Lab, University of Washington (2000)
25. Sattar, J., Bourque, E., Giguere, P., Dudek, G.: Fourier tags: Smoothly degradable fiducial markers for use in human-robot interaction. In: Fourth Canadian Conference on Computer and Robot Vision (CRV), Montreal, Quebec, Canada, pp. 165–174 (2007)

# Localized Earth Mover's Distance for Robust Histogram Comparison

Kwang Hee Won and Soon Ki Jung

School of Computer Science and Engineering,
College of IT Engineering, Kyungpook National University,
1370 Sankyuk-dong, Buk-gu, Daegu 702-701, South Korea

**Abstract.** The Earth Mover's Distance (EMD) is a useful cross-bin distance metric for comparing two histograms. The EMD is based on the minimal cost that must be paid to transform one histogram into the other. But outlier noise in the histogram causes the EMD to be greatly exaggerated. In this paper, we propose the localized Earth Mover's Distance (LEMD). The LEMD separates noises from meaningful transportation of data by specifying local relations among bins, and gives a predefined penalty to those noises, according to the applications. An extended version of the tree-based transportation simplex algorithm is proposed for LEMD. The localized property of LEMD is formulated similarly to the original EMD with the thresholded ground distance, such as EMD-hat [7] and FastEMD [8]. However, we show that LEMD is more stable than EMD-hat for noise-added or shape-deformed data, and is faster than FastEMD that is the state of the art among EMD variants.

## 1 Introduction

The Earth Mover's Distance (EMD) is an actively used cross-bin distance metric because of its robustness, especially in the task of comparison between two histogram-based descriptors. There are many variants involved in reducing the high computation cost [4,7,8] and resolving the non-metric property [7] of EMD. EMD and EMD-variants show their robustness in many computer vision tasks, such as image retrieval [11], human pose recognition [6], visual tracking [13], etc.

The EMD is obtained by measuring the minimum total cost to transform one histogram into the other, and is expressed as a minimization problem in Equation (1) by Rubner *et al.* [11].

$$EMD(P,Q) = \min_{\{f_{ij}\}} \frac{\sum_{ij} f_{ij} d_{ij}}{\sum_{ij} f_{ij}},$$
$$s.t. \sum_j f_{ij} \le P_i, \sum_i f_{ij} \le Q_j, \sum_{i,j} f_{ij} = \min\left(\sum_i P_i, \sum_j Q_j\right), f_{ij} \ge 0, \quad (1)$$

where $P$ and $Q$ are two given histograms, $f_{ij}$ represents the flow amount transported from bin $i$ to bin $j$, and $d_{ij}$ represents its ground distance. The cost

R. Kimmel, R. Klette, and A. Sugimoto (Eds.): ACCV 2010, Part I, LNCS 6492, pp. 478–489, 2011.

for transporting the unit amount from one bin to another is determined by its ground distance. During the optimization process, if some amount fails to find its destination among nearby bins, EMD often transfers this locally remained amount to distant bins at high cost. It is because EMD does not consider outlier noises, which do not have their corresponding pairs in the histograms. As a result, EMD is unstable for histograms which have noise or geometric distortion in them.

In this paper, we propose localized EMD (LEMD, in short), which can handle the transference of noises and meaningful data separately. LEMD is computed by transforming the source bins to the destination bins using only locally connected edges. During the optimization process, if some amounts fail to find their destinations locally, then they are considered as noises and are removed on their own node with a predefined cost (for the case of consumer, the required amounts are created on the node itself). This self-creation/removal concept is represented clearly in Equation (2), where $\theta$ specifies locality and $\theta/2$ is the penalty for noises.

$$\begin{aligned} &LEMD_\theta(P,Q) = \\ &\min\left(\sum_{ij} f_{ij}d_{ij} + \frac{\theta}{2}\left(\sum_i\left(P_i - \sum_j f_{ij}\right) + \sum_j\left(Q_j - \sum_i f_{ij}\right)\right)\right), \qquad (2) \\ &s.t. \quad \sum_j f_{ij} \le P_i, \sum_i f_{ij} \le Q_j, \ f_{ij} \ge 0. \end{aligned}$$

In this formulation, the optimization process increases the flow $f_{ij}$ as much as possible for $d_{ij} \le \theta$. The second term of the optimization is the remaining amounts after the flow maximization, and is treated as noises. The penalty for noises is $\theta/2$. So the LEMD has the same result as the original EMD, with the thresholded ground distance $\theta$. It can also handle non-normalized histograms like EMD-hat. In order to separate locality and the penalty for the noises, we introduce another notation, $\text{LEMD}_{\theta_1,\theta_2}(P,Q)$, where the flow result is the same as that of $\text{LEMD}_{\theta_1}(P,Q)$ but the penalty is $\theta_2/2$.

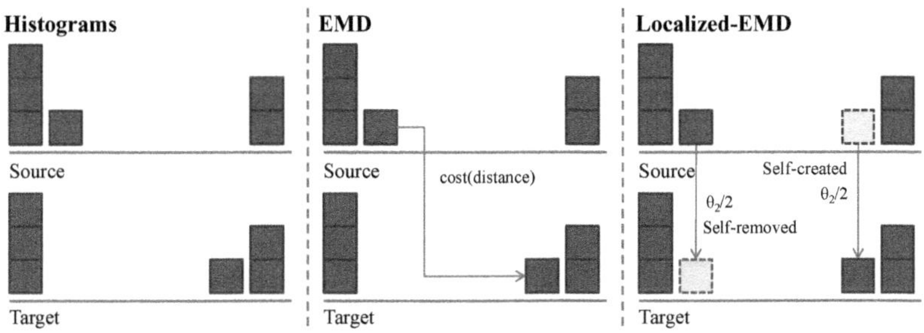


**Fig. 1.** EMD and LEMD

The conceptual example of EMD and LEMD is illustrated in Fig. 1. The EMD interprets one unit amount of the second bin in the source histogram as matched data of right side bin in the target histogram, but LEMD treats them as noise and removes(or creates)them on the bin itself with the fixed cost, $\theta_2/2$. This simple example can be expanded to natural images and region descriptors (like SIFT) which are represented in multi-dimensional histograms. The SIFT descriptor [5] is generated from a 2-dimensional area which is divided into 16 cells. Each cell contains gradient distribution (it is a 1D-circular histogram, actually). So, if the same feature is captured from an optical sensor with a slightly different camera pose, the amount of a bin of specific gradient angle in a cell can be transferred to the bins of close angles in its adjacent cells.

In order to show the useful properties of LEMD, we perform two simple experiments with the video clip obtained from [10]. Two reference images are chosen that one is the 120th frame of the video clip and the other is a quite dissimilar to the first one as shown in Fig. 2d. The first experiment is for the noise-added images in

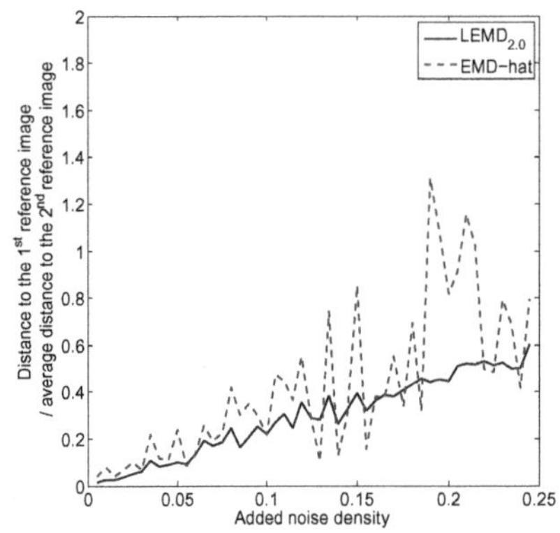


(a) Result of the first experiment.

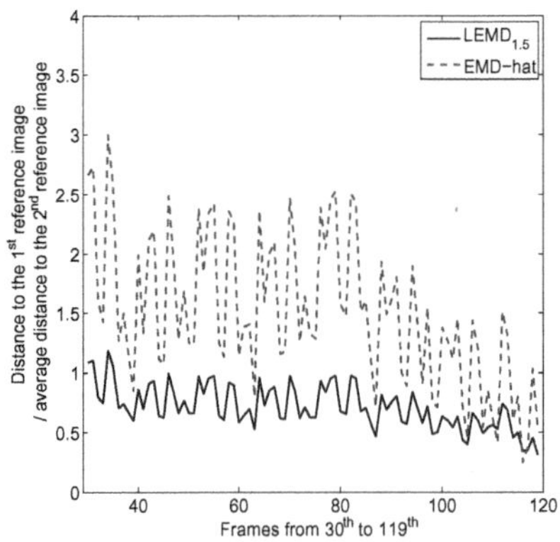


(b) Result of the second experiment.

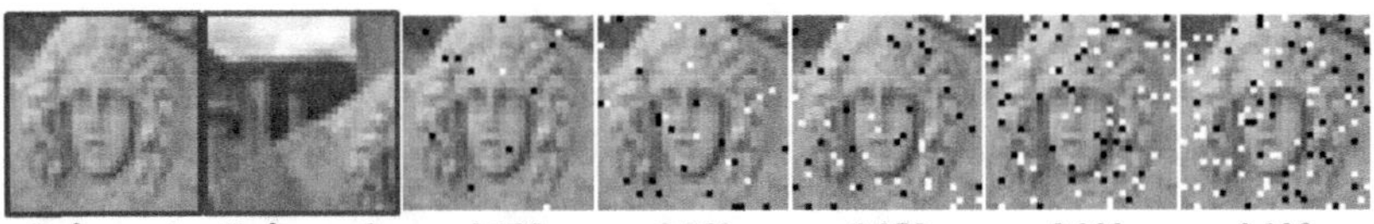


(c) Noise added images.

(d) Video frames with geometric view transform.

**Fig. 2.** EMD and LEMD for the noise-added or view changed images which the synthetic noises are added to the first reference image as shown in Fig. 2a and 2c. The second experiment is for the image sequence changing the view direction as shown in Fig. 2b and 2d. The graphs illustrate the distances using EMD or LEMD metric from the reference image, but the distances are normalized by the average distance from the second reference image to all image subjects. We use resized 30 x 30 pixel images as histograms in the first experiment and use SIFT like histograms in the second experiment. LEMD gives more stable result than EMD under the distortion by noise or geometric transformation.

Another issue about EMD is its heavy computation cost. We propose several approaches to reduce the computation cost. First, we simplify the flow network. The optimization is performed on a bipartite graph, which has $2N$ nodes and $N^2$ edges where $N$ is the number of bins in the histogram. The complexity of the network to compute LEMD is specified with $M$ and $k$. The number of nodes, $M$ is the same as the number of non-zero bins in a signature and $M$ is always smaller than $N$. The number of edges on one node, $k$ is specified by $\theta_1$ and is not increased according to $N$. Second, we construct an efficient configuration of the flow network. Our algorithm is about ten times faster than FastEMD[8] that is the state of the art among EMD variants, though it has similar network complexity to ours. LEMD uses a modified tree-based simplex that partitions the graph into multiple trees and the self-creation or removal concept is implemented with two dummy nodes positioned on the root. It reduces the height of the trees so that the tree-based operations have low complexity. Moreover, LEMD gives a lower bound for matching problems to trim off the unpromising candidates. In our experiment, a high percentage of candidates are excluded before the optimization. LEMD is easily adapted to higher dimensional histograms without weakening either the robustness or computational efficiency for above mentioned characteristics.

The remainder of this paper consists of the following sections. In section 2, we briefly introduce the previous researches about EMD-variants. In section 3, we describe the optimization process for LEMD and its characteristics including two lower bounds. In section 4, we show the experimental setup and results to compare with other EMD-variants. In section 5, we conclude the paper with a statement of intended future works.

## 2 Related Works

Many EMD variants have been developed to reduce the heavy computational cost and to resolve the non-metric property for the non-normalized histograms. To enhance computation speed, many variants impose constraints on their formulations or on specific applications, and have the approximation instead of the exact computation. Shirdhonkar and Jacobs proposed an approximated EMD on wavelet transformed histograms [12]. It requires a preprocessing step and the shape of histograms are restricted only to which are able to be decomposed with wavelet coefficient. Ling and Okada proposed EMD-L1 [2] which uses $L_1$ ground distance and it reduces the complexity of the flow network. But it can be applied only to regular grid. Grauman and Darrell proposed a fast contour matching algorithm using EMD concept but it requires to transform the target data and hard to be applied to other applications. Others deal with only low dimensional data for linear time complexity algorithm, one of them is $\mathrm{SIFT_{DIST}}$[7].

For the non-metric property, Pele and Werman proposed EMD-hat, which is metric [7]. The flow network to compute EMD-hat forms the signatures from the bin-to-bin differences of the histograms and adds an additional sink node to deal with the difference of total amounts of the two histograms as shown in Fig. 3. The cost of sending quantities to the additional sink node is given as $\theta$ times the maximum cost. However, EMD-hat also requires optimization (minimization) in its equation, so the computation cost is not smaller than that of the original EMD. So, they designed $\mathrm{SIFT_{DIST}}$ to show that EMD-hat can be used in applications with some modification. $\mathrm{SIFT_{DIST}}$ is designed for SIFT descriptors, and EMD-hat was applied to gradient histogram (1D-circular) of each rectangular cell. The flow network to compute EMD-hat in each SIFT cell has three types of edges (zero-, one-, and two-cost edge). They accomplished linear-time computation with outperforming results among its competitor algorithms, but they did not fully utilize the three dimensional structure of bins. Also, it is required to design different simplified versions of EMD-hat for different applications.

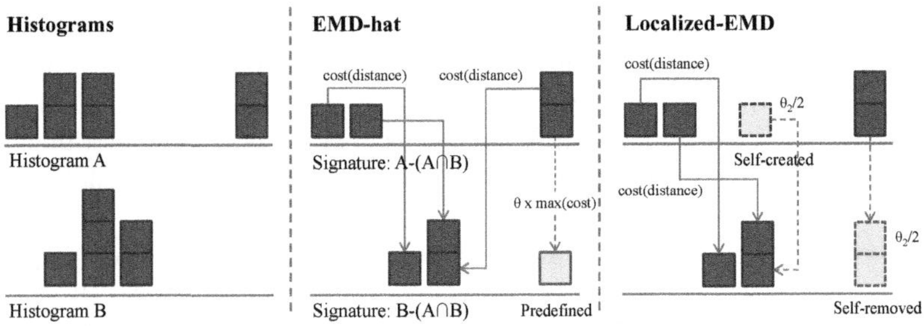


**Fig. 3.** EMD-hat and LEMD. EMD-hat has one dummy to adjust the different amounts of two histograms and its cost is same with maximum of all ground distances. While LEMD has two dummies for locally remained amount of each sides (supplier and demander) and the cost to these dummy nodes are fixed and not expansive than locally maximum cost.

On the other hand, LEMD fully utilizes the structures of input histograms and LEMD is easily applied to various dimensional histograms by defining the ground distance and adjusting two parameters properly. Besides, EMD and EMD-hat often produce unstable results for noise-added histograms, while LEMD can handle noise by giving them the appropriate penalty according to the purposes of the applications. Fig. 4 shows the relations of $L_1$ distance, LEMD and EMD-hat. If the LEMD has $\theta$ (both $\theta_1$ and $\theta_2$) of $1-\epsilon(0 < \epsilon \ll 1)$, then LEMD is close to a half of $L_1$ distance between two histograms and LEMD is getting closer to EMD-hat in accordance with the increase of $\theta$. Under the same maximum distance,

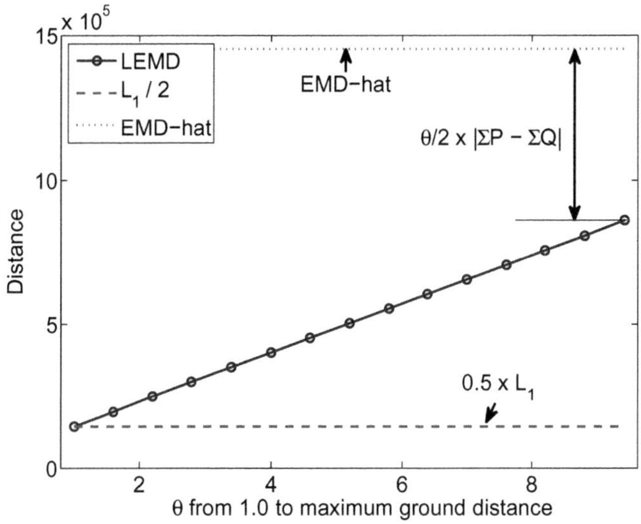


**Fig. 4.** $L_1$ distance, LEMD (of $\theta$ from 1.0 to maximum ground distance) and EMD-hat

LEMD is smaller than EMD-hat by as much as $\theta/2$ times of the difference between total amounts, $\sum P_i$ and $\sum Q_j$ of input histograms.

Pele and Werman proposed FastEMD [8] by using thresholded ground distance. Their main reason to suggest FastEMD is to make EMD-hat faster. In its implementation, they removed edges which have larger ground distance than the threshold and reduced the complexity of flow network from $N^2$ to $kN$, where $k$ is constant and $N$ varies according to the number of bins. They solved a min-cost-flow problem with the shortest path algorithm in the flow network with two special nodes - one is a sink node to balance the total amount of source and target histograms, and the other is a transference node to connect any two nodes from source to target with a fixed cost.

The complexity of flow graph to compute LEMD is not much different with FastEMD, but we approached from different concept, especially for noise added (or geometrically deformed) histograms, to improve the performance of EMD. The flow network to solve LEMD is designed with a filtered set of edges by $\theta_1$, and has two additional nodes (one dummy supplier and one dummy consumer) to implement self-creation/removal concepts. Two dummy nodes handle (create/remove) locally detected noises with cost of $\theta_2/2$ and the difference of the total amounts of two histograms, simultaneously.

We employ a transportation simplex model using tree structure. However, $\theta_1$ breaks a basic variable tree into multiple trees in our formulation. So, we modified the intermediate process to optimize the basic feasible solution. FastEMD and LEMD are still impractical for matching problems of thousands candidates but LEMD is about ten times faster than FastEMD in our experiments (discussed in section 4) and provides a lower bound to trim off the unpromising candidates effectively for various problems, such as image matching based on feature descriptors.

## 3 Computation of LEMD and Lower Bounds

LEMD is designed to handle the effect of noise and the transference of data to nearby bins in histograms separately. The quantities which cannot find their corresponding bins in local relations are considered as noise. As we explained below Equation (2), LEMD has two parameters: $\theta_1$ and $\theta_2$. The first parameter, $\theta_1$, specifies local relations for each of the bins and $\theta_2$ represents the penalty of noises. The formulation is the optimization problem on the bipartite graph, which has two more dummy nodes as shown in Fig. 5.

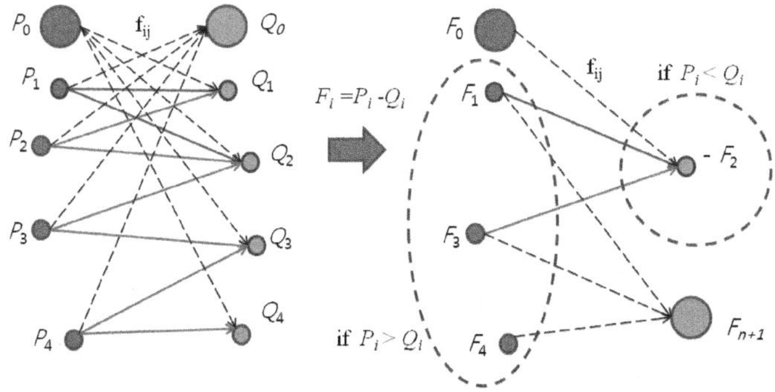


**Fig. 5.** Bipartite graph of signatures for LEMD: (left side) $P$ and $Q$ are two histograms and each node has edges. The dash lines represent from-, to-dummy edges. $P_0$ and $Q_0$ denote the dummy supplier and consumer, respectively. (right side) $F_{i(1 \le i \le 4)}$ is the signature of $P$ and $Q$. $F_0$ is the dummy supplier and $F_{n+1}$ is the dummy consumer.

We employed the transportation simplex algorithm [1] to find the minimum. The detailed description of transportation simplex is not the scope of this paper and for the readers whom do not have background about transportation simplex, the literature [1] will be helpful. We perform the optimization on the graph obtained from the signature of two histograms. For the first step, we find a basic feasible solution using Russell's initialization [1] because this initial solution converges faster than other initializations. The resulting basic variables representing the basic feasible solution are formed into the tree structure. In our problem, multiple trees are generated because the edges between two distant bins are not used. To determine whether the current solution is optimal, $\Delta$-values are computed for all non-basic variables [1]:

$$\Delta = \text{cost} - u - v, \quad \text{where } u = \text{cost} - v. \tag{3}$$

$u$ (or $v$) represents the added distance (total amount) when we add unit flow on the path from/to the root and $u$ (or $v$) is determined by starting from the root ($u = 0$) of a basic variable tree in original transportation simplex. However, in our case, $u$ (or $v$)-values have to be computed from the dummy nodes in order to evaluate the delta value of the non-basic variable, which connects two trees, properly. To avoid loss of generality, we reform the multiple trees to have a dummy node as the root. $\Delta$ represents the added distance when we add flow on that edge(non-basic variable).

If all $\Delta$-values are non-negative, then the current solution is optimal. If not, the non-basic variable which has the minimum (largest negative) delta is selected as an entering variable and its corresponding edge is added to improve the evaluation of given objective function. By adding a non-basic variable to the basic variable trees, a cycle is formed. Cycles can be formed inside a single tree and also be formed between trees which have dummy nodes in them as shown in Fig. 6. Those two-types of cycles are canceled by augmenting minimum-flow along the cycle. The basic variable which is removed from the tree is the leaving variable.

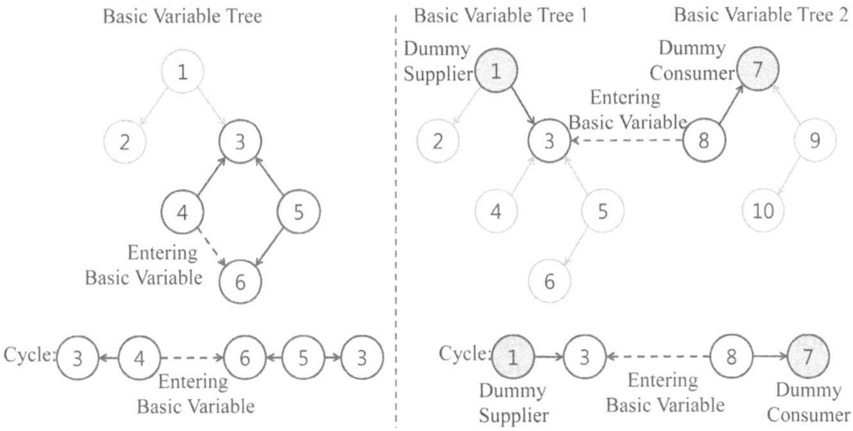


**Fig. 6.** Basic variable tree and cycles in a single tree (left) and dummy-cycle (right) If a sequence of edges connects between the dummy supplier and the dummy consumer, then this sequence does the same function with a cycle in the single tree

The overall process to compute LEMD consists of the following steps:

**Step 1.** Initialize multiple basic variable trees and a non-basic variable list from the signature of two histograms.
**Step 2.** Compute $\Delta$ for all non-basic variables on the list
**Step 3.** Select an Entering variable that has negative largest delta value.
**Step 4.** Find a cycle that contains the Entering variable.
**Step 5.** Find the minimum flow on the cycle and canceling it (including the update of tree structures) by adding the same flow on the cycle.
**Step 6.** Repeat Step 2 to 5 until all delta values are not negative.

Computation cost of LEMD is closely related with the depth of each tree. For example, finding a cycle and canceling it contains depth based retrieval and reformulation of sub-trees. With the same number of nodes, our multiple tree structure has lower depth than single tree. More over the root-positioned dummy nodes make our trees even shorter because two dummy nodes have the largest number of edges. Above reasons make our algorithm faster than FastEMD which have the same number of nodes with LEMD and linearly increasing edges according to the number of nodes.

The optimization process is time consuming, even though our flow graph is less complex than EMD. So, we suggest a lower bound which can be obtained from the above data structure. They effectively trim off the unpromising candidates, which cannot be closer than the current closest candidate. The lower bound is obtained by computing the total amounts of each positive and negative signed signature (Equation (4)):

$$\begin{aligned} LB_1(P,Q) &= d_{min} \cdot f_{in} + \frac{\theta_2}{2} f_d, \\ s.t.\ f_{in} &= \frac{1}{2}\left(\left(\sum_i P_i + \sum_j Q_j\right) - \left|\sum_i P_i - \sum_j Q_j\right|\right), \\ f_d &= \left|\sum_i P_i - \sum_j Q_j\right|, \quad \forall i,j\ (0 < d_{min} \leq d_{ij}). \end{aligned} \tag{4}$$

The minimum of two absolute amounts of signatures, $f_{in}$, is weighted by minimum edge cost (except zero-cost), $d_{min}$, and the difference of two absolute amounts, $f_d$, is weighted by $\theta/2$, the penalty for noises. The interpretation of this lower bound is that all data is matched with the nearest bins and noises are minimized. The computation time of this lower bound is linear and it is much faster than the original LEMD. The lower bound is useful for various applications, such as image matching using feature descriptors. This is because they compare one histogram with many candidates to find the closest one. We will show the efficiency of the lower bound in section 4 by performing a simple experiment.

## 4 Experimental Results

In this section, we show some experimental results that prove the robustness of LEMD by comparing it to other EMD-variants. We mainly compared our algorithm with FastEMD, our competitor algorithm, because other EMD-variants have constraints on their formulation for specific application or are limited for only low dimensional histograms. EMD and EMD-hat are excluded for the computation costs are too large to evaluate. The performances of other variants are found on other literatures [4,7,8,11].

For the first experiment, we compared our algorithm with EMD-hat with thresholded ground distance (FastEMD), EMD-L1, $L_2$ distance that is bin to bin distance, in the task of image query. The descriptor of image is gradient histogram

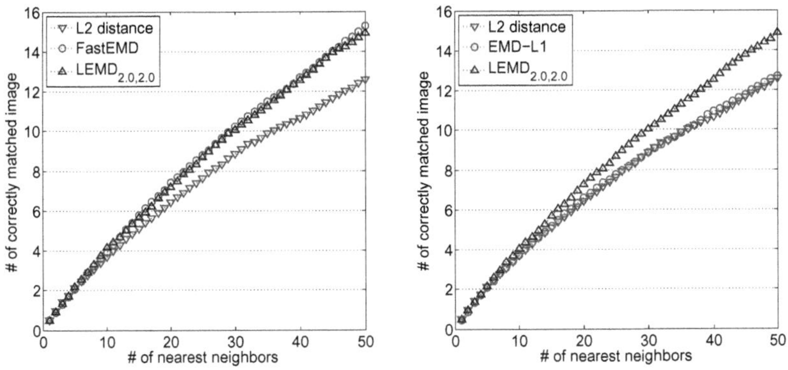


**Fig. 7.** Accuracy of Image Query (left) $L_2$, $LEMD_{2.0}$, FastEMD (right) EMD-L1, $LEMD_{2.0}$

that is obtained from 5 by 5 regions and each region contains circular histogram of 8 directions. For the orientations the opposite directions are considered the same (0 and 180, for example) and the smaller distance of clockwise and count clockwise is chosen. Image samples are taken from Corel Data Set 1000 that are consist of 10 classes and each class contains 100 images. For FastEMD, we used the latest implementation (version 2) in the webpage of the author [9] and the threshold values are set as the same with LEMD. For EMD-L1, we also obtained the authors implementation from [3]. We prepared ten-fold validation sets. Sample images of each class are divided into 10 sets and one set of each class is chosen for query images and other 9 sets are database. The accuracy is obtained by comparing the classes of 50-closest neighbors and that of the corresponding query image. The average accuracy of 10-fold validation sets is shown in Fig 7.

In Fig 7, we plotted EMD-L1 separately with FastEMD because EMD-L1 requires normalized histograms as its input and is implemented only for 1, 2 and 3D histograms. So, we modified LEMD according to the setting of EMD-L1 for fair comparison. In Table 1, the computation times are measured only for the distance computations. LEMD is much faster than FastEMD and close to EMD-L1. LEMD gives similar accuracy with FastEMD but the computation time of LEMD is one-tenth of FastEMD. The lower bound ($LEMD_{2.0}^{LB}$ in Table 1) makes LEMD even faster by trimming off 53.4% and 51.4% of candidates while sustaining the same accuracy. This shows that LEMD will give the best performance among them with given time and computation resources.

We compared LEMD values with different $\theta$. Five sets of query images and target images from the first experiment are used. The image descriptors are the same form. We increased $\theta$ from 1.0(minimum edge cost) to maximum ground distance that is close to EMD-hat. Another plot shows the computation time of LEMDs with the corresponding $\theta$s.

The result shows the EMD(EMD-hat for non-normalized histograms) is less effective distance measure for our image query task and LEMD requires a proper tuning of $\theta$-value to apply other applications.

**Table 1.** Computation time with different distance measure for image query task

| | L2 | EMD-L1 | $LEMD_{2.0}$ | $LEMD_{2.0}^{LB}$ | FastEMD |
|---|---|---|---|---|---|
| 1st graph of Fig 7(sec.) | 7.7 | - | 1,464.4 | 747.8 | 12,029.0 |
| 2nd graph of Fig 7(sec.) | 7.5 | 731.5 | 1,352.4 | 713.3 | - |

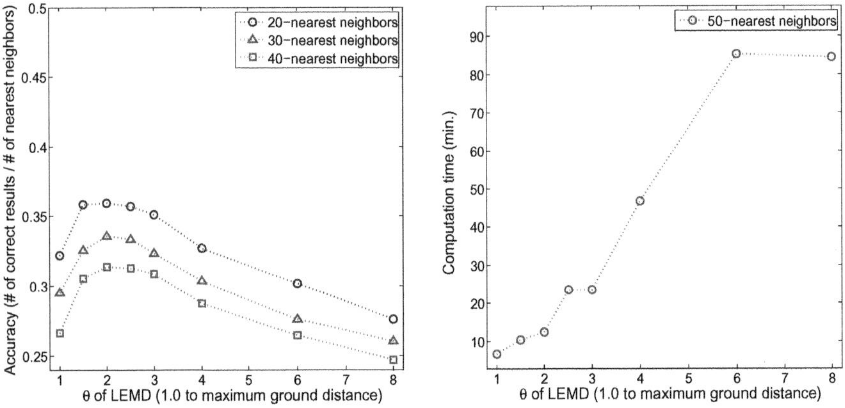


**Fig. 8.** LEMD with different $\theta$. (left) accuracy for different number of neighbors according to various $\theta$.(right) computation time in min. for 50-nearest neighbors.

## 5 Conclusion and Discussion

In this paper, we proposed a localized EMD (LEMD). LEMD is more stable than EMD and EMD-hat when the histograms have noises and geometrical transformation on them. The reason is that the LEMD separates noise terms from data and weights them regularly using self creation/remove concept. Also, the LEMD is faster than EMD-hat with thresholded ground distance and provides a lower bound to improve its performance in some applications, such as matching problems. LEMD can be used in various applications which require robustness regarding noises and have reasonable computation complexity. Our intended future work is to study more characteristics of LEMD, especially related to two parameters. And it is required to enhance the speed of EMD-like distance measures for histogram comparison.

**Acknowledgement.** The authors wish to thank Prof. Ram. Nevatia and Dr. Sung Chun Lee, University of Southern California, for helpful discussions and generous support on this work. This work was also partially supported by the Dual Use Technology Center (DUTC) projects through ARVision Inc.

## References


1. Hillier, F.S., Lieberman, G.J.: Introduction to Mathematical Programming. McGraw-Hill, New York (1990)
2. Ling, H., Okada, K.: EMD-$L_1$: An efficient and robust algorithm for comparing histogram-based descriptors. In: Leonardis, A., Bischof, H., Pinz, A. (eds.) ECCV 2006. LNCS, vol. 3953, pp. 330–343. Springer, Heidelberg (2006)
3. Ling, H., Okada, K., http://www.ist.temple.edu/~hbling/publication.htm
4. Ling, H., Okada, K.: An efficient earth mover's distance algorithm for robust histogram comparison. IEEE Trans. Pattern Anal. Mach. Intell. 29, 840–853 (2007)
5. Lowe, D.G.: Object recognition from local scale-invariant features. In: ICCV 1999: Proceedings of the International Conference on Computer Vision, Washington, DC, USA, vol. 2, pp. 1150–1157. IEEE Computer Society, Los Alamitos (1999)
6. Mori, G.: Guiding model search using segmentation. In: ICCV 2005: Proceedings of the Tenth IEEE International Conference on Computer Vision, Washington, DC, USA, pp. 1417–1423. IEEE Computer Society, Los Alamitos (2005)
7. Pele, O., Werman, M.: A linear time histogram metric for improved SIFT matching. In: Forsyth, D., Torr, P., Zisserman, A. (eds.) ECCV 2008, Part III. LNCS, vol. 5304, pp. 495–508. Springer, Heidelberg (2008)
8. Pele, O., Werman, M.: Fast and robust earth mover's distances. In: ICCV 2009: Proceedings of the International Conference on Computer Vision (2009)
9. Pele, O., Werman, M., http://www.cs.huji.ac.il/~ofirpele/fastemd/
10. Pollefeys, M., http://www.inf.ethz.ch/personal/pomarc/
11. Rubner, Y., Tomasi, C., Guibas, L.J.: The earth mover's distance as a metric for image retrieval. Int. J. Comput. Vision 40, 99–121 (2000)
12. Shirdhonkar, S., Jacobs, D.: Approximate earth mover's distance in linear time. In: IEEE Computer Society Conference on Computer Vision and Pattern Recognition (2008)
13. Zhao, Q., Yang, Z., Tao, H.: Differential earth mover's distance with its applications to visual tracking. IEEE Trans. Pattern Anal. Mach. Intell. 32, 274–287 (2010)

# Geometry Aware Local Kernels for Object Recognition

Dimitri Semenovich and Arcot Sowmya

The University of New South Wales, Australia

**Abstract.** Standard learning techniques can be difficult to apply in a setting where instances are sets of features, varying in cardinality and with additional geometric structure. Kernel-based classification methods can be effective in this situation as they avoid explicitly representing the instances. We describe a kernel function which attempts to establish correspondences between local features while also respecting the geometric structure. We generalize some of the existing work on context dependent kernels and demonstrate a connection to popular graph kernels. We also propose an efficient computation scheme which makes the new kernel applicable to instances with hundreds of features. The kernel function is shown to be positive semidefinite, making it suitable for use in a wide range of learning algorithms.

## 1 Introduction

Our work is motivated by sparse object representations [1,2] which remain popular in computer vision. This usually entails images being described by a collection of local affine-invariant descriptors computed over certain subsampled regions. Such a setting makes the application of standard machine learning techniques somewhat difficult as instances consist of sets (usually with additional geometric structure) of feature vectors, where the cardinality of these sets may vary and the correspondence between the features is unknown.

One way to address this is to use a learning method that abstracts representation of instances through the use of a kernel function [3], such as support vector machines. In this context a kernel function $k(x, x')$ can be thought of as a similarity measure which will provide high values when two images $x$ and $x'$ share similar structure and appearances and low otherwise, while remaining invariant to transformations. It must also satisfy two additional requirements in order to guarantee convergence of learning algorithms to a globally optimal solution [3]: the kernel function must be symmetric, that is, $k(x, x') = k(x', x)$, and positive semi-definite (*p.s.d.*); the latter is also known as the Mercer condition. The challenge is to define a kernel that captures the semantics inherent in the natural images but at the same time is reasonably efficient to compute.

In this work we investigate the design of kernel functions for recognition based on local features.

**Related work.** Kernel-based object recognition methods were initially global [4,5] - each image was mapped to a fixed-length vector, such as a colour histogram

R. Kimmel, R. Klette, and A. Sugimoto (Eds.): ACCV 2010, Part I, LNCS 6492, pp. 490–503, 2011.

or some other global descriptor [6] and a similarity between these vectors was defined. Later on local kernels that can handle variable length and unordered data were developed to deal with objects which cannot be easily represented by fixed length vectors, e.g. graphs or collections of interest points. While global kernels are simple to define and evaluate, they are less flexible than local kernels when it comes to handling invariance which is particularly important for object recognition applications.

Local kernels applied to object recognition can be categorized into two broad groups - those based on a distance, such as KL divergence, defined between parametric distributions or histograms of local features and those based on the summation kernel (1) or, more generally, the $R$-convolution kernel [7] of Haussler. We do not discuss the "principal angles" kernel of Wolf and Shashua [8] and related work as it is more suitable for representing collections of correlated objects, such as successive video frames, rather than sets of local features.

In the first category, Kondor and Jebara [9] fit a normal distribution to each set of vectors and the kernel value between two sets is defined as the Bhattacharya affinity between the respective distributions. Gaussian assumption is needed in order to provide a closed form solution. Similarly, Moreno et al. [10] fit normal distributions to vector sets and then compare them using KL divergence as the distance measure, but unlike [9] the resulting kernel is not positive semidefinite.

Grauman and Darrell [11] have introduced a "pyramid match" kernel which does not make distributional assumptions and instead maps feature vectors to a multiresolution histogram which is computed by binning data points into discrete regions of increasingly larger size. The kernel is then defined as a weighted histogram intersection, which was shown to be positive semidefinite [12]. It is efficient to evaluate with linear complexity in the number of local features. A similar idea is to use a "bag of words" representation, where the local features are clustered and each image is reduced to a histogram of cluster centres to which a standard kernel is then applied. This approach is particularly effective when combined with a spatial pyramid [13]. The second group of local kernels can be interpreted as modifications to the simple summation kernel (for the relationship with histogram kernels see [14]) where a minor kernel $k'$ is computed over all pairs of local features $x \in A$ and $y \in B$:

$$k(A, B) = \sum_{x \in A} \sum_{y \in B} k'(x, y) \tag{1}$$

The discriminative ability of the summation kernel is limited by the fact that all possible matches between local features are combined with equal weight, ensuring that "correct" matches are swamped by the bad ones. To get around this Wallraven et al. [15] and Froelich [16] proposed versions of an "optimal assignment" or "max" kernel which only includes highest values of $k'$ for each $x \in A$ and $y \in B$ in the above sum. Here $\pi_A$ and $\pi_B$ denote permutations of the respective sets:

$$k(A, B) = \begin{cases} \max_{\pi_B} \sum_{i=1}^{|A|} k'(x_i, \pi_B(i)), \text{ if} |A| \geq |B| \\ \max_{\pi_A} \sum_{i=1}^{|B|} k'(\pi_A(i), y_i), \text{ otherwise} \end{cases}$$

Despite claims [15,16] these kernels are not in fact positive semidefinite [17,18]. Statistical bounds on the positive semidefinite property of a variant of the "max" kernel have been discussed [19].

Lyu [17] approximated the "max" kernel while maintaining the $p.s.d$ property by introducing an exponent $|p| > 1$ into (1) in order to increase prominence given to the largest terms. To further improve the quality of matching a circular-shift invariant term $k_\theta$ was added capturing information about local geometry in the form of relative angles between neighbouring features:

$$k(A,B) = \sum_{x \in A} \sum_{y \in B} (k'(x,y) k_\theta(\theta_x, \theta_y))^p \tag{2}$$

Parsana et al [20] have observed that well matching local features should also have neighbouring points which are similar. They propose two variations to the summation kernel, one that averages the minor kernel values over local neigborhoods:

$$k(A,B) = \sum_{x \in A} \sum_{y \in B} k'(x,y) \sum_{p \in \mathcal{N}_x} \sum_{q \in \mathcal{N}_y} k'(p,q) \tag{3}$$

and another that uses graph spectra to compare the shapes of neighborhoods.

Sahbi et al [21] have proposed context-dependent kernel which is similar to (3) in that it uses neighborhood criteria to capture object geometry. They motivate it as a fixed point of an energy function. We recover the kernels of [20,21] in our framework.

We should also point out a close relationship between summation kernels and a situation when a standard kernel is applied to a "bag of words" representation. Let us assume that images $A$ and $B$ have associated "bag of words" histograms $H(A)$ and $H(B)$. Then a linear kernel applied to $H(A)$ and $H(B)$ would be as follows, with $h(.)$ representing histogram mappings of individual features:

$$k_{BoW}(A,B) = H(A)^\top H(B) = \left(\sum_{x \in A} h(x)\right)^\top \left(\sum_{y \in B} h(y)\right) = \sum_{x \in A} \sum_{y \in B} k_\delta(x,y) \tag{4}$$

and $k_\delta$ defined as a very simple local kernel:

$$k_\delta(x,y) = \begin{cases} 1, \text{ if } x \text{ and } y \text{ belong to the same cluster} \\ 0, \quad \text{otherwise} \end{cases}$$

Beyond the use of kernels for object recognition with local features there also exists related but largely independent literature on applying kernel methods to general graphs. These are usually instances of the $R$-convolution kernel [7] relying on decomposing graphs into substructures such as paths and subtrees. In particular, random walk kernels proposed by Gärtner [22,23] look at all possible walks and sum a minor kernel over all possible walks of the graphs. The computation can be achieved in polynomial time by inverting or exponentiating the adjacency matrix of the direct product graph. A generalised framework for graph kernels has been proposed [24]. Graph kernels have previously been applied to recognition from point clouds [25].

**Motivation and contribution.** The first category of kernels suffers from considerable drawbacks. Some methods make strong and unrealistic probabilistic assumptions in order to model the sets of local features. They also generally lack a mechanism to take into account the geometric structure of data,which is particularly relevant for object recognition.

The performance of the second type of kernels is strongly affected by the quality of alignment between the features. mainly when images contain redundant and repeatable structures. A naive greedy matching kernel will not perform well when the features are not sufficiently discriminative (and not at all if we are dealing with point sets) and the local neighborhood graphs used in [20, 21] are quite restrictive. The main contribution of this work is a kernel function for local features that respects the global geometry of the objects in a principled way based on pairwise correspondences. We also show that kernels already proposed [20,21] are special cases of our framework and establish a strong conection to kernels on general graphs.

## 2 Geometry Aware Local Kernel

A feature is a local interest point $x$ for which we define $\phi(x) \in \mathbb{R}^2$ and $\psi(x) \in \mathbb{R}^n$ to be its 2-D coordinates and the associated $n$ dimensional descriptor (e.g. SIFT or shape context coefficients) respectively. Let $\mathcal{X}$ be the set of all features, then our goal is to construct a kernel function $k(A, B) = \langle \Phi(A), \Phi(B) \rangle$ defined on all finite subsets $A, B \subset \mathcal{X}$ where $\Phi$ is some map taking finite subsets of $\mathcal{X}$ to a reproducing kernel Hilbert space $\mathcal{H}$.

### 2.1 Definition

Given two sets of local features $A$ and $B$ let $C = A \times B$ be a set of all possible assignments $(i, j)$, where $i \in A$ and $j \in B$. We wish to construct a correspondence $L \subset C$ that respects the geometric structure present in the data.

Now for each pair of assignments $(a, b)$, where $a = (i, j) \in A \times B$ and $b = (i', j') \in A \times B$, we define a score that measures how compatible the pairs of features $(i, i') \in A \times A$ are with $(j, j') \in B \times B$ (see Figure 1). We require these pairwise affinities to increase with the quality of the match and to form a valid (i.e. symmetric and $p.s.d$) kernel $k_p(\cdot, \cdot)$. We store the kernel values for every pair of assignments $(a, b) \in C$ in a $|C| \times |C|$ matrix $M_\times$ as follows:

$$[M_\times]_{ab} = k_p\left((i, i'), (j, j')\right) \quad (5)$$

$M_\times$ describes how well the pairwise geometry of two local features $(i, i')$ is preserved after mapping them to $(j, j')$. The assignments $a = (i, j)$ can be seen as nodes of a graph, with the pairwise scores $[M_\times]_{ab}$ as edge weights and $M_\times$ as its adjacency matrix. While $M_\times$ need not be positive and symmetric, it will have these properties for most practical choices of $k_p(\cdot, \cdot)$.

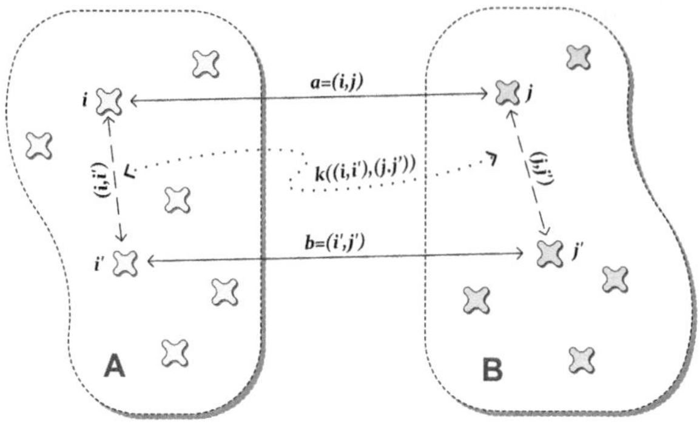


**Fig. 1.** Illustration to clarify the definitions in Section 2.1. Blue and red crosses represent local features.

Leordeanu and Hebert [26] have observed that "good" assignments should form a well-connected cluster in the graph defined by $M_\times$, i.e. good assignments will all be consistent with other "good" assignments. The correspondence problem can then be reduced to finding a cluster $L$ of assignments $(i,j)$ that maximizes the intracluster score

$$r(u) = \frac{u^\top M_\times u}{u^\top u} \tag{6}$$

where $u$ is a cluster membership indicator vector:

$$u(a) = \begin{cases} 1, \text{ if } a \in L \\ 0, \text{ otherwise} \end{cases}$$

The choice of $r(u)$ can be justified by noting that $u^\top M_\times u$ is the sum of all the edge weights in the resulting subgraph and $u^\top u$ is the the number of its nodes. This means that $r(u)$ represents a per-node measure of self-consistency for the cluster and would discourage the inclusion of outliers. While integer optimization of $r(u)$ is not tractable, if we ignore both integer and one-to-one correspondence constraints on $u$ we are left with the standard Rayleigh quotient which is maximised by the leading eigenvector $\xi$ of $M_\times$. If $M_\times$ is non-negative then by Perron-Frobenius theorem, the elements of $\xi$ will also be non-negative and we can interpret $\xi(a)$ as the strength of association of $a$ with the cluster $L$. Using the algorithm of Leordeanu and Hebert [26] as motivation, our geometry aware kernel is defined as follows:

$$k^n_\times(A,B) = \mathsf{vec}(K_1)^\top M^n_\times \mathsf{vec}(K_2) \tag{7}$$

where $K_1$, $K_2$ are $|A| \times |B|$ matrices obtaied by evaluating arbitrary kernels $k_m : \mathcal{X} \times \mathcal{X} \to \mathbb{R},\ m \in \{1,2\}$ on the local features $i \in A,\ j \in B$:

$$[K_m]_{ij} = k_m(i,j) \tag{8}$$

and $\mathsf{vec}$ is the standard column stacking operator with $[K_m]_{*p}$ denoting the $p$-th column of the matrix $K_m$:

$$\mathsf{vec}(K) = \begin{pmatrix} [K_m]_{*1} \\ \vdots \\ [K_m]_{*|B|} \end{pmatrix} \tag{9}$$

If $\rho(M_\times)$, the spectral radius of $M_\times$, is equal to 1 then there exists a limit:

$$\lim_{n\to\infty} k^n(A, B) = \lim_{n\to\infty} \mathsf{vec}(K_1)^\top M_\times^n \mathsf{vec}(K_2) \tag{10}$$

$$= \mathsf{vec}(K_1)^\top (\xi\xi^\top)\mathsf{vec}(K_2) \tag{11}$$

$$= (\xi^\top \mathsf{vec}(K_1))(\xi^\top \mathsf{vec}(K_2)) \tag{12}$$

Note that $\xi^\top \mathsf{vec}(K_1)$ is simply matrix notation for the summation kernel (1), but with each term weighted by $\xi(a)$ which is the quality of correspondence derived from pairwise geometry of local features. Unfortunately it is not possible to ensure that $\rho(M_\times) = 1$ without losing the *p.s.d.* property of $k_\times^n(\cdot,\cdot)$, so we cannot adopt (12) as the definition. In practice $M_\times$ tends to have large eigengap, and so (7) provides a good approximation to (12) for small $n$.

## 2.2 Special Cases

Let $P$ and $Q$ be $|A| \times |A|$ and $|B| \times |B|$ adjacency matrices for the neighbourhood graphs of $A$ and $B$ respectively:

$$[P]_{ii'} = \begin{cases} 1, \text{ if } \ \|\phi(i) - \phi(i')\|_2 < \epsilon \\ 0, \text{ otherwise} \end{cases} \tag{13}$$

We can use them to define $M_\times$ as follows (transposition isn't strictly necessary as both $P$ and $Q$ are symmetric):

$$M_\times = Q^\top \otimes P^\top \tag{14}$$

where $\otimes$ denotes the Kronecker product:

$$A \otimes B = \begin{pmatrix} a_{11}B \; a_{12}B \cdots \\ a_{21}B \; a_{22}B \\ \vdots \qquad\quad \ddots \end{pmatrix} \tag{15}$$

This is consistent with (5) as:

$$[M_\times]_{ab} = [Q]_{ii'}[P]_{jj'} \tag{16}$$

and the right-hand side is of the form $f(x)f(x')$, which is a valid kernel [3] (embedding into a one dimensional space). We can then recover the neighbourhood kernel (3) of [20]:

$$k_{\mathcal{N}}(A,B) = \mathsf{vec}(K)^\top (Q^\top \otimes P^\top)\mathsf{vec}(K) \tag{17}$$

The equivalence can be verified by simple algebra.

Sahbi et al. define their "context dependent" kernel [21] iteratively as follows, with starting point being an arbitrary kernel:

$$K_n^{CDK} = \exp(-D + P^\top K_{n-1}^{CDK} Q - I) \tag{18}$$

$$= K' \circ \exp(P^\top K_{n-1}^{CDK} Q) \tag{19}$$

Here $K_n^{CDK}$ is an $|A| \times |B|$ matrix such that:

$$[K_1^{CDK}]_{ij} = k^{local}(i,j) \tag{20}$$

and exp is applied elementwise rather than denoting traditional matrix exponentiation and $\circ$ is Hadamard (elementwise) matrix product. $D$ is a matrix of distances between feature descriptors. The value of the kernel between two sets $A$ and $B$ is then the followin with $\mathsf{e}$ being a vector of all ones:

$$k_n^{CDK}(A,B) = \mathsf{e}^\top \mathsf{vec}(K_n^{CDK}) \tag{21}$$

To show the connection with our kernel we apply a standard result relating $\mathsf{vec}$ and the Kronecker product:

$$(A \otimes B)\mathsf{vec}(P) = \mathsf{vec}(B^\top P A) \tag{22}$$

But first we need to define our kernel iteratively ($L_0 = K_2$ from (7)):

$$L_l = \mathsf{vec}^{-1}\left((Q^\top \otimes P^\top)\mathsf{vec}(L_{l-1})\right) \tag{23}$$

$$= P^\top L_{l-1} Q \tag{24}$$

$$k_\times^n(A,B) = \mathsf{vec}(K_1)^\top \mathsf{vec}(L_n) \tag{25}$$

Clearly the kernels (23) and (18) are closely related and we can extend CDK to use the full matrix $M_\times$.

## 2.3 Relation to Graph Kernels

Vishwanathan et al. [24] have proposed a framework for graph kernels that generalizes many of the previous results (e.g. [22,23]), by defining

$$k_G(G_1,G_2) = p_\times^\top [\sum_{n=0}^{\infty} \lambda_n W_\times^n] q_\times \tag{26}$$

Here $W_\times$ is a kernel matrix evaluated between the (attributed) edges of the graphs $G_1$ and $G_2$ and $p_\times = p \otimes p'$, $q_\times = q \otimes q'$ are Kronecker products of starting and stopping probabilities for $G_1$ and $G_2$. Our proposed kernel is closely related and can be recovered in this setting if we treat feature sets $A$ and $B$

as fully connected graphs with each edge $(i, i')$ having attributes in the form $(\phi(i), \phi(i'), \psi(i), \psi(i'))$and $k_1(\cdot,\cdot)$, $k_2(\cdot,\cdot)$ are of the form $f(i)(f(j)$.

The advantage of our approach is that it is motivated in terms of alignment between local features which can provide additional intuition and aid in the selection of minor kernels. Also in practice the values of $\lambda_n$ in (26) must be very small for larger $n$ in order to ensure convergence, which leads to higher powers of $W_\times$ having negligible contribution. In our experiments we show that using higher values of $n$ helps performance.

### 2.4 Mercer Condition

Let $\mathcal{X}$ be an input space and let $k : \mathcal{X} \times \mathcal{X} \to \mathbb{R}$ be symmetric and continuous. Kernel $k$ is positive semidefinite, if and only if any Gram matrix constructed by restricting $k$ to any finite subset of $\mathcal{X}$ is positive semidefinite, namely all its eigenvalues are non-negative. A Mercer kernel $k$ guarantees that there exists a reproducing kernel Hilbert space (RKHS) $\mathcal{H}$ and a function $\Phi : \mathcal{X} \to \mathcal{H}$ such that $k(x, x') = \langle \Phi(x), \Phi(x')\rangle_{\mathcal{H}}$ for all $x, x' \in \mathcal{X}$.

**Proposition 1 ( [3]).** *Mercer kernels are closed under addition, multiplication and exponentiation.*

**Lemma 1.** *Kernel between local features* $k_l^n : (\mathcal{X} \times \mathcal{X}) \to \mathbb{R}$ *such that* $k_l^n(i,j) = \left[\mathsf{vec}^{-1}(M_\times^n \mathsf{vec}(K_2))\right]_{ij}$ *is a p.s.d. kernel.*

*Proof.* We show that this is a special case of an $R$-convolution kernel [7]. We define a relation for all local features belonging to the same image $A$ as $R(i', i)$ iff $i \in A \wedge i' \in A$. $R^{-1}(i)$ is then the set of all possible decompositions of local feature $i$ into pairs $(i, i')$, where $i'$ is another feature in the same image. $R$ is finite, as there is only a finite number of features per image. We define a $R$-convolution kernel $k_R$ as follows:

$$k_R(i,j) = \sum_{R^{-1}(i)} \sum_{R^{-1}(j)} k_p((i,i'),(j,j'))k_2(i',j') \tag{27}$$

It is positive semidefinite as shown in [7]. Simple algebra shows that $k_R(\cdot,\cdot)$ is equal to $k_l^1(\cdot,\cdot)$ and therefore the latter is also positive semidefinite. Result for $k_l^n(\cdot,\cdot)$ follows by induction.

**Lemma 2.** *As defined in (7), kernel* $k_\times^n(\cdot,\cdot)$ *is positive semidefinite.*

*Proof.* By Lemma 1 $k_l^n(\cdot,\cdot)$ is $p.s.d$ and then so is $k_1(i,j)k_l^n(i,j)$ by closure under multiplication. By substituting into (1) we obtain $k_\times^n(\cdot,\cdot)$ which is therefore also $p.s.d.$

### 2.5 Efficient Computation

If the kernel $k_p(\cdot,\cdot)$ (5) corresponds to an inner product in a finite dimensional RKHS and the mapping $\Phi_p : (\mathcal{X} \times \mathcal{X}) \to \mathcal{H}_p$ can be constructed explicitly, it

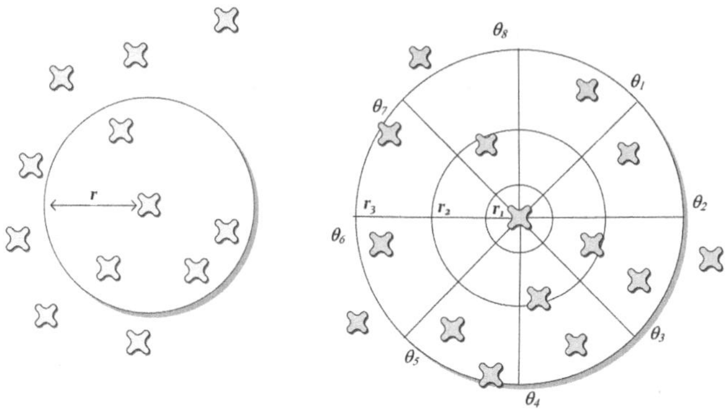


**Fig. 2.** One way to construct a finite dimensional $\mathcal{H}_p$. Instead of looking at a local neighborhood around an interest point as a whole (left) divide it into multiple bins and construct a separate neighborhood graph for each bin (right). This allows us to capture geometric relations such as distance and angle just as the original definition (7) and yet provides an impressive computational improvement (Section 2.5).

is possible to exploit the block structure of $M_\times$ to considerably speed up kernel computation. Let $\mathcal{H}_p$ be $d$ dimensional and $\pi_i : \mathcal{H}_p \to \mathbb{R}$ a projection from $\mathcal{H}_p$ to the $i$-th coordinate. Then after some simple algebra we can show that:

$$M_\times = \sum_{i=1}^{d} \pi_i(\Phi_p(B))^\top \otimes \pi_i(\Phi_p(A))^\top \tag{28}$$

where $[\Phi_p(A)]_{ij} = \Phi_p([A]_{ij})$ or the vector valued result of mapping the edge $(i, i')$ (see Figure 1.) to $\mathcal{H}_p$ and $\pi_i$ is applied elementwise. We can now rewrite our iterative definition of $k_\times^n(\cdot\,,\cdot)$ given in (23) as:

$$L_n = \mathsf{vec}^{-1}\left(M_\times \mathsf{vec}(L_{n-1})\right) \tag{29}$$

$$= \sum_{i=1}^{d} \pi_i(\Phi_p(A))^\top L_{n-1} \pi_i(\Phi_p(B)) \tag{30}$$

While normal computation of a single iteration would take $O(m^4)$ time, where $m$ is the number of interest points per image, the efficient version takes only $O(dm^2)$ time, making the kernel practical for images with hundreds of features.

In Figure 2 we show one possible way to construct a finite dimensional $\mathcal{H}_p$ inspired by shape context descriptor [27]. Instead of looking at a local neighborhood (see (13)) around an interest point as a whole (Figure 2 *left*) divide it into multiple bins and construct a separate neighborhood graph for each bin (Figure 2 *right*):

$$M_\times = \sum_{r,\theta} P_{r,\theta}^\top \otimes Q_{r,\theta}^\top \tag{31}$$

**Fig. 3.** Sample images from Caltech101 dataset. This figure shows the feature pairs with highest kernel values (in order of decreasing intensity) for a simple local kernel (*left*) and after 3 iterations of the geometry aware kernel (12) (*right*). We can see that most of the spurious matches have been supressed. The bottom row shows that it may be possible to form multiple consistent cliques.

This allows us to capture geometric relations such as distance and angle, just as the original definition (7), and yet provides a considerable computational improvement.

# 3 Performance

We conducted experiments on standard MNIST [28], Olivetti (OLR faces) [29] and Caltech101 [30] datasets. Due to computational complexity we were only able to evaluate the kernel with the full $M_{\times}$ for Gram matrices of dimension not exceeding $400 \times 400$. To allow a comparison with other popular methods, however, we provide results of the efficient version of the kernel with shape context neighborhood structure for the full Caltech101 dataset.

## 3.1 Experiments and Results

**Full pairwise consistency scores.** For these experiments we used the entire Olivetti dataset (400 images), for MNIST 40 random examples per class were drawn, for Caltech101 we took the first 50 images from the following 6 categories: *airplane, rear cars, faces, motorbikes, watches, ketches*. For MNIST handwritten digits interest points were sampled from each contour and local features constructed by using 30 coefficients of the shape context descriptor [27]. For Olivetti and Caltech101 images we used the DoG detector and the standard 128 dimensional SIFT descriptors [1]. Descriptors were clustered into 1,000 "visual words" using k-means for the bag of words kernel ("Gaussian BoW").

**Table 1.** Results on MNIST, Olivetti and Caltech101 datasets. We report 10-fold cross validation accuracy (percentages) and the standard deviations on both datasets using "one-vs-all" SVM. "Neighbourhood" is the formulation in (17), "Context" refers to the shape-context neighborhood structure described in Figure 2, while for "pairwise" we have used pairwise consistency scores.

| | Dataset | | |
|---|---|---|---|
| **Kernel** | MNIST (SUBSET) | OLIVETTI (ORL - FACES)I | CALTECH101 (6 CATEGORES) |
| GAUSSIAN BOW | 87.2 ± 3.7 | 95.1 ± 3.1 | 86.9 ± 4.6 |
| SUMMATION | 91.2 ± 3.9 | 97.9 ± 1.8 | 90.4 ± 4.4 |
| NEIGHBOURHOOD [20] | 90.1 ± 4.3 | 97.5 ± 2.1 | 91.0 ± 4.7 |
| GK-CONTEXT | 94.5 ± 2.1 | 98.7 ± 1.9 | 93.0 ± 2.8 |
| GK-PAIRWISE | 96.4 ± 2.2 | 99.2 ± 1.1 | 94.9 ± 1.7 |

**Table 2.** Results on the full Caltech101 dataset with 15 and 30 training images per class with (L=2) and without (L=0) a two level spatial pyramid. We report accuracy rates and their standard deviations over 5 random training/testing splits using "one-vs-all" SVM. "GK-Context" refers to the efficient computation scheme using shape-context neighborhood structure described in Figure 2.

| | Caltech101 (L=0) | | Caltech101 (L=2) | |
|---|---|---|---|---|
| **Kernel** | 15 TRAINING | 30 TRAIING | 15 TRAINING | 30 TRAIING |
| GAUSSIAN BOW | 35.6 ± 0.8 | 43.5 ± 0.9 | 45.6 ± 0.8 | 57.5 ± 0.9 |
| PMK [11] | 50.0 ± 0.9 | 58.2 | - | - |
| SPM [13] | - | 41.5± 01.2 | 56.4 | 64.4 ± 0.5 |
| EMK [14] | 46.6 ± 0.9 | 54.5 ± 0.8 | 60.5 ± 0.9 | 70.3 ± 0.8 |
| GK-CONTEXT | 57.3 ± 1.1 | 67.1 ± 0.8 | 61.7 ± 0.7 | 72.1 ± 0.9 |

Table 1. shows 10-fold cross validation accuracy on three datasets using "one-vs-all" SVM. "GK-Neighborhood" corresponds to (17), "GK-Context" refers to the shape-context neighborhood structure described in Figure 2 while for "GK-Pairwise" we defined

$$k_p(\cdot,\cdot) = k_{Gauss}(\|\phi(i)-\phi(i')\|, \|\phi(j)-\phi(j')\|)k_{pol}(\psi(i),\psi(j))k_{pol}(\psi(i'),\psi(j'))$$

Here $k_{Gauss}$ and $k_{pol}$ are Gaussian and third degree polynomial kernels respectively. Local kernels were Gaussian. An implementation will be made available.

**Caltech101.** Following the frequently applied settings we used 15 and 30 images for training per category and tested on the remaining images. For the local features we again used SIFT descriptors but this time instead of detecting the interest points, they were evaluated over $16 \times 16$ patches on a dense regular grid. K-means clustering was applied to identify 1,000 "visual words" for the bag of words kernel.

We used the same shape context neighborhood structure ("GK-Context") as in the other experiments. To show the effectiveness of our approach we compare the results with and without a spatial pyramid [13]. Table 2. shows classification accuracy using a "one-vs-all" SVM.

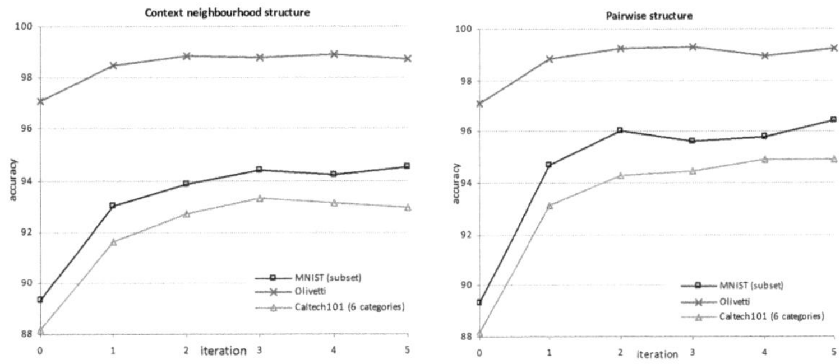


**Fig. 4.** Average classification accuracy (percent) as a function of the number of iterations. On the left are results for the shape context neighborhood structure (GK-Context) and on the right for the pairwise correspondences (GK-Pairwise). There are no significant improvements after the third iteration.

### 3.2 Discussion

Our kernel consistently outperforms the baseline summation and bag of words kernels demonstrating the importance of information about the geometry of objects when working with local features. In all cases performance improved for several iterations (Figure 4).

The efficient implementation ("context") still shows good prediction results, while being feasible to evaluate for larger datasets. Further computational improvements are available if we assume that the local kernels $k_1$ and $k_2$ correspond to projections to finite dimensional spaces (e.g. bag of words).

Our method performs considerably better than the other kernels in the literature [13, 14] without the use of spatial pyramid and has the strong advantage of being invariant to translations (the neighbourhood structure can also be modified to be invariant to rotations).

With the introduction of spatial pyramid it still performs better than other methods but the margin decreses as the spatial pyramid is particularly effective on the Caltech101 dataset where the objects are centered and mostly at the same orientation.

## 4 Conclusion

In this paper we have described a new kernel that is designed for object recognition with local features. This kernel function is motivated from spectral relaxation of the pointset matching problem. We proposed an efficient computation technique and established a strong connection between kernels on local features and kernels on general graphs. The proposed kernel was validated on standard datasets with encouraging results.

Future work may include computational improvements, similar to the recent work of Bo and Sminchisescu [14] on approximating the summation kernel.

## References

1. Lowe, D.: Distinctive image features from scale-invariant keypoints. International Journal of Computer Vision 60, 91–110 (2004)
2. Lazebnik, S., Schmid, C., Ponce, J.: Semi-local affine parts for object recognition. In: British Machine Vision Conference, BMVC (2004)
3. Schölkopf, B., Smola, A.: Learning with Kernels. MIT Press, Cambridge (2002)
4. Chapelle, O., Haffner, P., Vapnik, V.N.: Support vector machines for histogram-based image classification. IEEE Transactions on Neural Network 10, 1055–1064 (1999)
5. Demirkesen, C., Cherifi, H.: A comparison of multiclass svm methods for real world natural scenes. In: International Conference on Advanced Concepts for Intelligent Vision Systems (2008)
6. Oliva, A., Torralba, A.: Building the gist of a scene: the role of global image features in recognition. Progress in brain research 155, 23–36 (2006)
7. Haussler, D.: Convolution kernels on discrete structures. Technical Report UCS-CRL-99-10, University of California at Santa Cruz, Santa Cruz, CA, USA (1999)
8. Wolf, L., Shashua, A.: Learning over sets using kernel principal angles. Journal of Machine Learning Research 4, 913–931 (2003)
9. Kondor, R.I., Jebara, T.: A kernel between sets of vectors. In: ICML (2003)
10. Moreno, P.J., Ho, P., Vasconcelos, N.: A kullback-leibler divergence based kernel for svm classification in multimedia applications. In: NIPS (2003)
11. Grauman, K., Darrell, T.: The pyramid match kernel: Efficient learning with sets of features. Journal of Machine Learning Research 8, 725–760 (2007)
12. Odone, F., Barla, A., Verri, A.: Building kernels from binary strings for image matching. IEEE Transactions on Image Processing 14, 169–180 (2005)
13. Lazebnik, S., Schmid, C., Ponce, J.: Beyond bags of features: Spatial pyramid matching for recognizing natural scene categories. In: CVPR (2006)
14. Bo, L., Sminchisescu, C.: Efficient match kernels between sets of features for visual recognition. In: NIPS (2009)
15. Wallraven, C., Caputo, B., Graf, A.B.A.: Recognition with local features: the kernel recipe. In: ICCV (2003)
16. Fröhlich, H., Wegner, J.K., Sieker, F., Zell, A.: Optimal assignment kernels for attributed molecular graphs. In: ICML (2005)
17. Lyu, S.: Mercer kernels for object recognition with local features. In: CVPR (2005)
18. Vert, J.P.: The optimal assignment kernel is not positive definite. Technical report, arXiv (2008)
19. Boughhorbel, S., Tarel, J.P., Fleuret, F.: Non-mercer kernels for svm object recognition. In: BMVC (2004)
20. Parsana, M., Bhattacharya, S., Bhattacharya, C., Ramakrishnan, K.R.: Kernels on attributed pointsets with applications. In: NIPS (2007)
21. Sahbi, H., Audibert, J.Y., Rabarisoa, J., Keriven, R.: Context-dependent kernel design for object matching and recognition. In: CVPR (2008)
22. Gärtner, T.: Exponential and geometric kernels for graphs. In: NIPS Workshop on Unreal Data: Principles of Modeling Nonvectorial Data (2002)
23. Gärtner, T., Flach, P., Wrobel, S.: On graph kernels: Hardness results and efficient alternatives. In: COLT (2003)
24. Vishwanathan, S., Borgwardt, K., Kondor, I., Schraudolph, N.: Graph kernels. Journal of Machine Learning Research 9, 1–37 (2008)
25. Bach, F.: Graph kernels between point clouds. In: ICML (2008)

26. Leordeanu, M., Hebert, M.: A spectral technique for correspondence problems using pairwise constraints. In: ICCV (2005)
27. Belongie, S., Malik, J., Puzicha, J.: Shape matching and object recognition using shape contexts. IEEE Transactions on Pattern Analysis and Machine Intelligence 24, 509–522 (2002)
28. LeCun, Y., Bottou, L., Bengio, Y., Haffner, P.: Gradient-based learning applied to document recognition. Proceedings of the IEEE 86, 2278–2324 (1998)
29. Samaria, F., Harter, A.: Parameterisation of a stochastic model for human face identification. In: Proceedings of 2nd IEEE Workshop on Applications of Computer Vision (1994)
30. Fei-Fei, L., Fergus, R., Perona., P.: Learning generative visual models from few training examples: an incremental bayesian approach tested on 101 object categories. In: CVPR Workshop on Generative-Model Based Vision (2004)

# Author Index

GPSR Compliance

*The European Union's (EU) General Product Safety Regulation (GPSR) is a set of rules that requires consumer products to be safe and our obligations to ensure this.*

*If you have any concerns about our products, you can contact us on ProductSafety@springernature.com*

In case Publisher is established outside the EU, the EU authorized representative is:

Springer Nature Customer Service Center GmbH
Europaplatz 3
69115 Heidelberg, Germany

**Batch number: 10682617**

Printed by Printforce, the Netherlands